NUMERICAL ALGORITHMS WITH FORTRAN

Springer
*Berlin
Heidelberg
New York
Barcelona
Budapest
Hong Kong
London
Milan
Paris
Santa Clara
Singapore
Tokyo*

Gisela Engeln-Müllges is Professor of Numerical Mathematics and Computer Science in the Department of Mechanical Engineering and Prorector for Research and Development at the Fachhochschule Aachen. She advises engineers at academia and in industry on the application of numerical methods and has published numerous books.

Frank Uhlig is Professor of Mathematics at Auburn University. He has been awarded two Fulbright Grants and has organized several research conferences. His research activity and publications are in linear algebra, matrix theory, numerical analysis, geometry, Krein spaces, graph theory, mechanics, and inverse problems.

GISELA ENGELN-MÜLLGES FRANK UHLIG

Numerical Algorithms with Fortran

With CD-ROM

With 46 Figures

 Springer

Gisela Engeln-Müllges
Fachhochschule Aachen
Department of
Mechanical Engineering
Goethestraße 1
D-52064 Aachen, Germany

Frank Uhlig
Auburn University
Department of Mathematics
218 Parker Hall
Auburn, AL 36849-5310, USA

Translated by M. Schon and F. Uhlig
CD-ROM with Fortran 77 Programs by J. Dietel, G. Dubois, G. Engeln-Müllges,
T. Eul, M. Grönheim, E. Heyne, I. Hinze, V. Krüger, K. Niederdrenk, G. Palm,
E. Pohl, R. Reuter, H.-J. Rheinbach, H. Turowski et al.
Transcription to Fortran 90 by A. Sohn

Cataloging-in-Publication Data applied for
Die Deutsche Bibliothek – CIP-Einheitsaufnahme
Numerical algorithms with Fortran/Gisela Engeln-Müllges;
Frank Uhlig. – Berlin; Heidelberg; New York, Barcelona;
Budapest; Hong Kong; London; Milan; Paris; Santa Clara;
Singapore; Tokyo: Springer.
ISBN 3-540-60529-0
NE: Engeln-Müllges, Gisela; Uhlig, Frank
Buch. – 1996

ISBN 3-540-60529-0 Springer-Verlag Berlin Heidelberg New York

The use of general descriptive names, trademarks, etc. in this publication does not imply, even in the absence of a specific statement, that such names are exempt from the relevant protective laws and regulations and therefore free for general use.
Cover Design: Künkel + Lopka Werbeagentur, Ilvesheim
Typesetting: Camera ready by authors
Printed on acid-free paper SPIN 10519378 33/3142 – 5 4 3 2 1 0

Preface

This book was begun about 20 years ago jointly by Gisela Engeln–Müllges and Fritz Reutter at the Technical University in Aachen (RWTH), Germany. Its first edition in 1974 was entitled "Formelsammlung zur Numerischen Mathematik mit FORTRAN-Programmen". It was based on courses in numerical analysis taught mainly to engineering students at the RWTH.

Professor Fritz Reutter (1911 – 1990) actively continued working on the text portion of this book until about 1978, and he served as a valued advisor until its third German edition in 1981. After 1983, his grave illness precluded him from developing this book further.

From the fourth to the seventh German editions (1984 – 1993), now entitled "Numerik-Algorithmen mit FORTRAN 77-Programmen", the book increased almost threefold in volume. This was due in part to the inclusion of new subject areas such as two–dimensional splines and numerical cubature, as well as to the presentation of new methods in the text and program portions of this book, such as Akima and Renner subsplines, Bézier and B–splines, and Runge–Kutta embedding formulas.

This book on **"Numerical Algorithms with FORTRAN"** is a translation of the seventh German edition (1993) with a few updates and changes of emphasis implemented by the new coauthor, Frank Uhlig of Auburn University. There are slight differences between this English edition and the current German edition, especially in chapters 1, 4, 7, 8 and 9.

Our book has a twofold purpose: for one it gives an informal introduction to the mathematical and computational principles of over 150 numerical algorithms, complete with many illustrations, as well as practical guidelines to their usage. Moreover a CD provides easy access to widely tested computer codes for these and other algorithms together with a battery of test programs and test examples and additional software. The wide range of quoted literature serves to broaden the treated subjects and is meant to help our users.

The book is not meant to be a textbook on 'Numerical Analysis': we include no single proof, hardly any examples and no problems. Our emphasis when writing this book lay in explaining the principles of various methods, and in describing efficient and proven algorithms for the standard methods of Numerical Analysis, as well as documenting and developing them. Moreover this book presents many not easily accessible and newly developed algorithms which prove to be of great merit in numerical computations.

The following topics are covered in 18 chapters:

- Computer representation of numbers, error and conditioning analysis, stability of algorithms.

- Methods to solve nonlinear and algebraic equations.

- Direct and iterative algorithms for solving linear systems of equations, both dense and sparse, structured and unstructured.

- Systems of nonlinear equations.

- Eigenvalues and eigenvectors of matrices.

- Linear and nonlinear approximation.

- Polynomial and rational approximation.

- Interpolating polynomial splines and cubic fitting splines for generating smooth curves.

- Two–dimensional splines, surface splines, Bézier splines, and B–splines.

- Akima and Renner subsplines.

- Numerical differentiation.

- Numerical integration.

- Numerical cubature.

- Initial and boundary value problems for ordinary differential equations.

Many algorithms in this book can only be found in the primary literature and are not available in programmed form elsewhere. Several algorithms were developed specifically for this book. These include: specific surely converging methods for solving nonlinear equations; methods for solving linear systems of equations with many specially structured system matrices as they appear naturally for splines e.g., in general or symmetric banded form with and without sparsity; Shepard interpolation; one– and multidimensional splines of several

types; Akima and Renner splines with and without rounding of corners and applications for finding arc length; numerical quadrature formulas of all stable orders, partially implemented adaptively; cubature for rectangular and triangular regions; stable adaptive initial value problem solvers and embedding methods up to order 9; effective methods of automatic step size control.

Our programs have proved their worth in the field. They have been widely applied and tested and are kept up to date continuously. We also keep developing and implementing new variants of existing methods. Clearly all of this cannot be detailed completely in the text portion. For more information on current developments and new additional software, we refer our users to our WWW sites below. To illustrate, our programs include versions of the FFT for node numbers not a power of two. Likewise surface splines for interpolation triples $(x_i, y_i, z_i = f(x_i, y_i))$ as described in the text part of the book have been augmented in the program section by surface splines for 4–tuples (x_i, y_i, z_i, f_i) to interpolate functions $w = f(x, y, z)$ as well.

All our programs supply the user with detailed error listings, should such be encountered. Many of our algorithms then propose alternate methods. The programs themselves are completely annotated. Our Numerical Library is used in industry, universities and research institutions. Over the last few years, we have issued more than 3000 individual site licenses and more than 50 campus licenses. To enhance this Numerical Library, we have developed additional software packages such as 2D/3D graphics programs, a symbolic language interpreter, and a program to plot implicit planar curves $F(x, y) = 0$ on a rectangle for example.

With the users of our previous German editions we have developed a steady and pleasant exchange of ideas over the years that has helped us stay up to date. We would be delighted if this exchange would continue with our users in English now. We shall welcome all comments and suggestions heartily.

We are indebted to many authors for their research work and to many programmers who have realized our programs. Especially we thank Andreas Beumers in Aachen, who has used his expertise to typeset this book in LaTeX and has produced the final printer's copy. Most of the drawings were done by Björn Terwege in PiCTeX in Aachen.
The editing work on two continents would not have been possible without Rosie Torbert, Darryl Hankerson and Greg Harris at Auburn, and e-mail.
An early English translation was started by Maria Schon of Aachen; Hans Neubarth, Doris Eggermann and Jörg Lindemann read the proofs.
We thank Reinhard Wodicka for many useful hints and discussions and cooperation to the sections on B–splines and Akima and Renner subsplines.
Finally we thank Springer Verlag in Heidelberg for supporting this book. Special mention here is due to H. Engesser for his support and help.

Almost none of our work would have been possible to accomplish without the continuous help and encouragement from our friends and families.

February 1996

Gisela Engeln–Müllges

Frank Uhlig

Information

on campus and site licenses for the Numerical Library in the programming languages FORTRAN 77/90, Turbo Pascal, ANSI C, QuickBASIC, as well as on further software packages can be obtained by FAX or e–mail:

Prof. Dr. Gisela Engeln–Müllges
Kesselstraße 88
D–52076 Aachen–Lichtenbusch

FAX: +49 2408 7812
e–mail: engeln–muellges@fh–aachen.de

Frank Uhlig
706 Cary Dr.
Auburn, AL 36830
USA

FAX: (334) 844 6555
e–mail: uhligfd@mail.auburn.edu

Further information together with current updates are to be found on the WWW at:

http://www.fh–aachen.de/home/engeln–muellges/num–alg.html or
http://www.auburn.edu/num–alg.html

Notations

\Rightarrow	implies, logical implication
\Leftrightarrow	if and only if, iff
$:=$	set equal to by definition
$\overset{!}{=}$	equality to be shown or obtained
$<,\ \leq$	less than, less than or equal to
$>,\ \geq$	greater than, or greater than or equal to
$a \ll b$	a is significantly smaller than b
$\approx,\ \doteq$	approximately
\equiv	identity
\sim	proportional to
$\{a_1, a_2, \ldots\}$	set with elements a_1, a_2, \ldots
$\{x \mid \ldots\}$	set of all x for which \ldots is true
\in	element of
\notin	not an element of
\subseteq	contained in, a subset of
\subset	properly contained in, a proper subset of
$\not\subset$	not a subset of
\mathbb{N}	set of positive integers 1, 2, 3, ...
\mathbb{N}_0	set of nonnegative integers 0, 1, 2, ...
\mathbb{Z}	set of integers ..., $-2, -1, 0, 1, 2, \ldots$
\mathbb{Q}	set of rational numbers
\mathbb{R}	set of real numbers
\mathbb{C}	set of complex numbers
$\mathbb{R}^+, \mathbb{R}^-$	set of positive or negative real numbers, respectively
(a, b)	open interval from a to b, $a < b$
$[a, b]$	closed interval from a to b, $a \leq b$
$[a, b)$	half open interval from a to b (open on the right), $a < b$

$(a, b]$	half open interval from a to b (open on the left), $a < b$				
$n!$	n factorial, $n! := 1 \cdot 2 \cdot 3 \cdots n$ for $n \in \mathbb{N}$ with $0! := 1$				
$\binom{n}{k}$	n choose k, $\binom{n}{k} := \frac{n!}{k!(n-k)!}$ for $k \in \mathbb{N}_0$, $k \leq n$, $n \in \mathbb{N}$				
$\prod_{i=1}^{n} a_i$	the product $a_1 \cdot a_2 \cdot a_3 \ldots a_n$				
$\sum_{i=1}^{n} a_i$	the sum $a_1 + a_2 + \ldots + a_n$				
\int_{a}^{b}	definite integral from a to b				
i	imaginary unit, $i := \sqrt{-1}$				
e	Euler constant, $e = 2.718\ 281\ 828\ 459\ 045\ 235\ 36\ldots$				
π	π, $\pi = 3.141\ 592\ 653\ 589\ 793\ 238\ 46\ldots$				
$	a	$	magnitude of a, $	a	:= \begin{cases} a & \text{for} \quad a \geq 0 \\ -a & \text{for} \quad a < 0 \end{cases}$
$\| \cdot \|$	norm of \cdot				
$\{a_k\}$	sequence of a_k				
$\lim_{k \to \infty} a_k$	limit of a_k for $k \to \infty$				
$\max \{f(x) \mid x \in [a, b]\}$	maximum of the function $f(x)$ for $x \in [a, b]$				
$\min \{	M_i	\mid i = 1, 2, \ldots, m\}$	minimum over $	M_i	$ for $i = 1, 2, \ldots, m$
(x, y)	ordered pair				
(x_1, x_2, \ldots, x_n)	n tupel				
$f : I \to \mathbb{R}$	a real valued function f from I to \mathbb{R}				
$x \to f(x), x \in D$	x is mapped to $f(x)$ for $x \in D$				
$f', f'', f''', f^{(4)}, \ldots, f^{(n)}$	first, second, third, fourth, \ldots, n^{th} derivative of f				
$C[a, b]$	set of continuous functions on $[a, b]$				
$C^n[a, b]$	set of n fold continously differentiable function on $[a, b]$				
\mathbb{R}^n	n dimensional real space				
$A = O(h^q)$	$	A/h^q	\leq C$ for $h \to 0, C = const$		
sign (a)	sign of a				
$\mathbf{x}, \mathbf{y}, \mathbf{z}, \ldots$	vectors				
$\mathbf{A}, \mathbf{B}, \mathbf{C}, \ldots$	matrices				
$\mathbf{0}$	zero vector				
$\mathbf{x} \times \mathbf{y}$	cross product of two vectors \mathbf{x} and \mathbf{y}				
\mathbf{I}	identity matrix				
\mathbf{A}^T	transpose matrix of \mathbf{A}				

$\mathbf{x}^T = (x_1, x_2, \ldots, x_n)$ transpose vector of $\mathbf{x} = \begin{pmatrix} x_1 \\ \vdots \\ x_n \end{pmatrix}$

\mathbf{A}^{-1} inverse matrix of a square matrix \mathbf{A}

$\det(\mathbf{A})$ determinant of a square matrix \mathbf{A}

[ABC] see quote in the bibliography under [ABC]

w.l.o.g. without loss of generality

Contents

Chapter 1

Computer Numbers, Error Analysis, Conditioning, Stability of Algorithms and Operations Count

1.1 Definition of Errors

In general a numerical procedure will only give an approximate value A for a desired number a. This problem leads us to study error phenomena first.

DEFINITION 1.1 (*True* and *absolute error*).
If A is an approximate value of a number a, then the difference

$$\Delta_a = a - A$$

is called the *true error* of A, while its absolute value

$$|\Delta_a| = |a - A|$$

is called the *absolute error* of A.

In most cases the number a is unknown. Thus it is usually impossible to determine either a true or an absolute error of the approximate value A of a. Consequently one often tries to find an upper bound $\delta_a > 0$ instead for the absolute error $|\Delta_a|$ of A, so that $|\Delta_a| \leq \delta_a$.

DEFINITION 1.2 (*Error bound for the absolute error, maximum absolute error*).
If $|\Delta_a|$ is the absolute error of an approximate value A of a, and if $\delta_a > 0$

is an upper bound for $\mid \Delta_a \mid$, i.e.,

(1.1) $$\mid \Delta_a \mid \le \delta_a,$$

then δ_a is called an *error bound for the absolute error* of A.

Since $\mid \Delta_a \mid = \mid a - A \mid \le \delta_a$, we have

(1.2) $$A - \delta_a \le a \le A + \delta_a, \quad \text{or} \quad a \in [A - \delta_a, A + \delta_a].$$

In order to be able to compare different approximations A of a, we introduce the relative error:

DEFINITION 1.3 (*Relative error*).
If $|\Delta_a|$ is the error of an approximate value A for the number a, where both A and a are nonzero, then the quotient

$$|\epsilon_a| := \frac{|\Delta_a|}{|a|} = \frac{|a - A|}{|a|} \text{ for } a \ne 0, \text{ or } |\epsilon_a| := \frac{|\Delta_a|}{|A|} = \frac{|a - A|}{|A|} \text{ for } A \ne 0,$$

is called the *true relative error* of A.
For estimating the accuracy of approximations A of a, the magnitude of the true relative error $|\epsilon_a|$ is much more useful and this is customarily called the *relative error* of A.

DEFINITION 1.4 (*Percentage error*).
If A is an approximate value for $a \ne 0$, then

$$\left| \frac{a - A}{a} \right| \cdot 100$$

is called the *percentage error* of A, and any σ_a with

$$\left| \frac{a - A}{a} \right| \cdot 100 \le \sigma_a$$

is called an *error bound of the percentage error*.

1.2 Decimal Representation of Numbers

For every integer a, there is exactly one decimal representation in powers of ten of the form

$$a = \pm(a_n 10^n + a_{n-1} 10^{n-1} + \ldots + a_1 10^1 + a_0 10^0) = \pm \sum_{k=0}^{n} a_k 10^k$$

with coefficients $a_k \in \{0,1,2,3,4,5,6,7,8,9\}$ and a nonnegative integer n. To get the *decimal representation* of a one writes down the *digits* a_k in decreasing order in the following fashion:

(1.3) $$a = \pm a_n a_{n-1} \ldots a_1 a_0.$$

If a is not an integer, then a has a representation in decreasing powers of ten with at least one term of the form

$$b_k 10^{-k} \quad \text{with} \quad k \in \mathbb{N} \text{ and } b_k \in \{1,2,3,4,5,6,7,8,9\}.$$

To separate the nonnegative powers of ten from the negative powers, one inserts a *decimal point* between a_0 and b_1 in the *decimal fraction* of a:

(1.4) $$a = \pm \left(\sum_{k=0}^{n} a_k 10^k + \sum_{k=1}^{\infty} b_k 10^{-k} \right) = \pm a_n a_{n-1} \cdots a_1 a_0 \, . \, b_1 b_2 \cdots b_t \cdots .$$

The integers b_k, $0 \le b_k \le 9$, to the right of the decimal point are called *decimals*. If for a given number a there is an index j with $b_j \ne 0$ while for all subsequent indices $b_{j+1} = b_{j+2} = \cdots = 0$, then a is said to have a finite decimal representation, otherwise a has an infinite decimal representation. The following is well known:

LEMMA 1.5 Every rational number p/q with relatively prime $p \in \mathbb{Z}$ and $q \in \mathbb{N}$ can be represented by a finite or an infinite periodic decimal fraction, each irrational number can be represented by an infinite, nonperiodic decimal fraction.

DEFINITION 1.6 (*Significant digits*).
All digits of a decimal representation (1.3) or (1.4), beginning with the first nonzero digit, are called *significant digits*.

DEFINITION 1.7 (*Normalized decimal floating point representation*).
Every real number $a \neq 0$ can be represented in the form

$$
\begin{aligned}
a &= \pm(.d_1 d_2 \ldots d_z d_{z+1} \ldots) \cdot 10^k \\
&= \pm m \cdot 10^k \quad \text{with} \quad 0.1 \leq m < 1 \quad \text{and} \quad k \in \mathbb{Z}.
\end{aligned}
$$

(1.5)

(1.5) is called the *normalized (decimal) floating point representation* of a.
m is its *mantissa*, k the *exponent*. If the mantissa has t significant digits
for some $t \in \mathbb{N}$, then it is said to carry t digits.

A computer has only a finite set of machine numbers available for its computations. Its mantissas are usually constricted in length to between 7 and 9 digits, and to twice that in *double precision* arithmetic. Moreover, the exponent $k \in \mathbb{Z}$ is usually bounded by $|k| \leq 100$ or 178 or some finite number. Thus if two large numbers are added or multiplied, the result might *overflow* the reservoir of numbers that can be represented in the computer. Similarly algebraic operations might result in a positive number that is smaller than the smallest positive number representable by the computer and we encounter *underflow*.
Overflow is in general fatal for the execution of a program. The standard corrective action for underflow is to continue computations with the value zero.

Often a decimal fraction a must be approximated by a machine number A which carries less digits than a. This is achieved by *rounding a*.

DEFINITION 1.8 (*Correct decimals*).
A number a in the form (1.4) with more than t decimals is correctly
rounded to the machine number A with t decimals, if

$$
\mid a - A \mid \leq \frac{1}{2} \cdot 10^{-t}.
$$

In this case the approximate value A for a has t *correct decimals*.

DEFINITION 1.9 (*Statistically correct rounding*).
If
$$
\mid a - A \mid = \frac{1}{2} \cdot 10^{-t},
$$
we round A to the lower value, provided b_t (1.4) is even; and we round A
up if b_t is odd.

DEFINITION 1.10 (*Correct digits*).
If A has been obtained from a by correct rounding to t decimals, then the digits in A that are associated with the power 10^{-t} or higher are called *correct digits* of A. Leading zeros are omitted.

The last digit of an approximate number should ideally be a correct digit.

THEOREM 1.11. Let a have the representation (1.5). If A is a t-digit decimal floating point number which was derived from a by rounding to a t-digit mantissa, then:

1) for the absolute error of A we have

$$|\Delta_a| = |\, a - A| \leq \frac{1}{2} \cdot 10^{k-t} \quad , \text{ and}$$

2) the relative error ϵ_a satisfies for $a \neq 0$

$$|\epsilon_a| = \left| \frac{a - A}{a} \right| \leq 5 \cdot 10^{-t}, \text{ or } |\epsilon_a| = \left| \frac{a - A}{A} \right| \leq 5 \cdot 10^{-t}$$

if $A \neq 0$.

For programming applications the following *mixed test* is quite useful. It helps insure that either the absolute or relative error of an approximate value A for a is bounded appropriately:

(1.6) $|a - A| \leq |A| \cdot \text{RELERR} + \text{ABSERR}, \quad \text{ABSERR} \geq 0, \text{ RELERR} \geq 0.$

If $\text{RELERR} = 0$ and $\text{ABSERR} > 0$, then (1.6) checks on the absolute error and if $\text{ABSERR} = 0$ and $\text{RELERR} > 0$, then (1.6) checks the relative error. It would not make much sense to have $\text{RELERR} \cdot \text{ABSERR} \neq 0$. However, $\text{RELERR} + \text{ABSERR} > 0$ is necessary in (1.6).

The *machine constant* ϱ is the smallest positive number, which – when added to 1 – supplies a value different from 1 on the specific computer:

DEFINITION 1.12 (*Machine constant* or *elementary rounding error*).
If a computer uses floating point numbers with t-digit mantissas for powers of 10, then

$$\varrho = 5 \cdot 10^{-t}$$

> is called the *machine constant* or the *elementary rounding error* for the computer.

Knowledge of the machine constant ϱ for the computer in use is extremely important when executing numerical algorithms:
A program designed to find a zero of a given function f for example should be well satisfied with having found a value x_0 for which $|f(x_0)| \approx c \cdot \varrho$ for $c \le 4$ or 6. For requiring more accuracy would certainly be unrealistic due to the unavoidable rounding errors of the computations. Similarly, while division by zero is clearly impossible, a numerical algorithm ought not even divide by a number close to zero. In Gauß algorithm programs for example, see chapter 4.5, there is a need to check whether the pivot element satisfies $|a_{ii}| > c \cdot \varrho$ for $c \ge 2$, 4 or 6 instead of just checking that $a_{ii} \ne 0$ in order to obtain meaningful results. We shall not mention these considerations again in the text part of the book, but emphasize that all the programs include judicious use of the machine constant ϱ to avoid the kind of problems just mentioned.

In chapter 1.3.3 we shall deal explicitly with the rounding errors of arithmetic operations in the course of computations by means of *differential error analysis*.

1.3 Sources of Errors

In numerical computations, errors come from different sources. The total error (or accumulated error) of a computation from input to output is generally composed of:

1) Input errors,

2) Procedural errors (truncation error, discretization error etc.),

3) Propagation errors, and

4) Computational errors.

1.3.1 Input Errors

Input errors depend directly on the quality of the input data. Poor measurements, poor data transfer etc. will affect the input and hence the output of any algorithm. For our purposes, however, these errors are unavoidable . And we have to structure our numerical procedures in such a way as to limit the effects of input errors, see the remarks on numerical stability of an algorithm in section 1.3.4.

1.3.2 Procedural Errors

Many algorithms in numerical analysis are designed to solve an approximate problem if a given problem has no known or usable explicit solution.

This approximate problem must be formulated in such a way that

a) it can be solved numerically, and

b) its solution does not differ excessively from the one for the given problem.

Note that for a properly formulated approximate problem, its exact solution will be an approximate solution for the given problem. We call the difference between these two solutions the *procedural error*. This procedural error depends highly on the given problem and on the selected problem approximation. The procedural error is not concerned with input or computational errors in the procedure itself.

1.3.3 Error Propagation and the Condition of a Problem

The error in the output of a problem which is caused by the error in the input data is called the *propagation error*. To model the effects of propagation errors, one can view the output y of a procedure as a real-valued function f of the input data:

$$y = f(x_1, x_2, \ldots, x_n) =: f(\boldsymbol{x}) \qquad \text{with} \qquad \boldsymbol{x} = (x_1, x_2, \ldots, x_n)^T.$$

If the input x_i is known only approximately as X_i, then f will yield an approximate output

$$Y = f(X_1, X_2, \ldots, X_n) =: f(\boldsymbol{X}) \quad \text{for } \boldsymbol{X} = (X_1, X_2, \ldots, X_n)^T.$$

In the following, we will give an upper bound for the output error $\Delta_y = y - Y$, given the input errors $\Delta_{x_i} = x_i - X_i$ of the x_i :

THEOREM 1.13.

Let $G = \{\boldsymbol{x} \mid |x_i - X_i| \leq \delta_{x_i}, \ i = 1, \ldots, n\}$ and assume $\boldsymbol{x} \in G$ and $\boldsymbol{X} \in G$. If f has continuous first partial derivatives f_{x_i} in G, then there is an $\bar{\boldsymbol{x}} = (\bar{x}_1, \bar{x}_2, \ldots, \bar{x}_n)^T \in G$ with \bar{x}_i between x_i and X_i for $i = 1, \ldots, n$, so that the *propagation error* is

$$\Delta_y = y - Y = f(\boldsymbol{x}) - f(\boldsymbol{X}) = \sum_{i=1}^{n} \frac{\partial f(\overline{\boldsymbol{x}})}{\partial x_i} \Delta_{x_i}, \text{ where } \Delta x_i = x_i - X_i.$$

If $y = f(\boldsymbol{x}) \neq 0$ and $x_i \neq 0$ for all i, then the *relative propagation error* is

$$\epsilon_y := \frac{\Delta y}{y} = \sum_{i=1}^{n} \frac{x_i}{f(\boldsymbol{x})} \frac{\partial f(\bar{\boldsymbol{x}})}{\partial x_i} \epsilon_{x_i},$$

where $\epsilon_{x_i} = \dfrac{\Delta x_i}{x_i} = \dfrac{x_i - X_i}{x_i}$.

To determine the propagation error caused by an input error is the task of *error analysis*. Procedures f with small partial derivatives f_{x_i} are always preferred. However, for a given input \boldsymbol{x}, the relative propagation error strongly depends on \boldsymbol{x} and $f(\boldsymbol{x})$, i.e., on the problem itself. One should in fact view the factors

$$\kappa_i := \frac{x_i}{f(\boldsymbol{x})} \frac{\partial f(\bar{\boldsymbol{x}})}{\partial x_i}$$

as *amplification factors* of the relative input errors ϵ_{x_i}. One calls the problem of computing $f(\boldsymbol{x})$ from \boldsymbol{x} a *well-conditioned* problem for the algorithm f, if all its amplification numbers are reasonably bounded. The numbers κ_i are often called the *condition numbers* of the problem, see e.g., chapter 4.15 for the condition of matrices. Ideally, there will be no deterioration of accuracy if $|\kappa_i| \leq 1$ for all i. In reality, one would rejoice if one could assert that $|\kappa_i| \leq 100$ for all i. For then, the output $f(\boldsymbol{x})$ would have lost only about 2 digits of accuracy from the given input accuracy. If $|\kappa_i| \leq 10^4$ for all i, then 4 digits of accuracy would be lost when computing $f(\boldsymbol{x})$ from \boldsymbol{x}, so that using double precision arithmetic would be warranted in this case.

With large problems, an evaluation of propagation errors and condition numbers is very complicated and rarely possible. In such cases, it is appropriate to use statistical error estimations, see [HENR72] vol.2, p.381.

In the following, we shall list the propagation errors for the basic arithmetic operations.

1) *Propagation error of a sum* .
 Let

$$y = f(x_1, x_2) = x_1 + x_2.$$

Clearly $f_{x_i} = 1$ and by theorem 1.13

$$\epsilon_y \doteq \frac{x_1}{x_1 + x_2}\epsilon_{x_1} + \frac{x_2}{x_1 + x_2}\epsilon_{x_2} = \sum_{i=1}^{2}\kappa_i\epsilon_{x_i}.$$

If the inputs x_1 and x_2 have the same sign, then addition is a well conditioned operation since $0 \le \kappa_i = \frac{x_i}{x_1 + x_2} < 1$. If, however, x_1 and x_2 have opposite signs and are nearly equal in size, then their addition is badly conditioned. This effect is called *cancellation* and algorithms should be designed so that cancellation is avoided for all input data.

2) *Propagation error of a product.*
Let

$$y = f(x_1, x_2) = x_1 x_2.$$

Here $f_{x_1} = x_2$ and $f_{x_2} = x_1$ making $\kappa_i \approx 1$ and thus $\epsilon_y \doteq \epsilon_{x_1} + \epsilon_{x_2}$. Hence multiplication is globally well-conditioned.

3) *Propagation error of a quotient.*
Let

$$y = f(x_1, x_2) = \frac{x_1}{x_2} \quad \text{for } x_2 \ne 0.$$

Here $f_{x_1} = 1/x_2$ and $f_{x_2} = -x_1/x_2^2$. Thus $\kappa_1 \approx 1$ while $\kappa_2 \approx -1$. Hence $\epsilon_y \doteq \epsilon_{x_1} - \epsilon_{x_2}$ and division is well-conditioned.

4) *Propagation error of powers.*
Let

$$y = f(x_1) = x_1^p \text{ for } p > 0, \text{ and } x_1 > 0 .$$

Here $f_{x_1} = px_1^{p-1}$ and thus $\kappa_1 \approx p$ giving $\epsilon_y \doteq p\epsilon_{x_1}$. This makes roots globally well-conditioned and powers moderately ill-conditioned for large p. This is one of the reasons that one cannot expect reasonable results when naively trying to evaluate a polynomial

$$P_n(x) = a_n x^n + \ldots + a_1 x + a_0$$

at x_0 by summing terms of the form $a_k x_0^k$. One should use the well-conditioned Horner scheme instead, which alternates multiplications and additions, but does not form powers directly, see chapter 3.2.

1.3.4 The Computational Error and Numerical Stability of an Algorithm

If one has chosen a method for solving a problem numerically, one is still free to fine–tune the algorithm. An *algorithm* is defined as a finite set of exactly pre-scribed instructions which are to be carried out in a predetermined sequence upon the given initial data in order to find the desired output. During these calculations, an *accumulated computational error* arises due to the accumula-tion of *local computational errors*. These local computational errors originate from certain operations: Irrational numbers like π, e, $\sqrt{2}$ are replaced by finite decimal expansions (machine numbers), giving *truncation* or *rounding errors*. There are *rounding errors* within the arithmetic procedures. Sufficiently small quantities may be neglected due to underflow. Leading correct digits may be cancelled when subtracting numbers of near equal size, etc. If the number of operations of an algorithm is very high - as it is for most algorithms - , then there is always a danger of coming up with completely false results.

Algorithms that avoid these pitfalls to a large degree are generally called *numerically stable*. There are several different notions of what exactly con-stitutes numerical stability of an algorithm. We shall give three versions of a definition:

1) If all rounding errors of an algorithm are of the same size as the *un-avoidable error* $f_x \cdot \Delta_x$ for the input error Δ_x, then Bauer [BAUE65] calls the algorithm numerically stable.

2) Stewart [STEW73], p. 76, distinguishes between the theoretical algo-rithm f and its numerical realization f^*. The algorithm f^* is called numerically stable, if for each input x there is a nearby x^* so that $f(x^*)$ is close to $f^*(x)$. With this definition, originated by A. W. Kahan, the computed output of a stable algorithm approximates the exact solution for well-conditioned problems. If f^* is numerically stable in Stewart's sense, then for an ill-conditioned problem, there is an x^* near x for which $f(x^*)$ and $f^*(x^*)$ differ about equally, and most likely very much, from $f(x)$.

3) Another often used definition for numerical stability requires the exis-tence of an x^* near x so that $f^*(x^*) = f(x)$. This notion is usually employed in *backwards error analysis*, see [WILK64].

Instable algorithms are of no use in practice. For instance: Algorithms of the general form $y_{n+1} = ay_n + by_{n-1}$, where a, b are constant and $n \in \mathbb{N}$, are stable if a and b are chosen in such a way that

$$\left| \frac{a}{2} \pm \sqrt{\frac{a^2}{4} + b} \right| < 1 \quad .$$

But even when employing a stable algorithm, it is of little use to work with a very exact procedure if - say - the computation is executed in very low precision, for then there will be large computational errors.

1.4 Operations Count, et cetera

While numerically stable algorithms and well-conditioned problems are usually necessary to be able to compute satisfactory results, they are by no means sufficient: A computer is a finite time and finite storage medium. Computer operations consist of general overhead, specific overhead, memory writing and retrieval, and arithmetic operations. In a computer all arithmetic operations such as powers, roots, evaluating sin x etc. are reduced to additions and multiplications by intrinsically defined procedures. And the latter, multiplication, takes the bulk of operational time. For this reason, different algorithms are generally compared with each other in terms of their *operations count*, i.e., the number of multiplies required in their execution. Adds typically take only a fraction of the time and are generally ignored.

Some attention is usually given to minimize the amount of data read in and out of memory. For example, it is better to program:

$$A = X(1)$$
$$B = (A + 2.0) * A$$

than to program

$$B = X(1) * X(1) + X(1) * 2.0,$$

for in the latter command, $x(1)$ must be retrieved from memory three times unless the compiler has optimized this inefficienct code.

Besides the numerical stability of the chosen algorithm, a well-conditioned problem and a relatively low operations count, the last barrier to a successful computation is the size of the computer memory and the know-how to make do with the memory at hand. If a problem is too big, maybe it can be broken into smaller subproblems, such as solving a huge linear system of equations by using block matrix methods instead, see chapter 4.16. Or another method could be used as, for example, an iterative linear equation solver from chapter 5 instead of a direct method from chapter 4.

The art of computing rests on recognizing the inherent dangers of all algorithms and its success relies on balancing their ill effects in such a way that the output is meaningful. Moreover it needs to work within the constraints of time and space inherent in all computational environments.

Bibliography for Chapter 1

[ATKI89], 1; [BAUE65]; [BJÖR74], 1,2; [BRON85], P.96-100; [CONT80], 1; [ENGE87], Chap.1; [HENR72] vol.2, Chap.15,16; [NOBL65] I,I; [RICE83], Chap.3; [STET76], 1-3; [STEW73]; [STOE91], 1; [TOERN79] vol.1, 1.7, 1.8; [WERN82], I §6; [WILK64].

Chapter 2

Nonlinear Equations in One Variable

2.1 Introduction

If f is a real-valued function defined on a closed interval $I = [a, b]$, then a number $\xi \in I$ is called a *zero* or a *root* of the function f, or a *solution* of the equation

$$(2.1) \qquad\qquad f(x) = 0,$$

if $f(\xi) = 0$.

If f is an *algebraic polynomial* of the form

$$(2.2) \qquad f(x) \equiv P_n(x) = \sum_{j=0}^{n} a_j x^j \quad \text{with } a_j \in \mathbb{R} \text{ and } a_n \neq 0,$$

then the equation (2.1) is called *algebraic* and the integer n is called its *degree*. Every equation (2.1) which is not algebraic is generally called *transcendental*, e.g. $\ln x - 1/x = 0$ or $x - \sin x = 0$ are transcendental equations.

In this chapter, we shall describe algorithms for determining simple and multiple roots ξ of f in I. In doing so, we shall distinguish between classical iteration procedures (iterative methods, Newton's method, regula falsi, Steffensen method) and the so-called inclusion methods (bisection, Pegasus method, methods of Anderson-Björck and King). The inclusion procedures converge with *certainty* if the iteration begins with an interval in which f changes sign, since inclusion is maintained throughout the course of their calculations. These procedures can also be applied in case of roots of even order if one makes use of theorem 2.3 first. In general, inclusion procedures with a high order of convergence are preferred over the classical iterative procedures (see the suggestions in section 2.9). Algorithms to determine some or all roots of algebraic polynomials are found in chapter 3.

2.2 Definitions and Theorems on Roots

DEFINITION 2.1. The root ξ of a function $f \in C[a,b]$ is called a *j-fold root* for $j \in \mathbb{N}$ if f can be written in the form

$$f(x) = (x - \xi)^j g(x)$$

with a continuous function g for which $g(\xi) \neq 0$. If $j = 1$, then ξ is called a *simple root* of f. If $j > 1$, it is called a *multiple root*.

If ξ is a root of odd order, then f has a sign change at $x = \xi$. If ξ is of even order, then the graph of f touches the x-axis at ξ and f does not change sign there.

In the following we shall denote the set of p *times continuously differentiable functions* f defined on an interval $I = [a,b]$ by the symbol $C^p[a,b]$ or $C^p(I)$.

THEOREM 2.2. Let $f \in C^j[a,b]$. Then $\xi \in I = [a,b]$ is a j-fold root of f if and only if

$$f(\xi) = f'(\xi) = \ldots = f^{(j-1)}(\xi) = 0, \quad \text{while} \quad f^{(j)}(\xi) \neq 0.$$

THEOREM 2.3. If $\xi \in I$ is a j-fold root of f, and if f is sufficiently often continuously differentiable on I, then ξ is a simple root of the function g defined by

$$g(x) = \frac{f(x)}{f'(x)}.$$

THEOREM 2.4 (*Bolzano's intermediate value theorem*).
If f is continuous in $I = [a,b]$ and $f(a) \cdot f(b) < 0$, then f has at least one root $\xi \in I$.

2.3 General Iteration Procedures

2.3.1 How to Construct an Iterative Process

Instead of an equation of the form (2.1), consider an equation of the form

$$(2.3) \qquad\qquad \varphi(x) = x.$$

In the context of finding zeros of a continuous function f, equation (2.3) is interpreted as a *fixed point equation* where φ is a real-valued continuous function on I. $\xi \in I$ is called a solution of (2.3) or a *fixed point* of φ if $\varphi(\xi) = \xi$.

Equations of the form (2.1) and (2.3) are indeed equivalent:

LEMMA 2.5. If f and g are continuous functions on a closed interval I, and if $g(x) \neq 0$ for all $x \in I$, then the equations (2.1) and (2.3) with

$$(2.4) \qquad\qquad \varphi(x) := x - f(x)g(x)$$

have the same solutions in the interval I, i.e., the two equations are equivalent.

Each suitable choice of g supplies an equation (2.3) equivalent to (2.1). It is often possible to change an equation (2.1) into various equations of the form (2.3) by solving (2.1) for x in different ways, see the example at the end of this section.

If we are given a fixed point equation of the form (2.3) on an interval I, then starting from an *initial value* $x^{(0)} \in I$, a sequence $\{x^{(\nu)}\}$ can be constructed by setting

$$(2.5) \qquad\qquad x^{(\nu)} := \varphi(x^{(\nu-1)}) \ \text{ for } \nu = 1, 2, \dots.$$

This sequence can only be constructed if for every $\nu = 1, 2, \dots$

$$x^{(\nu)} = \varphi(x^{(\nu-1)}) \in I.$$

If the sequence $\{x^{(\nu)}\}$ converges, i.e., if

$$\lim_{\nu \to \infty} x^{(\nu)} = \xi \in I,$$

then ξ is a solution of equation (2.3). In this case

$$\xi = \lim_{\nu \to \infty} x^{(\nu)} = \lim_{\nu \to \infty} x^{(\nu+1)} = \lim_{\nu \to \infty} \varphi\left(x^{(\nu)}\right) = \varphi\left(\lim_{\nu \to \infty} x^{(\nu)}\right) = \varphi(\xi).$$

Such a procedure is usually called an *iterative procedure*. The definition (2.5) is called the *iteration rule*. The procedure advances in *iteration steps*. The function φ is called its *iteration function*. The sequence $\{x^{(\nu)}\}$ is called *iteration sequence*. The iteration steps for $\nu = 1, \ldots, N$ with $N \in \mathbb{N}$ give the algorithm of the iteration procedure for the initial value $x^{(0)}$:

$$\begin{aligned}
x^{(0)} &= \text{initial value,} \\
x^{(1)} &= \varphi(x^{(0)}), \\
x^{(2)} &= \varphi(x^{(1)}), \\
&\vdots \\
x^{(N)} &= \varphi(x^{(N-1)}).
\end{aligned}$$

Without using lemma 2.5 directly, one can often construct equivalent fixed point equations by algebraic manipulations of the given f. For example, to solve $f(x) = \ln x - 1/x = 0$, one can form the equivalent equations $x = 1/\ln x$ or $x = e^{1/x}$. Thus $\varphi_1(x) = 1/\ln x$ and $\varphi_2(x) = e^{1/x}$ are two different iteration functions for the same root-finding problem. We shall look at this example again in chapter 2.4.2 and determine the order of convergence for each of the methods φ_1 and φ_2 experimentally there.

2.3.2 Existence and Uniqueness of Solutions

THEOREM 2.6 (*Existence theorem*).
For a finite closed interval I, the equation $\varphi(x) = x$ has at least one solution $\xi \in I$, if

(i) φ is continuous on I,

(ii) $\varphi(x) \in I$ for all $x \in I$.

In order to determine whether the solution of (2.3) is unique, one generally relies on a so-called *Lipschitz condition* for φ:

If there exists a number L with $0 \le L < 1$, so that

$$(2.6) \qquad |\varphi(x) - \varphi(x')| \le L|x - x'| \quad \text{for all } x, x' \in I,$$

then (2.6) is a Lipschitz condition for the function φ. The constant L is called its *Lipschitz constant* on I. A function φ that fulfills a Lipschitz condition (2.6) is called *Lipschitz-bounded*. A differentiable function φ for which

(2.7) $|\varphi'(x)| \leq L < 1$

holds for all $x \in I$ is always Lipschitz-bounded.

THEOREM 2.7 (*Uniqueness theorem*).
The equation $\varphi(x) = x$ has at most one solution $\xi \in I$ if φ is Lipschitz-bounded in the interval I.

Since a Lipschitz-bounded function φ is continuous in I, we obtain from theorem 2.6:

THEOREM 2.8 (*Existence and uniqueness theorem*).
Let I be a finite closed interval. Then the equation $\varphi(x) = x$ has exactly one solution $\xi \in I$ if

(i) $\varphi(x) \in I$ for all $x \in I$,

(ii) φ is Lipschitz-bounded on I.

2.3.3 Convergence and Error Estimates of Iterative Procedures

THEOREM 2.9 (*Fixed point theorem*).
Let $I = [a, b]$ and assume that the function φ satisfies

(i) φ is a mapping of an interval I into itself $(\varphi : I \rightarrow I)$, i.e., $\varphi(x) \in I$, for all $x \in I$,

and

(ii) φ is Lipschitz-bounded on I.

Then φ has exactly one fixed point $\xi \in I$ with $\varphi(\xi) = \xi$. This fixed point ξ can be reached iteratively from any starting point $x^{(0)} \in I$ by setting

$$x^{(\nu)} = \varphi(x^{(\nu-1)}) \quad \text{for } \nu = 1, 2, \ldots,$$

i.e., $\lim_{\nu \to \infty} x^{(\nu)} = \xi$.

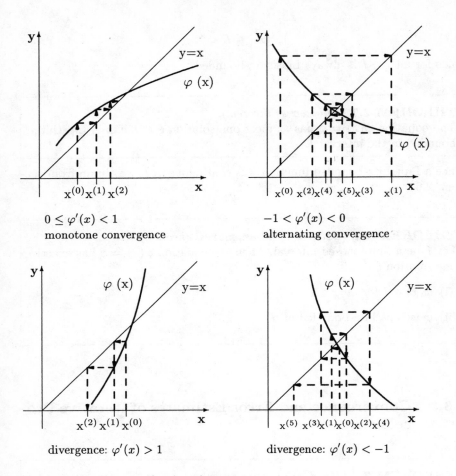

Figure 2.1: Convergence of an iterative process

THEOREM 2.10 (*Theorem of Collatz*):
If $|\varphi'(x)| \leq L < 1$ for $|x - x^{(0)}| \leq r$ and if $|\varphi(x^{(0)}) - x^{(0)}| \leq (1 - L)r$,
then all iterated values $x^{(\nu)} = \varphi(x^{(\nu-1)})$ lie in the circle $|x - x^{(0)}| \leq r$ and
converge towards the unique solution ξ of $\varphi(x) = x$.

An approximate iterative solution $x^{(\nu)}$ will differ from the exact solution ξ by
the error $\Delta^{(\nu)} := x^{(\nu)} - \xi$, where we assume for the moment that no compu-
tational errors have been made in obtaining $x^{(\nu)}$. Let us find a bound α for
the absolute error $|\Delta^{(\nu)}|$:

1) The *a posteriori estimate* of the error

(2.8) $$|\Delta^{(\nu)}| = |x^{(\nu)} - \xi| \le \frac{L}{1-L}|x^{(\nu)} - x^{(\nu-1)}| = \alpha$$

gives an estimate derived from two successive iterates $x^{(\nu-1)}$ and $x^{(\nu)}$.

2) The *a priori estimate* of the error

(2.9) $$|\Delta^{(\nu)}| = |x^{(\nu)} - \xi| \le \frac{L^{\nu}}{1-L}|x^{(1)} - x^{(0)}| = \beta$$

gives an estimate based on knowing $x^{(0)}$ and the first iterate $x^{(1)}$ only. Note that $\alpha \le \beta$.

The a priori error estimate of (2.9) can already be used after the first iteration. It is often used to estimate the number ν of iteration steps required so that $x^{(\nu)}$ will lie within a given distance of the fixed point ξ. The a posteriori error estimate in (2.8) can be computed only during the course of the calculations. It supplies a better bound than the a priori error estimate and therefore it is often preferred. If one wants to obtain rapid convergence, the function φ and the associated interval I should be selected in such a way that $L < 1/5$, see [DÖRI69].

If $0 < L \le 1/2$, then in (2.8) $|x^{(\nu)} - \xi| \le |x^{(\nu)} - x^{(\nu-1)}|$. Hence the absolute error of $x^{(\nu)}$ is lower (for $L < 1/2$) or at most equal (for $L = 1/2$) to the absolute difference of the last two approximations $x^{(\nu)}$ and $x^{(\nu-1)}$ of ξ.
For $1/2 < L < 1$ the absolute error of $x^{(\nu)}$ can, however, exceed $|x^{(\nu)} - x^{(\nu-1)}|$, so that the iteration sequence does not say anything by itself about the error. In the case of *alternating convergence*, i.e, if $-1 < \varphi'(x) \le 0$, the error estimate

$$|x^{(\nu)} - \xi| \le \frac{1}{2}|x^{(\nu)} - x^{(\nu-1)}|$$

can be used. For $L = 1/3$ this is identical with (2.8); for $L > 1/3$, it is slightly better than (2.8).

Error estimates without using a Lipschitz constant L.

In practice a method by J.B. Kioustelidis in [KIOU78] is recommended. In general, it gives more precise bounds than the a posteriori error estimate. Besides, it does not assume knowledge of the Lipschitz constant L for φ:

Let $x^{(\nu)}$ be an iterative approximation of a root ξ of odd order and assume for $\varepsilon > 0$ that

(2.10) $$f(x^{(\nu)} - \varepsilon)f(x^{(\nu)} + \varepsilon) < 0.$$

Then by the intermediate value theorem (theorem 2.4), $f(\xi) = 0$ with $\xi \in (x^{(\nu)} - \varepsilon, x^{(\nu)} + \varepsilon)$. Hence $|x^{(\nu)} - \xi| < \varepsilon$.

This idea is normally implemented as follows: Choose an ε as the break-off criterion for the iteration, e.g. $\varepsilon = 10^{-k}$ with $k \in \mathbb{N}$. Then examine the condition (2.10) with computations carried out in double precision. If (2.10) is satisfied, ε is an upper bound for the absolute error. To obtain an even better upper bound, carry out the calculations once more with a smaller ε_1 (e.g. with $\varepsilon_1 = 10^{-k-1}$). If (2.10) is satisfied for ε_1, continue in an analogous way until (2.10) is not satisfied for an $\varepsilon_j = 10^{-k-j}$. Then $\varepsilon_{j-1} = 10^{-k-j+1}$ is the most exact error bound thus obtainable.

If ξ is a zero of even order of f, then due to theorem 2.3, one must employ the function $g = f/f'$ instead of f for the error estimate. In this case ξ is only a simple zero of g and (2.10) becomes:

$$g(x^{(\nu)} + \varepsilon)g(x^{(\nu)} - \varepsilon) < 0.$$

Computational error.

If $\varepsilon^{(\nu)}$ is the local computational error of the ν^{th} iterate that was generated by computing $x^{(\nu)} = \varphi(x^{(\nu-1)})$ and if $|\varepsilon^{(\nu)}| \leq \varepsilon$ for $\nu = 1,2,...$, then the accumulated computational error $r^{(\nu)}$ at the ν^{th} iteration step satisfies

$$|r^{(\nu)}| \leq \frac{\varepsilon}{1 - L} \quad \text{for the Lipschitz constant } L \text{ if } 0 \leq L < 1.$$

Note that the error bound $\varepsilon/(1 - L)$ is independent of the iteration step count ν. Hence the algorithm (2.5) is stable, see the definitions in chapter 1.3.4.

Since the total error is composed of the procedural error and the computational error, ideally these two errors should be allowed to have the same order of magnitude. Then, given L and (2.9), we have

$$\frac{L^\nu}{1 - L}|x^{(1)} - x^{(0)}| \approx \frac{\varepsilon}{1 - L}.$$

Thus the number $\nu = \nu_0$ of iteration steps needed to achieve the given accuracy level ϵ is

(2.11) $$\nu_0 \approx \left(\log \frac{\varepsilon}{|x^{(1)} - x^{(0)}|} \right) / \log L.$$

2.3.4 Practical Implementation

To solve a given equation $f(x) = 0$ iteratively one proceeds as follows:

ALGORITHM 2.11.

Task: Find a solution ξ of the equation $f(x) = 0$.

1^{st} *step*: Transform $f(x) = 0$ into an equivalent equation of the form $\varphi(x) = x$.

2^{nd} *step*: Find an interval I that surely contains a zero of f; refer to "Determination of a starting point" below.

3^{rd} *step*: Examine, whether the function φ fulfills the conditions of theorem 2.9 for all $x \in I$; see "Test for convergence" below.

4^{th} *step*: Formulate the iteration rule $x^{(\nu)} = \varphi(x^{(\nu-1)})$ after selecting an arbitrary starting point $x^{(0)} \in I$.

5^{th} *step*: Calculate the iteration sequence $\{x^{(\nu)}\}$, $\nu = 1, 2, \ldots$.
The iteration is to be continued until for a $\nu \in \mathbb{N}$ and given error bounds $\delta_i > 0$:

(2.12) $|x^{(\nu)} - x^{(\nu-1)}| \leq \delta_1 |x^{(\nu)}|$ (relative iteration error), or

(2.13) $|x^{(\nu)} - x^{(\nu-1)}| \leq \delta_2$ (absolute iteration error), or

(2.14) $|f(x^{(\nu)})| \leq \delta_3$ (absolute functional error).

The first condition is preferred to the second. In fact we recommend replacing (2.12) and (2.13) by a mixed error test such as (1.6) which accounts for both.

6^{th} *step*: Estimate the error, see section 2.3.3.

Determination of a starting point.

To find a starting point for the iteration, it is neccessary to find an interval I in which the conditions of theorem 2.9 hold. The following heuristic approaches might help to find such an I.

a) *Plotting:* We can write the given function f as the difference $f = f_1 - f_2$ of two functions f_1 and f_2 and solve the equation $f(x) \equiv f_1(x) - f_2(x) = 0$ by solving $f_1(x) = f_2(x)$. Here we determine those x-values for which the graphs of f_1 and f_2 intersect. In order to succeed, we must express $f = f_1 - f_2$ in such a way that f_1 and f_2 are both easy to plot. One should use functions f_1, f_2 that can be found in tables or which can be drawn easily. An interval around the perceived

intersection of the graphs of f_1 and f_2 will serve as an initial guess for I, and $x^{(0)} \in I$ as a starting point for the iteration.

b) *Rough estimates:* Evaluating f at a small number of points makes it often possible to find an interval $I = [a, b]$ with $f(a) \cdot f(b) < 0$ if the zero of f has odd order.

c) *Estimating the iteration function:* Here one uses the equivalent equation $\varphi(x) = x$. If φ is bounded, or $|x| = |\varphi(x)| \leq r$, then one can choose $I = [-r, +r]$ for the interval I. Due to the continuity of φ, it might be possible to find bounds $a \leq x = \varphi(x) \leq b$. Then set $I = [a, b]$. If the conditions of theorem 2.9 are satisfied in I, any value $x^{(0)} \in I$ may be chosen as a starting point.

Test of convergence.

In most practical cases it is very difficult to check the sufficiency conditions for convergence in theorem 2.9, such as the Lipschitz conditions (2.6) or (2.7) for a given interval I and function φ. This difficulty alone makes the inclusion procedures much more attractive for zeros of odd order.

With a general iteration procedure, one should first try to verify that

$$\text{(i)} \quad |\varphi'(x^{(0)})| < 1 \quad \text{and} \quad \text{(ii)} \quad \varphi(x^{(0)}) \in I$$

for the chosen starting point $x^{(0)} \in I$. Then iterate and after each iteration step, re-examine whether any of the conditions

(i) $|\varphi'(x^{(\nu-1)})| < 1,$ or

$$|x^{(\nu)} - x^{(\nu-1)}| = |\varphi(x^{(\nu-1)}) - \varphi(x^{(\nu-2)})| < |x^{(\nu-1)} - x^{(\nu-2)}|, \quad \text{or}$$

(ii) $\varphi(x^{(\nu-1)}) \in I$

are satisfied. Additionally one inserts a break-off criterion (2.12), (2.13) or (2.14) into the algorithm. These will be satisfied for sufficiently large ν if φ satisfied a Lipschitz condition.

2.4 Order of Convergence of an Iterative Procedure

2.4.1 Definitions and Theorems

For iterative procedures, the order of convergence can be used as a measure for the computational work required.

DEFINITION 2.12 (*Order of convergence*).
The iteration sequence $\{x^{(\nu)}\}$ converges of at least p^{th} order towards ξ if there is a constant $0 \le M < \infty$ so that for $p \in \mathbb{R}$, $p \ge 1$

(2.15) $$\lim_{\nu \to \infty} \frac{|x^{(\nu)} - \xi|}{|x^{(\nu-1)} - \xi|^p} = M.$$

In this case the iterative procedure $x^{(\nu)} = \varphi(x^{(\nu-1)})$ is called a procedure of at least p^{th} order. It is exactly of order p if $M \ne 0$.

(2.15) thus shows that the error of the ν^{th} approximation $x^{(\nu)}$ equals about M times the p^{th} power of the error of the $(\nu-1)^{th}$ approximation. The rate of convergence increases as the order of convergence does. If $p = 1$, one speaks of *linear convergence*; for $p = 2$ we have *quadratic convergence* and, in general, if $p > 1$, we speak of *superlinear convergence*.

THEOREM 2.13. Assume $\varphi \in C^p[a,b]$. If $\lim_{\nu \to \infty} x^{(\nu)} = \xi$ with $\varphi(\xi) = \xi$ and
$$\varphi'(\xi) = \varphi''(\xi) = \ldots = \varphi^{(p-1)}(\xi) = 0, \quad \text{while } \varphi^{(p)}(\xi) \ne 0,$$
then $\mathrm{x}^{(\nu)} = \varphi(\mathrm{x}^{(\nu-1)})$ is an iterative procedure of order p with

$$M = \frac{1}{p!} |\varphi^{(p)}(\xi)| \le \frac{1}{p!} \max_{x \in I} |\varphi^{(p)}(x)| \le M_1.$$

If p = 1, then $M = |\varphi'(\xi)| < 1$.

Moreover, to increase the order of convergence one can always make use a theorem proved in [COLL68], p.231.

THEOREM 2.14. If $x^{(\nu)} = \varphi_1(x^{(\nu-1)})$ and $x^{(\nu)} = \varphi_2(x^{(\nu-1)})$ are two iterative procedures with orders of convergence p_1 and p_2, respectively, then the iteration rule

$$x^{(\nu)} := \varphi_1(\varphi_2(x^{(\nu-1)}))$$

has at least the order of convergence $p_1 \cdot p_2$.

By applying theorem 2.14, one can construct iteration procedures of an arbitrarily high order of convergence. If e.g. $x^{(\nu)} = \varphi(x^{(\nu-1)})$ is an iteration procedure of order $p > 1$, define the iteration functions $\varphi_s(x)$ by

$$\varphi_1(x) = \varphi(x), \ \varphi_s(x) = \varphi(\varphi_{s-1}(x)) \qquad \text{for} \qquad s = 2, 3, \ldots.$$

Then φ_s is an iterative procedure with order of convergence p^s.

Special case: *Quadratic convergence.*

If the iteration function φ gives a quadratically convergent procedure, then the conditions of theorem 2.9 can be simplified:

THEOREM 2.15. If $\varphi \in C^2[a,b]$ and the equation $\varphi(x) = x$ has a solution ξ in the interior of I with $\varphi'(\xi) = 0$, then there exists an interval $I_r = \{x \mid |x - \xi| \le r\}$ for some $r > 0$, so that the sequence $\{x^{(\nu)}\}$ defined by $x^{(\nu)} = \varphi(x^{(\nu-1)})$, $x^{(0)} \in I_r$, converges towards ξ at least quadratically for an arbitrary starting point $x^{(0)} \in I$.

Put more simply, under the assumptions (i) and (ii) of the above theorem, the sequence $\{x^{(\nu)}\}$ converges quadratically towards the fixed point ξ whenever $x^{(0)}$ has been chosen close enough to ξ.

One might think that the above theorem is only of theoretical importance. However it is easily possible (due to the freedom in transforming $f(x) = 0$ to an equivalent equation $\varphi(x) = x$) to construct iteration functions φ in such a way that $\varphi'(\xi) = 0$, see section 2.5.

2.4.2 Determining the Order of Convergence Experimentally

Let $\varphi : I \to I$ be an iteration function, $\xi \in I$ its fixed point and let $x^{(\nu)}$ be defined by the iteration rule $x^{(\nu)} = \varphi(x^{(\nu-1)})$. In this case, we define

$$p_\nu := \frac{\log |x^{(\nu)} - \xi|}{\log |x^{(\nu-1)} - \xi|}.$$

Then the iterative procedure $\mathrm{x}^{(\nu)} = \varphi(x^{(\nu-1)})$ has the order of convergence

$$p = \lim_{\nu \to \infty} p_\nu.$$

We can use this to determine the order of convergence of any iteration as follows: Express the iteration function φ in terms of f and use a simple function g with a known zero in the formulation (2.4) of φ. Iterate with φ for the known root of g and form the sequences $\{x^\nu\}$, $\Delta_\nu = x^{(\nu)} - \xi$ and p_ν. A test of Newton's method and his modified method for double roots applied to solving $f(x) = 1 - sin \ x$ easily yields $p_{\text{Newton}} = 1$ and

$p_{\mathrm{mod. Newton}} = 2$ experimentally, as indicated by theorems 2.16 and 2.17 below. Note that the iteration functions φ_1 and φ_2, introduced as an example for solving $f(x) = \ln x - 1/x = 0$ in chapter 2.3.1, can be rewritten as $\varphi_1(x) = x - \frac{xf(x)}{\ln x}$ and $\varphi_2(x) = e^{\ln x - f(x)}$ and these iterations experimentally yield orders of convergence $p_1 = 1 = p_2$. This indicates that for finding roots of specific functions f one may not expect to do better with custom-made iterations than with the standard procedures that we shall describe next.

To obtain a reliable estimate of the order of convergence for an iteration with φ, one should calculate the sequence $\{p_\nu\}$ for $\nu = 1, 2, \ldots$, until the break-off condition

$$|p_\nu - p_{\nu-1}| \le |p_\nu| \cdot \mathrm{RELERR} + \mathrm{ABSERR}$$

is satisfied for given bounds $\mathrm{RELERR} > 0$, $\mathrm{ABSERR} = 0$ or $\mathrm{RELERR} = 0$, $\mathrm{ABSERR} > 0$. The p_ν should be computed in double precision.

2.5 Newton's Method

2.5.1 Finding Simple Roots

If $f \in C^2[a, b]$ has a *simple root* ξ in (a,b), then $f(\xi) = 0$ and $f'(\xi) \neq 0$. Using (2.4) with $g(x) = 1/f'(x)$, the iteration function φ becomes

(2.16) $$\varphi(x) = x - \frac{f(x)}{f'(x)}.$$

And the iteration rule is given as

(2.17) $$x^{(\nu)} = \varphi(x^{(\nu-1)}) = x^{(\nu-1)} - \frac{f(x^{(\nu-1)})}{f'(x^{(\nu-1)})} \quad \text{for } \nu = 1, 2, \ldots .$$

Figure 2.2: Newton's method

Since $\varphi'(\xi) = 0$, the procedure converges at least quadratically by theorem 2.13. We can make use of theorem 2.15 to examine its convergence:

There must be an interval

$$(2.18) \qquad I_r(x) := \{x \mid |x - \xi| \le r\} \subset [a,b] \text{ for } r > 0,$$

in which the iteration function φ in (2.16) satisfies a Lipschitz condition

$$(2.19) \qquad |\varphi'(x)| = |f(x)f''(x)/f'^2(x)| \le L < 1,$$

while $f'(x) \ne 0$ for $x \in I_r$.

THEOREM 2.16. Let $f \in C^2[a,b]$. If f is three times differentiable and has a simple zero in the open interval (a,b), then there exists an interval I_r (2.18) so that the iteration $x^{(\nu)} = \varphi(x^{(\nu-1)})$ converges at least quadratically for φ defined in (2.17) for any starting point $x^{(0)} \in I_r$. On this interval I_r, the Lipschitz condition (2.19) is satisfied if f' does not vanish on I_r. Moreover the error estimates (2.8) and (2.9) hold and

$$|x^{(\nu+m)} - \xi| \le \frac{1}{M_1}\left(M_1|x^{(\nu)} - \xi|\right)^{2^m} \text{ for all } \nu, m = 0,1,2,\dots,$$

provided

$$\max_{x \in I}(|f''(x)| \ / \ 2\min_{x \in I}|f'(x)|) \le M_1.$$

If ξ is a root of f of multiplicity j, then $\varphi'(\xi) = 1 - 1/j$, so that for $j \ge 2$ the quadratic convergence of Newton's method no longer holds. The simple root-finding method of Newton can still be used here, but it would be advisable to use Newton's method for multiple roots (section 2.5.3) instead if the order j of the root is known. If the order is not known, we recommend the modified Newton's method of section 2.5.3, or a combination of both.

In (2.19), $f'(x)$ must not vanish in I_r. Moreover, $|f'(x)|$ should be bounded below, and $|f''(x)|$ and $|f(x)|$ should be bounded above. [DÖRI69], p. 28, gives a stronger error estimate provided $L \le 1/2$, that requires $f'(x) \ne 0$ and (2.19) to be satisfied only for one point $x^{(0)} \in I$.

Theorem 2.16 insures the convergence of Newton's method only for initial values $x^{(0)}$ which are sufficiently close to ξ. However, the following statement of a global nature holds ([HENR72], vol. 1, p.107):

Let $f \in C^2[a,b]$. Assume that

a) $f(a) \cdot f(b) < 0$,

b) $f'(x) \neq 0$ for all $x \in [a,b]$,

c) $f''(x) \geq 0$ (or ≤ 0) for all $x \in [a,b]$, and

d) if c is that end-point of $[a,b]$, at which $|f'(x)|$ assumes its smaller

value, then $\left| \dfrac{f(c)}{f'(c)} \right| \leq b - a$.

Under these assumptions, Newton's method converges for each $x^{(0)} \in [a,b]$ towards the unique solution $\xi \in [a,b]$ of $f(x) = 0$.

2.5.2 A Damped Version of Newton's Method

In analogy to the damped Newton method for nonlinear systems in section 6.2.1.2 and algorithm 6.5, we can describe the damped Newton's method for single equations. For $\nu = 1,2,...$, do:

(i) Compute $\Delta x^{(\nu)} := -f(x^{(\nu-1)})/f'(x^{(\nu-1)})$.

(ii) Compute an integer j so that
$$j := \min\{ \, i \mid i \geq 0, |f(x^{(\nu-1)} + 1/2^i \Delta x^{(\nu)})| < |f(x^{(\nu-1)})|\}.$$

If the condition cannot be fulfilled for $0 \leq i \leq i_{max}$, where i_{\max} is a given threshold, then the computations continue with $j = 0$.

(iii) $x^{(\nu)} := x^{(\nu-1)} + \Delta x^{(\nu)}/2^j$.

2.5.3 Newton's Method for Multiple Zeros; a Modified Newton's Method

Assume that the given function f is sufficiently often differentiable in $I = [a,b]$. If f has a zero ξ of multiplicity j in (a,b), then

$$f(\xi) = f'(\xi) = f''(\xi) = \ldots = f^{(j-1)}(\xi) = 0 \text{ and } f^{(j)}(\xi) \neq 0.$$

THEOREM 2.17 (*Newton's method for multiple roots*).
Let $f \in C^{j+1}[a,b]$ have a zero ξ in (a,b) with multiplicity $j \geq 2$. Then the iteration rule

$$x^{(\nu)} = x^{(\nu-1)} - j \frac{f(x^{(\nu-1)})}{f'(x^{(\nu-1)})} = \varphi(x^{(\nu-1)}) \text{ for } \nu = 1, 2, \ldots$$

converges to ξ in an interval I_r of type (2.18) at least quadratically.

To apply theorem 2.17 assumes the knowledge of the order j of the zero. However, j is known precisely only very rarely. Even for multiple roots whose order is unknown, it is possible to design a quadratically convergent iterative Newton method: Simply apply (2.17) to $g := f/f'$. This yields

$$(2.20) \begin{cases} x^{(\nu)} = x^{(\nu-1)} - \dfrac{g(x^{(\nu-1)})}{g'(x^{(\nu-1)})} = x^{(\nu-1)} - j(x^{(\nu-1)}) \dfrac{f(x^{(\nu-1)})}{f'(x^{(\nu-1)})} \quad \text{with} \\[2mm] j(x^{(\nu-1)}) := \dfrac{1}{1 - \dfrac{f(x^{(\nu-1)}) f''(x^{(\nu-1)})}{f'^2(x^{(\nu-1)})}} \qquad \text{for} \quad \nu = 1, 2, \ldots . \end{cases}$$

By theorem 2.3 we have:

THEOREM 2.18 (*Modified Newton's method for multiple roots*).
Assume f can be differentiated sufficiently often in $I = [a,b]$ and has a zero ξ of multiplicity $j \geq 2$ in (a,b), then the iterative procedure (2.20) converges quadratically for each $x^{(0)} \in I_r$ defined in (2.18). Moreover

$$\lim_{\nu \to \infty} x^{(\nu)} = \xi \qquad \text{and} \qquad \lim_{\nu \to \infty} j(x^{(\nu)}) = j,$$

where j is the order of the zero $\xi \in [a,b]$ of f.

2.6 Regula Falsi

2.6.1 Regula Falsi for Simple Roots

We assume that the function f is continuous in $I = [a,b]$ and has a simple zero ξ in (a,b). To use Newton's method we have to assume that f is differentiable. However, calculating f' can be very difficult in practice, see chapter 14. The *regula falsi* is an iterative method that works well for functions that are not even differentiable. It requires two initial values $x^{(0)}$, $x^{(1)}$, see also the "inclusion procedures" in section 2.8.

One-point formula with storage:

$$(2.21) \qquad x^{(\nu)} = x^{(\nu-1)} - \frac{x^{(\nu-1)} - x^{(\nu-2)}}{f(x^{(\nu-1)}) - f(x^{(\nu-2)})} f(x^{(\nu-1)}),$$

$$\text{provided } f(x^{(\nu-1)}) - f(x^{(\nu-2)}) \neq 0 \text{ for } \nu = 2, 3, \ldots$$

Thus the regula falsi (2.21) finds the zero of the line through the points $(x^{(\nu-1)}, f(x^{(\nu-1)}))$ and $(x^{(\nu-2)}, f(x^{(\nu-2)}))$ and replaces the role of the derivative $f'(x^{(\nu-1)})$ in Newton's method with the difference quotient.

Figure 2.3: Regula Falsi

If $f(x^{(\nu)}) = 0$, the procedure stops at $x^{(\nu)}$. Essential for the convergence of the regula falsi is that the initial values $x^{(0)}$, $x^{(1)}$ are sufficiently close to the zero ξ. Here is a result regarding its *convergence* ([STUM82], p.43):

THEOREM 2.19. If the function $f \in C^2[a, b]$ satisfies

$$|f'(x)| \geq m \text{ and } |f''(x)| \leq M \text{ for all } x \in (a, b), \text{ and } m, M > 0,$$

then there is an interval $I_r(\xi) \subset (a, b)$ for $r > 0$, so that $\xi \in I_r$ is the only zero of f and for each pair of initial values $x^{(0)}$, $x^{(1)} \in I_r$ with $x^{(0)} \neq x^{(1)}$ the procedure converges towards the zero ξ .

The order of convergence for the regula falsi as defined in (2.21) is known to be $p = (1 + \sqrt{5})/2 \approx 1.62$. Again it is possible to employ procedure (2.21) for the approximate calculation of multiple zeros. However, its relatively high order of convergence is then lost. The *modified regula falsi* below has a convergence order of $p \approx 1.62$ even in the case of multiple zeros. To learn more about the efficiency of the regular falsi, see section 2.9 .

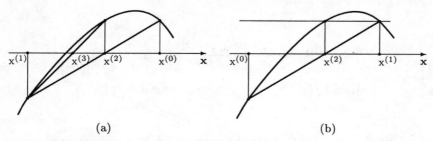

(a) (b)

Figure 2.4: Regula falsi, (a) convergence, (b) no convergence for same initial
 values but with a different numbering,

REMARK: For the same set of initial values, the convergence behaviour of
the method may be affected by their numbering.

2.6.2 Modified Regula Falsi for Multiple Zeros

If ξ is a zero of multiplicity j of f and if $|f^{(j+1)}(x)|$ is bounded in a neighbor-
hood of ξ, then ξ is a simple zero of the function

$$(2.22) \qquad h(x) = f^2(x)/(f(x + f(x)) - f(x))$$

and $|h''(x)|$ is bounded in a neighborhood of ξ. If, in the iteration rule (2.21),
one uses the function h from (2.22) instead of f, the *modified regula falsi*
converges with order $p = (1 + \sqrt{5})/2$ towards the multiple zero ξ of f. For a
proof see [KIOU79].

2.6.3 Simplest Version of the Regula Falsi

The simplest such iteration rule would be as follows:

$$(2.23) \qquad x^{(\nu)} = x^{(\nu-1)} - \frac{x^{(\mu)} - x^{(\nu-1)}}{f(x^{(\mu)}) - f(x^{(\nu-1)})} f(x^{(\nu-1)})$$

$$\text{for } \mu = 0, 1, 2, \ldots, \ \nu = 2, 3, \ldots.$$

Here $\mu < \nu - 1$ is the largest integer for which $f(x^{(\mu)}) \neq 0$, $f(x^{(\nu-1)}) \neq 0$
and $f(x^{(\mu)})f(x^{(\nu-1)}) < 0$. To start (2.23), one must find two pints $x^{(0)}$, $x^{(1)}$
with $f(x^{(0)})f(x^{(1)}) < 0$. The simple iteration formula (2.23) has only $p = 1$
as its order of convergence. However, it provides a very easy error estimate
since the required solution ξ always lies between any two values $x^{(s)}$ and $x^{(t)}$
of the sequence $\{x^{(\nu)}\}$ for which $f(x^{(s)})f(x^{(t)}) < 0$. If f is continuous on
$I = [a, b]$ and has a zero in I, then if $x^{(0)}$ and $x^{(1)}$ satisfy $f(x^{(0)})f(x^{(1)}) < 0$,
the sequence (2.23) will converge, for a proof see [COLL68], p.240.

2.7 Steffensen Method

2.7.1 Steffensen Method for Simple Zeros

Assume that the equations $\varphi(x) = x$ and $f(x) = 0$ are equivalent in $I = [a, b]$ and that φ is differentiable. Let ξ be the only solution of $f(x) = 0$ or $x = \varphi(x)$ in (a, b). According to theorem 2.9, the iteration procedure with iteration function φ converges towards ξ , provided that

(i) $|\varphi'(x)| \leq L < 1$, and (ii) $\varphi(x) \in I$ for all $x \in I$.

Using the iteration function φ, one can construct an iterative process which converges quadratically towards ξ for both $|\varphi'(x)| < 1$ and $|\varphi'(x)| > 1$. This procedure is called the *Steffensen procedure*. Compared with Newton's method, it has the advantage of not using derivatives while having the same order of convergence. This procedure is useful especially in cases where $|\varphi'|$ is very close to 1 or when $|\varphi'| > 1$.

THEOREM 2.20 (*Steffensen procedure for simple zeros*).
Let $\varphi \in C^3[a, b]$ and assume ξ is the only solution of the equation $\varphi(x) = x$ in (a, b). If $\varphi'(\xi) \neq 1$, then the iteration rule

$$x^{(\nu)} \quad = \quad x^{(\nu-1)} - \frac{(\varphi(x^{(\nu-1)}) - x^{(\nu-1)})^2}{\varphi(\varphi(x^{(\nu-1)})) - 2\varphi(x^{(\nu-1)}) + x^{(\nu-1)}} =: \Phi(x^{(\nu-1)}),$$

$$\nu \quad = \quad 1, 2, \ldots$$

converges at least quadratically towards ξ for each initial value $x^{(0)} \in I$.

Its disadvantage is the fact that, by assuming $\varphi'(\xi) \neq 1$, it can only be used for simple zeros. For alternate methods see sections 2.8, 2.9.

2.7.2 Modified Steffensen Method for Multiple Zeros

The following *modified Steffensen procedure* is designed for multiple zeros; it yields both the zero and its multiplicity (see [ESSE75]).

THEOREM 2.21 (*Modified Steffensen method for multiple zeros*).
If φ is sufficiently often differentiable in $I = [a, b]$ and if the equation $\varphi(x) = x$ has only one solution ξ in (a, b) with multiplicity $j \geq 2$, then the

iteration procedure

$$x^{(\nu)} = x^{(\nu-1)} - j(x^{(\nu-1)}) \frac{(x^{(\nu-1)} - \varphi(x^{(\nu-1)}))^2}{z(x^{(\nu-1)})} =: \Phi(x^{(\nu-1)}), \ \nu = 1, 2, \ldots$$

with

$$j(x^{(\nu-1)}) := \frac{(z(x^{(\nu-1)}))^2}{(z(x^{(\nu-1)}))^2 + (x^{(\nu-1)} - \varphi(x^{(\nu-1)}))(z(x^{(\nu-1)}) + \varphi(2x^{(\nu-1)} - \varphi(x^{(\nu-1)})) - x^{(\nu-1)})}$$

and

$$z(x^{(\nu-1)}) := x^{(\nu-1)} - 2\varphi(x^{(\nu-1)}) + \varphi(\varphi(x^{(\nu-1)}))$$

converges at least quadratically towards ξ for any initial value $x^{(0)} \in I$.
Moreover $\lim_{\nu\to\infty} j(x^{(\nu)}) = j$.

To avoid an accumulation of rounding errors one should compute in double
precision when using this modified procedure.

2.8 Inclusion Methods

In this section, we treat inclusion methods that will always converge and that
can be used for zeros of odd order as well as for zeros of even order. With zeros
of even order, on account of theorem 2.3, the procedures must be employed
not on f, but instead on the function $g(x) = f(x)/f'(x)$. Inclusion procedures
are generally preferred to algorithms derived from the regula falsi in chapters
2.6, 2.7 due to their more certain convergence and higher order of convergence.
Only the bisection method converges linearly here:

2.8.1 Bisection Method

Let $f \in C[a, b]$ be given with $f(a)f(b) < 0$. Then we can determine a zero
$\xi \in [a, b]$ of odd order by means of repeated interval bisection.

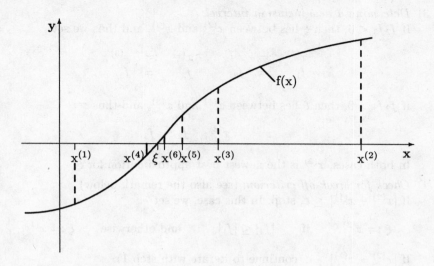

Figure 2.5: Bisection method

REMARK: The algorithm below always refers to just three points $x^{(1)}$, $x^{(2)}$ and $x^{(3)}$. This is because the end points of the inclusion interval are always designated by $x^{(1)}$, $x^{(2)}$ in the computations.

ALGORITHM 2.22 (*Bisection method*).

 Given: (i) $f \in C[a, b]$ with $f(a)f(b) < 0$,
 (ii) an error bound $\varepsilon > 0$, $\varepsilon \in \mathbb{R}$.

 Solve for: $\xi \in (a, b)$ with $f(\xi) = 0$.

Initial values: $x^{(1)} := a, x^{(2)} := b$ (inclusion interval),
 $f_1 := f(x^{(1)}), f_2 := f(x^{(2)})$.

The iteration steps are carried out as follows:

1) *Bisection of the inclusion interval*:

$$x^{(3)} := x^{(2)} + 0.5(x^{(1)} - x^{(2)}).$$

2) *Calculate the new functional value*: $f_3 = f(x^{(3)})$. If $f_3 = 0$, the calculation has ended; $\xi = x^{(3)}$.
Otherwise continue with 3).

3) *Determine a new inclusion interval:*
If $f_2 f_3 < 0$, then ξ lies between $x^{(2)}$ and $x^{(3)}$, and thus we set

$$x^{(1)} \ := \ x^{(2)} \quad , \quad x^{(2)} \ := \ x^{(3)},$$
$$f_1 \ := \ f_2 \quad , \quad f_2 \ := \ f_3.$$

If $f_2 f_3 > 0$, then ξ lies between $x^{(1)}$ and $x^{(3)}$, and thus set

$$x^{(2)} := x^{(3)}, \quad f_2 := f_3.$$

In both cases, $x^{(2)}$ is the newest best approximation for ξ.

4) *Check for break-off criterion:* (see also the remark below)
If $|x^{(2)} - x^{(1)}| \le \varepsilon$, stop. In this case, we set

$$\xi := x^{(2)}, \quad \text{if} \quad |f_2| \le |f_1|, \quad \text{and otherwise} \quad \xi := x^{(1)}.$$

If $|x^{(2)} - x^{(1)}| > \varepsilon$, continue to iterate with step 1).

Instead of the break-off criterion $|x^{(2)} - x^{(1)}| \le \varepsilon$, the mixed test (1.6)

$$|x^{(2)} - x^{(1)}| \le |x^{(2)}| \cdot \text{RELERR} + \text{ABSERR}$$

is recommended with given bounds RELERR and ABSERR for the relative and absolute errors (see section 1.2).

2.8.2 Pegasus Method

If f is continuous on the closed interval $[a,b]$ and if $f(a)f(b) < 0$, then according to theorem 2.4, f has at least one zero of odd order in (a,b). The *Pegasus method*, a variant of the regula falsi, calculates one of these zeros by repeated minimization of the root inclusion interval. It always converges if initially $f(a)f(b) < 0$. The order of convergence is $p \approx 1.642\ldots$ in the case of a simple zero (see [DOWE72], [DOWE71]).

ALGORITHM 2.23 (*Pegasus Method*).

Given: (i) $f \in C[a,b]$ with $f(a)f(b) < 0$,
 (ii) an error bound $\varepsilon > 0$, $\varepsilon \in \mathbb{R}$.

Solve for: Zero $\xi \in [a, b]$ of f, approximated in such a way that the differ-
ence between the last two iterative values is smaller than ϵ.

Initial values: $x^{(1)} := a$, $x^{(2)} := b$; Compute
$$f_1 := f(x^{(1)}), \ f_2 := f(x^{(2)}).$$

1) *Secant step:* Compute the slope s_{12} of the line joining the points
$(x^{(1)}, f_1)$, $(x^{(2)}, f_2)$:
$$s_{12} = \frac{f_2 - f_1}{x^{(2)} - x^{(1)}}.$$

Find the intersection $x^{(3)}$ of this secant with the x-axis:
$$x^{(3)} = x^{(2)} - f_2/s_{12}.$$

2) *Calculate:* $f_3 = f(x^{(3)})$.
If $f_3 = 0$, the calculation stops with $\xi := x^{(3)}$.
Otherwise continue with 3).

3) *Determine a new inclusion interval:*
If $f_3 f_2 < 0$, the zero ξ lies between $x^{(2)}$ and $x^{(3)}$.
Set
$$x^{(1)} := x^{(2)}, \qquad x^{(2)} := x^{(3)}, \ \text{and} \qquad f_1 := f_2, \qquad f_2 := f_3.$$

If $f_3 f_2 > 0$, the zero ξ lies between $x^{(1)}$ and $x^{(3)}$ *(modified step)*:
Assign a new functional value $g f_1$ at $x^{(1)}$ where $g = f_2/(f_2 + f_3)$.
Then $0 < g < 1$ in the modified step, and
$$x^{(2)} := x^{(3)}, \qquad f_1 := g f_1, \qquad f_2 := f_3.$$

4) *Check for break-off condition:*
If $|x^{(2)} - x^{(1)}| \leq \varepsilon$, the iteration stops.
Take
$$\xi := x^{(2)}, \qquad \text{if} \qquad |f_2| \leq |f_1|, \qquad \text{otherwise take} \qquad \xi := x^{(1)}.$$

If $|x^{(2)} - x^{(1)}| > \varepsilon$, continue the iteration in 1) with the values $x^{(1)}, x^{(2)}$
and f_1, f_2 from 3).

Instead of testing for the absolute error in 4), one should use the mixed
test
$$|x^{(2)} - x^{(1)}| \leq |x^{(2)}| \cdot \text{RELERR} + \text{ABSERR}$$
(see section 1.2) which allows a choice between the absolute or relative
error test.

Geometric interpretation for the modified step.

The two lines connecting the two pairs of points $(x_2, f_2 + f_3)$ and (x_1, f_1), and (x_2, f_2) and (x_1, f_1^*) intersect on the x-axis if $f_1^* := f_1 \cdot f_2 / (f_2 + f_3)$.

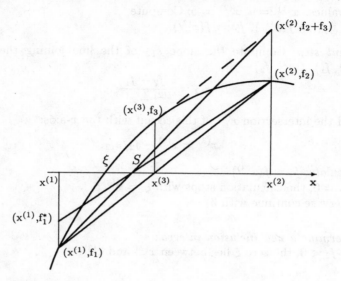

Figure 2.6: Pegasus method
 The line connecting $(x^{(1)}, f_1)$ and $(x^{(2)}, f_2 + f_3)$ intercepts the
 x–axis at S. The line connecting $(x^{(2)}, f_2)$ and S determines the
 point $(x^{(1)}, f_1^*)$.

2.8.3 Anderson-Björck Method

The Anderson-Björck method (see [ANDE73]) works essentially like the Pegasus method. Only in the third step of the algorithm, when determining a new inclusion interval, one modifies the functional value f_1 at $x^{(1)}$ differently. The effectivity E of this procedure lies between 1.682 and 1.710.

ALGORITHM 2.24 (*Anderson-Björck Method*).

Given: (i) $f \in C[a, b]$ with $f(a)f(b) < 0$,
 (ii) an error bound $\varepsilon > 0, \varepsilon \in \mathbb{R}$.

Task: Find a zero $\xi \in (a, b)$ of f, approximated in such a way that the
 difference between the last two iterated values is smaller than ε.

Start: Take $x^{(1)} := a$, $x^{(2)} := b$ as initial values and compute
$f_1 := f(x^{(1)})$, $f_2 := f(x^{(2)})$.

The iteration steps are as follows:

1) *Secant step*:
 Compute the slope of the line connecting $(x^{(1)}, f_1)$ and $(x^{(2)}, f_2)$:

 $$s_{12} = \frac{f_2 - f_1}{x^{(2)} - x^{(1)}}, \text{ and set } x^{(3)} = x^{(2)} - f_2/s_{12} \text{ as in Algorithm 2.23.}$$

2) *Calculate*: $f_3 = f(x^{(3)})$.
 If $f_3 = 0$, the calculation ends with $\xi := x^{(3)}$, otherwise continue with
 3).

3) *Determine a new inclusion interval*:
 If $f_3 f_2 < 0$, the zero lies between $x^{(2)}$ and $x^{(3)}$, hence we set:

 $$
 \begin{array}{llll}
 x^{(1)} & := & x^{(2)} & , \quad x^{(2)} \quad := \quad x^{(3)}, \\
 f_1 & := & f_2 & , \quad f_2 \quad := \quad f_3.
 \end{array}
 $$

 If $f_3 f_2 > 0$ and the zero lies between $x^{(1)}$ and $x^{(3)}$, we assign a new
 functional value $g f_1$ at $x^{(1)}$ *(modified step)*:
 If $1 - f_3/f_2 \leq 0$, take $g = 0.5$, otherwise use
 $g = 1 - f_3/f_2$ and set

 $$x^{(2)} := x^{(3)}, \qquad f_1 := g f_1, \qquad f_2 := f_3.$$

4) *Check for break-off condition*:
 If $|x^{(2)} - x^{(1)}| \leq \varepsilon$, stop the iteration.
 Set
 $$\xi := x^{(2)}, \quad \text{if} \quad |f_2| \leq |f_1|, \quad \text{and otherwise} \quad \xi := x^{(1)}.$$

 If $|x^{(2)} - x^{(1)}| > \varepsilon$, the iteration is continued in 1) with the new values
 $x^{(1)}, x^{(2)}, f_1, f_2$ from the third step.

The break-off criterion in the fourth step ought again to be replaced by
the mixed error test (1.6)

$$|x^{(2)} - x^{(1)}| \leq |x^{(2)}| \cdot \text{RELERR} + \text{ABSERR}$$

(see section 1.2).

Geometric interpretation:

Replacing f_1 by gf_1 in the third step of this algorithm can be interpreted geometrically as follows: For the three points

$$(x^{(1)}, f_1), (x^{(3)}, f_3), (x^{(2)}, f_2),$$

we form the interpolating parabola and use its tangent at the middle point $(x^{(3)}, f_3)$. If this tangent intersects the x-axis between $x^{(3)}$ and $x^{(1)}$, this intersection point is taken to be the next approximation of the desired zero.

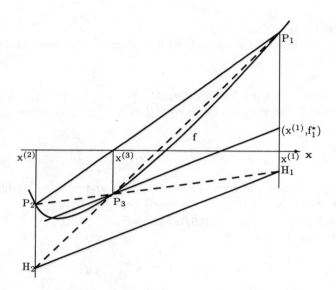

Figure 2.7: Anderson–Björck method. The lines P_2P_3 and P_1P_3 determine the points H_1 and H_2. The line parallel to H_1H_2 passing through P_3 is the tangent to the interpolating parabola through P_1, P_2, P_3 at P_3.

In this algorithm, the tangent is determined by the point $(x^{(3)}, f_3)$ and its intersection point $(x^{(1)}, f_1^*)$ with the line $x = x^{(1)}$.
According to the Anderson-Björck method, we have

$$f_1^* = gf_1 \qquad \text{with} \qquad g = \frac{s_{23}}{s_{12}} \qquad \text{and} \qquad s_{23} = \frac{f_2 - f_3}{x^{(2)} - x^{(3)}}.$$

If $g > 0$, the points $(x^{(1)}, f_1)$ and $(x^{(1)}, f_1^*)$ lie on the same side of the x-axis, and the point of intersection of the tangent to the parabola with the x-axis is between $x^{(3)}$ and $x^{(1)}$.

Due to the secant step 1), the points $(x^{(1)}, f_1)$, $(x^{(2)}, f_2)$ and $(x^{(3)}, 0)$ are collinear, and thus:

$$s_{12} = \frac{f_1 - f_2}{x^{(1)} - x^{(2)}} = \frac{f_2}{x^{(2)} - x^{(3)}}.$$

Hence we obtain g as:

$$g = \frac{s_{23}}{s_{12}} = \frac{f_2 - f_3}{x^{(2)} - x^{(3)}} \frac{x^{(1)} - x^{(2)}}{f_1 - f_2} = \frac{f_2 - f_3}{x^{(2)} - x^{(3)}} \frac{x^{(2)} - x^{(3)}}{f_2} = \frac{f_2 - f_3}{f_2} = 1 - \frac{f_3}{f_2}.$$

Now $f_3 f_2 > 0$ implies $g < 1$. If $g \leq 0$, the tangent does not intersect the x-axis between $x^{(3)}$ and $x^{(1)}$. In this case we take $g = 0.5$; compare also with the Illinois Method below.

2.8.4 The King and the Anderson-Björck-King Methods, the Illinois Method

The King method (see [KING73]) differs from the Pegasus method only slightly. In it two secant iteration steps are never performed consecutively. Instead each secant iteration is followed by a modified step.

The Anderson-Björck-King method combines the Anderson-Björck method with the method of King, where two secant steps may never follow each other. Here the modified step is performed according to the Anderson-Björck method.

Common to the inclusion procedures of Pegasus, Anderson-Björck and King is the secant step. The modified step is realized in different ways. For Pegasus, $g = f_2/(f_2 + f_3)$; for Anderson-Björck, $g = 1 - f_3/f_2$ and, in case $g \leq 0$, set $g = 0.5$. In the Illinois method g is always set to be 0.5.

2.8.5 Zeroin Method

The Zeroin method (see [DEKK69] and [FORS77]) is an ingenious combination of the bisection method ($p = 1$), the secant method ($p = 1.618$) and of inverse quadratic interpolation ($p = 2$). It is an inclusion procedure that operates on three iterative approximations of the desired zero. It chooses the procedure which is most appropriate in each situation depending on geometric considerations.

After running many tests comparing the Pegasus method, the Anderson-Björck method, the King method, and the Anderson-Björck-King method with the Zeroin procedure, we found that all these methods require about the same number of functional evaluations. The Anderson-Björck-King method generally needs the lowest number of function evaluations; the Zeroin procedure is superior with functions f of the form $f(x) = x^n - a$, whenever n is large.

2.9 Efficiency of the Methods and Aids for Decision Making

By means of the *efficiency index* E of Traub (see [TRAU81], App. C), iterative methods can be compared very easily. If H is the *Horner number*, defined as the number of functional evaluations required per iteration step, and p is the order of convergence of an iterative method, then the efficiency index E is given by

$$E := p^{1/H}.$$

The following table gives a list of the orders of convergence and the efficiency indices of the methods proposed above for the computation of simple and multiple zeros of real valued functions of one variable.
The bigger E, the more efficient the procedure will be in a neighborhood of the zero.

Method	Order of Convergence p	Horner number H	Efficiency E	Inclusion method
Newton's method for simple zeros	2	2	1.414	No
Newton's method for multiple zeros	2	2	1.414	No
Modified Newton's method	2	3	1.260	No

Method	Order of Convergence p	Horner number H	Efficiency E	Inclusion method
Bisection method	1	1	1	Yes
Simple version of the regula falsi	1	1	1	Yes
Regula falsi for simple zeros	1.618	1	1.618	No
Illinois method	1.442	1	1.442	Yes
Pegasus method	1.642	1	1.642	Yes
Anderson-Björck method	1.682...1.710	1	1.682...1.710	Yes
King method	1.710...1.732	1	1.710...1.732	Yes
Anderson-Björck-King method	1.710...1.732	1	1.710...1.732	Yes

The best efficiency is obtained for the Pegasus method, the Anderson-Björck method or the methods of King and Anderson-Björck-King. Besides, as inclusion methods, these methods will surely converge to a zero of f.

If one wants to compute individual *polynomial* roots, then one can also use Newton's method together with a Horner scheme for computing functional values and their derivatives effectively, see chapter 3.2. For transcendental equations *neither* the regula falsi *nor* Newton's method can be recommended.

Bibliography for Chapter 2

[ANDE73]; [ATKI89], 2; [BJÖR79], 6.; [BÖHM85]; [COLL68], §§17-19; [COLL73] I, 1.5-1.7; [CONT80], 2.1-2.4; [DOWE71]; [DOWE72]; [ENGE87], chap.2; [ESSE75]; [HÄMM91], 8.; [HENR72] vol. 1, chap. 4; [HILD78], 10.; [ISAA66], 3.1-3.2; [KING73]; [KIOU78]; [KIOU79]; [MAES88], 5.; [MEIN79], §8; [NIED87], chap.7; [NOBL65] I, 2.1-2.4; [PATR75]; [RALS67], V.; [SCHW89], 5.; [SHRA85]; [STET76], 5.; [STOE91]; [STUM82], 2; [TRAU81]; [WERN82], I, II §§1-5; [ZURM65], §1.

Chapter 3

Roots of Polynomials

3.1 Preliminary Remarks

We shall consider algebraic equations of the form

$$(3.1) \qquad P_n(x) = \sum_{j=0}^{n} a_j x^j = 0, \qquad a_j \in \mathbb{C}, \, a_n \neq 0.$$

The fundamental theorem of algebra states that a polynomial equation (3.1) has exactly n complex solutions x_k if each is counted with its multiplicity α_k. Every algebraic polynomial P_n (3.1) can thus be factored into n linear factors

$$(3.2) \qquad P_n(x) = a_n(x - x_1)(x - x_2)\ldots(x - x_n)$$

with $x_i \in \mathbb{C}$. If the linear factor $(x - x_k)$ arises exactly α_k times, then x_k is an α_k-fold solution of (3.1). One can write (3.2) in the form

$$P_n(x) = a_n(x - x_1)^{\alpha_1}(x - x_2)^{\alpha_2}\ldots(x - x_m)^{\alpha_m},$$

with distinct $x_i \in \mathbb{C}$, $i = 1, \ldots, m$, $m \leq n$, and $\alpha_1 + \alpha_2 + \ldots + \alpha_m = n$.

For $a_j \in \mathbb{R}$, complex solutions of (3.1) can only appear in pairs of complex conjugate numbers, i.e., if $x = \alpha + i\beta$ is a solution, then $\bar{x} = \alpha - i\beta$ is also a solution of (3.1) and it has the same multiplicity. The degree of an equation (3.1) with real coefficients that has no real roots, must therefore be even. Alternatively, each algebraic equation with real coefficients of odd degree has at least one real root.

The Horner scheme is used to determine functional values of P_n and its derivatives. It can also be used for calculating deflated polynomials, once a root or a pair of complex conjugate roots of P_n is known. It is the main tool used for computing zeros of polynomials.

To calculate all the zeros of a polynomial (3.1), the Muller, Bauhuber, or the Jenkins-Traub methods of chapters 3.3.2 to 3.3.4 are recommended.

3.2 The Horner Scheme

The *Horner scheme* is used for calculating functional values of a polynomial P_n and its derivatives at a given number x_0. It is well conditioned and quite efficient with regard to computational time.

3.2.1 First Level Horner Scheme for Real Arguments

In order to compute $P_n(x_0)$ for $x_0 \in \mathbb{R}$ and a given polynomial

$$P_n(x) = a_n^{(0)} x^n + a_{n-1}^{(0)} x^{n-1} + \ldots + a_1^{(0)} x + a_0^{(0)} \quad \text{with} \quad a_j^{(0)} \in \mathbb{R},$$

we can rewrite $P_n(x_0)$, $x_0 \in \mathbb{R}$, in the following form

$$(3.3) \qquad P_n(x_0) = (\ldots ((a_n^{(0)} x_0 + a_{n-1}^{(0)}) x_0 + a_{n-2}^{(0)}) x_0 + \ldots) x_0 + a_0^{(0)}.$$

By setting

$$(3.4) \qquad \begin{cases} a_n^{(1)} & := \quad a_n^{(0)} \\ a_j^{(1)} & := \quad a_{j+1}^{(1)} x_0 + a_j^{(0)} \qquad \text{for } j = n-1, n-2, \ldots, 1, 0, \end{cases}$$

and using (3.3), we obtain

$$P_n(x_0) = a_0^{(1)}.$$

The calculations in (3.4) can be carried out by hand in the following arrangement, which is the so-called *Horner scheme*:

P_n	$a_n^{(0)}$	$a_{n-1}^{(0)}$	$a_{n-2}^{(0)}$	\cdots	$a_1^{(0)}$	$a_0^{(0)}$
$x = x_0$	0	$a_n^{(1)} x_0$	$a_{n-1}^{(1)} x_0$	\cdots	$a_2^{(1)} x_0$	$a_1^{(1)} x_0$
\sum	$a_n^{(1)}$	$a_{n-1}^{(1)}$	$a_{n-2}^{(1)}$	\cdots	$a_1^{(1)}$	$\boxed{a_0^{(1)} = P_n(x_0)}$

Note that in the top row all the coefficients of P_n appear in decreasing order. Any zero coefficients must be included.

If we divide $P_n(x)$ by $(x - x_0)$ for $x_0 \in \mathbb{R}$ and $x \neq x_0$, then

(3.5) $$P_n(x) = (x - x_0)P_{n-1}(x) + P_n(x_0),$$

where

$$P_n(x_0) \quad = \quad a_0^{(1)}, \text{ and}$$
$$P_{n-1}(x) \quad = \quad a_n^{(1)}x^{n-1} + a_{n-1}^{(1)}x^{n-2} + \ldots a_2^{(1)}x + a_1^{(1)}.$$

The coefficients of the polynomial P_{n-1} are identical to the coefficients $a_j^{(1)}$ which appear in the bottom row of the Horner scheme for $x = x_0$. They are in fact the coefficients that were defined in (3.4).

Deflation.

If x_0 is a zero of P_n, then (3.5) implies that

$$P_n(x) = (x - x_0)P_{n-1}(x).$$

And the coefficients of the *deflated polynomial* P_{n-1} are the numbers $a_j^{(1)}$ of the Horner scheme for P_n evaluated with $x = x_0$.

3.2.2 First Level Horner Scheme for Complex Arguments

If the given polynomial P_n has complex coefficients or if we want to evaluate $P_n(x_0)$ for a complex argument x_0, we can still use the Horner scheme. In this case, we have to split the calculations into a real and an imaginary computation for each coefficient.

If P_n has real coefficients, then $P_n(x_0)$ can be computed over the reals for $x_0 \notin \mathbb{R}$ by using a two-rowed Horner scheme. For $x_0 \notin \mathbb{R}$ we form

$$(x - x_0)(x - \bar{x}_0) = x^2 - px - q.$$

Clearly $p = x_0 + \bar{x}_0$ and $q = -x_0 \cdot \bar{x}_0$ are both real, where \bar{x}_0 denotes the complex conjugate of x_0.

Dividing P_n by $(x^2 - px - q)$ yields

$$(3.6) \quad \begin{cases} P_n(x) &= (x^2 - px - q)P_{n-2}(x) + b_1^{(1)}x_0 + b_0^{(1)} \quad \text{with} \\ P_{n-2}(x) &= b_n^{(1)}x^{n-2} + b_{n-1}^{(1)}x^{n-3} + \ldots + b_3^{(1)}x + b_2^{(1)}. \end{cases}$$

Hence the coefficients $b_k^{(1)}$ of P_{n-2} are real and given by

$$(3.7) \quad \begin{cases} b_n^{(1)} &= a_n^{(0)}, \\ b_{n-1}^{(1)} &= a_{n-1}^{(0)} + p\, b_n^{(1)}, \\ b_k^{(1)} &= a_k^{(0)} + p\, b_{k+1}^{(1)} + q\, b_{k+2}^{(1)} \quad \text{for } k = n-2, n-1, \ldots, 1, \text{ and} \\ b_0^{(1)} &= a_0^{(0)} + q\, b_2^{(1)}. \end{cases}$$

The calculations of (3.7) can be carried out in the following real *two-rowed Horner scheme*:

P_n	$a_n^{(0)}$	$a_{n-1}^{(0)}$	$a_{n-2}^{(0)}$	\cdots	$a_2^{(0)}$	$a_1^{(0)}$	$a_0^{(0)}$
q	0	0	$qb_n^{(1)}$	\cdots	$qb_4^{(1)}$	$qb_3^{(1)}$	$qb_2^{(1)}$
p	0	$pb_n^{(1)}$	$pb_{n-1}^{(1)}$	\cdots	$pb_3^{(1)}$	$pb_2^{(1)}$	0
\sum	$b_n^{(1)}$	$b_{n-1}^{(1)}$	$b_{n-2}^{(1)}$	\cdots	$b_2^{(1)}$	$b_1^{(1)}$	$b_0^{(1)}$

Since $x_0^2 - px_0 - q = 0$, for $x = x_0$ (3.6) implies

$$(3.8) \qquad\qquad P_n(x_0) = b_1^{(1)}x_0 + b_0^{(1)},$$

and we have in fact computed the required functional value.

If $x_0 \notin \mathbb{R}$ is a zero of P_n then by (3.8), $b_0^{(1)} = 0$ and $b_1^{(1)} = 0$.

Deflation of polynomials with real coefficients for complex roots.

If $x_0 \notin \mathbb{R}$ is a zero of P_n, then according to (3.6)

$$P_n(x) = (x^2 - px - q)P_{n-2}(x).$$

The coefficients of the deflated polynomial P_{n-2} are the real numbers $b_k^{(1)}$ at the bottom of the two-rowed Horner scheme.

3.2.3 Complete Horner Scheme for Real Arguments

Since the first level Horner scheme not only supplies the functional value $P_n(x_0)$ but also the coefficients $a_j^{(1)}$ of the deflated polynomial P_{n-1}, it can be used inductively to compute the k^{th} derivative $P_n^{(k)}(x_0)$ of the polynomial P_n for $k = 1, \ldots, n$ at $x_0 \in \mathbb{R}$: From (3.5) and the product rule for differentiation

$$P_n'(x) = P_{n-1}(x) + (x - x_0)P_{n-1}'(x).$$

Thus for $x = x_0$:

$$P_n'(x_0) = P_{n-1}(x_0),$$

and $P_n'(x_0)$ can be found by forming another Horner scheme below the third row of the first level Horner scheme set up to evaluate $P_{n-2}(x_0)$.

Continuing in this way one can obtain

$$P_{n-k}(x_0) = \frac{1}{k!}P_n^{(k)}(x_0) \quad \text{for} \quad k = 1, \ldots, n.$$

By completing the Horner scheme thus, one can find all coefficients $P_{n-k}(x_0)$ of the *Taylor series* for P_n at $x = x_0$:

$$P_n(x) = \tilde{P}_n(x - x_0) = \sum_{k=0}^{n}(x - x_0)^k P_{n-k}(x_0) = \sum_{k=0}^{n}(x - x_0)^k \frac{1}{k!}P_n^{(k)}(x_0).$$

COMPUTATIONAL SCHEME 3.1 (*Complete Horner scheme*).

$$
\begin{array}{c|cccccc}
P_n & a_n^{(0)} & a_{n-1}^{(0)} & a_{n-2}^{(0)} & \cdots & a_1^{(0)} & a_0^{(0)} \\
x = x_0 & 0 & a_n^{(1)}x_0 & a_{n-1}^{(1)}x_0 & \cdots & a_2^{(1)}x_0 & a_1^{(1)}x_0 \\
\end{array}
$$

$$
\begin{array}{c|ccccc}
P_{n-1} & a_n^{(1)} & a_{n-1}^{(1)} & a_{n-2}^{(1)} & \cdots & a_1^{(1)} \\
x = x_0 & 0 & a_n^{(2)}x_0 & a_{n-1}^{(2)}x_0 & \cdots & a_2^{(2)}x_0 \\
\end{array}
\qquad \boxed{a_0^{(1)} = P_n(x_0)}
$$

$$
\begin{array}{c|cccc}
P_{n-2} & a_n^{(2)} & a_{n-1}^{(2)} & a_{n-2}^{(2)} & \cdots \\
\cdots & \cdots & \cdots & \cdots & \cdots \\
\end{array}
\qquad \boxed{a_1^{(2)} = \tfrac{1}{1!}P_n'(x_0)} \quad = P_{n-1}(x_0)
$$

$$
\begin{array}{c|cc}
P_1 & a_n^{(n-1)} & a_{n-1}^{(n-1)} \quad \cdots \\
x = x_0 & 0 & a_n^{(n)}x_0 \\
\end{array}
$$

$$
\begin{array}{c|c}
P_0 & a_n^{(n)} \\
& \\
x = x_0 & 0 \\
\end{array}
\qquad \boxed{a_{n-1}^{(n)} = \tfrac{1}{(n-1)!}P_n^{(n-1)}(x_0)} \quad = P_1(x_0)
$$

$$
\boxed{a_n^{(n+1)} = \tfrac{1}{n!}P_n^{(n)}(x_0)} \qquad = P_0(x_0)
$$

with $\quad a_n^{(k)} = a_n^{(k-1)}, \quad a_j^{(k)} = a_{j+1}^{(k)}x_0 + a_j^{(k-1)}$ for $j = 0, \ldots, n-1$ and $k = 1, \ldots, n+1$.

Operations count.

Forming the Taylor series for P_n at x_0 by means of the complete Horner scheme requires $(n^2+n)/2$ operations. The standard method of differentiation, calculating the derivatives and dividing by $k!$, takes $n^2 + 2n - 2$ operations. For $n \geq 3$ the complete Horner scheme thus saves more than half of the operations. Moreover the staggered formulation $a_j^{(1)} = a_{j+1}^{(1)}x_0 + a_j^{(0)}$ of the higher level coefficients in (3.4) is known to be extremely stable, see [WILK84]. Thus the complete Horner scheme is much preferred over a naive evaluation of $P_n(x_0) = a_n^{(0)}x_0^n + \ldots + a_0^{(0)}x_0$ by forming powers of x_0 and summing.

3.2.4 Applications

The Horner scheme can be used

(1) to calculate the values of a polynomial P_n and its derivatives at $x = x_0$ easily, quickly and with very small rounding errors,

(2) to form the Taylor series for a polynomial at $x = x_0$, and

(3) to deflate a polynomial for a known root x_0.

If for example one wants to determine a root of a polynomial P_n by Newton's method, see chapter 2.5, one should always use a Horner scheme to calculate $P_n(x^{(\nu)})$, $P'_n(x^{(\nu)})$ and $P''_n(x^{(\nu)})$.

If an iterative root-finding procedure finds a zero x_1 of P_n sufficiently well, one can divide P_n by $(x - x_1)$ and apply the iterative procedure again to the deflated polynomial P_{n-1}. Thus one can successively find all zeros of P_n and avoid the possibility of calculating a zero twice. However, the zeros of the deflated polynomials could differ more and more from the zeros of the initial polynomial P_n. For this reason, Wilkinson recommends in [WILK64], p. 70-83, to always perform deflation with the zero of least magnitude of P_j. One should thus work with a method which chooses its approximation for a polynomial root in such a way that the iteration converges towards the zero of least norm (see Muller's method, section 3.3.2). If this requirement is met, all the computed zeros have an accuracy which is essentially determined by their own condition number and not by the accuracy of the previously determined zeros. After all zeros of P_n have been found numerically, Wilkinson recommends to use those zeros as initial values for a second pass of the same algorithm with the given polynomial P_n. This enhances the accuracy of the computed roots, especially if their condition has been worsened by deflation.

3.3 Methods for Finding all Solutions of Algebraic Equations

3.3.1 Preliminaries

If sufficiently close approximations for the zeros of a polynomial are known, one can easily construct sequences of approximations that converge towards the true zeros. One problem thus is how to obtain close initial approximations of the zeros.

If one wants to calculate all real zeros of a real polynomial P_n by using one of the iterative procedures of Chapter 2, one can for example proceed by any of the following methods:

1) Find an interval I which contains all zeros. This can be done, in the following way:

 If $P_n(x) = x^n + a_{n-1}x^{n-1} + ... + a_1x + a_0$ is the given polynomial and $A = \max_{0 \le k \le n-1} |a_k|$, then all the zeros of P_n lie in a circle around the origin in the complex plane with radius $r = A + 1$ by Gershgorin's Theorem, see chapter 7.1, applied to the companion matrix for P_n. Therefore I can be taken as $[-r, +r]$.

 If P_n is a polynomial that has only real zeros such as an orthogonal polynomial (see section 8.1.2, in particular case 2), one can employ *Laguerre's theorem*:

 The zeros of P_n all lie within an interval whose end-points are given by the two solutions of the quadratic equation

 $$nx^2 + 2a_{n-1}x + \big(2(n-1)a_{n-2} - (n-2)a_{n-1}^2\big) = 0.$$

2) Count the number of the real zeros of P_n according to the sign change rules of Sturm and Descartes.

3) Determine the location of roots by interval splitting. Compute functional values and count the number of sign changes.

Using 3), it usually becomes possible to determine subintervals $I_k \subset I$ so that in each subinterval there is precisely one zero x_k of odd order. If such subintervals I_k are known, one can use Newton's method to calculate approximations for the roots $x_k \in I_k$. Here P_n and P_n' (or P_n, P_n' and P_n'') should be calculated using the Horner scheme.

References for 1) through 3) are [ZURM65], p. 46; [BRON85], p. 119-121.

The above methods are labor-intensive and very tedious in practice. One really needs methods that find all real and complex roots of a polynomial with real or complex coefficients quickly without previous knowledge of the location of the roots.

This is accomplished by *Muller's method* (see section 3.3.2) for polynomials with real coefficients.

For polynomials with complex coefficients we shall indicate two methods below, the *Jenkins-Traub method* and the *Bauhuber method*. in chapters 3.3.4

and 3.3.3. The Muller method can be extended to polynomials with complex coefficients. We shall only give a short description of the Jenkins-Traub method without desribing the algorithm completely.

3.3.2 Muller's Method

Muller's method [MULL56] finds all real and conjugate complex zeros of a real polynomial without any previous estimates for the roots. Let

$$P_n(x) = a_0 + a_1 x + a_2 x^2 + \ldots + a_n x^n = \sum_{j=0}^{n} a_j x^j \quad \text{for} \quad a_j \in \mathbb{R} \quad \text{with } a_n \neq 0.$$

Idea behind the method:

Initially, an approximate value $x_1^{(N)}$ is determined for the zero x_1 of P_n of smallest magnitude by means of Muller's iteration, see "Muller's iteration" below. After obtaining the coefficients for $P_{n-1} = P_n(x)/(x - x_1^{(N)})$ by a Horner scheme where one disregards the remainder, one determines an approximate value $x_2^{(N)}$ for the zero x_2 of P_{n-1} of least amount, again using Muller's iteration with $f \equiv P_{n-1}$ instead of $f \equiv P_n$ and x_2 instead of x_1. After finding $x_2^{(N)}$, one proceeds analogously. Thus one can obtain approximate values of all zeros of P_n, nearly ordered according to magnitude. Only nearly ordered, since it might be possible to obtain the 'second' zero first. In most test examples, however, the zeros will appear in an ordered fashion:

$$|x_1| \leq |x_2| \leq \ldots \leq |x_n|.$$

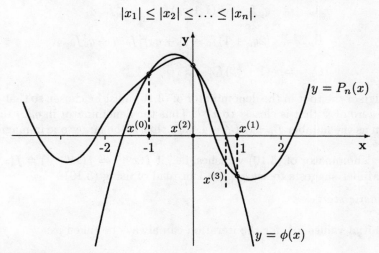

Figure 3.1: Muller's method

NOTE. Once one has found all zeros $x_i^{(N)}$ of P_n approximately by using Muller's method, one should employ a second root-finding method with the original polynomial P_n and take the approximate zeros as initial values for it. It is best to use Newton's method for this. One should calculate all zeros of P_n first by means of the described procedure before correcting them. According to tests carried out by Wilkinson (see [WILK64], II, 30.), it is not necessary to correct immediately before deflating P_n.

Muller's iteration.

For each triple of points $(x^{(k)}, f_k)$, $k = \nu-2$, $\nu-1$, ν, determine the quadratic interpolating polynomial Φ with $f_k := f(x^{(k)})$. Then find the zeroes of Φ. One of the zeros is chosen as a new approximate value $x^{(\nu+1)}$ for the desired zero x_1 of least magnitude of P_n. Write $f(x) \equiv P_n(x)$.

Then

(3.9) $$x^{(\nu+1)} = x^{(\nu)} + h_\nu q_{\nu+1}, \qquad \nu = 2,3,\dots$$

with

(3.10) $$q_{\nu+1} = \frac{-2C_\nu}{B_\nu \pm \sqrt{B_\nu^2 - 4A_\nu C_\nu}},$$

where the constants are defined by the following equations:

$$h_\nu = x^{(\nu)} - x^{(\nu-1)}, \qquad q_\nu = h_\nu/h_{\nu-1},$$
$$A_\nu = q_\nu f_\nu - q_\nu(1+q_\nu)f_{\nu-1} + q_\nu^2 f_{\nu-2},$$
$$B_\nu = (2q_\nu+1)f_\nu - (1+q_\nu)^2 f_{\nu-1} + q_\nu^2 f_{\nu-2},$$
$$C_\nu = (1+q_\nu)f_\nu.$$

The sign of the root in the denominator of (3.10) must be chosen so that $x^{(\nu+1)}$ is the zero of Φ that is closest to $x^{(\nu)}$. Thus the denominator in $q_{\nu+1}$ must be chosen as the number $B_\nu \pm \sqrt{B_\nu^2 - 4A_\nu C_\nu}$ that is the larger one in magnitude.

If the denominator of (3.10) vanishes, i.e., if $f(x^{(\nu)}) = f(x^{(\nu-1)}) = f(x^{(\nu-2)})$, then Muller suggests to set $q_{\nu+1} = 1$ instead of using (3.10).

Automatic start.

The initial values $x^{(i)}$ for the iteration can always be taken as:

$$x^{(0)} = -1, \qquad x^{(1)} = 1, \qquad x^{(2)} = 0$$

for all $f = P_n$. Instead of evaluating $f(x^{(i)}) = P_n(x^{(i)})$, Muller has suggested to truncate P_n with coefficients a_i to the quadratic polynomial $P_2(x) = a_0 + a_1 x + a_2 x^2$ and set

$$f_0 = P_2(x^{(0)}) = a_0 - a_1 + a_2,$$
$$f_1 = P_2(x^{(1)}) = a_0 + a_1 + a_2, \quad \text{and}$$
$$f_2 = P_2(x^{(2)}) = P_n(x^{(2)}) = a_0.$$

However, the true polynomial values at -1, 1 and 0 can also be used.

Break-off criterion.

The iteration (3.9) can be stopped, if for a given $\varepsilon > 0$

$$\frac{|x^{(\nu+1)} - x^{(\nu)}|}{|x^{(\nu+1)}|} < \varepsilon.$$

If this is the case for $\nu = N - 1$, then $x_1^{(N)} := x^{(N)}$ is the desired approximation of a root x_1 of f.

Complex conjugate roots.

The radicand in (3.10) can be negative for two reasons:

(1) A real solution of the equation $f(x) \equiv P_n(x) = 0$ is being approximated by a sequence of conjugate complex numbers. The imaginary parts of the sequence $\{x^{(\nu)}\}$ and the imaginary parts of the associated polynomial values $f(x^{(v)})$ tend to zero in this case.

(2) x_1 is a complex root. With x_1 its complex conjugate \bar{x}_1 is also a root of P_n. In this case, the division $P_n/[(x - x_1^{(N)})(x - \bar{x}_1^{(N)})]$ supplies a real polynomial P_{n-2} of degree $n - 2$ where one disregards the remainder (see section 3.2.2). One can then use Muller's iteration on P_{n-2} and obtain an approximation for its zero of smallest magnitude.

Convergence.

It has proven impossible to establish any result on global convergence of the method. However, it has been proved that convergence does occur if the method begins in a sufficiently close neighborhood of a simple or double zero. With the modifications indicated in [MULL56], p.210, Muller obtained convergence in all test cases with his modified starting procedure. The modification consists of examining in each case whether $|f(x^{(\nu+1)})/f(x^{(\nu)})| \leq 10$. If not,

$q_{\nu+1}$ should be halved, and $h_{\nu+1}, x^{(\nu+1)}$ and $f(x^{(\nu+1)})$ are calculated anew.

Order of convergence.

In [MULL56], the order of convergence is established as $p = 1.84$ for simple zeros, while double zeros have $p = 1.23$.

3.3.3 Bauhuber's Method

Bauhuber's method [BAUH70] computes all real and complex roots of a polynomial P_n with complex coefficients.

Principle of the procedure.

For an arbitrary starting point $x^{(0)} \in \mathbb{R}$, a sequence of approximations $\{x^{(\nu)}\}$, $\nu = 1, 2, ...$, is constructed so that the sequence of absolute values $\{|P_n(x^{(\nu)})|\}$ is monotonically decreasing:

$$|P_n(x^{(0)})| > |P_n(x^{(1)})| > \dots .$$

Then Newton's method is employed. The iteration is stopped if $|P_n(x^{(\nu+1)})| < \varepsilon$ for a given $\varepsilon > 0$. If for one ν

(3.11) $$|P_n(x^{(\nu)})| \leq |P_n(x^{(\nu+1)})|,$$

then $x^{(\nu+1)}$ must be excluded from the sequence of iterates $\{x^{(\nu)}\}$. In this case one has to determine a new $x^{(\nu+1)}$ by using a two-dimensional search called *spiralization* , carried out in the complex plane, so that

$$|P_n(x^{(\nu+1)})| < |P_n(x^{(\nu)})|.$$

Then the iteration is continued. An extrapolation technique will then improve the sequence of approximate values. If $x^{(N)}$ is an approximate root, then one calculates the deflated polynomial $P_{n-1} = P_n(x)/(x - x^{(N)})$ by means of a Horner scheme. Disregarding the remainder, one can apply the same procedure to the polynomial P_{n-1} of degree $n - 1$. Continuing in this fashion, one can obtain all roots of the polynomial P_n.

Basic idea underlying spiralization.

Let $\nu + 1$ be the index for which (3.11) holds for the first time in the sequence $\{x^{(\nu)}\}$. Then there must be an x_{s+1} inside the circle centered at $x^{(\nu)}$ with radius $r = |x^{(\nu+1)} - x^{(\nu)}|$ for which

$$|P_n(x_{s+1})| < |P_n(x^{(\nu)})|.$$

Such an x_{s+1} can be determined using a spiral polygon to search the disk from the outside inwards. For this, we use a complex factor $q = q_1 + iq_2$ with q_1, q_2 real and $|q| < 1$. Bauhuber has suggested the factor $q = 0.1 + 0.9i$, other choices are possible.

Algorithm for the spiralization.

With the initial values

$$
\begin{aligned}
x_0 &:= x^{(\nu+1)} \quad \text{satisfying} \quad (3.11) \\
\Delta x_0 &:= x^{(\nu+1)} - x^{(\nu)} \\
q &:= 0.1 + 0.9i,
\end{aligned}
$$

we calculate the following quantities for $k = 0$:

$$
(3.12) \quad
\begin{cases}
\Delta x_{k+1} &= q\Delta x_k \\
x_{k+1} &= x^{(\nu)} + \Delta x_{k+1} \\
\Delta P &= |P_n(x^{(\nu)})| - |P_n(x_{k+1})|.
\end{cases}
$$

If $\Delta P \leq 0$, then k is increased by one and the terms in (3.12) are calculated anew. We continue in this way until $\Delta P > 0$ for some $k = s$. Then we replace $x^{(\nu+1)}$ by x_{s+1}. The iteration is then continued with Newton's method.

3.3.4 The Jenkins-Traub Method

The *Jenkins-Traub method* ([JENK70], [TRAU66]) is an iterative method that determines the zero of smallest magnitude of a polynomial P_n with complex coefficients. It is globally convergent for all initial values $x^{(0)} \in (-\infty, |x_i|_{min}]$, where $|x_i|_{min}$ denotes the smallest absolute value of a root of P_n. This convergence is at least quadratic. This method also handles the case of two or more zeros of the same magnitude. Depending on whether the zero of least size is simple, double or multiple, the algorithm is modified automatically by internal logical branching. After the zero(s) of least size is (are) determined, a lower degree polynomial can be determined by deflation. This procedure can then be applied inductively to find approximate values for all of the roots of P_n.

3.3.5 The Laguerre Method

The *Laguerre method* is globally convergent. For real polynomials P_n it converges for every real starting value $x^{(0)}$. If P_n has simple zeros it has the order of convergence $p = 3$. Since the method requires three functional evaluations $P_n(x^{(\nu)}), P_n'(x^{(\nu)})$ and $P_n''(x^{(\nu)})$ in each iteration step, its Horner number is $H = 3$ and in case of simple roots, its efficiency is

$$E = p^{1/H} = \sqrt[3]{3} \doteq 1.442 \quad .$$

The iteration rule is:

$$x^{(\nu+1)} = x^{(\nu)} - \frac{nP_n(x^{(\nu)})}{P_n'(x^{(\nu)}) \pm \sqrt{S(x^{(\nu)})}} \quad \text{for } \nu = 0, 1, 2, \ldots$$

with

$$S(x) := (n-1)[(n-1)(P_n'(x))^2 - nP_n(x)P_n''(x)].$$

The sign in the denominator is chosen so that $|x^{(\nu+1)} - x^{(\nu)}|$ becomes as small as possible. Thus it should be chosen as the sign of $P_n'(x^{(\nu)})$.

If the zeros x_i of P_n are ordered monotonically

$$x_1 \leq x_2 \leq \cdots \leq x_n,$$

then the method converges for $x^{(0)} \in (x_{i-1}, x_i)$ towards either x_i or x_{i-1}. If $x^{(0)} < x_1$, it converges towards x_1, and for $x^{(0)} > x_n$ it converges towards x_n. As soon as a zero is found, the polynomial is deflated with a Horner scheme, and then one continues as usual.

We recommend to employ the first approximations of the roots as initial values for a second iteration using Newton's method on the original polynomial P_n. Otherwise accuracy might decrease due to the errors introduced by deflation, see also the note for Muller's method in section 3.3.2.

3.4 Hints for Choosing a Method

General hints on selecting an appropriate method for finding the roots of an algebraic polynomial were given in section 3.3.1. For the individual merits of each method see the various sections.

Bibliography for Chapter 3

[ATKI89], 2; [BAUH70]; [BJÖR69], 6.9; [BÖHM85]; [CARN69], 3.2-3.4, 3.9; [COLL73], I, 1.3-1.4; [CONT80], 2.6-2.7; [FORD77]; [HENR72] vol.1, chap.7.8; [ISAA66], 3.3-3.4; [NIED87], 7.5; [NOBL65], I, 2.5-2.7; [RALS78], 8; [RALS60] vol.1, VI 21; [STIE63], 4.5-4.6; [WERN82], II §§3,6-8; [WILK64], chap.2; [WILK84]; [ZURM65], §2

Chapter 4

Direct Methods for Solving Systems of Linear Equations

4.1 The Problem

In this chapter and the next, we consider the problem of finding values for n unknown quantities x_i so that m given linear equations

$$(4.1) \qquad \begin{cases} a_{11}x_1 + a_{12}x_2 + \ldots + a_{1n}x_n & = b_1, \\ a_{21}x_1 + a_{22}x_2 + \ldots + a_{2n}x_n & = b_2, \\ \vdots & \quad \vdots \\ a_{m1}x_1 + a_{m2}x_2 + \ldots + a_{mn}x_n & = b_m \end{cases}$$

are simultaneously satisfied. In formula (4.1) the coefficients $a_{ik} \in \mathbb{R}$ and the right hand sides $b_i \in \mathbb{R}$ are given numbers for $i = 1, \ldots, m$ and $k = 1, \ldots, n$. Written in matrix notation, (4.1) becomes

$$(4.2) \qquad \boldsymbol{A}\boldsymbol{x} = \boldsymbol{b} \quad \text{for} \quad \boldsymbol{x} \in \mathbb{R}^n \quad \text{and} \quad \boldsymbol{b} \in \mathbb{R}^m, \quad \text{where}$$

$$\boldsymbol{A} = (a_{ik}) = \begin{pmatrix} a_{11} & a_{12} & \cdots & a_{1n} \\ a_{21} & a_{22} & \cdots & a_{2n} \\ \vdots & & & \vdots \\ a_{m1} & a_{m2} & \cdots & a_{mn} \end{pmatrix}, \quad \boldsymbol{x} = \begin{pmatrix} x_1 \\ x_2 \\ \vdots \\ x_n \end{pmatrix} \quad \text{and} \quad \boldsymbol{b} = \begin{pmatrix} b_1 \\ b_2 \\ \vdots \\ b_m \end{pmatrix}.$$

Note that the first index i of a_{ik} indicates the row in which the element a_{ik} occurs in \boldsymbol{A}, while the second index k of a_{ik} denotes the column in which a_{ik} occurs in \boldsymbol{A}.

A vector \boldsymbol{x} with components x_i, $i = 1, \ldots, n$, that solves (4.1) or (4.2) is called a *solution* of the linear system. In Chapter 4.14 overdetermined systems $\boldsymbol{A}\boldsymbol{x} = \boldsymbol{b}$ with $m \geq n$ are treated. Otherwise we restrict ourselves to real square systems with n equations and n unknowns, i.e., to the case where $\boldsymbol{A}\boldsymbol{x} = \boldsymbol{b}$ for an (n, n) real matrix \boldsymbol{A}, $\boldsymbol{x} \in \mathbb{R}^n$ and $\boldsymbol{b} \in \mathbb{R}^n$.

Complex linear systems can be dealt with analogously by either using complex arithmetic or alternatively by solving a real system twice the original size by observing that the complex system

$$Cw = e \quad \text{for} \quad C = A + iB \in \mathbb{C}^{nn}, \; w = x + iy \in \mathbb{C}^{n}, \; \text{and} \; e = c + id \in \mathbb{C}^{n},$$

where A, B are real matrices and x, y, c and d are real vectors, is equivalent to the $(2n, 2n)$ real system

$$\begin{pmatrix} A & -B \\ B & A \end{pmatrix} \begin{pmatrix} x \\ y \end{pmatrix} = \begin{pmatrix} c \\ d \end{pmatrix}.$$

One distinguishes between *direct* and *iterative methods* for finding the numerical solution of systems of linear equations. Direct methods supply the exact solution if one disregards rounding errors. Iterative methods compute successively better approximate solutions of a linear system beginning with an initial vector. Iterative methods will be described in chapter 5.

This chapter deals with direct methods, specifically with Gauß algorithm, the Gauß-Jordan method, Cholesky decomposition and procedures for systems with band matrices, pivoting strategies etc.

Bibliography for Section 4.1

[ENGE87], 3.1, 3.2; [MAES84], 2.1; [STET76], 6.

4.2 Definitions and Theoretical Background

We shall first assemble those definitions and theorems for real matrices that we need in this chapter. We note, however, that most of the following concepts also apply or can be easily generalized to complex matrices, see e.g. [LANC85], or [GOLU89], 1, 2.

DEFINITION 4.1 (*Leading principal submatrix, leading principal minor*).
For $k \leq n$ the (k, k) matrix A_k, formed by deleting rows and columns

$k + 1, \ldots, n$ from \boldsymbol{A}_{nn} is called the *leading principal submatrix* of order k. Its determinant is called the k^{th} *leading principal minor* of \boldsymbol{A}.

DEFINITION 4.2 (*Nonsingular, strongly nonsingular, rank*).
A matrix \boldsymbol{A}_{nn} is called *nonsingular*, if $\det \boldsymbol{A} \neq 0$. We shall call \boldsymbol{A}_{nn} *strongly nonsingular*, if all leading principal minors are different from zero:

$$\det(\boldsymbol{A}_k) \neq 0 \quad \text{for all} \quad k = 1, \ldots, n.$$

If $\det \boldsymbol{A} = 0$, \boldsymbol{A} is called *singular*. The *rank* of a matrix \boldsymbol{A}_{nn} is defined as the size of the largest square submatrix of \boldsymbol{A} that is nonsingular.

DEFINITION 4.3 (*Triangular matrix*) .
An (n, n) matrix $\boldsymbol{L} = (l_{ik})$ is called *lower triangular* if $l_{ik} = 0$ for $k > i$. It is called *unit lower triangular*, if it is triangular and $l_{ii} = 1$ for all i.
An (n, n) matrix $\boldsymbol{R} = (r_{ik})$ is called *upper triangular* if $r_{ik} = 0$ for $i > k$. It is called *unit upper triangular* if moreover $r_{ii} = 1$ for all i.

DEFINITION 4.4 (*Permutation matrix*) .
An (n, n) matrix \boldsymbol{P} is called a *permutation matrix* if there is exactly one 1 and $n - 1$ zero entries in each of its rows and columns.

NOTE. A permutation matrix \boldsymbol{P} can be obtained from the (n, n) identity matrix \boldsymbol{I}_n by a rearrangement of its rows or columns. Note that left-multiplication of \boldsymbol{A}_{nn} by a permutation matrix \boldsymbol{P} in which the i^{th} and k^{th} rows of \boldsymbol{I}_n are interchanged, results in the interchange of rows i and k of \boldsymbol{A} in $\boldsymbol{P}\boldsymbol{A}$.

THEOREM 4.5. (*LR factorization*).
Every strongly nonsingular (n, n) matrix \boldsymbol{A} can be factored uniquely into a product $\boldsymbol{A} = \boldsymbol{L}\boldsymbol{R}$ where

$$L = \begin{pmatrix} 1 & & & \\ l_{21} & 1 & & \\ \vdots & & \ddots & \\ l_{n1} & l_{n2} & \cdots & 1 \end{pmatrix}, \quad R = \begin{pmatrix} r_{11} & r_{12} & \cdots & r_{1n} \\ & r_{22} & \cdots & r_{2n} \\ & & \ddots & \vdots \\ & & & r_{nn} \end{pmatrix}$$

are unit lower triangular and upper triangular, respectively.

Such a factorization of A is called an LR *factorization* of A.

One can drop the requirement that $\det(A_k) \neq 0$ for all k in the LR factorization of A, if the rows of the nonsingular matrix A may be permuted first:

THEOREM 4.6. For every nonsingular (n, n) matrix A there exists an (n, n) permutation matrix P, so that PA has an LR factorization

$$PA = LR,$$

where L is unit lower triangular and R is upper triangular. L and R are uniquely determined by P and A. Moreover $\det P = (-1)^k$ where k denotes the number of row interchanges in P when compared with I_n. And

$$\det A = (-1)^k \det R = (-1)^k r_{11} r_{22} \cdots r_{nn}.$$

Thus every nonsingular matrix A_{nn} can be transformed to a strongly nonsingular matrix PA by left-multiplication with a permutation matrix P, or by an interchange of rows of A.

DEFINITION 4.7 (*Transpose matrix, symmetric matrix*).
For $A_{nn} = (a_{ik})$ the matrix obtained from A by writing rows as columns (and columns as rows) is called the *transpose* of A and denoted by $A_{nn}^T := (a_{ki})$. An (n, n) matrix $A = (a_{ik})$ is called *symmetric* if

$$a_{ik} = a_{ki}$$

for all $i, k = 1, \ldots, n$.
Thus $A = A^T$ for symmetric matrices A.

DEFINITION 4.8 (*Orthogonal matrix*).
A real (n, n) matrix U is called *orthogonal* if

$$U^T U = I \quad \text{or} \quad U^{-1} = U^T.$$

THEOREM 4.9. (*Householder matrix*).
If $v \neq 0 \in \mathbb{R}^n$ and I_n is the (n, n) identity matrix, then the *Householder matrix*

$$H := I_n - \frac{2}{\|v\|^2} v v^T$$

is a symmetric and orthogonal (n, n) matrix, i.e., $H^T H = H^2 = I_n$.

DEFINITION 4.10 (*Band matrix*).
An (n, n) matrix $A = (a_{ik})$, $i, k = 1, \ldots, n$ is called a *band matrix*, if its entries vanish outside of a band parallel to the main diagonal. Let m_ℓ be the number of subdiagonals of A and m_r be the number of superdiagonals of A that are not all zero. Then $a_{ik} = 0$ for $i - k > m_\ell$ and $k - i > m_r$ where $0 \leq m_\ell \leq n - 2$ and $0 \leq m_r \leq n - 2$. The number $m = m_\ell + m_r + 1$ is the *bandwidth* of A.

A matrix A of bandwidth m can have at most m non-zero elements in any one row. The following are special band matrices:

diagonal matrices	for $m_\ell = m_r = 0$,
bidiagonal matrices	for $m_\ell = 1$, $m_r = 0$ or $m_\ell = 0$, $m_r = 1$,
tridiagonal matrices	for $m_\ell = m_r = 1$,
five-diagonal matrices	for $m_\ell = m_r = 2$.

DEFINITION 4.11 (*Cyclically tridiagonal matrix*).
A matrix $A_{nn} = (a_{ik})$, $i, k = 1, \ldots, n$ is called *cyclically tridiagonal*, if

$$a_{ik} = 0 \quad \text{for} \quad 1 < |i - k| < n - 1.$$

DEFINITION 4.12 (*Diagonally dominant and strictly diagonally dominant matrix*).
A square matrix $A = (a_{ik})$, $i, k = 1, \ldots, n$ is called *diagonally dominant*, if

$$|a_{ii}| \geq \sum_{\substack{k=1 \\ k \neq i}}^{n} |a_{ik}| \quad \text{for all} \quad i = 1, \ldots, n,$$

and for at least one index i the strict inequality holds.
A matrix A is called *strictly diagonally dominant*, if

$$|a_{ii}| > \sum_{\substack{k=1 \\ k \neq i}}^{n} |a_{ik}| \quad \text{for all} \quad i = 1, \ldots, n.$$

DEFINITION 4.13 (*Positive definite and positive semidefinite matrices*).
A symmetric (n, n) matrix $A = (a_{ik}) = A^T$, $i, k = 1, \ldots, n$ is called *positive definite* if

$$Q(x) := x^T A x > 0 \quad \text{for all} \quad x \neq 0, \ x \in \mathbb{R}^n.$$

It is called *positive semidefinite*, if

$$Q(x) \geq 0 \quad \text{for all} \quad x \in \mathbb{R}^n.$$

If A is positive definite then $a_{ii} > 0$ for all i.

Equivalent conditions for positive definiteness of a symmetric matrix $A_{nn} = (a_{ik}) = A^T$ are:

1) All leading principal minors of A are positive:

 $$\det(A_k) > 0 \quad \text{for} \quad k = 1, \ldots, n.$$

2) The LR factorization $A = LR$ in theorem 4.5 leads to an R with $r_{ii} > 0$ for all i.

3) A factorization $A = LDL^T$ with a unit lower triangular matrix L and a diagonal matrix $D = (d_{ik})$ makes $d_{ii} > 0$ for all i.

A symmetric matrix $A = A^T$ is positive definite if:

1) It is strictly diagonally dominant with positive diagonal entries $a_{ii} > 0$ for all i.

2) It is diagonally dominant with positive diagonal entries $a_{ii} > 0$ for all i and $a_{ik} < 0$ for all $i \neq k$.

3) It is tridiagonal and diagonally dominant with $a_{ii} > 0$ for all i and $a_{ik} \neq 0$ for $|i - k| = 1$.

THEOREM 4.14. A strictly diagonally dominant (n, n) matrix A_{nn} is strongly nonsingular.

THEOREM 4.15. Every symmetric, strongly nonsingular matrix A_{nn} can be factored uniquely as LDL^T with a unit lower triangular matrix L and a diagonal matrix D.

THEOREM 4.16. Every positive definite matrix A_{nn} is strongly nonsingular and can be factored into a product LDL^T as in theorem 4.15 where all diagonal entries of D are positive.

THEOREM 4.17. Every positive definite matrix A_{nn} can be factored uniquely into the product LL^T with a lower triangular matrix $L = (\ell_{ik})$, where $\ell_{ii} > 0$ for all i.

Bibliography for Section 4.2

[GOLU89], 1, 2; [LANC85]; [MAES84], 2.1; [TÖRN79] vol.1, 1.1-1.3; [ZURM65], 6.1 .

4.3 Solvability Conditions for Systems of Linear Equations

A linear system $Ax = b$ with an (m, n) matrix A is called *homogeneous*, if $b = 0$. Otherwise it is called *inhomogeneous*.

A homogeneous system $Ax = 0$ is always solvable. It has $n - \text{rank}A$ linearly independent solutions y_i. The set of all solutions is the space of all linear combinations

$$x = c_1 y_1 + c_2 y_2 + \ldots + c_{n-r} y_{n-r} \text{ for } c_i \in \mathbb{R}$$

where $r = \text{rank}A$. If rank $A = n$, then $Ax = 0$ has only the trivial solution $x = 0$.

For square systems of the form (4.2), the above remarks can be expressed via the determinant of A_{nn} thus:

a) If $\det A \neq 0$, then $Ax = 0$ has only the trivial solution $x = 0$.

b) If $\det A = 0$ and A has rank r, then the homogeneous system has exactly $n - r$ linearly independent solutions.

For an inhomogeneous system $Ax = b$ with $b \neq 0$, solvability is governed by:

THEOREM 4.18. An inhomogeneous linear system $Ax = b \neq 0$ is solvable if and only if the rank of the *augmented* $(m, n + 1)$ *matrix* (A, b) is the same as the rank of the (m, n) matrix A. Any solution x of $Ax = b$ is the sum of any solution x_h of the homogeneous system $Ax = 0$ and a solution x_i of the inhomogeneous system.

For square systems matrices A in (4.2), we can use the determinant function to illustrate solvability of linear systems $Ax = b$:

a) If $\det A \neq 0$, then there is exactly one solution to $Ax = b$, namely $x = A^{-1}b$. (Here clearly rank $(A, b) = $ rank $A = n$).

b) If $\det A = 0$ and if the system is solvable, i.e. if rank $(A, b) = $ rank $(A) = r < n$, then the solution is not uniquely determined. Any linear combination of the $n - r$ linearly independent solutions of $Ax = 0$ can be added to a specific solution of the inhomogeneous system and this sum will solve $Ax = b$.

Bibliography for Section 4.3

[ATKI89] ch. 7; [NIEM87], 4.1; [STIE63], p.20; [TÖRN79], vol.1, 1.4; [ZURM65], 5.5, 5.6; [ZURM84], 8.

4.4 The Factorization Principle

The direct methods rely mainly on factoring the system matrix A of the linear system $Ax = b$. Usually the (n, n) matrix A is factored into a product LR of a lower triangular matrix L and an upper triangular matrix R, provided this can be done. Using the product LR instead of A in the linear system $Ax = b$ changes the problem into the equivalent system $Rx = c$ where $c = L^{-1}b$. The aim of all direct methods is to find fast and accurate ways to update the right hand side b to $c = L^{-1}b$ and the system matrix A to $L^{-1}A = R$. If the method used cannot factor A directly or if such a factorization is not warranted due to instability of the method for the given input matrix A, one standardly tries to exchange rows in A and thus to replace A with PA and factor $PA = LR$ where P is a permutation matrix, see chapter 4.5.2 on pivot strategies.

Theorems 4.5 and 4.6 are the motivation for the following algorithms for solving linear systems by means of a triangular factorization.

ALGORITHM 4.19 *(Factorization without row interchanges).*

Given: $Ax = b$ with A_{nn} strongly nonsingular and $b \in \mathbb{R}^n$.
Find : Solution $x \in \mathbb{R}^n$.

1^{st} step: Find a lower triangular matrix L so that $L^{-1}A = R$ is upper triangular.

2^{nd} step: Update b to $c = L^{-1}b$.

3^{rd} step: Find x by solving $Rx = c$ using backsubstitution.

ALGORITHM 4.20 (*Factorization with row interchanges*).

Given: $Ax = b$ with a nonsingular matrix A_{nn} and $b \in \mathbb{R}^n$.

Find : Solution $x \in \mathbb{R}^n$.

1^{st} step: Determine a suitable permutation matrix P.

2^{nd} step: Find a lower triangular matrix L so that $L^{-1}PA = R$ is upper triangular.

3^{rd} step: Update b to $c = L^{-1}Pb$.

4^{th} step: Find x by solving $Rx = c$ using backsubstitution.

The direct methods to be mentioned in this chapter are variants of the two algorithms just described. They differ only in so far as special properties of the system matrix A in $Ax = b$ are used to reduce the amount of computations or intermediate storage. For the moment we shall deal only with systems of n equations in n unknowns. In section 4.14 we will treat overdetermined systems.

Bibliography for Section 4.4

[BJÖR74],5.; [FORS67], 9; [MAES84], 2.3; [RICE83], 6.1; [SCHE89], 3; [SCHW89],1 .

4.5 Gauß Algorithm

4.5.1 Gauß Algorithm with Column Pivot Search

The idea behind *Gauß algorithm with column pivot search* is to transform a linear system of the form (4.1) with $m = n$ into a triangular system

$$(4.3) \quad \begin{cases} r_{11}x_1 + & r_{12}x_2 + & \cdots + & r_{1n}x_n & = c_1, \\ & r_{22}x_2 + & \cdots + & r_{2n}x_n & = c_2, \\ & & \ddots & \vdots & \\ & & & r_{nn}x_n & = c_n, \end{cases}$$

by eliminating the lower triangular coefficients of A one column at a time after possibly exchanging rows to bring the largest element of a column into

its diagonal position. If $r_{11}r_{22}\cdots r_{nn} \neq 0$, the system (4.3) can be solved recursively for x_i, $i = n, n-1, \ldots, 1$ by what is called *backsubstitution*.

Details.

Clearly, the solution x of a linear system does not depend on the order of the equations in (4.1). Therefore one can interchange equations or rows of A in such a way that the element of largest size in the first column of A appears in the (1,1) position. This is called *column pivot search*, see section 4.5.2. Call the matrix thus obtained $A^{(0)} = PA = (a_{ik}^{(0)})$ and set $b^{(0)} = Pb = (b_i^{(0)})$. Then (4.1) is equivalent to

$$(4.4) \qquad \sum_{k=1}^{n} a_{ik}^{(0)} x_k = a_i^{(0)} \quad \text{for } i = 1, \ldots, n.$$

Since $\det A \neq 0$, the first column of A must be different from zero, i.e., the element of largest size in the first column of A cannot be zero, making $a_{11}^{(0)} \neq 0$. To eliminate x_1 from the equations $2, \ldots, n$, one multiplies the first equation of (4.4) by $-a_{i1}^{(0)}/a_{11}^{(0)}$ and adds it to the i^{th} equation successively for $i = 2, \ldots, n$. After this first elimination step we obtain the equivalent system:

$$(4.5) \qquad \left\{ \begin{array}{ll} a_{11}^{(0)}x_1+ \quad a_{12}^{(0)}x_2 \quad +\ldots \quad +a_{1n}^{(0)}x_n = \quad b_1^{(0)}, \\ \qquad\qquad \tilde{a}_{22}^{(1)}x_2 \quad +\ldots \quad +\tilde{a}_{2n}^{(1)}x_n = \quad \tilde{b}_2^{(1)}, \\ \qquad\qquad \vdots \qquad\qquad\qquad\qquad\qquad \vdots \\ \qquad\qquad \tilde{a}_{n2}^{(1)}x_2 \quad +\ldots \quad +\tilde{a}_{nn}^{(1)}x_n = \quad \tilde{b}_n^{(1)}, \end{array} \right.$$

where

$$\tilde{a}_{ik}^{(1)} = \left\{ \begin{array}{ll} 0 & \text{for } k = 1 \text{ and } i = 2, \ldots, n, \\ a_{ik}^{(0)} - a_{1k}^{(0)} \dfrac{a_{i1}^{(0)}}{a_{11}^{(0)}} & \text{otherwise, and} \end{array} \right.$$

$$\tilde{b}_i^{(1)} = \quad b_i^{(0)} - b_1^{(0)} \dfrac{a_{i1}^{(0)}}{a_{11}^{(0)}} \quad \text{for } i = 2, \ldots, n.$$

The system (4.5) thus consists of one equation in all n unknown x_1, x_2, \ldots, x_n and $n - 1$ equations in the $n - 1$ unknown x_2, \ldots, x_n only.

Another analogous elimination can be applied to the last $n - 1$ equations in (4.5): First, one interchanges rows so that the element of largest size among the $\tilde{a}_{i2}^{(1)}$ for $i = 2, \ldots, n$ appears in the (2,2) position. Denote the resulting system by:

$$(4.6) \quad \begin{cases} a_{11}^{(0)} x_1 + a_{12}^{(0)} x_2 + \ldots + a_{1n}^{(0)} x_n = b_1^{(0)}, \\ \qquad\qquad a_{22}^{(1)} x_2 + \ldots + a_{2n}^{(1)} x_n = b_2^{(1)}, \\ \qquad\qquad\qquad\qquad \vdots \qquad\qquad\qquad \vdots \\ \qquad\qquad a_{n2}^{(1)} x_2 + \ldots + a_{nn}^{(1)} x_n = b_n^{(1)}. \end{cases}$$

Since $\det \boldsymbol{A} \neq 0$, in theory we must have $a_{22}^{(1)} \neq 0$. Proceeding from the third equation down to the n^{th} equation of (4.6) analogously, the elements of the j^{th} row are updated for $j = 3, \ldots, n-1$ as:

$$\tilde{a}_{ik}^{(2)} = \begin{cases} 0 & \text{for} \quad k = 2 \quad \text{and} \quad i = 3, \ldots, n, \\ a_{ik}^{(1)} - a_{2k}^{(1)} \dfrac{a_{i2}^{(1)}}{a_{22}^{(1)}} & \text{otherwise, and} \end{cases}$$

$$\tilde{b}_i^{(2)} = b_i^{(1)} - b_2^{(1)} \dfrac{a_{i2}^{(1)}}{a_{22}^{(1)}} \quad \text{for} \quad i = 3, \ldots, n.$$

This elimination process of the lower triangular entries of \boldsymbol{A} is repeated analogously for the columns $j = 3, \ldots, n-1$ with updated entries defined as

$$(4.7) \begin{cases} \tilde{a}_{ik}^{(j)} = \begin{cases} 0 & \text{for} \quad k = 1, \ldots, j \quad \text{and} \quad i = j+1, \ldots, n, \\ a_{ik}^{(j-1)} - a_{jk}^{(j-1)} \dfrac{a_{ij}^{(j-1)}}{a_{jj}^{(j-1)}} & \text{otherwise, and} \end{cases} \\ \\ \tilde{b}_i^{(j)} = b_i^{(j-1)} - b_j^{(j-1)} \dfrac{a_{ij}^{(j-1)}}{a_{jj}^{(j-1)}} \quad \text{for } i = j+1, \ldots, n. \end{cases}$$

Having eliminated column j below its diagonal entry, the equations numbered $j+1$ to n must be permuted in such a way that the element of largest size among the $\tilde{a}_{i,j+1}^{(j)}$ for $j+1 \leq i \leq n$ is found in position $(j+1, j+1)$. We shall call the elements of the newly updated equations $j+1$ to n by $a_{ik}^{(j)}$ and $b_i^{(j)}$. Then after $n-1$ such elimination steps we obtain an equivalent system with a triangular matrix:

$$(4.8) \begin{cases} a_{11}^{(0)} x_1 + a_{12}^{(0)} x_2 + a_{13}^{(0)} x_3 + \ldots + a_{1n}^{(0)} x_n = b_1^{(0)}, \\ \qquad\quad a_{22}^{(1)} x_2 + a_{23}^{(1)} x_3 + \ldots + a_{2n}^{(1)} x_n = b_2^{(1)}, \\ \qquad\qquad\qquad a_{33}^{(2)} x_3 + \ldots + a_{3n}^{(2)} x_n = b_3^{(2)}, \\ \qquad\qquad\qquad\qquad \ddots \quad \vdots \qquad\quad \vdots \\ \qquad\qquad\qquad\qquad\qquad\qquad a_{nn}^{(n-1)} x_n = b_n^{(n-1)}. \end{cases}$$

Thus $r_{ik} = a_{ik}^{(i-1)}$ and $c_i = b_i^{(i-1)}$. And (4.8) is in triangular form (4.3). From (4.8) one can compute the x_i recursively by *backsubstitution*:

$$(4.9)\begin{cases} x_n = \dfrac{b_n^{(n-1)}}{a_{nn}^{(n-1)}}, \quad \text{and} \\[3mm] x_j = \dfrac{b_j^{(j-1)}}{a_{jj}^{(j-1)}} - \sum_{k=j+1}^{n} \dfrac{a_{jk}^{(j-1)}}{a_{jj}^{(j-1)}} x_k \quad \text{for} \quad j = n-1, n-2, \ldots, 2, 1. \end{cases}$$

A careful counting of the number of divisions and multiplications performed in (4.7) shows that the LR factorization of A requires $\frac{n(n^2-1)}{3}$ operations, while the updating of b to $c = L^{-1}Pb$ requires $\frac{n(n-1)}{2}$ operations. Since backsubstitution takes $\frac{n(n+1)}{2}$ operations, a linear system $Ax = b$ can be solved in a total of $\frac{n^3}{3} + n^2 - \frac{n}{3} \approx \frac{n^3}{3}$ operations.

Since $\det A \neq 0$, theoretically none of the computed diagonal elements $a_{jj}^{(j-1)}$ can vanish. If, after an elimination step, it should turn out impossible to find a nonzero pivot element $a_{jj}^{(j-1)}$, then $\det A$ would equal 0 numerically and A would be numerically singular. In this case, the resulting partial triangular form for A could theoretically tell whether there is a solution and how many linearly independent ones there are, see [ENGE87], section 3.3.5. But this answer would be very unreliable numerically, and in this situation a solution to $Ax = b$ should best be found using least square methods as described in chapter 8.1.3.4.

As the rank of A equals the number of nonzero diagonal entries r_{ii} in R in theory, one might be falsely tempted to try to determine the rank of a singular matrix A via its triangular factorization. Rounding errors, however, will almost never create a true zero on the diagonal and thus almost all singular matrices will appear numerically nonsingular if using Gauß reduction. To determine the rank of a matrix reliably, one should use the *singular value decomposition* (SVD), see [GOLU89], Chapters 5 and 8.

Since $PA = LR$ and $\det P = (-1)^k$, where k is the number of the row interchanges effected by P,

$$\det A = (-1)^k r_{11} r_{22} \ldots r_{nn}.$$

Let us illustrate the steps of Gauß algorithm in a formal scheme for $n = 3$ next:

COMPUTATIONAL SCHEME 4.21 (*Gauß algorithm for $n = 3$*).

Rows	$a_{ik}^{(j)}, \tilde{a}_{ik}^{(j)}$			$b_i^{(j)}, \tilde{b}_i^{(j)}$	Operations
$1^{(0)}$	$a_{11}^{(0)}$	$a_{12}^{(0)}$	$a_{13}^{(0)}$	$b_1^{(0)}$	(Given a matrix \boldsymbol{A}
$2^{(0)}$	$a_{21}^{(0)}$	$a_{22}^{(0)}$	$a_{23}^{(0)}$	$b_2^{(0)}$	and right hand side \boldsymbol{b}.
$3^{(0)}$	$a_{31}^{(0)}$	$a_{32}^{(0)}$	$a_{33}^{(0)}$	$b_3^{(0)}$	Here $h^{(i)}$ shall mean
					the "h^{th} row at step i".)
$\tilde{2}^{(1)}$	0	$\tilde{a}_{22}^{(1)}$	$\tilde{a}_{23}^{(1)}$	$\tilde{b}_2^{(1)}$	$2^{(0)} - \dfrac{a_{21}^{(0)}}{a_{11}^{(0)}}1^{(0)}$
$\tilde{3}^{(1)}$	0	$\tilde{a}_{32}^{(1)}$	$\tilde{a}_{33}^{(1)}$	$\tilde{b}_3^{(1)}$	$3^{(0)} - \dfrac{a_{31}^{(0)}}{a_{11}^{(0)}}1^{(0)}$
$2^{(1)}$	0	$a_{22}^{(1)}$	$a_{23}^{(1)}$	$b_2^{(1)}$	Interchange rows from
$3^{(1)}$	0	$a_{32}^{(1)}$	$a_{33}^{(1)}$	$b_3^{(1)}$	$\tilde{2}^{(1)}, \tilde{3}^{(1)}$ to $2^{(1)}, 3^{(1)}$,
					so that
					$\lvert a_{22}^{(1)} \rvert = \max(\lvert\tilde{a}_{22}^{(1)}\rvert, \lvert\tilde{a}_{32}^{(1)}\rvert)$
$\tilde{3}^{(2)} = 3^{(2)}$	0	0	$\tilde{a}_{33}^{(2)} = a_{33}^{(2)}$	$\tilde{b}_3^{(2)} = b_3^{(2)}$	$3^{(1)} - \dfrac{a_{32}^{(1)}}{a_{22}^{(1)}}2^{(1)}$

The rows $1^{(0)}$, $2^{(1)}$ and $3^{(2)}$ make up the final upper triangular system (4.8). From (4.8) we can use backsubstitution (4.9) to determine the solution x. The indicated interchange of rows is of course not necessary if $\lvert\tilde{a}_{22}^{(1)}\rvert \geq \lvert\tilde{a}_{32}^{(1)}\rvert$. In that case $\tilde{a}_{2i}^{(1)} = a_{2i}^{(1)}$ and $\tilde{a}_{3i}^{(1)} = a_{3i}^{(1)}$ for $i = 2, 3$.

4.5.2 Pivot Strategies

Each equation in the original system (4.1) represents a hyperplane

$$\sum_{i=1}^{n} a_{ji}x_i = b_j$$

in \mathbb{R}^n. To solve $\boldsymbol{A}\boldsymbol{x} = \boldsymbol{b}$ for nonsingular \boldsymbol{A}_{nn} thus amounts to finding the unique point of intersection of these n hyperplanes. Gauß elimation changes the rows in \boldsymbol{A} to those of \boldsymbol{R} in (4.8). When we use a small pivot $a_{ij}^{(j-1)}$ to eliminate the entries in the j^{th} column of $\boldsymbol{A}^{(j)}$ below its diagonal, we essentially change the geometry of the hyperplane intersection for $\boldsymbol{A}^{(j)}$ to one where the hyperplanes intersect at smaller angles. Hence the point of intersection becomes less precise, see [STOE91], 4.5. This is the main reason for a pivot search. Gauß algorithm is known to be numerically stable if at step j one searches for the maximal element in the whole submatrix to the right and below the (j, j) position in $\boldsymbol{A}^{(j-1)}$. In this *complete pivot search* one would then swap both rows and columns in $\boldsymbol{A}^{(j-1)}$. However this search and double swap is very expensive in computer overhead time and is usually not necessary. A *column pivot search*, also called a *partial pivot search* leads to a stable algorithm for most problems and is generally preferred.

In many cases, a pivot search is only effective if the rows and columns of \boldsymbol{A} have approximately the same size, i.e., if \boldsymbol{A} is *equilibrated* or *scaled*:

(4.10)
$$
\begin{aligned}
\text{row sums} \quad & r_i \quad := \quad \sum_{j=1}^{n} |a_{ij}| \\
\text{column sums} \quad & c_k \quad := \quad \sum_{j=1}^{n} |a_{jk}| \\
& r_i \approx c_k \quad \text{for all} \quad i, k = 1, \dots, n.
\end{aligned}
$$

Explicit scaling of a matrix is time consuming and introduces rounding errors of its own. Hence one uses *implicitly scaled pivot searches* instead. In elimination step j in chapter 4.5.1 one searches for the index i where $\left|\frac{\tilde{a}_{ij}^{(j-1)}}{r_i}\right|$ is maximal and uses the pivot $a_{jj}^{(i-1)}$ for which

(4.11)
$$\frac{|a_{jj}^{(j-1)}|}{r_j} := \max_{i=j,\dots,n}\left(\frac{|\tilde{a}_{ij}^{(j-1)}|}{r_i}\right).$$

4.5.3 Computer Implementation of Gauß Algorithm

"Gauß algorithm" is the historical name for algorithm 4.19. Without row interchanges it obtains the triangular factorization $\boldsymbol{A} = \boldsymbol{L}\boldsymbol{R}$ with $\boldsymbol{L} = (\ell_{ij})$, $\boldsymbol{R} = (r_{ij})$ from (4.8):

$$r_{ij} = \begin{cases} a_{ij}^{(i-1)} & , \quad i \leq j \\ 0 & , \quad i > j \end{cases} , \quad \ell_{ij} = \begin{cases} a_{ij}^{(j-1)}/a_{jj}^{(j-1)} & , \quad i > j \\ 1 & , \quad i = j \\ 0 & , \quad i < j \end{cases} .$$

And for $c = L^{-1}b = (c_1, \ldots, c_n)^T$ we have $c_i = b_i^{(i-1)}$ for $i = 1, \ldots, n$ in (4.8). The solution x of $Ax = b$ is given by backsubstitution as

$$x_n = \frac{c_n}{r_{nn}},$$

$$x_i = \frac{1}{r_{ii}} \left(c_i - \sum_{j=i+1}^{n} r_{ij}x_j \right) \quad \text{for } i = n-1, n-2, \ldots, 1.$$

During the elimination process, the elements of A can be overwritten row by row with the elements of the factorization matrices L and R. The nonzero ℓ_{ij} are situated on and below the main diagonal. On the diagonal $\ell_{ii} = 1$ and there is no need to store these l_{ii}. Instead the diagonal elements r_{ii} and r_{ij} for $i \leq j$ can be stored in the full upper triangle of A. Similarly c can be overwritten onto the right hand side vector b.

In the following algorithm, we shall formulate a triangular factorization of A with column pivot search which overwrites the elements of A with those of L and R.

ALGORITHM 4.22 (*Triangular matrix factorization with column pivot search*).

Given: A nonsingular (n, n) matrix A.

 Task: Find a triangular factorization of $PA = LR$ where P is a permutation matrix, L is unit lower triangular and R is upper triangular. Overwrite A with L and R.

1^{st} step: Initialize the pivot selection vector $p = (p_1, \ p_2, \ldots, p_n)^T$ by setting $p_i = i$ for all i.

2^{nd} step: For every $j = 1, \ldots, n-1$ do

 2.1 Determine $i_0 \geq j$ with $|a_{i_0,j}| = \max\{|a_{ij}|, \ i = j, \ldots, n\}$ (Pivot search), interchange p_{i_0} with p_j and the i_0^{th} row of A with the j^{th} row. If $a_{jj} = 0$, A is numerically singular, and the procedure must be stopped. Else:

 2.2 For each $i = j + 1, \ldots, n$ do:

2.2.1 Overwrite a_{ij} with a_{ij}/a_{jj}.

2.2.2 For $k = j + 1, \ldots, n$ do:

Replace a_{ik} by $a_{ik} - a_{jk}a_{ij}$.

This yields: $\boldsymbol{PA} = \boldsymbol{LR}$ with $\boldsymbol{P} = (e_{p_1}, e_{p_2}, \ldots, e_{p_n})^T$,

$$
\boldsymbol{L} = \begin{pmatrix} 1 & 0 & \cdots & 0 \\ a_{21} & 1 & & 0 \\ \vdots & & \ddots & \vdots \\ a_{n1} & \cdots & a_{n,n-1} & 1 \end{pmatrix} \text{ and } \boldsymbol{R} = \begin{pmatrix} a_{11} & a_{12} & \cdots & a_{1n} \\ 0 & a_{22} & \cdots & a_{2n} \\ \vdots & & \ddots & \vdots \\ 0 & \cdots & & a_{nn} \end{pmatrix}.
$$

Here \boldsymbol{e}_{p_j} denotes the p_j^{th} unit vector in \mathbb{R}^n with a 1 in the p_j^{th} position and zeros everywhere else.

ALGORITHM 4.23 (*Gauß algorithm with column pivot search*).

Given: $\boldsymbol{Ax} = \boldsymbol{b}$ with $\det \boldsymbol{A} \neq 0$.

Task: Find the solution \boldsymbol{x}.

1^{st} step: Determine the pivot vector \boldsymbol{p} and the triangular matrices \boldsymbol{L} and \boldsymbol{R} according to algorithm 4.22. (2 steps)

2^{nd} step: Calculate $\boldsymbol{c} = (c_1, c_2, \ldots, c_n)^T$ by updating $\boldsymbol{Pb} = \boldsymbol{Lc}$. Here

$$
\boldsymbol{Pb} = (b_{p_1}, b_{p_2}, \ldots, b_{p_n})^T \text{ and}
$$

$$
c_1 = b_{p_1}, \quad c_i = b_{p_i} - \sum_{j=1}^{i-1} a_{ij}c_j \quad \text{for} \quad i = 2, \ldots, n.
$$

3^{rd} step: Calculate the solution \boldsymbol{x} of $\boldsymbol{Ax} = \boldsymbol{b}$ by backsubstitution from $\boldsymbol{Rx} = \boldsymbol{c}$:

$$
x_n = \frac{c_n}{a_{nn}},
$$

$$
x_i = \frac{1}{a_{ii}} \left(c_i - \sum_{j=i+1}^{n} a_{ij}x_j \right) \quad \text{for} \quad i = n-1, n-2, \ldots, 1.
$$

The previous two algorithms can be adapted to an implicitly scaled column pivot search. In this case, one simply searches for the index i_0 that satisfies (4.11) of chapter 4.5.2 in step 2.1 of algorithm 4.22.

4.5.4 Gauß Algorithm for Systems with Several Right Hand Sides

If we have to solve a linear system with one nonsingular matrix A for m right hand sides $b_j, j = 1, \ldots, m$, we can write

$$AX = B \quad \text{with} \quad X = (x_1, x_2, \ldots, x_m) \quad \text{and} \quad B = (b_1, b_2, \ldots, b_m).$$

Here A is the (n, n) matrix common to all m linear systems, X is an (n, m) matrix made up of the columns of the solution vectors x_j of $Ax_j = b_j$ for $j = 1, \ldots, m$, and B is the (n, m) matrix formed with the right hand sides b_j as its columns. To solve $AX = B$, the triangular factorization of A has to be computed only once. And updating and backsubstitution has to be performed m-times. Hence $n^3/3 - n/3 + mn^2$ operations are required, compare with chapters 4.5.1 and 4.6, and the table in chapter 4.17.

ALGORITHM 4.24 (*Gauß algorithm with column pivot search for m right hand sides*).

Given: $AX = B$ with A nonsingular and $B = (b_1, \ldots, b_m)$, $b_i \in \mathbb{R}^n$.

Task: Find the solution matrix $X = (x_1, x_2, \ldots, x_m)$ with $x_i \in \mathbb{R}^n$.

1^{st} step: Factor $PA = LR$ according to algorithm 4.22.
2^{nd} step: Update $PB = LC$.
3^{rd} step: Backsubstitution $RX = C$ gives X.

Bibliography for Section 4.5

[BJÖR74], 5.32; [COLL73] I, 2.3; [CONT80], 5.3; [ENGE87], 3.2; [FADD59], §16; [GOLU89]; [HÄMM91], 17a; [ISAA66], 2.1; [McCR72], 4.3; [MAES84], 2.2; [NIEM87], 4.3; [NOBL65] I, 4.1-4.3; [SCHW89], 1.1; [SPEL85], 1.6/1.7; [STOE91], 4.1; [TÖRN79] vol.1, 4; [WERN82], III §1; [WILK64], pp.119, 198; [ZURM65], §5.1; [ZURM84],§6.1.

4.6 Matrix Inversion via Gauß Algorithm

To find the inverse of a nonsingular (n, n) matrix A, we must solve the following n linear systems with right hand sides e_i:

$$Ax_i = e_i \quad \text{for} \quad i = 1, \dots, n.$$

Here $\det A \neq 0$, and e_i denotes the i^{th} unit vector in \mathbb{R}^n. If we combine the n right hand sides e_i to form the identity matrix I_n and the n unknown solution vectors x_i to form the columns of a matrix X, then we have to solve the matrix equation $AX = I_n$, see section 4.5.4, with $m = n$ and B replaced by I_n.

By definition of an inverse, $X = A^{-1}$. Thus the n solution vectors x_i of the n systems $Ax_i = e_i$ form the columns of A^{-1}. In order to obtain A^{-1}, we can solve the n systems simultaneously via Gauß algorithm. Partial pivoting is again recommended. Finding the inverse thus will take $(4n^3/3) - (n/3)$ operations.

ALGORITHM 4.25 (*Finding A^{-1} via Gauß algorithm with partial column pivot search*).

Given: A nonsingular (n, n) matrix A.
 Find: $X = A^{-1}$ so that $AX = I$.

1^{st} step: Factor $PA = LR$ according algorithm 4.22
2^{nd} step: Update $P = LC$.
3^{rd} step: Backsubstitute $RX = C$, the solution X is A^{-1}.

NOTE. This procedure is intended only in case that A^{-1} is explicitly needed. It should never be used to solve m systems $Ax_i = y_i$, $i = 1, \dots, m$ for any $m < n$, by computing $x_i = A^{-1}y_i$. The reason for this is that matrix inversion alone takes $4n^3/3 - n/3 + mn^2$ operations in contrast to $n^3/3 - n/3 + mn^2$ total operations if one employs the Gauß algorithm for m right hand sides as detailed in section 4.5.4.

Bibliography for Section 4.6

[CONT80], 5.5; [ENGE87], §3.3.6; [FADD59], §21; [GOLU89]; [NOBL65], 5.3; [STIE63], p.24; [ZURM65], p.147; [ZURM84], §6.6.

4.7 Linear Equations with Symmetric Strongly Nonsingular System Matrices

If the matrix $A = A^T$ is symmetric, then, due to symmetry, one only needs to store its lower or upper triangular elements and the diagonal elements of A. In the following, we shall describe algorithms which are based on theorems 4.15 - 4.17 for solving linear systems $Ax = b$ with symmetric strongly non-singular matrices A and specifically for positive definite matrices for which $\det(A_k) \neq 0$ for all $k = 1, \ldots, n$, recall definition 4.2. The procedures of this chapter will maintain symmetry since they do not use a column pivot search.

4.7.1 The Cholesky Decomposition

The basic algorithm of this chapter relies on the LDL^T factorization of strongly nonsingular real symmetric matrices $A = A^T$.

ALGORITHM 4.26 (LDL^T *factorization for symmetric strongly non-singular A*).

Given: $Ax = b$ for a symmetric strongly nonsingular matrix A and a right hand side b.

Find: Solution x.

1^{st} step: Factor $A = LDL^T$ with a unit lower triangular matrix L and a diagonal matrix D.

2^{nd} step: Update:
$$y = L^{-1}b, w = D^{-1}y.$$

3^{rd} step: Backsubstitution $L^T x = w$.

This procedure is analogous to algorithm 4.19. It allows for negative diagonal entries in the matrix D, but no zero entries can be allowed in order to be able to perform the updating step.

If for a linear system $Ax = b$ the system matrix A is symmetric ($a_{ik} = a_{ki}$ for all i, k) and positive definite ($x^T Ax > 0$ for all $x \neq 0$), then one can use the standard *Cholesky decomposition*. For large n it uses only half as many operations as Gauß algorithm and needs only about half as much storage. We shall give two forms of the Cholesky decomposition: the standard one factors $A = L \cdot L^T$, the second one factors $A = LDL^T$ for a diagonal matrix D and a lower triangular or a unit lower triangular L, respectively. Both versions of the Cholesky decomposition are benign, should the system matrix A not be numerically positive definite or not strongly nonsingular: the first one will stop when it is asked to find a square root of a nonpositive real number. The second one need not stop in this case. For numerically positive definite A, all entries d_{ii} in D with $A = LDL^T$ will be positive, but for general symmetric matrices A the diagonal entries d_{ii} may assume any distribution. If, in the second version, one d_{ii} is computed to be zero, then we can find no x with $Ax = b$, or with $LDL^T x = b$, since backsubstitution would force us to divide by zero. If, however, all $d_{ii} \neq 0$, then $Ax = b$ can be solved using the known LDL^T factorization of A. It can be easily seen that the leading principal minors of A can be expressed as $\det(A_k) = d_{11} \cdots d_{kk}$ for $k = 1, \ldots, n$ if $A = LDL^T$. Moreover the sign distribution of the set $\{d_{ii}\}$ gives the *inertia* of the symmetric strongly nonsingular matrix A, i.e., the number of positive and negative eigenvalues of A is given by the number of positive and negative entries among the $\{d_{ii}\}$, respectively. If all $d_{ii} > 0$, A is numerically positive definite. The second, root-free version of the Cholesky decomposition is preferred numerically because it avoids calculating square roots.

First version of Cholesky: $A = LL^T$.

Factor $A = LL^T$ where $L = (\ell_{ik})$ is lower triangular with $\ell_{ii} > 0$. Transform the system $Ax = b$ into an equivalent system $L^T x = c$ via the following steps:

1) Factor $A = LL^T$.

2) Update $c = L^{-1}b$.

3) Backsubstitute $L^T x = c$.

ALGORITHM 4.27. *(Linear system solver for positive definite A using the standard Cholesky decomposition).*

Given: A linear system $Ax = b$ with a symmetric positive definite (n, n) matrix A.

Find: Solution x.

1^{st} step: (Factor $A = LL^T$)
 For each $j = 1, \ldots, n$ do:

$$1.1 \quad \ell_{jj} = \sqrt{a_{jj} - \sum_{i=1}^{j-1} \ell_{ji}^2}$$

1.2 For each $k = j + 1, \ldots, n$ do:

$$\ell_{kj} = \frac{1}{\ell_{jj}} \left[a_{kj} - \sum_{i=1}^{j-1} \ell_{ki} \cdot \ell_{ji} \right].$$

2^{nd} step: (Update $c = L^{-1}b$)
 For each $j = 1, \ldots, n - 1$ do:
 For each $k = j+1, \ldots, n$ do:
 $b_k := b_k - \ell_{kj} \cdot b_j$.
 For each $j = 1, \ldots, n$ do:
 $c_j := b_j / \ell_{jj}$.

3^{rd} step: (Backsubstitution $x = (L^T)^{-1}c$)
 3.1 $x_n := \frac{c_n}{\ell_{nn}}$
 3.2 For each $i = n - 1, \ldots, 1$ do:

$$x_i := \frac{1}{\ell_{ii}} \left(c_i - \sum_{k=i+1}^{n} \ell_{ki} x_k \right).$$

Then the determinant of A is:

$$\det A = \det L \det(L^T) = (\ell_{11} \ell_{22} \cdots \ell_{nn})^2.$$

Root-free version of the Cholesky decomposition: $A = LDL^T$

If A is strongly nonsingular and symmetric we can factor $A = LDL^T$ so that D is a diagonal matrix and L is a unit lower triangular matrix. Then the system $Ax = b$ is equivalent to the system $L^T x = c$, where $c = L^{-1}b$. The factorization requires $n^3/6 + O(n^2)$ operations. If A is not positive definite, i.e., if not all diagonal entries d_{ii} of D are positive, then, on occasion, the

LDL^T factorization of A may be unstable.

Implementation.

We shall use the following notation to describe D and L explicitly:

$$D = \begin{pmatrix} d_1 & & & & \\ & d_2 & & & \\ & & \ddots & & \\ & & & d_{n-1} & \\ & & & & d_n \end{pmatrix}, \quad L = \begin{pmatrix} 1 & & & \cdots & 0 \\ \ell_{21} & 1 & & \cdots & \\ \vdots & & \ddots & \ddots & \vdots \\ & & & 1 & \\ \ell_{n1} & \cdots & & \ell_{n,n-1} & 1 \end{pmatrix}.$$

ALGORITHM 4.28. (*Linear system solver via root-free Cholesky*).

Given: $Ax = b$ with a symmetric strongly nonsingular matrix A and a
 right hand side b.

 Find: Solution $x \in \mathbb{R}^n$.

1^{st} step: (Factor $A = LDL^T$)
 For each $i = 1, \ldots, n-1$ do:
$$d_i = a_{ii} - \sum_{k=1}^{i-1} d_k \ell_{ik}^2$$
 For each $j = i+1, \ldots, n$ do:
$$\ell_{ji} = (a_{ji} - \sum_{k=1}^{i-1} d_k \ell_{jk} \ell_{ik})/d_i$$
$$d_n = a_{nn} - \sum_{k=1}^{n-1} d_k \cdot \ell_{nk}^2$$

2^{nd} step: (Update $c = L^{-1}b$, rename $b = D^{-1}c$)

 2.1 For each $i = 1, \ldots, n-1$ do:
 For each $j = i+1, \ldots, n$ do:
$$b_j := b_j - \ell_{ji} b_i$$

 2.2 For each $i = 1, \ldots, n$ do:
$$b_i := b_i/d_i$$

3^{rd} step: (Backsubstitution $x = (L^T)^{-1}b$)
 $x_n = b_n$.

For each $i = n-1, n-2, \ldots, 2, 1$ do:
$$x_i := b_i - \sum_{j=i+1}^{n} \ell_{ji} x_i.$$

If, during the factorization, one compares the coefficients of A and LDL^T somewhat differently, it is possible to save $\frac{n(n-1)}{2}$ operations in algorithm 4.28 (for this see [MAES84], p. 77):

ALGORITHM 4.29.

Given: Linear system $Ax = b$ with a symmetric strongly nonsingular matrix A.

Find: Solution x.

1^{st} step: (Factor $A = LDL^T$)
 For each $j = 1, \ldots, n$ do:

 1.1 For each $i = 1, \ldots, j-1$ do:

 1.1.1 $h = a_{ji}$

 1.1.2 $\ell_{ji} = h/d_i$

 1.1.3 For each $k = i+1, \ldots, j$ do:
 a_{jk} is replaced by $a_{jk} - h\ell_{ki}$

 1.2 $d_j = a_{jj}$

2^{nd} step: (Update $Lz = b, Dc = z$)
 For each $j = 1, \ldots, n$ do:

 2.1 $z_j = b_j$

 2.2 For each $i = 1, \ldots, j-1$ do:
 $z_j := z_j - \ell_{ji} z_i$

 2.3 $c_j = z_j/d_j$

3^{rd} step: (Backsubstitution $L^T x = c$)
 For each $j = n, \ldots, 1$ do:

 3.1 $x_j = c_j$

 3.2 For each $i = j+1, \ldots, n$ do:
 $x_j := x_j - \ell_{ij} x_i$

The determinant of A is:

$$\det A = \det(L)\det D \det(L^T) = \det D = d_1 d_2 \cdots d_n.$$

The inertia of A is given by the number of positive and negative entries among the d_i.

REMARK. Note that neither version of the Cholesky decomposition can be used for symmetric matrices that are not strongly nonsingular: the nonsingular symmetric matrix $A = \begin{pmatrix} 0 & 1 \\ 1 & 0 \end{pmatrix}$ with $\det A_1 = 0$ cannot be factored as $A = LL^T$ or $A = LDL^T$ as can be easily verified. Thus the linear equation solving algorithms 4.27-4.29, 4.33, 4.35 and 4.37 in chapters 4.7.1, 4.10.2, 4.11.2 and 4.12.2, respectively, designed for strongly nonsingular symmetric, tridiagonal, cyclically tridiagonal and five-diagonal matrices $A = A^T$, can only be used for such matrices. If $A = A^T$ is not numerically strongly nonsingular, one must use the corresponding LR style algorithms preceding those LDL^T algorithms mentioned above. Moreover, unless $A = A^T$ is positive definite, certain instabilities may arise on occasion with the root–free Cholesky decomposition, see specifically [GOLU 89], chapter 4.4.

4.7.2 The Conjugate Gradient Method

If A is positive definite, the *conjugate gradient method* can also be used to solve a linear system $Ax = b$. Here instead of the linear system $Ax = b$ itself, an equivalent minimizing problem is solved iteratively.

LEMMA 4.30. The problems

(i) solve $Ax = b$, and

(ii) minimize $F(x) = \frac{1}{2}(x^T A x - 2x^T b)$

are equivalent for a positive definite matrix A.

The proof becomes apparent by investigating the auxiliary function

$$E(x) = \frac{1}{2}(Ax - b)^T A^{-1}(Ax - b).$$

Since A^{-1} is positive definite if A is, $E(x) \geq 0$ for all $x \in \mathbb{R}^n$. Therefore $E(x)$ is minimized if and only if $Ax - b = 0$. We can rewrite $E(x)$ as

$$E(x) = \frac{1}{2}(x^T Ax - 2x^T b) + \frac{1}{2}\underbrace{b^T A^{-1} b}_{\geq 0} \quad .$$

Consequently $E(x)$ is minimized if and only if

$$F(x) := \frac{1}{2}(x^T Ax - 2x^T b) = \text{Min.} \iff Ax - b = 0.$$

The direction of steepest descent of any function F is that of its negative gradient. For F, the gradient is

$$\text{grad } F(x) = Ax - b.$$

The problem of solving $Ax = b$ thus consists of finding a point x where the gradient of F vanishes. This is done by the conjugate gradient method iteratively in n steps, moving from an $x_i \in \mathbb{R}^n$ to the point $x_{i+1} \in \mathbb{R}^n$ with $E(x_{i+1}) \leq E(x_i)$ along the direction of the gradient or conjugate gradient of $F(x_i)$ until $F(x_n) = 0$.

ALGORITHM 4.31 (*Conjugate Gradient Method*).

Given: A positive definite symmetric matrix A.
 Find: Solution x of the linear system $Ax = b$.

1^{st} step: Select a starting point $x_0 \in \mathbb{R}^n$ arbitrarily.

 Compute $d_0 = -g_0 = b - Ax_0$. If $g_0 = 0$, then x_0 is the solution, stop. Else:

2^{nd} step: For $k = 0, 1, \ldots, n-1$ calculate:

 2.1 $\alpha_k = -\dfrac{d_k^T g_k}{d_k^T A d_k}$.

 2.2 Set $x_{k+1} = x_k + \alpha_k d_k$, where $x_{k+1} - x_0 \in \text{span}\{d_0, \ldots, d_k\}$.

 2.3 Set $g_{k+1} = g_k + \alpha_k A d_k$,

 where $g_{k+1} \perp d_0, \ldots, d_k$ and $g_{k+1} \notin \text{span}\{d_0, \ldots, d_k\}$.

 If $\|g_{k+1}\|_\infty < \varepsilon$ for a given $\varepsilon > 0$, then accept x_{k+1} as the solution and stop. Else:

 2.4 Set $\beta_k = \dfrac{g_{k+1}^T A d_k}{d_k^T A d_k}$.

 2.5 Set $d_{k+1} = -g_{k+1} + \beta_k d_k$,

 where $d_{k+1} \perp A d_k$ and $d_{k+1} \notin \text{span}\{d_0, \ldots, d_k\}$.

NOTE: For $k = n - 1$, i.e., in the n^{th} step, span $\{d_0, d_1, \ldots, d_{n-1}\} = \mathbb{R}^n$. And thus, barring rounding errors, the precise solution has been found.

The conjugate gradient method can be considered as a direct method because it furnishes the exact solution except for rounding errors. It is often used as an iterative procedure because in general it furnishes a sufficiently good approximation of the solution after a few steps due to its rapid convergence.

The advantages of this method are

- - it can be easily vectorized and used on parallel processors,

- - it converges rapidly. Its speed, however, depends on the condition number of the matrix A; the better its condition number, the fewer iterative steps are needed.

A disadvantage is its high sensitivity to rounding errors.

The conjugate gradient method can be vectorized much more easily than Gauß algorithm due to the nature of its operations. It is particularly suited for large sparse matrices A. Per iteration step, one must calculate three dot products and one matrix-vector multiplication Ad_k. In sequential processors, these operations take about 97% of the computational time. Thus it is necessary to pay attention to their effective calculation.

NOTE: In the literature there is a variant of the CG method, the *CG method with preconditioning*. In this method, the number of iteration steps is lowered, while the amount of calculations per step is increased. This procedure is described e.g. in [BUNS85], pp. 156, [SCHW88], [SCHW89], [MAES84], pp. 132-133.

Geometric Interpretation of the CG Method for $n = 2$:

If $n = 2$, the equation $F(x, y) = z$ describes an elliptical paraboloid in \mathbb{R}^3, the level curves $F(x, y) = const$ are ellipses, and their horizontal projections are similar ellipses with the same center.
The conjugate gradient method can be viewed in the following steps:

1) For an arbitrary $x_0 \in \mathbb{R}^2$ determine the level curve $E(x) = E(x_0)$, which must be an ellipse.

2) Construct the tangent t_0 of this ellipse at x_0 and its normal n_0.

3) Halve the segment of n_0 inside the ellipse $E(x) = E(x_0)$ to obtain x_1.

4) For the similar ellipse containing x_1, construct the diameter that is conjugate to the direction of $t_1 = n_0$. It connects the two points on the

elliptical level curve $E(x) = E(x_1)$ whose tangents are parallel to the direction of $t_1 = n_0$.

5) Halve this conjugate diameter to find x_2.

The following picture illustrates the steps for $A = \begin{pmatrix} 2 & 0 \\ 0 & 3 \end{pmatrix}$, $b = \begin{pmatrix} 4 \\ 1 \end{pmatrix}$ and $x_0 = \begin{pmatrix} 1 \\ 1 \end{pmatrix}$. Then $x_1 = \begin{pmatrix} 1.8 \\ 0.2 \end{pmatrix}$ and $x_2 = \begin{pmatrix} 2 \\ 1/3 \end{pmatrix}$.

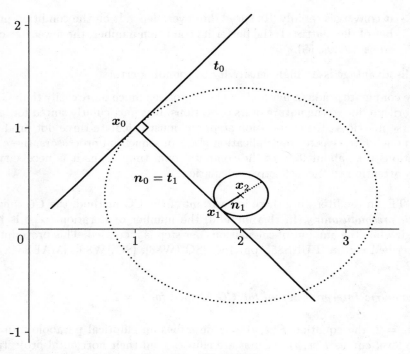

Figure 4.1: Geometric Interpretation of the CG Method

Note in the above drawing, that the first iteration $x_0 \rightarrow x_1$ proceeds along the gradient n_0 of $F(x_0)$. The subsequent iterations, i.e., $x_1 \rightarrow x_2$ in case $n = 2$, proceed along a modified gradient according to step 2.5 of the algorithm. This corresponds precisely to the direction n_1 of the ellipse diameter that is conjugate to its tangent t_1 at x_1 for the solidly drawn level curve $E(x) = E(x_1)$. And thus arises the name for this method of Hestenes and Stiefel.

Bibliography for Section 4.7

[GOLU89] ch. 3, 4; [ISAA73], p.57/58; [MAES84]; [NIEM87], 6.3; [SCHW72], p.36 ff; [STOE91], 8.7; [WERN82], p.149/150; [ZURM65], p.99; [ZURM84], §6.7.

4.8 The Gauß-Jordan Method

The *Gauß-Jordan method* is a modification of Gauß algorithm. Here the elimination rule (4.7) of Chapter 4.5.1 is used not only below the diagonal position, i.e., for $i > j$, but also above the diagonal for $i < j$:

$$(4.7\text{.G-J}) \quad \begin{cases} \tilde{a}_{ik}^{(j)} = \begin{cases} 0 \quad \text{for } k = 1, \ldots, j \ \text{ and } \ i = 1, \ldots, j-1, j+1, \ldots, n, \\[2mm] a_{ik}^{(j-1)} - a_{jk}^{(j-1)} \dfrac{a_{ij}^{(j-1)}}{a_{jj}^{(j-1)}} \quad \text{otherwise, and} \end{cases} \\[6mm] \tilde{b}_i^{(j)} = b_i^{(j-1)} - b_j^{(j-1)} \dfrac{a_{ij}^{(j-1)}}{a_{jj}^{(j-1)}} \quad \text{for } i = 1, \ldots, j-1, j+1, \ldots, n. \end{cases}$$

Pivoting is only done partially in column j to find the element of maximal modulus among the \tilde{a}_{ij}^{j-1} for $i \geq j$. Thus finally all but the diagonal entries in (4.8) are zeroed out, i.e., (4.8) becomes $a_{ii}^{(i-1)} x_i = b_i^{(i-1)}$ for $i = 1, \ldots, n$ and the backsubstitution (4.9) becomes rather trivial:

$$x_i = \frac{b_i^{(i-1)}}{a_{ii}^{(i-1)}} \quad \text{for all } i.$$

We note that this method takes $\frac{n^3}{2} + O(n^2)$ operations, i.e., asymptotically more than Gauß algorithm does, but it saves a little work in the backsubstitution step.

Bibliography for Section 4.8

[ENGE87], 3.5; [ISAA73], S.52/53; [McCA67], 5.7A; [McCR72], p.205 ff; [NIEM87] 4.4.

4.9 The Matrix Inverse via Exchange Steps

The *exchange method*, also called the *method of pivotization*, computes the inverse matrix A^{-1} by means of inverting a system of linear equations

$$(4.12) \qquad Ax = y \quad \text{for } A = (a_{ik}^{(0)}), \text{ where } i, k = 1, \ldots, n,$$

with $x^T = (x_1, x_2, \ldots, x_n), y^T = (y_1, y_2, \ldots, y_n)$, and $\det A \neq 0$. The solution of (4.12) is obtained after n exchanges of variables in the form $x = A^{-1}y$.

Principle of the Procedure.

A standard substitution method is used here schematically. The solution x is obtained step by step by exchanging one arbitrary variable y_i with an x_k. Each of these exchanges is called a *pivot step*; hence the method is called *pivotization*. The column of A that corresponds to x_k is the *pivot column*, the row belonging to y_i is the *pivot row*. Pivot row and column intersect in the *pivot element*. An exchange of x_k and y_i is possible only if the corresponding pivot element a_{ik} is different from zero. After n such pivot steps A^{-1} is obtained. The elements of the matrix obtained after the j^{th} pivot step shall be designated by $a_{ik}^{(j)}$ for $j = 1, \ldots, n$.

Rules for a pivot step.

1) The pivot element is replaced by its reciprocal: $a_{ik}^{(j)} = 1/a_{ik}^{(j-1)}$.

2) The other elements of the pivot column are divided by the pivot element: $a_{sk}^{(j)} = a_{sk}^{(j-1)}/a_{ik}^{(j-1)}$ for $s \neq i$.

3) The other elements of the pivot row are divided by the negative pivot element: $a_{is}^{(j)} = -a_{is}^{(j-1)}/a_{ik}^{(j-1)}$ for $s \neq k$.

4) The remaining elements of the matrix are updated according to the rule

$$a_{sm}^{(j)} = a_{sm}^{(j-1)} - \frac{a_{sk}^{(j-1)} a_{im}^{(j-1)}}{a_{ik}^{(j-1)}} \quad \text{for all} \quad s \neq i, \, m \neq k,$$

Scheme of the first pivot step with pivot element $a_{ik}^{(0)}$:

$$
\begin{array}{c|cccc}
 & x_1 \ldots x_k \ldots x_m \ldots x_n \\
\hline
y_1 & a_{11}^{(0)} \ldots a_{1k}^{(0)} \ldots a_{1m}^{(0)} \ldots a_{1n}^{(0)} \\
\vdots & \vdots \quad \vdots \quad \vdots \quad \vdots \\
y_i & a_{i1}^{(0)} \ldots a_{ik}^{(0)} \ldots a_{im}^{(0)} \ldots a_{in}^{(0)} \\
\vdots & \vdots \quad \vdots \quad \vdots \quad \vdots \\
y_s & a_{s1}^{(0)} \ldots a_{sk}^{(0)} \ldots a_{sm}^{(0)} \ldots a_{sn}^{(0)} \\
\vdots & \vdots \quad \vdots \quad \vdots \quad \vdots \\
y_n & a_{n1}^{(0)} \ldots a_{nk}^{(0)} \ldots a_{nm}^{(0)} \ldots a_{nn}^{(0)}
\end{array}
\qquad
\begin{array}{c|cccc}
 & x_1 \ldots y_i \ldots x_m \ldots x_n \\
\hline
y_1 & a_{11}^{(1)} \ldots a_{1k}^{(1)} \ldots a_{1m}^{(1)} \ldots a_{1n}^{(1)} \\
\vdots & \vdots \quad \vdots \quad \vdots \quad \vdots \\
x_k & a_{i1}^{(1)} \ldots a_{ik}^{(1)} \ldots a_{im}^{(1)} \ldots a_{in}^{(1)} \\
\vdots & \vdots \quad \vdots \quad \vdots \quad \vdots \\
y_s & a_{s1}^{(1)} \ldots a_{sk}^{(1)} \ldots a_{sm}^{(1)} \ldots a_{sn}^{(1)} \\
\vdots & \vdots \quad \vdots \quad \vdots \quad \vdots \\
y_n & a_{n1}^{(1)} \ldots a_{nk}^{(1)} \ldots a_{nm}^{(1)} \ldots a_{nn}^{(1)}
\end{array}
$$

If, at one point of the calculation, further pivot steps are impossible because all elements that could serve as pivots are zero, then $\det \boldsymbol{A} = 0$ numerically, see [ENGE87], section 3.6.

To avoid an accumulation of rounding errors in each exchange step, one chooses the element of largest magnitude available for a pivot element for the exchange of x_k with y_i.

This exchange method, as well as Gauß algorithm for finding \boldsymbol{A}^{-1}, are only recommended when \boldsymbol{A}^{-1} is explicitly needed. The Gauß algorithm is preferred over the exchange method. See specifically algorithm 4.25 and chapter 4.5.4. The exchange method, however, plays an important part in linear programming ([STIE63], p.28).

Bibliography for Section 4.9

[ENGE87], 3.6; [SCHW89], 1.4; [STIE63], 1.3.

4.10 Linear Systems with Tridiagonal Matrices

4.10.1 Systems with Tridiagonal Matrices

A matrix $\boldsymbol{A} = (a_{ik})$ is called *tridiagonal*, if $a_{ik} = 0$ for $|i - k| > 1$ and all $i, k = 1, \ldots, n$. A linear system (4.1) or (4.2) with a tridiagonal matrix \boldsymbol{A} has the form

$$\begin{pmatrix} a_{11} & a_{12} & & & & \\ a_{21} & a_{22} & a_{23} & & & \\ & a_{32} & a_{33} & & a_{34} & \\ & & \ddots & \ddots & & \ddots \\ & & a_{n-1,n-2} & a_{n-1,n-1} & a_{n-1,n} \\ & & & a_{n,n-1} & a_{nn} \end{pmatrix} \begin{pmatrix} x_1 \\ x_2 \\ x_3 \\ \vdots \\ x_{n-1} \\ x_n \end{pmatrix} = \begin{pmatrix} b_1 \\ b_2 \\ b_3 \\ \vdots \\ b_{n-1} \\ b_n \end{pmatrix}.$$

The system $Ax = b$ can be transformed into an equivalent system $Rx = c$ by factoring $A = LR$. Here, L is a bidiagonal lower triangular matrix and R is a bidiagonal unit upper triangular matrix. Necessary for such a factorization without row interchanges is the strong nonsingularily of A, i.e.,

$$\det(A_k) \neq 0 \quad \text{for} \quad k = 1, \dots, n \quad \text{(see theorem 4.5)}.$$

If this is not satisfied for A, we have to employ a column pivot search. In this case the band width of A is increased in general (see section 4.13). Without pivoting the reduction of $Ax = b$ to $Rx = c$ is achieved by the steps:

1) Factor $A = LR$. This gives L and R.
2) Update $b = Lc$ to get c.
3) Backsubstitution $Rx = c$ gives x.

Implementation.

The entries of the matrices A, L, R and the vectors b and c are labelled as follows:

$$(4.13) \begin{cases} A = \begin{pmatrix} d_1 & f_1 & & & \\ e_2 & d_2 & f_2 & & \\ & \ddots & \ddots & \ddots & \\ & & e_{n-1} & d_{n-1} & f_{n-1} \\ & & & e_n & d_n \end{pmatrix}, R = \begin{pmatrix} 1 & \gamma_1 & & & \\ & 1 & \gamma_2 & & \\ & & \ddots & \ddots & \\ & & & 1 & \gamma_{n-1} \\ & & & & 1 \end{pmatrix}, \\[3em] L = \begin{pmatrix} \alpha_1 & & & \\ \beta_2 & \alpha_2 & & \\ & \ddots & \ddots & \\ & & \beta_n & \alpha_n \end{pmatrix}, c = \begin{pmatrix} c_1 \\ c_2 \\ \vdots \\ c_n \end{pmatrix}, b = \begin{pmatrix} b_1 \\ b_2 \\ \vdots \\ b_n \end{pmatrix}. \end{cases}$$

If we compare the coefficients of the given matrix A with those of the product LR we notice that $\beta_i = e_i$ for all i. For this reason, we can replace the β_i immediately with e_i in the following algorithm.

ALGORITHM 4.32 (*Linear equation solver for tridiagonal A*).

Given: $Ax = b$ with a strongly nonsingular tridiagonal matrix A.
 Find: Solution x.

1^{st} step: (Factor $A = LR$)

 1.1 $\alpha_1 = d_1$

 1.2 $\gamma_1 = f_1/\alpha_1$

 1.3 For each $i = 2, \ldots, n-1$ calculate:

 1.3.1 $\alpha_i = d_i - e_i\gamma_{i-1}$

 1.3.2 $\gamma_i = f_i/\alpha_i$

 1.4 $\alpha_n = d_n - e_n\gamma_{n-1}$

2^{nd} step: (Updating $b = Lc$)

 2.1 $c_1 = b_1/d_1$

 2.2 For each $i = 2, \ldots, n$ calculate:
 $c_i = (b_i - e_ic_{i-1})/\alpha_i$

3^{rd} step: (Backsubstitution $Rx = c$)

 3.1 $x_n = c_n$

 3.2 For each $i = n-1, \ldots, 1$ calculate:
 $x_i = c_i - \gamma_ix_{i+1}$.

The tridiagonal matrix A will be nonsingular or $\det A \neq 0$, if $|d_1| > |f_1| > 0$, $|d_i| \geq |e_i| + |f_i|$, $e_if_i \neq 0$ for $i = 2, \ldots, n-1$ and $|d_n| > |e_n| > 0$. In this case, A is a tridiagonal diagonally dominant matrix. ([CONT80], p. 184; [ISAA66], pp.58; [STUM82] 6.3).
For the determinant of A we have as before:

$$\det A = \det L \det R = \det L = \alpha_1\alpha_2\cdots\alpha_n.$$

If the system matrix A is symmetric and tridiagonal or cyclically tridiagonal, diagonally dominant or positive definite, then a Gauß algorithm is numerically stable without pivot search. Condition improving scaling and iterative refinement, see chapters 4.15.3 and 4.15.4, in general will not improve the accuracy of the solution (see [FORS67], 8, 10, 11; [SPÄT74/2], p.15, [WILK61]). For all other A, a pivot search is recommended. Then the band width might increase at most to twice its original size (see section 4.13).

4.10.2 Systems with Tridiagonal Symmetric Strongly Nonsingular Matrices

If $A = A^T$ is a tridiagonal strongly nonsingular symmetric (n, n) matrix, it is advisable to use a method like the Cholesky method. Here storage can be very much reduced. The transition from $Ax = b$ to the equivalent system $L^T x = c$ proceeds according to algorithm 4.29.

1) Factor $A = LDL^T$. This gives L and D.

2) Updating $Lz = b$ gives z, and $Dc = z$ gives c.

3) Backsubstition $L^T x = c$ gives x.

Implementation.

The elements of A, L, D, c, z, b and x are labelled as follows:

$$A = \begin{pmatrix} d_1 & f_1 & & & \\ f_1 & d_2 & f_2 & & \\ & \ddots & \ddots & \ddots & \\ & & f_{n-2} & d_{n-1} & f_{n-1} \\ & & & f_{n-1} & d_n \end{pmatrix}, \quad L = \begin{pmatrix} 1 & & & & \\ \gamma_1 & 1 & & & \\ & \gamma_2 & \ddots & \ddots & \\ & & & 1 & \\ & & & \gamma_{n-1} & 1 \end{pmatrix},$$

$$D = \begin{pmatrix} \alpha_1 & & & \\ & \alpha_2 & & \\ & & \ddots & \\ & & & \alpha_n \end{pmatrix}, \quad c = \begin{pmatrix} c_1 \\ c_2 \\ \vdots \\ c_n \end{pmatrix}, \quad z = \begin{pmatrix} z_1 \\ z_2 \\ \vdots \\ z_n \end{pmatrix}, \quad b = \begin{pmatrix} b_1 \\ b_2 \\ \vdots \\ b_n \end{pmatrix},$$

$$x = \begin{pmatrix} x_1 \\ x_2 \\ \vdots \\ x_n \end{pmatrix}.$$

ALGORITHM 4.33 (*Linear equation solver for tridiagonal symmetric strongly nonsingular* A).

Given: $Ax = b$ for a tridiagonal strongly nonsingular symmetric matrix A.

Find: Solution x.

1^{st} step: (Factor $A = LDL^T$)

1.1 $\alpha_1 = d_1$

1.2 $\gamma_1 = f_1/\alpha_1$

1.3 For each $i = 2, \ldots, n-1$ do:

 1.3.1 $\alpha_i = d_i - f_{i-1}\gamma_{i-1}$

 1.3.2 $\gamma_i = f_i/\alpha_i$

1.4 $\alpha_n = d_n - f_{n-1}\gamma_{n-1}$

2^{nd} step: (Update $\boldsymbol{L}^T \boldsymbol{z} = \boldsymbol{b}, \boldsymbol{D}\boldsymbol{c} = \boldsymbol{z}$)

2.1 $z_1 = b_1$

2.2 For each $i = 2, \ldots, n$ do:
$z_i = b_i - \gamma_{i-1}z_{i-1}$

2.3 For each $i = 1, \ldots, n$ do:
$c_i = z_i/\alpha_i$

3^{rd} step: (Backsubstitution $\boldsymbol{L}^T \boldsymbol{x} = \boldsymbol{c}$)

3.1 $x_n = c_n$

3.2 For each $i = n-1, \ldots, 1$ do:
$x_i = c_i - \gamma_i x_{i+1}$

And

$$\det \boldsymbol{A} = \det(\boldsymbol{LDL}^T) = \det(\boldsymbol{L})\det \boldsymbol{D} \det(\boldsymbol{L}^T) = \det \boldsymbol{D} = \alpha_1\alpha_2\cdots\alpha_n,$$

while the inertia of \boldsymbol{A} is given by the number of positive and negative entries among the diagonal entries α_i of \boldsymbol{D}.

Note specifically the remark at the end of chapter 4.7.1.

Bibliography for Section 4.10

[BERG86]; [CONT80], p.182/4; [ENGE87], 3.3.4; [ISAA66], 2.3.2; [MAES84], 2.4; [SCHW89], 1.3; [WILK61].

4.11 Linear Systems with Cyclically Tridiagonal Matrices

4.11.1 Systems with a Cyclically Tridiagonal Matrix

An (n, n) matrix $A = (a_{ik})$ is called *cyclically tridiagonal* if $a_{ik} = 0$ for $1 < |i - k| < n - 1$ and $i, k = 1, \ldots, n$.

Such a system $Ax = b$ can be solved as usual without a pivot search if A is strongly nonsingular:

1) Factor $A = LR$ to obtain L and R.

2) Update $b = Lc$ to obtain c.

3) Backsubstitute $Rx = c$ to find x.

Implementation.

The elements of A, L, R, b, x and c shall be labelled as follows:

$$
A = \begin{pmatrix} d_1 & f_1 & & & e_1 \\ e_2 & d_2 & f_2 & & \\ & \ddots & \ddots & \ddots & \\ & & e_{n-1} & d_{n-1} & f_{n-1} \\ f_n & & & e_n & d_n \end{pmatrix}, \quad
R = \begin{pmatrix} 1 & \gamma_1 & & & \delta_1 \\ & 1 & \gamma_2 & & \delta_2 \\ & & \ddots & \ddots & \vdots \\ & & & \gamma_{n-2} & \delta_{n-2} \\ & & & 1 & \gamma_{n-1} \\ & & & & 1 \end{pmatrix},
$$

$$
L = \begin{pmatrix} \alpha_1 & & & & \\ \beta_2 & \alpha_2 & & & \\ & \ddots & \ddots & & \\ & & \beta_{n-1} & \alpha_{n-1} & \\ \varepsilon_3 & \varepsilon_4 \cdots & \varepsilon_n & \beta_n & \alpha_n \end{pmatrix}, \quad
c = \begin{pmatrix} c_1 \\ c_2 \\ \vdots \\ c_n \end{pmatrix}, \quad
b = \begin{pmatrix} b_1 \\ b_2 \\ \vdots \\ b_n \end{pmatrix},
$$

$$
x = \begin{pmatrix} x_1 \\ x_2 \\ \vdots \\ x_n \end{pmatrix}.
$$

ALGORITHM 4.34 (*Linear system solver for a cyclically tridiagonal matrix*).

Given: $Ax = b$ with a cyclically tridiagonal strongly nonsingular matrix A.

Find: Solution x.

1^{st} step: (Factor $A = LR$)

 1.1 $\alpha_1 = d_1$
 1.2 $\gamma_1 = f_1/\alpha_1$
 1.3 $\delta_1 = e_1/\alpha_1$
 1.4 For each $i = 2, \ldots, n - 2$ do:
 1.4.1 $\alpha_i = d_i - e_i\gamma_{i-1}$
 1.4.2 $\gamma_i = f_i/\alpha_i$
 1.4.3 $\beta_i = e_i$
 1.4.4 $\delta_i = -\beta_i\delta_{i-1}/\alpha_i$
 1.5 $\alpha_{n-1} = d_{n-1} - e_{n-1}\gamma_{n-2}$
 1.6 $\beta_{n-1} = e_{n-1}$
 1.7 $\varepsilon_3 = f_n$
 1.8 For each $i = 4, \ldots, n$ do:
 $\varepsilon_i = -\varepsilon_{i-1}\gamma_{i-3}$
 1.9 $\gamma_{n-1} = (c_{n-1} - \beta_{n-1}\delta_{n-2})/\alpha_{n-1}$
 1.10 $\beta_n = e_n - \varepsilon_n\gamma_{n-2}$
 1.11 $\alpha_n = d_n - \displaystyle\sum_{i=3}^{n} \varepsilon_i\delta_{i-2} - \beta_n\gamma_{n-1}$

2^{nd} step: (Update $Lc = b$)

 2.1 $c_1 = b_1/\alpha_1$
 2.2 For each $i = 2, \ldots, n - 1$ do:
 $c_i = (b_i - c_{i-1}\beta_i)/\alpha_i$
 2.3 $c_n = \left(b_n - \displaystyle\sum_{i=3}^{n} \varepsilon_i c_{i-2} - \beta_n c_{n-1}\right)/\alpha_n$

3^{rd} step: (Backsubstitution $Rx = c$)

 3.1 $x_n = c_n$
 3.2 $x_{n-1} = c_{n-1} - \gamma_{n-1}x_n$
 3.3 For each $i = n - 2, \ldots, 1$ do:
 $x_i = c_i - \gamma_i x_{i+1} - \delta_i x_n$

And the determinant of A is:

$$\det A = \det L \det R = \det L = \alpha_1\alpha_2 \cdots \alpha_n.$$

4.11.2 Systems with Symmetric Cyclically Tridiagonal Strongly Nonsingular Matrices

Let $A = (a_{ik}) = A^T$ be cyclically tridiagonal strongly nonsingular and symmetric. To solve the linear system $Ax = b$ one can employ a method like the Cholesky decomposition. Compared with chapter 4.11.1, one can economize on storage space. With a known factorization $A = LDL^T$, the system $Ax = b$ can be replaced following algorithm 4.29 by the equivalent system $L^T x = c$:

1) Factor $A = LDL^T$ to find L and D.

2) Update $Lz = b$ to find z. Then solve $Dc = z$ to find c.

3) Use backsubstitution $L^T x = c$ to find x.

Implementation.

The elements of A, L, D, c, z, x, b are labelled as follows:

$$A = \begin{pmatrix} d_1 & f_1 & & & f_n \\ f_1 & d_2 & f_2 & & \\ & \ddots & \ddots & \ddots & \\ & & f_{n-2} & d_{n-1} & f_{n-1} \\ f_n & & & f_{n-1} & d_n \end{pmatrix}, \quad L = \begin{pmatrix} 1 & & & & \\ \gamma_1 & 1 & & & \\ & \gamma_2 & & & \\ & & \ddots & \ddots & \\ & & & \gamma_{n-2} & 1 \\ \delta_1 & \ldots & & \delta_{n-2} & \gamma_{n-1} & 1 \end{pmatrix},$$

$$D = \begin{pmatrix} \alpha_1 & & & \\ & \alpha_2 & & \\ & & \ddots & \\ & & & \alpha_n \end{pmatrix}, \quad c = \begin{pmatrix} c_1 \\ c_2 \\ \vdots \\ c_n \end{pmatrix}, \quad b = \begin{pmatrix} b_1 \\ b_2 \\ \vdots \\ b_n \end{pmatrix},$$

$$x = \begin{pmatrix} x_1 \\ x_2 \\ \vdots \\ x_n \end{pmatrix}, \quad z = \begin{pmatrix} z_1 \\ z_2 \\ \vdots \\ z_n \end{pmatrix}.$$

ALGORITHM 4.35

Given: $Ax = b$ with a cyclically tridiagonal strongly nonsingular symmetric matrix A.

 Find: Solution x.

1^{st} step: (Factor $A = LDL^T$)

1.1 $\alpha_1 = d_1$

1.2 $\gamma_1 = f_1/\alpha_1$

1.3 $\delta_1 = f_n/\alpha_1$

1.4 For each $i = 2, \ldots, n - 2$ do:

 1.4.1 $\alpha_i = d_i - f_{i-1}\gamma_{i-1}$

 1.4.2 $\gamma_i = f_i/\alpha_i$

 1.4.3 $\delta_i = -\delta_{i-1}f_{i-1}/\alpha_i$

1.5 $\alpha_{n-1} = d_{n-1} - f_{n-2}\gamma_{n-2}$

1.6 $\gamma_{n-1} = (f_{n-1} - f_{n-2}\delta_{n-2})/\alpha_{n-1}$

1.7 $\alpha_n = d_n - \sum\limits_{i=1}^{n-2} \alpha_i\delta_i^2 - f_{n-1}\gamma_{n-1}^2$

2^{nd} step: (Update $\boldsymbol{L}z = \boldsymbol{b}, \boldsymbol{D}c = z$)

 2.1 $z_1 = b_1$

 2.2 For each $i = 2, \ldots, n-1$ do:
 $z_i = b_i - z_{i-1}\gamma_{i-1}$

 2.3 $z_n = b_n - \sum\limits_{i=1}^{n-2} \delta_i z_i - \gamma_{n-1}z_{n-1}$

 2.4 For each $i = 1, \ldots, n$ do:
 $c_i = z_i/\alpha_i$

3^{rd} step: (Backsubstitution $\boldsymbol{L}^T\boldsymbol{x} = \boldsymbol{c}$)

 3.1 $x_n = c_n$

 3.2 $x_{n-1} = c_{n-1} - \gamma_{n-1}x_n$

 3.3 For each $i = n - 2, \ldots, 1$ do:
 $x_i = r_i - \gamma_i x_{i+1} - \delta_i x_n$

For the determinant of \boldsymbol{A} we have:

$$\det \boldsymbol{A} = \det(\boldsymbol{L}\boldsymbol{D}\boldsymbol{L}^T) = \det(\boldsymbol{L})\det \boldsymbol{D}\det(\boldsymbol{L}^T) = \det \boldsymbol{D} = \alpha_1\alpha_2\cdots\alpha_n,$$

while the inertia of \boldsymbol{A} is given by the number of positive and negative entries among the α_i.

Note specifically the remark at the end of chapter 4.7.1.

Bibliography for Section 4.11

[SPÄT74/2], p.19/21 .

4.12 Linear Systems with Five-Diagonal Matrices

4.12.1 Systems with Five-Diagonal Matrices

A matrix $A = (a_{ik})$ is called *five-diagonal* if $a_{ik} = 0$ for $|i - k| > 2$ and all $i, k = 1, \ldots, n$.

By factoring $A = LR$, where L is a tridiagonal lower triangular matrix and R is a tridiagonal unit upper triangular matrix, one can reduce the system $Ax = b$ with a five-diagonal matrix A to the equivalent system $Rx = c$ without row interchanges according to algorithm 4.19 - provided $\det(A_k) \neq 0$ for $k = 1, \ldots, n - 1$:

1) Factor $A = LR$ and obtain L and R,

2) Update $b = Lc$ to obtain c,

3) Backsubstitution $Rx = c$ gives x.

Implementation.

Again we need not store A, L or R in full matrix format but can economize storage by storing the necessary information vectorially as follows:

$$
A = \begin{pmatrix}
d_1 & e_1 & f_1 & & & & \\
h_2 & d_2 & e_2 & f_2 & & & \\
g_3 & h_3 & d_3 & e_3 & f_3 & & \\
 & \ddots & \ddots & \ddots & \ddots & \ddots & \\
 & & g_{n-2} & h_{n-2} & d_{n-2} & e_{n-2} & f_{n-2} \\
 & & & g_{n-1} & h_{n-1} & d_{n-1} & e_{n-1} \\
 & & & & g_n & h_n & d_n
\end{pmatrix}, \quad
b = \begin{pmatrix}
b_1 \\
b_2 \\
\vdots \\
b_{n-1} \\
b_n
\end{pmatrix},
$$

$$
c = \begin{pmatrix}
c_1 \\
c_2 \\
\vdots \\
c_{n-1} \\
c_n
\end{pmatrix}, \quad
L = \begin{pmatrix}
\alpha_1 & & & & & \\
\beta_2 & \alpha_2 & & & & \\
\varepsilon_3 & \beta_3 & \alpha_3 & & & \\
 & \ddots & \ddots & \ddots & & \\
 & & \varepsilon_{n-1} & \beta_{n-1} & \alpha_{n-1} & \\
 & & & \varepsilon_n & \beta_n & \alpha_n
\end{pmatrix},
$$

$$
R = \begin{pmatrix}
1 & \gamma_1 & \delta_1 & & & & \\
 & 1 & \gamma_2 & \delta_2 & & & \\
 & & 1 & \gamma_3 & \delta_3 & & \\
 & & & \ddots & \ddots & \ddots & \\
 & & & & 1 & \gamma_{n-2} & \delta_{n-2} \\
 & & & & & 1 & \gamma_{n-1} \\
 & & & & & & 1
\end{pmatrix}.
$$

ALGORITHM 4.36

Given: $Ax = b$ with a five-diagonal strongly nonsingular matrix A.

Find: Solution x.

1^{st} step: (Factor $A = LR$)

 1.1 $\alpha_1 = d_1$

 1.2 $\gamma_1 = e_1/\alpha_1$

 1.3 $\delta_1 = f_1/\alpha_1$

 1.4 $\beta_2 = h_2$

 1.5 $\alpha_2 = d_2 - \beta_2\gamma_1$

 1.6 $\gamma_2 = (e_2 - \beta_2\delta_1)/\alpha_2$

 1.7 $\delta_2 = f_2/\alpha_2$

 1.8 For each $i = 3, \ldots, n-2$ do:

 1.8.1 $\beta_i = h_i - g_i\gamma_{i-2}$

 1.8.2 $\alpha_i = d_i - g_i\delta_{i-2} - \beta_i\gamma_{i-1}$

 1.8.3 $\gamma_i = (e_i - \beta_i\delta_{i-1})/\alpha_i$

 1.8.4 $\delta_i = f_i/\alpha_i$

 1.9 $\beta_{n-1} = h_{n-1} - g_{n-1}\gamma_{n-3}$

 1.10 $\alpha_{n-1} = d_{n-1} - g_{n-1}\delta_{n-3} - \beta_{n-1}\gamma_{n-2}$

 1.11 $\gamma_{n-1} = (e_{n-1} - \beta_{n-1}\delta_{n-2})/\alpha_{n-1}$

 1.12 $\beta_n = h_n - g_n\gamma_{n-2}$

 1.13 $\alpha_n = d_n - g_n\delta_{n-2} - \beta_n\gamma_{n-1}$

 1.14 For each $i = 3, \ldots, n$ do:

 $\varepsilon_i = g_i$

2^{nd} step: (Update $b = Lc$)

 2.1 $c_1 = b_1/\alpha_1$

2.2 $c_2 = (b_2 - \beta_2 c_1)/\alpha_2$

2.3 For each $i = 3, \ldots, n$ do:
$$c_i = (b_i - \varepsilon_i c_{i-2} - \beta_i c_{i-1})/\alpha_i$$

3rd step: (Backsubstitution $\boldsymbol{R}\boldsymbol{x} = \boldsymbol{c}$)

3.1 $x_n = c_n$

3.2 $x_{n-1} = c_{n-1} - \gamma_{n-1} x_n$

3.3 For each $i = n - 2, \ldots, 1$ do:
$$x_i = c_i - \gamma_i x_{i+1} - \delta_i x_{i+2}$$

And the determinant of \boldsymbol{A} is:

$$\det \boldsymbol{A} = \det \boldsymbol{L} \det \boldsymbol{R} = \det \boldsymbol{L} = \alpha_1 \alpha_2 \cdots \alpha_n.$$

4.12.2 Systems with Five-Diagonal Symmetric Matrices

If $\boldsymbol{A} = \boldsymbol{A}^T$ is five-diagonal, strongly nonsingular and symmetric, one can employ a procedure equivalent to the Cholesky decomposition for solving $\boldsymbol{A}\boldsymbol{x} = \boldsymbol{b}$. Since \boldsymbol{A} is symmetric one can save about half of the storage requirements of Algorithm 4.36.

We factor $\boldsymbol{A} = \boldsymbol{L}\boldsymbol{D}\boldsymbol{L}^T$ where \boldsymbol{L} is a tridiagonal unit lower triangular matrix and \boldsymbol{D} is a diagonal matrix. Then the system $\boldsymbol{A}\boldsymbol{x} = \boldsymbol{b}$ is equivalent to the system $\boldsymbol{L}^T\boldsymbol{x} = \boldsymbol{c}$, as in algorithm 4.29:

1). Factor $\boldsymbol{A} = \boldsymbol{L}\boldsymbol{D}\boldsymbol{L}^T$. This determines \boldsymbol{L} and \boldsymbol{D}.

2) Updating $\boldsymbol{L}\boldsymbol{z} = \boldsymbol{b}$ gives \boldsymbol{z}, $\boldsymbol{D}\boldsymbol{c} = \boldsymbol{z}$ gives \boldsymbol{c},

3) Backsubstitution $\boldsymbol{L}^T\boldsymbol{x} = \boldsymbol{c}$ determines \boldsymbol{x}.

Implementation.

To use vectorial storage, we introduce the following notation:

$$\boldsymbol{A} = \begin{pmatrix} d_1 & f_1 & e_1 & & & & \\ f_1 & d_2 & f_2 & e_2 & & & \\ e_1 & f_2 & d_3 & f_3 & e_3 & & \\ & \ddots & \ddots & \ddots & \ddots & \ddots & \\ & & e_{n-4} & f_{n-3} & d_{n-2} & f_{n-2} & e_{n-2} \\ & & & e_{n-3} & f_{n-2} & d_{n-1} & f_{n-1} \\ & & & & e_{n-2} & f_{n-1} & d_n \end{pmatrix}, \quad \boldsymbol{b} = \begin{pmatrix} b_1 \\ b_2 \\ \vdots \\ b_{n-1} \\ a_n \end{pmatrix},$$

$$D = \begin{pmatrix} \alpha_1 & & & & \\ & \alpha_2 & & & \\ & & \alpha_3 & & \\ & & & \ddots & \\ & & & & \alpha_n \end{pmatrix}, \quad L = \begin{pmatrix} 1 & & & & & \\ \gamma_1 & 1 & & & & \\ \delta_1 & \gamma_2 & 1 & & & \\ & \delta_2 & \ddots & \ddots & & \\ & & & & 1 & \\ & & & \delta_{n-2} & \gamma_{n-1} & 1 \end{pmatrix},$$

$$c = \begin{pmatrix} c_1 \\ c_2 \\ \vdots \\ c_n \end{pmatrix}, \quad z = \begin{pmatrix} z_1 \\ z_2 \\ \vdots \\ z_n \end{pmatrix}.$$

ALGORITHM 4.37

Given: $Ax = b$ with a five-diagonal symmetric strongly nonsingular matrix A.

Find: Solution x.

1^{st} step: (Factor $A = LDL^T$)

 1.1 $\alpha_1 = d_1$

 1.2 $\gamma_1 = f_1/\alpha_1$

 1.3 $\delta_1 = e_1/\alpha_1$

 1.4 $\alpha_2 = d_2 - f_1\gamma_1$

 1.5 $\gamma_2 = (f_2 - e_1\gamma_1)/\alpha_2$

 1.6 $\delta_2 = e_2/\alpha_2$

 1.7 For each $i = 3, \ldots, n-2$ do:

 1.7.1 $\alpha_i = d_i - e_{i-2}\delta_{i-2} - \alpha_{i-1}\gamma_{i-1}^2$

 1.7.2 $\gamma_i = (f_i - e_{i-1}\gamma_{i-1})/\alpha_i$

 1.7.3 $\delta_i = e_i/\alpha_i$

 1.8 $\alpha_{n-1} = d_{n-1} - e_{n-3}\delta_{n-3} - \alpha_{n-2}\gamma_{n-2}^2$

 1.9 $\gamma_{n-1} = (f_{n-1} - e_{n-2}\gamma_{n-2})/\alpha_{n-1}$

 1.10 $\alpha_n = d_n - e_{n-2}\delta_{n-2} - \alpha_{n-1}\gamma_{n-1}^2$

2^{nd} step: (Update $Lz = b, Dc = z$)

 2.1 $z_1 = b_1$

 2.2 $z_2 = b_2 - \gamma_1 z_1$

2.3 For each $i = 3, \ldots, n$ do:
$$z_i = b_i - \gamma_{i-1} z_{i-1} - \delta_{i-2} z_{i-2}$$

2.4 For each $i = 1, \ldots, n$ do:
$$c_i = z_i / \alpha_i$$

3^{rd} step: (Backsubstitution $\boldsymbol{L}^T \boldsymbol{x} = \boldsymbol{c}$)

3.1 $x_n = c_n$

3.2 $x_{n-1} = c_{n-1} - \gamma_{n-1} x_n$

3.3 For each $i = n - 2, \ldots, 1$ do:
$$x_i = c_i - \gamma_i x_{i+1} - \delta_i x_{i+2}$$

For the determinant of \boldsymbol{A} we have:

$$\det \boldsymbol{A} = \det(\boldsymbol{L}) \det \boldsymbol{D} \det \boldsymbol{L}^T = \det \boldsymbol{D} = \alpha_1 \alpha_2 \cdots \alpha_n,$$

while the inertia of \boldsymbol{A} is given by the number of positive and negative entries on the diagonal of \boldsymbol{D}.

Note specifically the remark at the end of chapter 4.7.1.

4.13 Linear Systems with Band Matrices

A matrix $\boldsymbol{A} = (a_{ik})$ whose elements vanish outside of a band parallel to the main diagonal is called a *band matrix* (see definition 4.10).

When factoring such a matrix $\boldsymbol{A} = \boldsymbol{LR}$, the triangular matrices \boldsymbol{R} and \boldsymbol{L} will also be banded. Thus the number of operations will be bounded by functions of m_ℓ, the number of nonzero lower diagonals in \boldsymbol{A}, and m_r, the number of nonzero upper diagonals of \boldsymbol{A}, rather than the dimension n of \boldsymbol{A} itself.

Thus, algorithms 4.19 and 4.20 can be used and the factors \boldsymbol{L} and \boldsymbol{R} will also be banded. In the following we shall indicate three algorithms, the first of which transforms \boldsymbol{A} to an upper triangular matrix, i.e., the \boldsymbol{L} of the LR factorization of \boldsymbol{A} is a unit lower triangular matrix and \boldsymbol{R} is an upper triangular matrix. The second algorithm describes a transformation to lower triangular form, i.e., \boldsymbol{L} is a unit upper triangular matrix and \boldsymbol{R} is a lower triangular matrix in a reversed "LR" factorization of \boldsymbol{A}.

The third algorithm uses a condensed version of the matrix \boldsymbol{A} and reduces it to upper or lower triangular form depending on the amount of calculations involved. The condensing of \boldsymbol{A} is performed in such a way that the co-diagonals

of A become the columns of the condensed matrix. This algorithm was developed by E. Pohl. The necessary row interchanges will be stored in a vector p. p is defined by setting $p_i = i + k$ if row i was interchanged with row $i + k$. If $p_i = 0$ there is no interchange. The parameter SIG accounts for row changes. After the transformation of A to upper or lower triangular form

$$\det A = \text{SIG } r_{11}r_{22}\cdots r_{nn}$$

where $\text{SIG} = (-1)^k$ and k is the number of row interchanges.

In the following algorithms, A is overwritten with its LR factors L and R so that

$$\det A = \text{SIG } a_{11}a_{22}\cdots a_{nn}.$$

ALGORITHM 4.38 (*Transformation to upper triangular form*).

Given: $Ax = b$ with a band matrix $A = (a_{ik})$ and $b \in \mathbb{R}^n$. m_ℓ = number of lower co-diagonals, m_r = number of upper co-diagonals.

Find: Solution x.

1^{st} step: Factor $A = LR$ with a unit lower triangular matrix L and an upper triangular matrix R with pivoting.

 1.1 SIG := 1

 1.2 For each $i = 1, \ldots, n - 1$ do:

 1.2.1 $p_i := \nu$ if
$|a_{i+\nu,i}| = max\{|a_{i+k,i}| \mid k = 0, \ldots, \min\{m_\ell, n - i\}\}$

 1.2.2 If $\nu \neq 0$, then

 1.2.2.1 SIG := $-$ SIG

 1.2.2.2 For each $k = 0, \ldots, \min\{m_l + m_r, n - i\}$ replace $a_{i,i+k}$ by $a_{i+\nu,i+k}$

 1.2.3 For $k = 1, \ldots, \min\{m_\ell, n - i\}$ do:

 1.2.3.1 Replace $a_{i+k,i}$ by $a_{i+k,i}/a_{ii}$

 1.2.3.2 For $j = 1, \ldots, \min\{m_l + m_r, n - i\}$ replace $a_{i+k,i+j}$ by $a_{i+k,i+j} - a_{i+k,i}\, a_{i,i+j}$

 1.3 $p_n := 0$ (last row is not exchanged)

2^{nd} step: Update $b = Lc$, b is overwritten with c. The row exchanges during the factorization are taken into consideration.

 2.1 For $i = 1, \ldots, n - 1$ do:

2.1.1 If $p_i \neq 0$, replace b_i by b_{i+p_i}

2.1.2 For $k = 1, \ldots, \min\{m_\ell, n - i\}$ do:
 replace b_{i+k} by $b_{i+k} - a_{i+k,i}b_i$

3^{rd} step: Backsubstitute $\boldsymbol{R}\boldsymbol{x} = \boldsymbol{c}$, the solution \boldsymbol{x} is stored in \boldsymbol{b}.

3.1 Replace b_n by b_n/a_{nn}

3.2 For each $i = n - 1, \ldots, 1$ do:

 3.2.1 For $k = 1, \ldots, \min\{n - i, m_\ell + m_r\}$ do:
 Replace b_i by $b_i - a_{i,i+k}b_{i+k}$

 3.2.2 Replace b_i by b_i/a_{ii}

ALGORITHM 4.39 (*Transformation to lower triangular form*).

Given: $\boldsymbol{A}\boldsymbol{x} = \boldsymbol{b}$ with $\boldsymbol{A} = (a_{ik})$ a band matrix and $\boldsymbol{b} \in \mathbb{R}^n$.

Find: Solution \boldsymbol{x}.

1^{st} step: Factor $\boldsymbol{A} = \boldsymbol{L}\boldsymbol{R}$ with a unit upper triangular matrix \boldsymbol{L} and a lower triangular matrix \boldsymbol{R}.

1.1 SIG := 1

1.2 For each $i = n, \ldots, 2$ do:

 1.2.1 $p_i := \nu$ for
 $|a_{i+\nu,i}| = \max\{|a_{i+k,i}|, k = 0, \ldots, \max\{1 - i, -m_r\}\}$

 1.2.2 If $\nu \neq 0$, then

 1.2.2.1 SIG := − SIG

 1.2.2.2 For each $k = 0, \ldots, \max\{1 - i, -m_r - m_l\}$ do:
 Replace $a_{i,i+k}$ by $a_{i+\nu,i+k}$

 1.2.3 For $k = -1, \ldots, \max\{1 - i, -m_r\}$ do:

 1.2.3.1 Replace $a_{i+k,i}$ by $a_{i+k,i}/a_{ii}$

 1.2.3.2 For $j = -1, \ldots, \max\{1 - i, -m_r - m_l\}$ set
 $a_{i+k,i+j} := a_{i+k,i+j} - a_{i+k,i}a_{i,i+j}$

 1.3 $p_1 := 0$ (first row is not interchanged)

2^{nd} step: Update $\boldsymbol{b} = \boldsymbol{L}\boldsymbol{c}$, \boldsymbol{c} is stored in \boldsymbol{b}

2.1 For $i = n, \ldots, 2$ do:

 2.1.1 If $p_i \neq 0$, replace b_i by b_{i+p_i}

2.1.2 For $k = -1, \ldots, \max\{1 - i, -m_r\}$ set
$$b_{i+k} := b_{i+k} - a_{i+k,i}b_i$$

3^{rd} step: Backsubstitution $\boldsymbol{Rx} = \boldsymbol{c}, \boldsymbol{x}$ is stored in \boldsymbol{b}

3.1 $b_1 := b_1/a_{11}$

3.2 For each $i = 2, \ldots, n$ do:

3.2.1 For $k = -1, \ldots, \max\{1 - i, -m_l - m_r\}$ do:
Replace b_i by $b_i - a_{i,i+k}b_{i+k}$

3.2.2 Replace b_i by b_i/a_{ii}

ALGORITHM 4.40 (*Condensed sparse matrix, transformation to upper or lower triangular form*).

Given: $\boldsymbol{Ax} = \boldsymbol{b}$ with a band matrix $\boldsymbol{A} = (a_{ik})$ and $\boldsymbol{b} \in \mathbb{R}^n$.

Find: Solution \boldsymbol{x}.

1^{st} step: Condense the matrix \boldsymbol{A} to \boldsymbol{A}^* by setting $a^*_{\alpha,\beta} := a_{\alpha,\alpha+\beta-m_\ell-1}$
or $a_{ik} = a^*_{i,m_\ell+1+k-i}$ for $i = 1, \ldots, n$,
$k = \max\{1, i - m_\ell\}, \ldots, \min\{n, i + m_r\}$

The (n, n) matrix \boldsymbol{A} is replaced by \boldsymbol{A}^* with n rows and $m = m_\ell + m_r + 1$ columns.

NOTE concerning the storage of additional co-diagonals of \boldsymbol{A} in algorithm 4.40: The transformation to lower triangular form produces additional co-diagonals below the diagonal band of \boldsymbol{A}. According to the condensing step, described in step 1, these would appear in columns to the left of column 1 in \boldsymbol{A}^*. Since at the start the number of such columns is not known, the additional co-diagonals are stored to the right of \boldsymbol{A}^* so that the total number of columns of \boldsymbol{A}^* becomes:

$$m^* = m + \min\{m_\ell, m_r\}.$$

2^{nd} step: Factorization

2.1 SIG := 1
$$m^* = m_\ell + m_r + 1 + \min\{m_\ell, m_r\}$$

2.2 If $m_\ell \le m_r$, set
$$i_a := 1, i_e := n - 1, i_s := 1, k_a := 1, j_a := 1,$$

otherwise set
$$i_a := n, i_e := 2, i_s := -1, k_a := -1, j_a := -1$$

2.3 For each $i = i_a, i_a + i_s, \ldots, i_e - i_s, i_e$ do:

 2.3.1 If $m_\ell \leq m_r$, then set $k_e := \min\{m_\ell, n - i\}$,
 otherwise set $k_e := \max\{1 - i, -m_r\}$

 2.3.2 $p_i := \nu$ if
 $|a^*_{i+\nu, m_\ell+1-\nu}| = \max\{|a^*_{i+k, m_\ell+1-k}| \mid k = 0, \ldots, k_e\}$

 2.3.3 If $m_\ell \leq m_r$, then set $j_e := \min\{m_l + m_r, n - i\}$,
 otherwise set $j_e := \max\{1 - i, -m_r - m_\ell\}$

 2.3.4 If $\nu \neq 0$, then

 2.3.4.1 SIG := $-$ SIG

 2.3.4.2 For $k = 0, i_s, 2_{i_s}, \ldots, j_e - i_s, j_e$ do:

 2.3.4.2.1 $k_m := k + m_\ell + 1$

 2.3.4.2.2 If $k_m \leq 0$, then $k_m := k_m + m^*$

 2.3.4.2.3 Replace a^*_{i,k_m} by $a^*_{i+\nu, k_m-\nu}$

 2.3.5 For $k = k_a, k_a + i_s, \ldots, k_e - i_s, k_e$ do:
 Set $a^*_{i+k, m_\ell+1-k} := a^*_{i+k, m_\ell+1-k}/a^*_{i, m_\ell+1}$
 For $j = j_a, j_a + i_s, \ldots, j_e - i_s, j_e$ do:
 $j_k := j + m_\ell + 1 - k$
 $j_m := j + m_\ell + 1$
 If $j_k \leq 0$, then $j_k := j_k + m^*$
 If $j_m \leq 0$, then $j_m := j_m + m^*$
 Set $a^*_{i+k, j_k} := a^*_{i+k, j_k} - a^*_{i+k, m_\ell+1-k} a^*_{i, j_m}$

2.4 $p_{i_e+i_s} := 0$

3^{rd} step: Update:

3.1 $m^* := m_\ell + m_r + 1 + \min\{m_\ell, m_r\}$
 If $m_\ell \leq m_r$, then
 $i_a := 1, i_e := n - 1, i_s := 1, k_a := 1,$
 otherwise
 $i_a := n, i_e := 2, i_s := -1, k_a := -1$

3.2 For $i = i_a + i_s, \ldots, i_e - i_s, i_e$ do:

 3.2.1 If $p_i \neq 0$, replace b_i by b_{i+p_i}

 3.2.2 If $m_\ell \leq m_r$, then set $k_e := \min\{m_\ell, n - i\}$,
 otherwise set $k_e := \max\{1 - i, -m_r\}$

 3.2.3 For $k = k_a, k_a + i_s, \ldots, k_e - i_s, k_e$ do:
 $b_{i+k} = b_{i+k} - a^*_{i+k, m_\ell+1-k} b_i$

3.3 $b_{i_e+i_s} = b_{i_e+i_s}/a^*_{i_e+i_s, m_\ell+1}$

3.4 For $i = i_e, i_e - i_s, \ldots, i_a + i_s, i_a$ do:

 3.4.1 If $m_\ell \leq m_r$, then set $k_e := \min\{n - i, m_\ell + m_r\}$, otherwise set $k_e := \max\{1 - i, -m_\ell - m_r\}$

 3.4.2 For $k = k_a, k_a + i_s, \ldots, k_e - i_s, k_e$ do:

 3.4.2.1 $k_m := k + m_\ell + 1$

 3.4.2.2 If $k_m \leq 0$, then $k_m := k_m + m^*$

 3.4.2.3 $b_i := b_i - a^*_{i,k_m} b_{i+k}$

 3.4.3 $b_i := b_i / a^*_{i,m_\ell+1}$

For a dense matrix A, there is no computational difference in reducing it to upper or lower triangular form. When using row interchanges for band matrices, however, there is a difference as far as calculation time and storage efficiency is concerned if $m_\ell \neq m_r$. A transformation to upper triangular form using row interchanges produces m_ℓ additional upper co-diagonals, a transformation to lower triangular form produces m_r additional lower co-diagonals. For an upper triangular form, we must deal with m_ℓ rows per elimination step, for a lower triangular form, m_r rows are involved, so that the amount of computations depends on our choice of transformation.

Bibliography for Section 4.13

[BERG86]; [FORS67] §23; [GOLU89], Sec.5.3; [MAES84], p.68; [WEIS84], 6.5.

4.14 Solving Linear Systems via Householder Transformations

Solving linear systems by means of Householder transformations is particulary important because the condition number $cond_2(A)$ of the system matrix A is not altered during the factorization of A into the product of an orthogonal matrix Q and an upper triangular matrix R (see e.g. [STOE91], ch. 4.7).

ALGORITHM 4.41 (*Linear equations solver via orthogonal matrices*).

Given: $Ax = b$ for an (m, n) matrix A with rank $A = n \leq m$, $b \in \mathbb{R}^m$.

Find: Solution $x \in \mathbb{R}^n$.

1^{st} step: Factor $A = QR$ for an orthogonal (m, m) matrix Q and an upper triangular (m, n) matrix R. By orthogonality:
$Q^T = Q^{-1}$.

2^{nd} step: Update c to $c = Q^T b$.

3^{rd} step: Backsubstitution $Rx = c$ gives the solution $x \in \mathbb{R}^n$.

Householder transformations

Householder transformations shall be used to transform the given (m, n) matrix A of rank n into an (m, n) upper triangular matrix $R = (r_{ik})$ by successive left-multiplication of A with Householder matrices H_i, see theorem 4.9:

$$A_{q+1} = \underbrace{H_q H_{q-1} \ldots H_2 H_1}_{=:Q^T} A_1 =: Q^T A = R$$

with $A_{i+1} = H_i A_i$ for $i = 1, \ldots, q$ with $q = \min(m - 1, n)$ and $A_1 := A$.

This method gives a QR factorization of A:

$$A = QR \quad \text{since} \quad Q^T = Q^{-1}.$$

Implementation of Householder reductions:

We set

$$A_1 = (a_{ik}^{(1)}) := A = (a_{ik}) \quad \text{for} \quad i = 1, \ldots, m, \ k = 1, \ldots, n,$$

and write

$$A_1 = \begin{pmatrix} a_{11}^{(1)} & a_{12}^{(1)} & \cdots & a_{1n}^{(1)} \\ a_{21}^{(1)} & a_{22}^{(1)} & \cdots & a_{2n}^{(1)} \\ \vdots & \vdots & & \\ a_{m1}^{(1)} & a_{m2}^{(1)} & \cdots & a_{mn}^{(1)} \end{pmatrix} = (a_1^{(1)}, a_2^{(1)}, \ldots, a_n^{(1)}),$$

where $a_j^{(1)}$ denotes the j^{th} column of A_1.

We can form the (m, m) Householder matrix \boldsymbol{H}_1 as in theorem 4.9 by setting

$$\boldsymbol{H}_1 := \boldsymbol{I}_m - \frac{2}{\|\boldsymbol{v}_1\|^2} \boldsymbol{v}_1 \boldsymbol{v}_1^T$$

for

$$\boldsymbol{v}_1 = \begin{pmatrix} a_{11}^{(1)} + \operatorname{sign}(a_{11}^{(1)})\|a_1^{(1)}\| \\ a_{21}^{(1)} \\ \vdots \\ a_{m1}^{(1)} \end{pmatrix}.$$

Then

$$\boldsymbol{A}_2 = \boldsymbol{H}_1 \boldsymbol{A}_1 = \begin{pmatrix} a_{11}^{(2)} & a_{12}^{(2)} & \cdots & a_{1n}^{(2)} \\ 0 & a_{22}^{(2)} & \cdots & a_{2n}^{(2)} \\ \vdots & & & \\ 0 & a_{m2}^{(2)} & \cdots & a_{mn}^{(2)} \end{pmatrix} = \begin{pmatrix} a_{11}^{(2)} & a_{12}^{(2)} & \cdots & a_{1n}^{(2)} \\ 0 & & & \\ \vdots & & \tilde{\boldsymbol{A}}_2 & \\ 0 & & & \end{pmatrix},$$

where $\tilde{\boldsymbol{A}}_2$ is $m-1$ by $n-1$. Continuing in this way, we can compute

$$\boldsymbol{A}_{i+1} = \boldsymbol{H}_i \boldsymbol{A}_i \quad \text{for} \quad i = 1, \ldots, p$$

for the (m, n) matrix \boldsymbol{A}_i which has a trapezoidal block of zeros in its lower left corner:

$$\boldsymbol{A}_i = \begin{pmatrix} a_{11}^{(2)} & a_{12}^{(2)} & & & \cdots & a_{1n}^{(2)} \\ & a_{22}^{(3)} & & & \cdots & a_{2n}^{(3)} \\ & & \ddots & & & \\ & & & a_{i-1,i-1}^{(i)} & \cdots & a_{i-1,n}^{(i)} \\ & & & 0 & & \\ & & & \vdots & & \tilde{\boldsymbol{A}}_i \\ & & & 0 & & \end{pmatrix}, \text{ where}$$

$$\tilde{\boldsymbol{A}}_i = \begin{pmatrix} a_{ii}^{(i)} & \cdots & a_{in}^{(i)} \\ \vdots & & \vdots \\ a_{mi}^{(i)} & \cdots & a_{mn}^{(i)} \end{pmatrix}.$$

Here \boldsymbol{H}_i is the (m, m) Householder matrix

$$\boldsymbol{H}_i = \begin{pmatrix} \boldsymbol{I}_{i-1} & \boldsymbol{O} \\ \boldsymbol{O} & \tilde{\boldsymbol{H}}_i \end{pmatrix} \begin{matrix} \} & i-1 \text{ rows} \\ \} & m-i+1 \text{ rows} \end{matrix}.$$

\tilde{H}_i is an $(m - i + 1, m - i + 1)$ Householder matrix, constructed according to theorem 4.9 as

$$\tilde{H}_i = I_{m-i+1} - \frac{2}{\|v_i\|^2} v_i v_i^T,$$

where the vector $v_i \in \mathbb{R}^{m-i+1}$ is

$$v_i = \begin{pmatrix} a_{ii}^{(i)} + \text{sign}(a_{ii}^{(i)}) \| \mathbf{a}_i^{(i)} \| \\ a_{i+1,i}^{(i)} \\ \vdots \\ a_{mi}^{(i)} \end{pmatrix} \quad \text{for the column } a_i = \begin{pmatrix} a_{ii}^{(i)} \\ a_{i+1,i}^{(i)} \\ \vdots \\ a_{mi}^{(i)} \end{pmatrix} \in \mathbb{R}^{m-i+1}.$$

The (m, n) matrix A_{i+1} has a larger lower left trapezoidol block with zero entries:

$$A_{i+1} = \begin{pmatrix} a_{11}^{(2)} & a_{12}^{(2)} & & \cdots & & a_{1n}^{(2)} \\ & a_{22}^{(3)} & & \cdots & & a_{2n}^{(3)} \\ & & \ddots & & & \\ & & & a_{ii}^{(i+1)} & \cdots & a_{in}^{(i+1)} \\ & & & 0 & & \\ & & & \vdots & & \tilde{A}_{i+1} \\ & & & 0 & & \end{pmatrix},$$

and a possibly dense matrix

$$\tilde{A}_{i+1} = \begin{pmatrix} a_{i+1,i+1}^{(i+1)} & \cdots & a_{i+1,n}^{(i+1)} \\ \vdots & & \vdots \\ a_{m,i+1}^{(i+1)} & \cdots & a_{m,n}^{(i+1)} \end{pmatrix}.$$

If one combines the Householder transformations H_i used to obtain the QR factorization of A in algorithm 4.41 with the updating equation

$$Rx = Q^T Ax = Q^T b =: c,$$

then the first two steps of algorithm 4.41 may be amended to simultaneously compute R and c. Thus one can avoid forming Q explicitly.

ALGORITHM 4.42 (*Householder reduction to upper triangular form*).

Given: $Ax = b$ with an (m, n) matrix $A = (a_{ik})$ of rank $n \leq m$, $b \in \mathbb{R}^m$.
Find: Solution $x \in \mathbb{R}^n$.

1^{st} step: (Householder transformations to factor $R = Q^T A$ and update $c = Q^T b$ with $Q^T = H_q H_{q-1} \cdots H_1$)
For each $i = 1, \ldots, n$ do:

 1.1 Calculate:
$$r := \sum_{k=i}^{m} a_{ki}^2$$
$$\alpha := \sqrt{r} \text{ sign } (a_{ii})$$
$$ak := 1/(r + \alpha \cdot a_{ii})$$
$$a_{ii} := a_{ii} + \alpha$$

 1.2 (Update the matrix A and the right hand side b, considered as the $(n + 1)^{st}$ column of A)
$$d_i := -\alpha$$
For each $k = i + 1, \ldots, n + 1$ do:
$$f := 0$$
For each $j = i, \ldots, m$ do:
$$f := f + a_{jk} a_{ji}$$
$$f := f \cdot ak$$
For each $j = i, \ldots m$ do:
$$a_{jk} := a_{jk} - f \cdot a_{ji}$$

2^{nd} step: (Backsubstitution $Rx = c$)
For each $i = n, \ldots, 1$ do:
$$x_i := a_{i,n+1}$$
For each $k = i + 1, \ldots, n$ do:
$$x_i := x_i - a_{ik} x_k$$
$$x_i := x_i / d_i$$

Bibliography for Section 4.14

[GOLU89], 5.1, 5.2; [MAES84], 4.3; [NIEM87], 6.4; [STEW73], 5.3; [STOE91], 4.7.

4.15 Errors, Conditioning and Iterative Refinement

4.15.1 Errors and the Condition Number

The solution of a linear system of equations can generally not be obtained exactly via direct methods because of

1) rounding errors that occur during the computations and accumulate, and
2) input error that can be overly magnified due to ill-conditioning, see chapter 1.3.4.

If small changes in the initial data cause big changes in the solution, then the problem is called *ill-conditioned*, see chapter 1.5.

It is important to measure the quality of an approximation $x^{(0)}$ of the true solution of $Ax = b$. The residual vector

$$(4.14) \qquad\qquad r^{(0)} = b - Ax^{(0)}$$

gives some indication of this. If $x^{(0)}$ is close to the exact solution x, the vector $r^{(0)}$ must be very small. Conversely, however, $|r^{(0)}| < \varepsilon$ need not imply that $x^{(0)}$ is a good approximation for x; this can only be expected for the solutions of well-conditioned systems. Therefore the residual vector itself cannot be considered as a suitable measure for the quality of an approximate solution. Similarly, the magnitude of the determinant of A is insufficient to characterize badly conditioned systems. Several useful condition measures are given next:

a) The number

$$K_H(A) = \frac{|\det A|}{\alpha_1 \alpha_2 \cdots \alpha_n} \quad \text{with} \quad \alpha_i = \sqrt{a_{i1}^2 + a_{i2}^2 + \ldots + a_{in}^2}$$
$$\text{for} \quad i = 1, \ldots, n$$

is called the *Hadamard condition number* of the matrix A. A matrix A is considered ill-conditioned if $K_H(A) \ll 1$. Here the symbol \ll denotes "much less than".

Empirical evidence suggests that: $K_H(A) < 0.01$ implies ill-conditioning, while $K_H(A) > 0.1$ implies well-conditioning.
For $0.01 \leq K_H(A) \leq 0.1$ no precise assertion is in general possible.

For linear systems with $K_H = O(10^{-k})$, a change in the k^{th} digit of one coefficient of A can, but need not lead to a change of order $O(10^k)$ in the solution (see [CONT80], p.163ss.; [WILK64], pp.116., pp.133., pp.143.).

b) The number

$$cond(A) = \|A\| \, \|A^{-1}\|$$

is standardly called the *condition number* of A, see section 5.2 for a definition of a matrix norm $\|A\|$.

c) The number

$$\mu(A) = \frac{\max_i |\lambda_i|}{\min_i |\lambda_i|},$$

where λ_i are the eigenvalues of the matrix A (see chapter 7), is called the *spectral condition number* of A.

For the condition numbers $cond(A)$ and $\mu(A)$, big values indicate ill-conditioning of A.
None of these condition numbers, however, give complete information about the condition of a linear system.

Different condition measures can be found in [COLL68], p.81/82; [FADD59] p.149-159; [ISAA66], p.39/40; [SPEL85], p.39/40; [MAES84], [NIEM87], 6.1, 6.2.

One way to notice ill-conditioning of a linear system is, that during Gauß elimination the elements $a_{jj}^{(j-1)}$ of the system (4.8) successively lose more and more accuracy due to cancellation.

To summarize, a system $Ax = b$ is ill-conditioned if any one of the following holds for the system matrix A:

1) $K_H(A) < 0.01$;

2) $cond(A) \gg 1$;

3) $\mu(A) \gg 1$;

4) small changes of a coefficients a_{ik} cause large changes in the solution vector;

5) the coefficients $a_{jj}^{(j-1)}$ of the system (4.8) obtained by Gauß algorithm lose accuracy successively;

6) the elements of the inverse A^{-1} of A are large when compared with the elements of A itself;

7) slow convergence or oscillation of iterative refinement, see chapter 4.15.4.

4.15.2 Condition Estimates

Only the Hadamard condition number $K_H(A)$ can be computed in a reasonable amount of time if a factorization of A is known. The LR factorization easily gives $\det A$ and the α_i are quickly computed. The operations count for $K_H(A)$, given $A = LR$, is $O(n^2)$. To compute $cond(A) = \|A\| \|A^{-1}\|$ would take $O(\frac{4}{3}n^3)$ operations, see chapter 4.6, and besides, for ill-conditioned matrices A, the inverse A^{-1} could only be computed with large errors. Thus one customarily relies on condition number estimates.

Condition number estimate of Forsythe and Moler.

[FORS67], p. 13 gives an heuristic estimate for the condition number $cond(A) = \|A\| \|A^{-1}\|$ which generally furnishes very useful results. Set

$$cond(A) \approx \frac{\|z^{(1)}\|}{\|x^{(0)}\|} \cdot \frac{1}{\text{EPS}},$$

where $z^{(1)}$ is the correction vector of the first iterative refinement step, see section 4.15.4, $x^{(0)}$ is the original solution of $Ax = b$ by means of Gauß algorithm and EPS is the machine constant, see definition 1.12.

ALGORITHM 4.43 (*Condition estimate of Forsythe and Moler*).

Given: (i) A linear system $Ax = b$ with a nonsingular (n,n) matrix A,
 (ii) a solution $x^{(0)}$ of $Ax = b$ from Gauß algorithm, $PA = LR$,
 (iii) the machine constant EPS

 Find: An estimate of the condition number $cond(A) = \|A\| \|A^{-1}\|$.

1^{st} step: Calculate the residual vector $r^{(0)}$ in double precision:

$$r^{(0)} := b - Ax^{(0)}.$$

2^{nd} step: Solve $Az^{(1)} = r^{(0)}$ via the known triangular factorization of A.
3^{rd} step: Calculate

$$cond(A) \approx \frac{\|z^{(1)}\|}{\|x^{(0)}\|} \cdot \frac{1}{\text{EPS}}.$$

Condition number estimate of Cline et al.

Suppose that an LR factorization of A or PA is known. Using the factors L and R of A one can calculate an approximate value for $\|(LR)^{-1}\|_\infty$ and,

using a matrix norm that is invariant to row changes such as the $1, 2,$ and ∞ matrix norms are, $\|A^{-1}\| = \|(LR)^{-1}\|$ gives an estimate for $cond(A)$.

The following algorithm originated in a paper of Cline, Moler, Stewart and Wilkinson [CLIN79]. It is used in a modified form in LINPACK and explained extensively in [KIEL88], 5.4.

The algorithm is based upon the inequality

$$(1) \qquad\qquad \|A^{-1}\|_\infty \geq \|z\|_\infty/\|x\|_\infty$$

derived from $Az = x$ with $x \neq 0$. Alternatively,

$$(2) \qquad \|A^{-1}\|_\infty = \|(A^{-1})^T\|_1 = \|(LR^{-1})^T\|_1 \geq \|z\|_1/\|x\|_1$$

for $(LR)^T z = x$ with $x \neq 0$. In both representations, z and x must be chosen so that $\|z\|/\|x\|$ is a good approximation for $\|A^{-1}\|$. Since $(LR)^T z = R^T L^T z = x$, two triangular systems

$$\begin{aligned} R^T y &= x \\ L^T z &= y \end{aligned}$$

must be solved when using (2). If a column pivot search is used it is desirable to find x and y so that $\|y\|_1/\|x\|_1$ is maximized. Then $L^T z = y$ is solved and thus $\|z\|_1/\|x\|_1$ is calculated. The proof, as well as further optimization steps can be found in [KIEL88], 5.4.

ALGORITHM 4.44 (*Condition number estimate of Cline et al*).

Given: LR factorization of A or PA with a unit lower triangular matrix $L = (\ell_{ik})$ and a nonsingular upper triangular matrix $R = (r_{ik})$.

Find: An estimate for $cond_\infty(A) = \|A\|_\infty \|A^{-1}\|_\infty$

1^{st} step: Determine $x = (x_i)$ with $x_i = \pm 1$ and $y = (y_i) = (R^T)^{-1}x$ so that $\|y\|_1 = \sum\limits_{i=1}^{n} |y_i|$ is maximal.

This is achieved by:

 1.1 Set $x_1 := 1$, $y_1 := 1/r_{11}$ and $y_i = -r_{1i} \cdot y_1/r_{ii}$ for $i = 2, \dots, n$

 1.2 For each $k = 2, \dots, n$ do:

 1.2.1 $v := \dfrac{1}{r_{kk}}$

 1.2.2 $x_k := y_k - v, \quad y_k := y_k + v$

 1.2.3 SMI $:= |x_k|, \quad$ SPL $:= |y_k|$

1.2.4 For each $i = k + 1, \ldots, n$ do:
$$v := r_{ki}/r_{ii}$$
$$x_i := y_i - v \cdot x_k, \quad y_i := y_i - v \cdot y_k$$
$$\text{SMI} := \text{SMI} + |x_i|, \quad \text{SPL} := \text{SPL} + |y_i|$$

1.2.5 If SMI > SPL, set
$$y_i := x_i \text{ for } i = k, \ldots, n$$
$$x_k := -1,$$
otherwise
$$x_k := 1.$$

2^{nd} Step: Solve $\boldsymbol{L}^T \boldsymbol{z} = \boldsymbol{y}$ for \boldsymbol{z} by backsubstitution.

3^{rd} Step: Set $K_\infty := \|\boldsymbol{z}\|_1 / \|\boldsymbol{x}\|_1$.

4^{th} Step: Calculate $\quad cond_\infty(\boldsymbol{A}) \approx \|\boldsymbol{A}\|_\infty \cdot K_\infty \quad$ as a condition number estimate for \boldsymbol{A}.

K_∞ is used as an approximate value for $\|\boldsymbol{A}^{-1}\|_\infty$ in [KIEL88] and [NIEM87]. Due to the equivalence of (1) and (2), $\bar{K}_\infty := \|\boldsymbol{z}\|_\infty / \|\boldsymbol{x}\|_\infty$ can be used as well. As shown by extensive tests, this second number \bar{K}_∞ can give even better approximations for $cond(\boldsymbol{A})$.

Bibliography for Section 4.15.2

[CLIN79]; [CLIN82]; [CLIN83]; [FORS67], 13.; [GOLU89], Sec. 3.5.4; [KIEL88], 5.4; [NIEM87], 6.2.4; [STEW73], 4.4 .

4.15.3 Improving the Condition Number

(a) *Row scaling* (see [WERN82], p.160): Multiply each row of \boldsymbol{A} by a constant, i.e., replace the given system $\boldsymbol{A}\boldsymbol{x} = \boldsymbol{b}$ by

$$\boldsymbol{D}_1 \boldsymbol{A}\boldsymbol{x} = \boldsymbol{D}_1 \boldsymbol{b},$$

where \boldsymbol{D}_1 is non-singular and diagonal. According to Wilkinson, one can improve the condition number if one chooses \boldsymbol{D}_1 in such a way that all row vectors of the matrix $\boldsymbol{D}_1 \boldsymbol{A}$ have the same norm.

(b) *Column scaling* (see [BJÖR74], 5; [FORS67], 11): Here we multiply each column of A by a constant. This means that the scalings for the unknown x_k are altered. Thus instead of $Ax = b$ we solve

$$(AD_2)y = b \text{ and } D_2 y = x.$$

(c) *Joint row and column scaling*: We can equilibrate both the rows and columns of A by forming $D_1 A D_2$. Then

$$(D_1 A D_2)Y = D_1 b \text{ and } D_2 y = x.$$

(d) Replacing certain rows of A with linear combinations of the given rows can also lead to an improvement of the condition number. The condition number, however, can also be worsened by this procedure.

4.15.4 Iterative Refinement

If a linear systems $Ax = b$ has an approximate solution $x^{(0)}$ that is marred by rounding errors , it can be improved upon by a simple iterative procedure: Let $x^{(0)}$ be the approximate solution of the system $Ax = b$ obtained by means of Gauß algorithm. Define the residual vector $r^{(0)} = b - Ax^{(0)}$. Then find the solution vector $z^{(1)}$ for the system

$$(4.15) \qquad\qquad Az^{(1)} = b - Ax^{(0)} = r^{(0)}$$

and set $x^{(1)} = x^{(0)} + z^{(1)}$. Since b, A and $x^{(0)}$ are known, the residuum $r^{(0)}$ can be calculated. In order to calculate $z^{(1)}$, we can use the fact that the system matrix A has already been transformed to upper triangular form by means of Gauß algorithm. Thus we can quickly determine $z^{(1)}$ from (4.15); we only need to update the right hand side $r^{(0)}$. Instead of the exact solution $z^{(1)}$ of (4.15), we will again only find an approximate solution, so that instead of the exact x,

$$x^{(1)} = x^{(0)} + z^{(1)}$$

will be another approximation.
With $x^{(1)}$ we can form the residual vector $r^{(1)} = b - Ax^{(1)}$, and repeat the process.

The general rule for calculating the $(\nu + 1)^{st}$ correction vector $z^{(\nu+1)}$ is

$$Az^{(\nu+1)} = b - Ax^{(\nu)} = r^{(\nu)} \quad \text{for } \nu = 0, 1, 2, \ldots.$$

One continues this process until the components of the successively corrected solution vector do not change any more or until the *relative improvement* satisfies $\|z^{(\nu+1)}\|_\infty / \|x^{(\nu)}\|_\infty < \varepsilon$ for an $\varepsilon > 0$. Then we accept

$$x \approx x^{(\nu+1)} = x^{(\nu)} + z^{(\nu+1)}.$$

We recommend to calculate the residual vector $r^{(i)}$ with double precision. In practice, iterative refinement will only improve the solution for a small number (1 or 2 or 3 at the most) of iterations. For larger i, the $x^{(i)}$ may begin to oscillate and differ largely from the true solution if the condition of A is bad.

ALGORITHM 4.45 (*Iterative refinement*).

Given: $A = LR$ or $PA = LR$ and an approximate solution $x^{(0)}$ of
 $Ax = b$.

 Find: Iterative refinement $x^{(\nu+1)} \approx x$, the true solution.

For each $\nu = 0, 1, 2 \ldots$, do:

1^{st} step: Form $r^{(\nu)} = b - Ax^{(\nu)}$.
2^{nd} step: Solve $Az^{(\nu+1)} = r^{(\nu)}$ by updating $r^{(\nu)}$ and backsubstitution.
3^{rd} step: Set $x^{(\nu+1)} = x^{(\nu)} + z^{(\nu+1)}$.

The calculation is stopped if the relative improvement is better than a given bound ε.

A sufficient condition for the convergence of iterative refinement is known ([WILK64], p.155), but it is too time consuming to verify in practice. Convergence of iterative refinement goes hand in hand with a well conditioned system matrix A, see [McCA67], 5.8; [McCR72], 4.8; [STIE63], p.24/25; [ZURM65], p.163.

Bibliography for Section 4.15

[CONT80], 5.4; [ENGE87], 3.7; [FADD59], §15; [GOLU89]; [ISAA66], p.39/40; [NIEM87], p.138 ff.; [NOBL65] I, 4.4; [SCHW89], 1.2; [STOE91], 4.4 - 4.6; [WERN82], III §2.

4.16 Systems of Equations with Block Matrices

4.16.1 Preliminary Remarks

Consider a linear system $\boldsymbol{Ax} = \boldsymbol{b}$ with n equations and n unknowns. A *partition* of the (n, n) matrix $\boldsymbol{A} = (a_{ik})$ into *blocks* or *submatrices* can be visualized by certain horizontal and vertical lines separating the rows and columns of \boldsymbol{A} into blocks. Thus a *block matrix* consists of submatrices \boldsymbol{A}_{ik} of $\boldsymbol{A} : \boldsymbol{A} = (\boldsymbol{A}_{ik})$, where each \boldsymbol{A}_{ik} itself is a matrix.

If one partitions a square matrix \boldsymbol{A} into blocks so that the diagonal blocks \boldsymbol{A}_{ii} are square (n_i, n_i) matrices and the blocks \boldsymbol{A}_{ik} are matrices with n_i rows and n_k columns each where $\sum_{i=1}^{N} n_i = n$, then one can partition the vectors \boldsymbol{x} and \boldsymbol{b} correspondingly and write the system $\boldsymbol{Ax} = \boldsymbol{b}$ in block form:

$$(4.16) \quad \begin{pmatrix} \boldsymbol{A}_{11} & \boldsymbol{A}_{12} & \cdots & \boldsymbol{A}_{1N} \\ \boldsymbol{A}_{21} & \boldsymbol{A}_{22} & \cdots & \boldsymbol{A}_{2N} \\ \vdots & \vdots & \ddots & \vdots \\ \boldsymbol{A}_{N1} & \boldsymbol{A}_{N2} & \cdots & \boldsymbol{A}_{NN} \end{pmatrix} \begin{pmatrix} \boldsymbol{x}_1 \\ \boldsymbol{x}_2 \\ \vdots \\ \boldsymbol{x}_N \end{pmatrix} = \begin{pmatrix} \boldsymbol{b}_1 \\ \boldsymbol{b}_2 \\ \vdots \\ \boldsymbol{b}_N \end{pmatrix}.$$

Here $\sum_{i=1}^{N} n_i = n$ and $\sum_{k=1}^{N} \boldsymbol{A}_{ik} \boldsymbol{x}_k = \boldsymbol{b}_i$ for $i = 1, \ldots, N$.

We will only consider those partitions of \boldsymbol{A} whose diagonal blocks are square, because we can then work with the block system in much the same way as if the blocks were numbers. In general it is possible to employ the previous methods to solve systems (4.16) with block matrices. The only difference is that one must calculate with submatrices and vectors instead of numbers. Division by diagonal matrix elements clearly must be replaced by multiplication with the inverse of a diagonal block here. And a pivot search cannot be used.

To illustrate this process we shall explain Gauß algorithms for dense block systems as well as for block tridiagonal systems in this chapter. At the end we shall mention some literature on methods for special block systems.

4.16.2 Gauß Algorithm for Block Matrices

1st elimination step.

Formally the elimination proceeds analogous to section 4.5 without pivoting. Division by diagonal elements is replaced by multiplication with the inverse diagonal block $(A_{jj}^{(j-1)})^{-1}$. Multiplication of the first block row

$$1^{(0)}: \qquad A_{11}^{(0)}x_1 + A_{12}^{(0)}x_2 + \ldots + A_{1N}^{(0)}x_N = b_1^{(0)}$$

on the left by the matrix $-A_{i1}^{(0)}\left(A_{11}^{(0)}\right)^{-1}$ and addition to the i^{th} block row for $i = 2, \ldots, N$ yields

$$
\begin{aligned}
1^{(0)}: \quad & A_{11}^{(0)}x_1 + & A_{12}^{(0)}x_2 & + \ldots + & A_{1N}^{(0)}x_N = & b_1^{(0)}, \\
2^{(1)}: \quad & & A_{22}^{(1)}x_2 & + \ldots + & A_{2N}^{(1)}x_N = & b_2^{(1)}, \\
3^{(1)}: \quad & & A_{32}^{(1)}x_2 & + \ldots + & A_{3N}^{(1)}x_N = & b_3^{(1)}, \\
& \vdots & \vdots & & \vdots & \vdots \\
N^{(1)}: \quad & & A_{N2}^{(1)}x_2 & + \ldots + & A_{NN}^{(1)}x_N = & b_N^{(1)}.
\end{aligned}
$$

2nd elimination step.

Multiplication of block row $2^{(1)}$ by $- A_{i2}^{(1)}(A_{22}^{(1)})^{-1}$ on the left and addition to the i^{th} block row for $i = 3, \ldots, N$ yields:

$$
\begin{aligned}
1^{(0)}: \quad & A_{11}^{(0)}x_1 + & A_{12}^{(0)}x_2 + & A_{13}^{(0)}x_3 & + \ldots + & A_{1N}^{(0)}x_N = & b_1^{(0)}, \\
2^{(1)}: \quad & & A_{22}^{(1)}x_2 + & A_{23}^{(1)}x_3 & + \ldots + & A_{2N}^{(1)}x_N = & b_2^{(1)}, \\
3^{(2)}: \quad & & & A_{33}^{(2)}x_3 & + \ldots + & A_{3N}^{(2)}x_N = & b_3^{(2)}, \\
& \vdots & & \vdots & & \vdots & \vdots \\
N^{(2)}: \quad & & & A_{N3}^{(2)}x_3 & + \ldots + & A_{NN}^{(2)}x_N = & b_N^{(2)}.
\end{aligned}
$$

After $N - 1$ such eliminations, we obtain a block triangular system $Bx = c$, where B is a block upper triangular matrix of the form

$$
\begin{pmatrix}
A_{11}^{(0)} & & \cdots & & A_{1N}^{(0)} \\
& A_{22}^{(1)} & & & A_{2N}^{(1)} \\
& & \ddots & & \vdots \\
& & & & A_{NN}^{(N-1)}
\end{pmatrix}
$$

analogous to formula (4.8) for matrices.

By block backsubstitution $Bx = c$ can be solved for the x_i:

$$A_{NN}^{(N-1)} x_N = b_N^{(N-1)},$$

$$A_{jj}^{(j-1)} x_j = b_j^{(j-1)} - \sum_{k=j+1}^{N} A_{jk}^{(j-1)} x_k \text{ for } j = N-1, N-2, \ldots, 1.$$

As each $A_{jj}^{(j-1)}$ is square, these N systems can now be treated individually by means of Gauß algorithm (with pivoting) according to section 4.5.

If one solves differential or integral equations numerically using discretization methods, linear systems with block tridiagonal matrices will quite often arise. Hence we shall describe a special algorithm for such block systems.

4.16.3 Gauß Algorithm for Block Tridiagonal Systems

Let the linear equation (4.16) have a block tridiagonal system matrix A

$$A = \begin{pmatrix} D_1 & C_1 & & & \\ B_2 & D_2 & C_2 & & \\ & \ddots & \ddots & \ddots & \\ & & B_{N-1} & D_{N-1} & C_{N-1} \\ & & & B_N & D_N \end{pmatrix}.$$

Here we can factor the matrix A analogously to section 4.10 into block bidiagonal matrices. To solve $Ax = b$, one performs the following block steps (see also [ISAA66], p.61-64):

(1) $A_1 = D_1, \Gamma_1 = A_1^{-1} C_1$

(2) $A_i = D_i - B_i \Gamma_{i-1}$ for $i = 2, \ldots, N$
 $\Gamma_i = A_i^{-1} C_i$ for $i = 2, \ldots, N-1$

(3) $c_1 = A_1^{-1} b_1$
 $c_i = A_i^{-1} (b_i - B_i c_{i-1})$ for $i = 2, \ldots, N$

(4) $x_N = c_N$
 $x_i = c_i - \Gamma_i x_{i+1}$ for $i = N-1, N-2, \ldots, 1.$

4.16.4 Other Block Methods

(1) If $A = (A_{ik})$ is positive definite and if all diagonal blocks A_{ii} have the same size $n_1 = n_2 = \ldots = n_N = n/N$, one can employ a square root method which is analogous to the Cholesky decomposition of Chapter 4.7.22.

(2) If all blocks A_{ik} in A are square of the same size, a block method can be used that is a modification of the Gauß-Jordan method.

(3) An example for systems with block tridiagonal matrices can be found in [SCHW89], p.210. There block tridiagonal matrices whose diagonal blocks are diagonal matrices are treated.

(4) For block iterative and block relaxation methods see e.g. [ISAA66], p.63 ff.; [SCHW72], p.216 ff.

Bibliography for Section 4.16

[ISAA66], 2.4; [SCHE89]; [SCHW72], 5.2.3; [STOE91], 8.5; [YOUN71], 14.

4.17 The Algorithm of Cuthill-McKee for Sparse Symmetric Matrices

Large sparse symmetric matrices occur naturally in many applications, such as when solving boundary value problems for ordinary or partial differential equations via discretizations, when using the finite element method etc.
For solving large sparse and symmetric systems of linear equations we recommend the following algorithm:

1. Use the Cuthill-McKee algorithm to transform the sparse symmetric system matrix A (e.g. a stiffness matrix) into a band matrix with nearly optimal band width, which in general, however, is not of minimal band width.

2. Subsequently use the Rosen algorithm with the node numbering taken from Cuthill-McKee as an initial numbering. In general, this will help to lower the band width further. There are, however, examples in which the band width will not decrease.

3. Use specially adapted Gauß or Cholesky schemes to solve the given system directly as in our programs CUTGAU and CUTCHO.

In the following we shall describe the Cuthill-McKee algorithm in the version that E. Pohl has prepared for the 7^{th} German FORTRAN edition of this handbook. The theoretical background is documented in [WEIS90], 1.5.2 for example.

ALGORITHM 4.46 (*Cuthill-McKee algorithm*)

Given: The nonzero elements a_{ik} of a sparse symmetric $n \times n$ matrix A
with their row and column indices i, k.

To find: The Cuthill-McKee numbering for the matrix.

1. Condense the matrix A:

 1.1 Set $NV := 0$

 1.2 For $i = 1, ..., n$:

 　1.2.1 Set $IR_i := NV + 1$

 　1.2.2 For all nonzero elements in row i:

 　　1.2.2.1 Find the next nonzero element a_{ik}

 　　1.2.2.2 Set $NV := NV + 1$; $V_{NV} := a_{ik}$; $IC_{NV} := k$

 1.3 Set $IR_{n+1} := NV + 1$

2. Form the incidence graph for the matrix in condensed form:

 2.1 Set $\mu := 0$

 2.2 For $i = 1, ..., n$:

 　2.2.1 Set $INB_i := \mu + 1$

 　2.2.2 For $k = IR_i, ..., IR_{i+1} - 1$:

 　　2.2.2.1 If $IC_k \neq i$, then
 　　　　$\mu := \mu + 1$;
 　　　　$NEIGHB_\mu := IC_k$

 　2.2.3 $IDEG_i := \mu + 1 - INB_i$

 2.3 $INB_{n+1} := \mu + 1$

3. Compute the Cuthill-McKee numbering:

 3.1 Initialize:
 　　For $i = 1, ..., n$:

 　3.1.1 Mark the node i "false". Set $ICM_i := 0$.

 3.2 Set $NFOUND := 0$

 3.3 For $i = 1, ..., n$:

 　3.3.1 If the node i is labelled "false":

 　　3.3.1.1 Find a starting node $IROOT$ in the component of the
 　　　　graph that contains i whose level is as large as possible
 　　　　(via algorithm 4.47).

3.3.1.2 Compute the Cuthill-McKee numbering of this component for the initial node $IROOT$ and the initial number $ISTART := NFOUND + 1$.
Mark all nodes of this component (via algorithm 4.48).

3.3.1.3 Set $NFOUND := NFOUND +$ number of nodes of the new component.

3.4 The vector ICM now contains the Cuthill-McKee numbering for the whole graph: For $i = 1, ..., n$, the i^{th} element in ICM, ICM_i, denotes the original number of the node with Cuthill-McKee number i.
Compute the inverse permutation $ICMREV$:
For $i = 1, ..., n$:
Set $ICMREV_{ICM_i} := i$.

With this Cuthill-McKee permutation the linear system $\boldsymbol{Ax = b}$ becomes equivalent to $\boldsymbol{A^* x^* = b^*}$, where

$$
\begin{aligned}
a_{ik}^* &= a_{ICM_i, ICM_k}, \\
b_i^* &= b_{ICM_i}, \\
x_i^* &= x_{ICM_i}, \qquad i = 1, ..., n, \ k = 1, ..., n,
\end{aligned}
$$

or alternatively

$$
\begin{aligned}
a_{ik} &= a_{ICMREV_i, ICMREV_k}^*, \\
b_i &= b_{ICMREV_i}^*, \\
x_i &= x_{ICMREV_i}^*, \quad i = 1, ..., n, \ k = 1, ..., n.
\end{aligned}
$$

ALGORITHM 4.47 (*Subalgorithm for algorithm 4.46*)

Given: (a) The graph of the $n \times n$ matrix \boldsymbol{A} stored in the vectors $NEIGHB$ and INB from Algorithm 4.46, second step.

(b) The vector $IDEG$ with the levels of the nodes from algorithm 4.46, step 2.2.3.

(c) Certain nodes in $NEIGHB$ may carry labels from earlier passes of step 3.3.1 in Algorithm 4.46.

(d) The index $IROOT$ of an unmarked node.

To find: An initial node $IROOT$ whose component in the graph has a level number as large as possible.

1. Set $NLVOLD := 0$.
 $NLVOLD$ denotes the maximal level number thus far found.

2. Record the level numbers for the component of $IROOT$ in the vectors $LEVEL$ and ILV:

 2.1 Set $NLV := 0$; $LEVEL_1 := IROOT$; $\mu := 1$; $LEVEND := 0$.
 Mark the node $IROOT$.

 2.2 Set $NLV := NLV + 1$; $LEVBEG := LEVEND + 1$;
 $LEVEND := \mu$; $ILV_{NLV} := LEVBEG$.
 Here NLV denotes the number of levels found so far. The first level nodes consist precisely of $IROOT$. $LEVBEG$ denotes the initial index, while $LEVEND$ marks the terminal index of the highest level NLV in the vector $LEVEL$.

 2.3 Find all nodes of the next higher level $NLV + 1$. To do so, find all non–labelled neighbours of nodes of level NLV and record them in $LEVEL$:
 For $i = LEVBEG, ..., LEVEND$:

 2.3.1 For $j = INB_{LEVEL_i}, ..., INB_{LEVEL_i+1} - 1$:

 2.3.1.1 If the node $NEIGHB_j$ is unlabelled, we set
 $\mu := \mu + 1$; $LEVEL_\mu := NEIGHB_j$
 and mark the node $NEIGHB_j$.

 2.4 If $\mu > LEVEND$, i.e., if new nodes have been found in step 2.3, repeat step 2.2. Otherwise the level structure for the connected component of $IROOT$ has been found.

 2.5 Set $LVNODES := LEVEND$, which denotes the number of nodes of this component.
 Set $ILV_{NLV+1} := LVNODES + 1$.

 2.6 Reverse all node labels set in steps 2.1 to 2.3.

3. If $NLV \leq NLVOLD$, then the level structure has not been increased when compared with the previous run through step 2. Hence we can stop algorithm 4.47.
 Otherwise:

4. Set $NLVOLD := NLV$.

5. Search for a node of minimal degree on the last level, i.e., search for an index i_0 with $IDEG_{i_0} = \min\{IDEG_{LEVEL_i} \mid i = ILV_{NLV}, ..., ILV_{NLV+1} - 1\}$. Replace $IROOT$ by $LEVEL_{i_0}$.

6. Go back to step 2.

ALGORITHM 4.48 (*Subalgorithm for algorithm 4.46*)

Given: (a) The graph of an $n \times n$ matrix A described by the vectors $NEIGHB$ and INB from algorithm 4.46, second step.

(b) The vector $IDEG$ of degrees of the nodes from algorithm 4.46, step 2.2.3 as determined via algorithm 4.47.

(c) The starting node for a new component of the graph as determined by algorithm 4.47.

(d) The starting index $ISTART$ for the Cuthill-KcKee numbering of the component.

(e) Earlier runs of algorithm 4.46, step 3.3.1, may have labelled components of the graph. The Cuthill-McKee numbering of this component is stored in the vector ICM in the positions ICM_i for $i = 1, ..., ISTART - 1$. The nodes in this component are already marked. All others are unmarked.

To find: The Cuthill-McKee numbering of the component of $IROOT$.

1. Set $ICM_{ISTART} := IROOT$; $NEWEND := ISTART$; $LEVEND := ISTART - 1$. Mark the node $IROOT$.

2. Record the level structure for the component of $IROOT$ in ICM:

2.1 Set $LEVBEG := LEVEND + 1$; $LEVEND := NEWEND$. Here $LEVBEG$ denotes the starting index, while $LEVEND$ denotes the terminal index of the most recently found level inside the vector ICM. (The first level clearly contains precisely the node $IROOT$ itself.)

2.2 Find the nodes of the next higher level: to do so, find all previously unmarked neighbours of nodes of the previous level and store their indices in ICM:
For $i = LEVBEG, ..., LEVEND$:

2.2.1 Set $NEWBEG := NEWEND + 1$.
$NEWBEG$ points to the start of the new level nodes inside ICM. $NEWEND$ always denotes the index of the most recently found node in ICM.

2.2.2 For $j = INB_{ICM_i}, ..., INB_{ICM_i+1} - 1$:

2.2.2.1 If the node $NEIGHB_j$ has no label, we set:
$NEWEND := NEWEND + 1$;
$ICM_{NEWEND} := NEIGHB_j$
and then mark the node $NEIGHB_j$.

> 2.2.3 Sort the elements ICM_i for $i = NEWBEG, ..., NEWEND$ by increasing levels.
>
> 2.3 If $NEWEND > LEVEND$, i.e., if step 2.2 has produced new nodes, go to 2.1. Otherwise terminate algorithm 4.48.

Bibliography for Section 4.17

[MAES84]; [SCHW84]; [SCHW88]; [WEIS90], 1.5.2.

4.18 Recommendations for Selecting a Method

In spite of the numerous numerical methods known to solve linear systems, solving a linear system for large n is a very problematic numerical task. The main reasons for this difficulty are:

(1) the computational time, expecially the operations count $O(n^3)$;

(2) the required memory;

(3) the effects of rounding errors and ill-conditioning of the problem.

Re (1): Computation time can be estimated by the operations count for the algorithm. The following table gives operations counts for the solution of a linear system consisting of n equations with n unknowns and the various procedures. This comparison does not take into consideration the number of necessary additions and subtractions, nor does it account for retrieval and write times to access quantities in memory.

TABLE (*Operations count for systems of n equations with n unknowns*).

PROCEDURE	OPERATIONS COUNT
Gauß algorithm	$\frac{n^3}{3} + n^2 - \frac{n}{3}$
Cholesky decomposition	$\frac{n^3}{6} + O(n^2)$
Gauß-Jordan method	$\frac{n^3}{2} + n^2 + \frac{n}{2}$
Exchange steps	$n^3 + n^2$
Procedure for tridiagonal matrices	$5n - 4$
Procedure for cyclically tridiagonal matrices	$11n - 16$
Procedure for five-diagonal matrices	$11n - 16$
And for comparison only, see chapter 5:	
Iterative method (per iteration step)	$2n^2 - 2n$

Re (2): As far as the computer memory is concerned, there are two critical quantities for large n:

(a) the memory available for the storage of elements such as a_{ik} in primary memory, and

(b) the memory available in secondary memory such as on magnetic disks or diskettes, magnetic tapes, cassettes and the like.

The needed memory decreases if A has special properties, such as a band structure, or is sparse and symmetric. There is practically no need for a large memory if the a_{ik} can be calculated within the computer, i.e., if A is a "generated matrix" with a known generating function in a particular case, see also the following note.

Re (3): By choosing a proper sequence of the computations one can minimize the accumulation of rounding errors, provided they do not originate from an ill-conditioned problem. Thus one should work with scaled partial pivoting algorithms - unless the special structure of the system matrix guarantees well-conditionedness. The results can be improved with relatively little work by using iterative refinement for well conditioned system matrices A.

In general, it is impossible to examine either the condition number of the system matrix or the question whether any of the conditions for unique solvability are satisfied before one begins numerical calculations. For this reason, programs in this area should be written in such a way that the user is kept informed during the course of the calculations, such as by a 'stop' notice in case of an ill-conditioned A which makes itself known e.g. by a slow convergence of iterative refinement.

NOTE: When dealing with very large systems where the elements of A and b cannot all be placed into primary memory, not even in condensed form, one can employ the block methods of section 4.16. Systems of that kind appear mostly in the numerical solution of partial differential equations. If the matrices are sparse, as is often the case with boundary value problems for ordinary and partial differential equations using difference or finite-element methods, one should use specialized algorithms, see e.g. [MAES84]. There many algorithms can be found ready for programming. If the matrices are both sparse and symmetric, we recommend to use the Cuthill-McKee algorithm of section 4.17. Iterative algorithms are described in chapter 5.

Bibliography for Section 4.18

[CONT80], 5; [DONG79]; [ISAA66], 2; [MAES84], 2; [RICE83], 6.4-6.6; [SCHW84]; [SCHW88]; [YOUN71], 18.

Chapter 5

Iterative Methods for Linear Systems

5.1 Preliminary Remarks

Direct methods for solving linear systems, with their large number of operations proportional to n^3, have a tendency to accumulate rounding errors so that for a not ideally conditioned system matrix A, the solution can become entirely useless. In contrast, iterative methods are unaffected by rounding errors to a large extent, because each approximate solution with its inherent computational error can easily be improved upon in the following iteration step. Iterative methods typically require around n^2 operations for each iteration step, but, unfortunately, they do not converge for all solvable systems.

The procedures of this chapter such as the Jacobi and the Gauß-Seidel method converge only linearly and very slowly in most practical cases on account of a growing Lipschitz constant L as the dimension n increases. For this reason, iterative methods are superior to the direct methods only in a very few special cases, namely if A is sparse, very large and structured in such a way that - if one were to employ one of the direct methods - the matrices would exceed the available memory. Their convergence can in general be accelerated by using a relaxation method based on either the Jacobi or Gauß-Seidel method. Relaxation methods additionally require that at least some approximations to the eigenvalues of largest and smallest magnitude of the iteration matrix are known when using the Gauß-Seidel method, or that the eigenvalue of largest magnitude only be known approximately when using the Jacobi method.

5.2 Vector and Matrix Norms

Let \mathbb{R}^n be the n-dimensional real vector space and x a vector in \mathbb{R}^n. The *norm* function $\|..\| : \mathbb{R}^n \to \mathbb{R}^+$ assigns a nonnegative real number $\|x\|$ to $x \in \mathbb{R}^n$ and fulfills the following *vector norm axioms*:

1) $\|x\| > 0$ for all $x \in \mathbb{R}^n$ with $x \neq 0$ (positive definite),

2) $\|x\| = 0$ if and only if $x = 0$,

3) $\|\alpha x\| = |\alpha| \, \|x\|$ for all $x \in \mathbb{R}^n$ and $\alpha \in \mathbb{R}$ (homogeneous),

4) $\|x + y\| \leq \|x\| + \|y\|$ for all $x, y \in \mathbb{R}^n$ (triangle unequality).

Examples for vector norms are:

$$\|x\|_\infty := \max_{1 \leq i \leq n} |x_i| \qquad \text{(maximum norm)},$$

$$\|x\|_1 := \sum_{i=1}^{n} |x_i|, \qquad \text{and}$$

$$\|x\|_2 := \sqrt{\sum_{i=1}^{n} |x_i|^2} \qquad \text{(Euclidian norm)}.$$

For an (n, n) matrix with A the *matrix norm* $\|A\|$ of A is a nonnegative real number which satisfies the following *matrix norm axioms:*

1) $\|A\| \geq 0$ for all (n, n) matrices A (positive definite),

2) $\|A\| = 0$ if and only if $A = 0$,

3) $\|\alpha A\| = |\alpha| \, \|A\|$ for all matrices A and $\alpha \in \mathbb{R}$ (homogeneous),

4) $\|A + B\| \leq \|A\| + \|B\|$ for all (n, n) matrices A, B (triangle inequality), and

5) $\|AB\| \leq \|A\| \, \|B\|$ for all (n, n) matrices A, B (submultiplicative).

Examples of matrix norms are:

$$\|\boldsymbol{A}\|_\infty \quad := \quad \max_{1 \le i \le n} \sum_{k=1}^{n} |a_{ik}| \qquad \text{(row sum norm)},$$

$$\|\boldsymbol{A}\|_1 \quad := \quad \max_{1 \le k \le n} \sum_{i=1}^{n} |a_{ik}| \qquad \text{(column sum norm)},$$

$$\|\boldsymbol{A}\|_2 \quad := \quad \sqrt{\sum_{i,k=1}^{n} |a_{ik}|^2} \qquad \text{(Frobenius norm)}.$$

Matrix norms must be *compatible* with an underlying vector norm:

DEFINITION 5.1. A matrix norm is *compatible* with a vector norm if for every matrix \boldsymbol{A} and every vector \boldsymbol{x} the inequality

$$\|\boldsymbol{A}\boldsymbol{x}\| \le \|\boldsymbol{A}\|\,\|\boldsymbol{x}\|$$

holds.

The matrix norms $\|\boldsymbol{A}\|_j$ are *compatible* with the vector norms $\|\boldsymbol{x}\|_j$ for $j = 1, 2$ and ∞.

5.3 The Jacobi Method

Assume that a linear system $\boldsymbol{A}\boldsymbol{x} = \boldsymbol{b}$ with $\det \boldsymbol{A} \ne 0$ is given in the form (4.1).

To approximate the solution vector \boldsymbol{x}, one constructs a sequence $\{\boldsymbol{x}^{(\nu)}\}$ for $\nu = 1, 2, ...$, for which $\lim\limits_{\nu \to \infty} \boldsymbol{x}^{(\nu)} = \boldsymbol{x}$.

We can assume without loss of generality that none of the diagonal entries a_{jj} of \boldsymbol{A} is zero; otherwise interchange its rows. By solving the i^{th} equation of (4.1) for x_i, we can obtain an equivalent form for the system:

$$x_i = -\sum_{\substack{k=1 \\ k \ne i}}^{n} \frac{a_{ik}}{a_{ii}} x_k + \frac{b_i}{a_{ii}}, \quad i = 1, \dots, n.$$

If we set

$$(5.1) \qquad c_i = \frac{b_i}{a_{ii}}, \quad b_{ik} = \begin{cases} -\dfrac{a_{ik}}{a_{ii}} & \text{for} \quad i \ne k \\ 0 & \text{for} \quad i = k \end{cases},$$

we can rewrite (5.1) thus:

$$(5.2) \qquad \boldsymbol{x} = \boldsymbol{B}\boldsymbol{x} + \boldsymbol{c} \quad \text{with} \quad \boldsymbol{B} = (b_{ik}) \quad \text{and} \quad \boldsymbol{c} = \begin{pmatrix} c_1 \\ \vdots \\ c_n \end{pmatrix}.$$

Define a vector valued iteration function φ by

$$\varphi(\boldsymbol{x}) := \boldsymbol{B}\boldsymbol{x} + \boldsymbol{c}.$$

And construct the sequence $\{\boldsymbol{x}^{(\nu)}\}$ for an initial vector $\boldsymbol{x}^{(0)}$ by setting

$$(5.3) \quad \begin{cases} \boldsymbol{x}^{(\nu+1)} = \varphi(\boldsymbol{x}^{(\nu)}) = \mathbf{B}\boldsymbol{x}^{(\nu)} + \boldsymbol{c} \\[2mm] \text{with} \quad \boldsymbol{x}^{(\nu)} = \begin{pmatrix} x_1^{(\nu)} \\ x_2^{(\nu)} \\ \vdots \\ x_n^{(\nu)} \end{pmatrix} \quad \text{for } \nu = 0, 1, 2, \dots. \end{cases}$$

Expressed componentwise, this *Jacobi iteration* becomes

$$(5.4) \quad x_i^{(\nu+1)} = c_i + \sum_{k=1}^{n} b_{ik} x_k^{(\nu)} = \frac{b_i}{a_{ii}} - \sum_{\substack{k=1 \\ k \neq 1}}^{n} \frac{a_{ik}}{a_{ii}} x_k^{(\nu)}, \qquad \begin{array}{l} i = 1, \dots, n, \\ \nu = 0, 1, 2, \dots \end{array}$$

The matrix \boldsymbol{B} is called the *Jacobi iteration matrix* . It is useful to view the calculations in a scheme of the following form:

COMPUTATIONAL SCHEME 5.2 (*Jacobi iteration for $n = 3$*).

c_i	b_{ik}			$x_i^{(0)}$	$x_i^{(1)}$	\cdots
$\dfrac{b_1}{a_{11}}$	0	$-\dfrac{a_{12}}{a_{11}}$	$-\dfrac{a_{13}}{a_{11}}$	0		
$\dfrac{b_2}{a_{22}}$	$-\dfrac{a_{21}}{a_{22}}$	0	$-\dfrac{a_{23}}{a_{22}}$	0		
$\dfrac{b_3}{a_{33}}$	$-\dfrac{a_{31}}{a_{33}}$	$-\dfrac{a_{32}}{a_{33}}$	0	0		

NOTE: When implementing any of the iterative schemes mentioned in this chapter for large sparse and structured matrices A, it is of great advantage to form the iteration matrices and program the iteration rules like (5.4), (5.12), (5.16) or (5.18) directly in terms of the few nonzero entries of A, rather than to use the double 1 to n loops that describe these iterations theoretically. This takes little effort and must be done differently for each system matrix A.

With the concepts of vector and matrix norms, one can decide when the sequence $\{x^{(\nu)}\}$ converges:

THEOREM 5.3. Let $x \in \mathbb{R}^n$ be a solution of $\varphi(x) = x$. If for some vector norm

$$\|\varphi(x) - \varphi(x')\| \leq L\|\mathbf{x} - \mathbf{x}'\| \quad \text{with} \quad 0 \leq L < 1$$

is fulfilled for all $x, x' \in \mathbb{R}^n$, then for the iteration sequence $\{x^{(\nu)}\}$ defined by $x^{(\nu+1)} = \varphi(x^{(\nu)})$ for an arbitrary initial vector $x^{(0)} \in \mathbb{R}^n$ we have:

1) $\lim\limits_{\nu \to \infty} x^{(\nu)} = x$;

2) x is uniquely determined;

3) $\|x^{(\nu)} - x\| \leq \frac{L}{1-L}\|x^{(\nu)} - x^{(\nu-1)}\|$ (a posteriori error estimate), and

$$\leq \frac{L^\nu}{1-L}\|x^{(1)} - x^{(0)}\| \qquad \text{(a priori error estimate)}.$$

THEOREM 5.4. If either one of the following conditions is satisfied for the matrix A:

a) *row sum criterion*

(5.5) $\qquad \max_{1 \leq i \leq n} \sum\limits_{k=1}^{n} |b_{ik}| = \max_{1 \leq i \leq n} \sum\limits_{\substack{k=1 \\ k \neq i}}^{n} \left|\frac{a_{ik}}{a_{ii}}\right| \leq L_\infty < 1$, or

b) *column sum criterion*

(5.6) $\qquad \max_{1 \leq k \leq n} \sum\limits_{i=1}^{n} |b_{ik}| = \max_{1 \leq k \leq n} \sum\limits_{\substack{i=1 \\ i \neq k}}^{n} \left|\frac{a_{ik}}{a_{ii}}\right| \leq L_1 < 1$, or

c) *criterion of Schmidt and v. Mises*

(5.7) $\qquad \sqrt{\sum\limits_{i=1}^{n} \sum\limits_{k=1}^{n} |b_{ik}|^2} = \sqrt{\sum\limits_{i=1}^{n} \sum\limits_{\substack{k=1 \\ k \neq i}}^{n} \left|\frac{a_{ik}}{a_{ii}}\right|^2} \leq L_2 < 1,$

then the iteration sequence (5.3) or (5.4) converges towards the unique solution x of the linear system $\mathbf{A}x = b$ for any initial vector $x^{(0)} \in \mathbb{R}^n$ with the following error estimates

$$f_\infty^{(\nu)} \;:=\; \|x^{(\nu)} - x\|_\infty = \max_{1 \le i \le n} |x_i^{(\nu)} - x_i|$$

(5.8)
$$\le\; \frac{L_\infty}{1 - L_\infty} \max_{1 \le i \le n} |x_i^{(\nu)} - x_i^{(\nu-1)}| \qquad \text{(a posteriori)}$$

$$\le\; \frac{L_\infty^\nu}{1 - L_\infty} \max_{1 \le i \le n} |x_i^{(1)} - x_i^{(0)}| \qquad \text{(a priori)},$$

or

$$f_1^{(\nu)} \;:=\; \|x^{(\nu)} - x\|_1 = \sum_{i=1}^n |x_i^{(\nu)} - x_i|$$

(5.9)
$$\le\; \frac{L_1}{1 - L_1} \sum_{i=1}^n |x_i^{(\nu)} - x_i^{(\nu-1)}| \qquad \text{(a posteriori)}$$

$$\le\; \frac{L_1^\nu}{1 - L_1} \sum_{i=1}^n |x_i^{(1)} - x_i^{(0)}| \qquad \text{(a priori)},$$

or

$$f_2^{(\nu)} \;:=\; \|x^{(\nu)} - x\|_2 = \sqrt{\sum_{i=1}^n |x_i^{(\nu)} - x_i|^2}$$

(5.10)
$$\le\; \frac{L_2}{1 - L_2} \sqrt{\sum_{i=1}^n |x_i^{(\nu)} - x_i^{(\nu-1)}|^2} \qquad \text{(a posteriori)}$$

respectively

$$\le\; \frac{L_2^\nu}{1 - L_2} \sqrt{\sum_{i=1}^n |x_i^{(1)} - x_i^{(0)}|^2} \qquad \text{(a priori)}.$$

ALGORITHM 5.5 (*Jacobi iteration*).

Given: A linear system $\mathbf{A}x = b$.

Find: Solution x via Jacobi iteration.

1^{st} Step: The given system is transformed to the equivalent form (5.2).

2^{nd} Step: Examine whether one of the three sufficiency criteria of theorem 5.4 is fulfilled. If not, try to alter the system, by taking linear combinations of the given equations, so that one of the diagonal dominance criteria (5.5), (5,6) or (5.7) is satisfied. Otherwise calculate the solution by means of a direct method of chapter 4.

3^{rd} Step: If one of the criteria for convergence is satisfied, choose an arbitrary initial vector $x^{(0)}$; $x^{(0)} = 0$ or $x^{(0)} = b$ will do.

4^{th} Step: Form the iteration sequence $\{x^{(\nu)}\}$ according to the iteration rule (5.3) or (5.4). It is helpful to use the scheme 5.2 here. Continue iterating until one of the following three break-off criteria is satisfied:

 a) *Absolute error criterion:*

 $$\max_{1 \le i \le n} |x_i^{(\nu+1)} - x_i^{(\nu)}| < \delta \text{ for a given } \delta > 0.$$

 b) *Relative error criterion:*

 $$\max_{1 \le i \le n} |x_i^{(\nu+1)} - x_i^{(\nu)}| \le \max_{1 \le i \le n} |x_i^{(\nu+1)}|\varepsilon \text{ for a given } \varepsilon > 0.$$

 c) $\nu > \nu_0$, where ν_0 is the maximal number of iterations predicted by one of the a priori error estimates.

5^{th} Step: (Error estimate). If (5.5) is fulfilled, use the error estimate (5.8). If (5.5) is not fulfilled, but (5.6) is, use the error estimate (5.9). If only (5.7) is fulfilled, use error estimate (5.10), which is the weakest one.

NOTE. The absolute error criterion 4a) in the 4^{th} step of algorithm 5.5 amounts in fact to establishing convergence for the iteration. For if $0 \le L_\infty < 1$ and ν is sufficiently large, then criterion 4a) will always be satisfied.

5.4 The Gauß-Seidel Iteration

The *Gauß-Seidel iteration* differs from the Jacobi iteration only slightly, for in it the calculated $(\nu + 1)^{st}$ approximations of the leading coefficients $x_1, x_2, ..., x_{i-1}$ of x^ν are used when calculating the $(\nu + 1)^{st}$ approximation of x_i. For this method, the iteration rule (5.2) becomes:

$$\begin{cases}
\boldsymbol{x}^{(\nu+1)} = \boldsymbol{B}_r \boldsymbol{x}^{(\nu)} + \boldsymbol{B}_\ell \boldsymbol{x}^{(\nu+1)} + \boldsymbol{c} \quad \text{with} \\[2mm]
\boldsymbol{B}_r = \begin{pmatrix}
0 & b_{12} & b_{13} & \cdots & b_{1n} \\
0 & 0 & \cdot & & \vdots \\
\vdots & \vdots & \cdot & \cdot & \\
\cdot & \cdot & & \cdot & b_{n-1,n} \\
0 & 0 & \cdot & \cdots & 0
\end{pmatrix}, \\[2mm]
\boldsymbol{B}_\ell = \begin{pmatrix}
0 & \cdot & \cdots & & 0 \\
b_{21} & \cdot & & & \vdots \\
\vdots & & \cdot & \cdot & \\
\cdot & \cdot & \cdot & \cdot & \cdot \\
b_{n1} & b_{n2} & \cdots & b_{n,n-1} & 0
\end{pmatrix}, \quad \text{and}
\end{cases}$$

(5.11)

$$c_i = \frac{b_i}{a_{ii}}, \quad b_{ik} = \begin{cases} -\dfrac{a_{ik}}{a_{ii}} & \text{for} \quad i \neq k \\[2mm] 0 & \text{for} \quad i = k \end{cases}.$$

Written out componentwise for $i = 1, \ldots, n$ and $\nu = 0, 1, 2, \ldots$ we have:

(5.12)
$$\begin{aligned}
x_i^{(\nu+1)} &= c_i + \sum_{k=i+1}^{n} b_{ik} x_k^{(\nu)} + \sum_{k=1}^{i-1} b_{ik} x_k^{(\nu+1)} \\
&= \frac{b_i}{a_{ii}} - \sum_{k=i+1}^{n} \frac{a_{ik}}{a_{ii}} x_k^{(\nu)} - \sum_{k=1}^{i-1} \frac{a_{ik}}{a_{ii}} x_k^{(\nu+1)}.
\end{aligned}$$

The following *convergence criteria* insure convergence for the Gauß-Seidel method:

1) the row sum criterion (5.5); or
2) the column sum criterion (5.6); or
3) if \boldsymbol{A} is positive definite ($\boldsymbol{x}^T \boldsymbol{A} \boldsymbol{x} > 0$ for all $\boldsymbol{x} \neq \boldsymbol{0}$), then the Gauß-Seidel method converges.

It is helpful to visualize the calculations in a scheme of the following form:

COMPUTATIONAL SCHEME 5.6 (*Gauß-Seidel iteration for $n = 3$*).

c_i	\boldsymbol{B}_r, or b_{ik} for $k \geq i$			\boldsymbol{B}_ℓ, or b_{ik} for $k < i$			$\mathrm{x}_i^{(0)}$	$\mathrm{x}_i^{(1)}$	\ldots
$\dfrac{b_1}{a_{11}}$	0	$-\dfrac{a_{12}}{a_{11}}$	$-\dfrac{a_{13}}{a_{11}}$	0	0	0	0		
$\dfrac{b_2}{a_{22}}$	0	0	$-\dfrac{a_{23}}{a_{22}}$	$-\dfrac{a_{21}}{a_{22}}$	0	0	0		
$\dfrac{b_3}{a_{33}}$	0	0	0	$-\dfrac{a_{31}}{a_{33}}$	$-\dfrac{a_{32}}{a_{33}}$	0	0		

Here we do not write out an explicit formula because it is nearly identical to that of algorithm 5.5. Only in the second step one should consider different convergence criteria, and in the fourth step the iteration rule given by (5.12) should be used instead of (5.11).

5.5 A Relaxation Method using the Jacobi Method

The Jacobi method uses the iteration rule

$$(5.13) \qquad x^{(\nu+1)} = \mathbf{B}x^{(\nu)} + c \quad \text{for} \quad \nu = 0, 1, 2, \ldots$$

with an iteration matrix \boldsymbol{B}. It can be viewed as a correction on the last computed iterate:

$$(5.14) \qquad x^{(\nu+1)} = x^{(\nu)} + z^{(\nu)},$$

where

$$(5.15) \qquad z^{(\nu)} = c - \boldsymbol{B}_J x^{(\nu)} \text{ for } \boldsymbol{B}_J := \boldsymbol{I} - \boldsymbol{B}.$$

$z^{(\nu)}$ is called the *correction vector* . In a relaxation method, one tries to improve the iterate $x^{(\nu)}$ by $\omega z^{(\nu)}$ instead of $z^{(\nu)}$ where ω is called the *relaxation coefficient*. Thus, the iteration procedure (5.14) has the form

$$(5.16) \qquad x^{(\nu+1)} = x^{(\nu)} + \omega(c - \boldsymbol{B}_J x^{(\nu)}).$$

ω is chosen in such a way that the rate of convergence increases when compared with that of the Jacobi method.

If the iteration matrix B of Jacobi's method (5.13) has real eigenvalues

$$\lambda_1 \geq \lambda_2 \geq \ldots \geq \lambda_n \quad \text{with} \quad \lambda_1 \neq -\lambda_n,$$

then the relaxation coefficient

$$\omega = \frac{2}{2 - \lambda_1 - \lambda_n}$$

improves the rate of convergence of (5.16) when compared to that of Jacobi's method, see [WERN82], pp.188.

If $\omega < 1$, the iteration is called *underrelaxation*; if $\omega > 1$, it is called *overrelaxation*. In order to use a relaxation method optimally, one needs bounds for the eigenvalues of B. Methods for determining approximate eigenvalues of a matrix are given in chapter 7.

5.6 A Relaxation Method using the Gauß-Seidel Method

5.6.1 Iteration Rule

The iteration rule for the Gauß-Seidel method is

$$x^{(\nu+1)} = B_r x^{(\nu)} + B_\ell x^{(\nu+1)} + c \quad \text{for} \quad \nu = 0, 1, 2, \ldots.$$

We can rewrite this as

$$(5.17) \quad \begin{cases} x^{(\nu+1)} & = \quad x^{(\nu)} + z^{(\nu)} \quad \text{for} \\ z^{(\nu)} & = \quad c + B_\ell x^{(\nu+1)} - (I - B_r)x^{(\nu)}, \end{cases}$$

where B_r and B_ℓ denote the triangular matrices in (5.11).

If one replaces the correction vector $z^{(\nu)}$ by $\omega z^{(\nu)}$ in (5.17) for a relaxation coefficient ω, then one obtains the iteration rule for the method of *successive relaxation*

$$(5.18) \quad x^{(\nu+1)} = x^{(\nu)} + \omega(c + B_\ell x^{(\nu+1)} - (I - B_r)x^{(\nu)}).$$

The calculation of an optimum value for ω is difficult. It is possible to prove that the relaxation method (5.18) can only converge if $0 < \omega < 2$ (see [STOE91], p. 547).

The optimum overrelaxation factor for the *the successive overrelaxation method*, abbreviated customarily as *"SOR method"* is

$$\omega_{opt} = \frac{2}{1 + \sqrt{1 - \lambda_1^2}}$$

for a linear system with a positive definite, tridiagonal or block tridiagonal matrix A. Here λ_1 is the largest eigenvalue of the matrix $B = B_\ell + B_r$ (see [SCHW72], p.60, p.208/210, p.214). Matrices of this kind appear naturally in discretizations of elliptic boundary value problems. In this case the SOR method converges much faster with ω_{opt} than a relaxation method with Jacobi's method.

For linear systems with symmetric, but not block tridiagonal matrices A, as well as for skew symmetric matrices, there is a useful approximation for the optimal parameter ω in [NIET70].

5.6.2 Estimate for the Optimal Relaxation Coefficient, an Adaptive SOR Method

In [BUNS85], p. 137 ff and [HAGE81] an adaptive method is given which computes a near optimal relaxation coefficient ω_{opt} for the Gauß–Seidel method with relaxation. In order to speed up the method, at first a certain number l of steps ($l \geq 1$) is performed using the Gauß–Seidel method for a fixed relaxation coefficient ω. Then the near optimal relaxation coefficient is obtained from :

ALGORITHM 5.7 (see [BUNS85], p.137ff)

 Given: $Ax = b$, where A satisfies the hypothesis for the Gauß–Seidel method and we assume that all eigenvalues of the iteration matrix B are real.

 Find: approximate solution x.

 Set: $\omega := 1, q := 1, \nu := 0$.

Choose: accuracy bound $\varepsilon \in \mathbb{R}, \ \varepsilon > 0$,
 number of steps l between adjustments for the relaxation parameter; $l \geq 1, l \in \mathbb{N}$,
 starting vector $x^{(0)}$.

For each $\nu = 0, 1, 2, \ldots$ perform the following steps :

1. Compute $x^{(\nu+1)}$ according to (5.18). If ν is an integer multiple of l, continue with step 2, otherwise go to step 3.

2. To adjust the relaxation coefficient compute

$$q := \max_k \frac{|x_k^{(\nu+1)} - x_k^{(\nu)}|}{|x_k^{(\nu)} - x_k^{(\nu-1)}|}.$$

If $q > 1$, raise ν by 1 and go to step 1;
otherwise set

$$q := \max(q, \omega - 1)$$

and compute the new approximate relaxation coefficient as :

$$\omega := \frac{2}{1 + \sqrt{1 - \frac{1}{q}\left(\frac{q+\omega-1}{\omega}\right)^2}}.$$

3. If

$$\|x^{(\nu+1)} - x^{(\nu)}\|_\infty \le \varepsilon(1-q)\|x^{(\nu+1)}\|_\infty ,$$

stop the iteration with $x \approx x^{(\nu+1)}$;
otherwise set $\nu := \nu + 1$ and continue with step 1.

Bibliography for Chapter 5

[ATKI89], 8; [BUNS85]; [GOLU89], 10; [HAGE81]; [MEIS81], III; [NIEM87], 6.5; [RICE83], 6.3; [SCHE889], 6; [SCHW72], 2 and 5.2; [SPEL85], 1.10 and 1.11; [STOE91], 8; [TIKH77], chap. III; [TÖRN79] vol.1, 6; [WERN82], III §5; [YOUN71], 6-8.

Chapter 6

Systems of Nonlinear Equations

This chapter extends chapter 2 from solving one nonlinear equation in one unknown to systems of nonlinear equations. Assume that we are given a system of n nonlinear equations for an $n \in \mathbb{N}$ with $n \geq 2$:

(6.1)
$$
\begin{cases}
f_1(x_1, x_2, \ldots, x_n) = 0, \\
f_2(x_1, x_2, \ldots, x_n) = 0, \\
\vdots \qquad\qquad \vdots \\
f_n(x_1, x_2, \ldots, x_n) = 0.
\end{cases}
$$

Let D_f be a closed and bounded domain in \mathbb{R}^n, on which the functions $f_i(x_1, x_2, ..., x_n), i = 1, \ldots, n$, are defined. We assume that each f_i is continuous and real valued.

Setting

$$
x = \begin{pmatrix} x_1 \\ x_2 \\ \vdots \\ x_n \end{pmatrix} \quad \text{and} \quad f = \begin{pmatrix} f_1 \\ f_2 \\ \vdots \\ f_n \end{pmatrix},
$$

(6.1) can be expressed as

$$
f(x) = 0 \text{ where } f : D_f \subset \mathbb{R}^n \to \mathbb{R}^n.
$$

6.1 General Iterative Methods

As in chapter 2, we want to transform the nonlinear system $f(x) = 0$ into an equivalent system $\varphi(x) = x$ where $\varphi(x) = (\varphi_1(x), \varphi_2(x), \ldots, \varphi_n(x))^T$ and

$$
\varphi : D_\varphi \subseteq D_f \subset \mathbb{R}^n \to \mathbb{R}^n.
$$

$\bar{x} \in \mathbb{R}^n$ is called a *fixed point* of φ or a *solution* if $\varphi(x) = x$. Starting with a vector $x^{(0)} \in D_\varphi$, one constructs an *iteration sequence* $\{x^{(\nu)}\}$ according to the *iteration rule*

$$(6.2) \qquad x^{(\nu)} = \varphi(x^{(\nu-1)}) \quad \text{for} \quad \nu = 1, 2, \dots.$$

Here φ is a vector valued iteration function.

THEOREM 6.1. (*Fixed point theorem for systems*).

Let $D \subseteq D_\varphi$ be a closed and bounded domain in \mathbb{R}^n and let $\varphi(x) \in D$ for all $x \in D$, i.e., φ maps D into itself.

If there is a constant L with $0 \le L < 1$ and a vector norm $\|\cdot\|$, so that for all $x, x' \in D$ the Lipschitz condition (6.3) $\|\varphi(x) - \varphi(x')\| \le L\|x - x'\|$

is fulfilled, then:

(a) there is exactly one fixed point \bar{x} in D;

(b) the iteration (6.2) converges for any starting vector $x^{(0)} \in D$ towards \bar{x};

(c) the following error estimates hold:

$$\|x^{(\nu)} - \bar{x}\| \le \frac{L^\nu}{1-L}\|x^{(1)} - x^{(0)}\| \qquad \text{(a priori error estimate)},$$

$$\|x^{(\nu)} - \bar{x}\| \le \frac{L}{1-L}\|x^{(\nu)} - x^{(\nu-1)}\| \quad \text{(a posteriori error estimate)}.$$

For a proof see [HENR72] vol. 1, pp.131.

It is possible to formulate a condition analogous to $|\varphi'| \le L < 1$ for the one-dimensional case (refer to condition 2.7) via the *Jacobi matrix* for φ. If each φ_i possesses continuous partial derivatives with respect to each x_k, one can replace the Lipschitz condition (6.3) in theorem 6.1 with

$$(6.4) \qquad \|J_\varphi\| \le L < 1,$$

provided D is convex and closed. Here the *functional matrix* or *Jacobi matrix* J_φ is defined as

$$
\boldsymbol{J}_\varphi := \left(\frac{\partial \varphi_i}{\partial x_k} \right)_{\substack{i=1,\dots,n \\ k=1,\dots,n}} = \begin{pmatrix} \dfrac{\partial \varphi_1}{\partial x_1} & \dfrac{\partial \varphi_1}{\partial x_2} & \cdots & \dfrac{\partial \varphi_1}{\partial x_n} \\[2ex] \dfrac{\partial \varphi_2}{\partial x_1} & \dfrac{\partial \varphi_2}{\partial x_2} & \cdots & \dfrac{\partial \varphi_2}{\partial x_n} \\[2ex] \vdots & \vdots & & \vdots \\[2ex] \dfrac{\partial \varphi_n}{\partial x_1} & \dfrac{\partial \varphi_n}{\partial x_2} & \cdots & \dfrac{\partial \varphi_n}{\partial x_n} \end{pmatrix}.
$$

By choosing either one of the matrix norms from chapter 5.2, we have the following convergence criteria for (6.4):

Row sum criterion:

$$
(6.5) \qquad \|\boldsymbol{J}_\varphi\|_\infty = \max_{\substack{i=1,\dots,n \\ \boldsymbol{x} \in D}} \sum_{k=1}^{n} \left| \frac{\partial \varphi_i(\boldsymbol{x})}{\partial x_k} \right| \le L_\infty < 1,
$$

Column sum criterion:

$$
(6.6) \qquad \|\boldsymbol{J}_\varphi\|_1 = \max_{\substack{k=1,\dots,n \\ \boldsymbol{x} \in D}} \sum_{i=1}^{n} \left| \frac{\partial \varphi_i(\boldsymbol{x})}{\partial x_k} \right| \le L_1 < 1,
$$

Criterion of E. Schmidt and R. von Mises:

$$
(6.7) \qquad \|\boldsymbol{J}_\varphi\|_2 = \max_{\boldsymbol{x} \in D} \left(\sum_{i=1}^{n} \sum_{k=1}^{n} \left(\frac{\partial \varphi_i(\boldsymbol{x})}{\partial x_k} \right)^2 \right)^{1/2} \le L_2 < 1.
$$

If one of the criteria (6.5), ..., (6.7) holds, the associated error estimate of theorem 6.1 holds for the respective vector norm.

DEFINITION 6.2 *(Order of convergence).*
The iteration sequence $\{\boldsymbol{x}^{(\nu)}\}$ converges with at least p^{th} order towards $\bar{\boldsymbol{x}}$, if there exists a constant $0 \le M < \infty$, so that

$$
\lim_{\nu \to \infty} \frac{\|\boldsymbol{x}^{(\nu)} - \bar{\boldsymbol{x}}\|}{\|\boldsymbol{x}^{(\nu-1)} - \bar{\boldsymbol{x}}\|^p} = M < \infty.
$$

In this case the iteration $\boldsymbol{x}^{(\nu)} = \varphi(\boldsymbol{x}^{(\nu-1)})$ is said to have at least p^{th} order of convergence. It has the exact order of convergence p if $M \ne 0$.

Useful error estimates without using a Lipschitz constant (see [KIOU78]; [MOOR80]):

The following method is based upon a result of C. Miranda: Let

$$K := \left\{ x \in \mathbb{R}^n \,\middle|\, \max_{1 \le i \le n} |x_i - \tilde{x}_i| \le p \right\}$$

be an n-dimensional cube and let E_i^+ and E_i^- be defined as

$$E_i^{\pm} := \{ x \in K \,|\, x_i = \tilde{x}_i \pm p \} \ \text{for } i = 1, \ldots, n.$$

Thus each pair E_i^{\pm} is a pair of faces of K that are perpendicular to the x_i-axis. If $f : K \to R^n$ is continuous and if

$$f_i(E_i^+) > 0, \qquad\qquad f_i(E_i^-) < 0 \qquad\qquad \text{for } i = 1, \ldots, n, \ \text{i.e.,}$$

$$f_i(x) > 0, \ \text{for } x \in E_i^+ \quad \text{and} \quad f_i(y) < 0, \ \text{for } y \in E_i^- \quad \text{and all } i = 1, \ldots, n,$$

then K contains a zero \bar{x} of f.

This theorem has little practical use for continuous functions f. If f, however, is differentiable one can derive a useful error estimate from it, provided one has already computed an approximation \tilde{x} for the exact solution \bar{x}.

If one approximates f in a neighborhood of \tilde{x} by its Taylor series including only terms of first order, i.e., if

$$(6.8) \qquad\qquad f(x) \approx f(\tilde{x}) + J_f(\tilde{x})(x - \tilde{x}) \approx J_f(\tilde{x})(x - \tilde{x}),$$

where $J_f(\tilde{x}) := \left(\frac{\partial f_i(\tilde{x})}{\partial x_k} \right)_{\substack{i=1,\ldots,n \\ k=1,\ldots,n}}$ is the Jacobi matrix of f at \tilde{x} and $f(\tilde{x}) \approx 0$, then (6.8) implies

$$(6.9) \qquad\qquad J_f^{-1}(\tilde{x})\, f(x) \approx x - \tilde{x} \qquad \text{for} \qquad \tilde{x} \approx \bar{x}.$$

Now the mapping $h(x) := x - \tilde{x}$ satisfies the desired conditions of Miranda:

$$h_i(E_i^+) = p > 0, \quad \text{and} \quad h_i(E_i^-) = -p < 0 \ \text{for } i = 1, \ldots, n.$$

Due to (6.9), it is reasonable to expect that each mapping of the type

$$(6.10) \qquad\qquad g(x) := \left(\frac{\partial f_i(\tilde{x})}{\partial x_k} \right)^{-1} f(x) = J_f^{-1}(\tilde{x})\, f(x)$$

will also satisfy the desired conditions, and specifically

$$g_i(E_i^+) \approx p, \quad g_i(E_i^-) \approx -p \text{ for } i = 1, \ldots, n.$$

This will occur, if the approximation $\boldsymbol{g}(\boldsymbol{x}) \approx \boldsymbol{x} - \tilde{\boldsymbol{x}}$ is sufficiently good, i.e., if the cube K is sufficiently small. Since $\boldsymbol{g}(\boldsymbol{x})$ has the same zeros as \boldsymbol{f}, we have developed a method of error estimation for the approximate solution $\tilde{\boldsymbol{x}}$ of \boldsymbol{x} with $\boldsymbol{f}(\boldsymbol{x}) = \boldsymbol{0}$:

ALGORITHM 6.3 (*Error estimation without a Lipschitz constant*).

1^{st} Step: Compute the approximate Jacobi matrix $\boldsymbol{J}_f(\tilde{\boldsymbol{x}})$ at $\tilde{\boldsymbol{x}}$.

2^{nd} Step: Compute the inverse $\boldsymbol{J}_f^{-1}(\tilde{\boldsymbol{x}})$. (Practical experience shows that the computations of the first and second step do not have to be very accurate.)

3^{rd} Step: Construct $\boldsymbol{g}(\boldsymbol{x})$ according to (6.10) and check the conditions

(6.11) $g_i(\tilde{\boldsymbol{x}} + p\boldsymbol{e}_i) > 0, \quad g_i(\tilde{\boldsymbol{x}} - p\boldsymbol{e}_i) < 0 \text{ for } i = 1, \ldots, n,$

successively for $p := p_k = 0.5 \cdot 10^{-k}, k = k_0, k_0 + 1, \ldots, k_0 \in \mathbb{N}$. Here \boldsymbol{e}_i denotes the i^{th} unit vector.

4^{th} Step: If (6.11) is satisfied with $p = p_N$, but not for $p = p_{N+1}$, one should check the conditions

(6.12) $g_i(\boldsymbol{x}) > 0, \quad \boldsymbol{x} \in E_i^+, \quad \text{and } g_i(\boldsymbol{y}) < 0, \quad \boldsymbol{y} \in E_i^- \text{ for } i = 1, \ldots, n$

for all points $\boldsymbol{x}, \boldsymbol{y}$ of the planes E_i^+, E_i^-, respectively (see [MOOR80]).

If (6.12) holds, then $\max\limits_{1 \le i \le n} \|\tilde{x}_i - \bar{x}_i\| < p_N$. Since the verification of (6.12) is rather laborious, one might be satisfied most of the time with step 4^*:

4^{*th} Step: Check whether the conditions

$$g_i(\tilde{\boldsymbol{x}} + p\boldsymbol{e}_i) \approx +p, \quad g_i(\tilde{\boldsymbol{x}} - p\boldsymbol{e}_i) \approx -p \quad \text{for } i = 1, \ldots, n$$

hold. This is strong evidence (but no guarantee) that the solution of $\bar{\boldsymbol{x}}$ satisfies:

$$\max_{1 \le i \le n} |\tilde{x}_i - \bar{x}_i| < p_N.$$

Bibliography for Section 6.1

[BROS76], §11; [COLL68], §13.2; [ENGE87], 4.1; [HENR72], vol.1, §5.1-5.3; [IMSL87], §7.3; [STOE91], 5.1, 5.2; [STUM82], 9.1; [TÖRN79] vol.1, 7.1.

6.2 Special Iterative Methods

6.2.1 Newton Methods for Nonlinear Systems

6.2.1.1 The Basic Newton Method

Consider a nonlinear system (6.1) whose solution \bar{x} lies in the interior of D_f. Let each f_i possess continuous second partial derivatives in D_f and assume that for the Jacobi matrix

$$J_f := \left(\frac{\partial f_i}{\partial x_k} \right)_{\substack{i=1,\ldots,n \\ k=1,\ldots,n}} = \begin{pmatrix} \dfrac{\partial f_1}{\partial x_1} & \dfrac{\partial f_1}{\partial x_2} & \cdots & \dfrac{\partial f_1}{\partial x_n} \\ \dfrac{\partial f_2}{\partial x_1} & \dfrac{\partial f_2}{\partial x_2} & \cdots & \dfrac{\partial f_2}{\partial x_n} \\ \vdots & \vdots & & \vdots \\ \dfrac{\partial f_n}{\partial x_1} & \dfrac{\partial f_n}{\partial x_2} & \cdots & \dfrac{\partial f_n}{\partial x_n} \end{pmatrix}$$

we have $\det(J_f) \neq 0$ in D_f. Then there is a neighborhood $D \subset D_f$ of \bar{x} so that the conditions of theorem 6.1 are satisfied for the iteration function φ of the Newton method:

$$\varphi(x) := x - J_f^{-1}(x)f(x).$$

The iteration rule is:

$$(6.13) \qquad x^{(\nu)} \;=\; x^{(\nu-1)} - J_f^{-1}(x^{(\nu-1)})f(x^{(\nu-1)}), \qquad \text{or}$$

$$(6.14) \qquad x^{(\nu)} \;=\; x^{(\nu-1)} + \Delta x^{(\nu)} \quad \text{for } \nu = 1,2,\ldots, \text{ with}$$

$$(6.14') \qquad \Delta x^{(\nu)} = -J_f^{-1}(x^{(\nu-1)})f(x^{(\nu-1)}).$$

In order to avoid computing the explicit inverse of the Jacobi matrix, one will usually try to solve the following linear system instead:

$$(6.15) \qquad J_f(x^{(\nu-1)})\Delta x^{(\nu)} = -f(x^{(\nu-1)})$$

instead in each iteration step.

ALGORITHM 6.4 (*Newton's method*).

For each $\nu = 1, 2, \ldots$, do:

(i) Solve the linear system (6.15) to obtain $\Delta x^{(\nu)}$.

(ii) Compute $x^{(\nu)}$ from (6.14). Possible break-off criteria are:

(a) $\nu \geq \nu_{max}$ for a given $\nu_{max} \in \mathbb{N}$;

(b) $\|x^{(\nu)} - x^{(\nu-1)}\| \leq \|x^{(\nu)}\| \varepsilon_1$ for a given $\varepsilon_1 \in \mathbb{R},\quad \varepsilon_1 > 0$;

(c) $\|x^{(\nu)} - x^{(\nu-1)}\| \leq \varepsilon_2$ for a given $\varepsilon_2 \in \mathbb{R},\quad \varepsilon_2 > 0$;

(d) $\|f(x^{(\nu)})\| \leq \varepsilon_3$ for a given $\varepsilon_3 \in \mathbb{R}$.

In algorithm 6.4, the calculation of the inverse in (6.13) should be replaced by solving the linear system (6.15) for which the methods of chapters 4 or 5 are appropriate. Convergence is guaranteed if the iteration starts close enough to the exact solution \bar{x}.

Simplified Newton's method for systems.

To avoid having to solve a different linear system (6.15) in each iteration step, one can solve the approximate system

$$(6.16) \qquad\qquad J_f(x^{(0)}) \Delta x^{(\nu)} = -f(x^{(\nu-1)})$$

with the same matrix $J_f(x^{(0)})$ in all iteration steps. In this case, one has to form the LR factorization of J_f only once, see chapter 4.2; updating and backsubstitution are necessary for each new right hand side, as is the calculation of $x^{(\nu)}$ in (6.14).

Alternatively, one can periodically change the system matrix $J_f(x^{(0)})$ in (6.16) to a new Jacobi matrix $J_f(x^{(\nu_0)})$. This way the convergence rate is generally increased.

We note that the simplified Newton method converges quadratically as in the single equation case, see chapter 2.5.1.

6.2.1.2 Damped Newton Method for Systems

The damped Newton method is a variant of the quadratically convergent Newton method (see [CONT80], §5.2). Here the iterate $x^{(\nu)}$ of Newton's method is only accepted as the next iterate if for the Euclidian norm

$$\|\boldsymbol{f}(\boldsymbol{x}^{(\nu)})\|_2 < \|\boldsymbol{f}(\boldsymbol{x}^{(\nu-1)})\|_2.$$

Under the same assumptions as in section 6.2.1.1, we have:

ALGORITHM 6.5 (*Damped Newton method for systems*).

For $\nu = 1, 2, \ldots$ do:

(i) Compute $\Delta\boldsymbol{x}^{(\nu)}$ according to (6.15).

(ii) Compute j such that

$$j := \min\left\{ i \mid i \geq 0, \|\boldsymbol{f}(\boldsymbol{x}^{(\nu-1)}) + \frac{1}{2^i}\Delta\boldsymbol{x}^{(\nu)})\|_2 < \|\boldsymbol{f}(\boldsymbol{x}^{(\nu-1)})\|_2 \right\}.$$

(iii) Set $\boldsymbol{x}^{(\nu)} := \boldsymbol{x}^{(\nu-1)} + \Delta\boldsymbol{x}^{(\nu)}/2^j$.

Step (ii) shall be carried out only for $0 \leq i \leq i_{\max}$ where i_{\max} is given in advance. Should (ii) be impossible to fulfill, one continues computations with $j = 0$. This method is also quadratically convergent.

In practice, one can use the damped Newton method in two variants. One of them uses the Jacobi matrix, the other approximates the Jacobi matrix with the forward difference quotient matrix of f_i. The latter procedure enables us to estimate how many iteration steps $IUPD$ should be carried out with the same approximate Jacobi matrix. If this number $IUPD > 1$, the method corresponds to the damped simplified form of Newton's method.

Extensive tests have shown that the damped Newton method or the damped basic form of it are preferable to the basic Newton method, the Brown method, or the steepest descent method. It has also been demonstrated that the smoothing size i_{\max} should be adjusted for the problem. For the same problem and the same starting vector, the method might converge for one given i_{\max} and it might not do so for another i_{\max}. With different i_{\max} it might also converge towards different zeros of \boldsymbol{f}. Initially one should therefore start out with $i_{\max} = 4, IUPD = 1$ and a maximum of 1000 iterations.

6.2.2 Regula Falsi for Nonlinear Systems

We again consider the nonlinear system (6.1) and form the vectors

$$(6.17) \quad \delta f(x_j, \tilde{x}_j) = \frac{1}{x_j - \tilde{x}_j} \left(f(x_1, \ldots, x_j, \ldots, x_n) - f(x_1, \ldots, \tilde{x}_j, \ldots, x_n) \right)$$

for $j = 1, \ldots, n$ and the matrix

$$(6.18) \qquad Df(x, \tilde{x}) = (\delta f(x_1, \tilde{x}_1), \delta f(x_2, \tilde{x}_2), \ldots, \delta f(x_n, \tilde{x}_n)).$$

This corresponds to the Jacobi matrix of Newton's method where the derivatives are replaced by difference quotients. Thus D is an approximate Jacobi matrix.

If \bar{x} is a solution of (6.1) and if $x^{(\nu-2)}$ and $x^{(\nu-1)}$ are two approximations for \bar{x}, then one can calculate an approximation $x^{(\nu)}$ for each $\nu = 2, 3, \ldots$ according to the *iteration rule* of the *regula falsi*:

$$(6.19) \quad x^{(\nu)} = x^{(\nu-1)} - (Df(x^{(\nu-1)}, x^{(\nu-2)}))^{-1} f(x^{(\nu-1)}) = \varphi(x^{(\nu-1)}, x^{(\nu-2)}).$$

To start the regula falsi we always need two starting vectors $x^{(0)}, x^{(1)}$. Computing of the inverse Df^{-1} can be avoided as before by setting

$$(6.20) \qquad x^{(\nu)} \;=\; x^{(\nu-1)} + \Delta x^{(\nu)} \qquad \text{for}$$

$$\Delta x^{(\nu)} \;=\; -(Df(x^{(\nu-1)}, x^{(\nu-2)}))^{-1} f(x^{(\nu-1)}).$$

and obtaining $\Delta x^{(\nu)}$ by solving the linear system

$$Df(x^{(\nu-1)}, x^{(\nu)})\Delta x^{(\nu)} = -f(x^{(\nu-1)}).$$

Sufficient conditions for convergence are known, but of little use in practice. If, however, $\det J(\bar{x}) \neq 0$, the procedure converges if the starting vectors are close enough to \bar{x}. In this case, the order of convergence is $k = (1 + \sqrt{5})/2$, just as in the single equation case, see chapter 2.6.1.

Error estimate:

$$\|\bar{x} - x^{(\nu)}\| \;\leq\; \prod_{k=1}^{\nu-1} \left(\frac{s_k}{1-s_k}\right) \frac{1-2s_1}{1-3s_1} \|x^{(2)} - x^{(1)}\| \;\;\text{for}\;\; \nu = 2,3,\dots,$$

where $\quad s_1 \leq \dfrac{2}{7}, \quad s_2 = \dfrac{s_1}{1-s_1}$ and $s_k = \dfrac{s_{k-1}}{1-s_{k-1}} \dfrac{s_{k-2}}{1-s_{k-2}}$ for $k \geq 3$.

6.2.3 Method of Steepest Descent for Nonlinear Systems

Assume that a given nonlinear system (6.1) has a solution \bar{x}. Let

$$(6.21) \qquad\qquad Q(x) := \sum_{i=1}^{n} f_i^2(x) \quad \text{for } x \in \mathbb{R}^n.$$

Then $Q(x) = 0$ if and only if $f_i(x) = 0$ for $i = 1,\dots,n$. Finding solutions \bar{x} for which $Q(x) = 0$, is therefore equivalent to the solving the system (6.1).

From (6.21), the gradient of Q can be easily computed:

$$\nabla Q(x) = \operatorname{grad} Q(x) = \begin{pmatrix} Q_{x_1} \\ Q_{x_2} \\ \vdots \\ Q_{x_n} \end{pmatrix}, \quad \text{where } Q_{x_i} := \frac{\partial Q}{\partial x_i} \text{ for } i = 1,\dots,n.$$

Thus we can form an iterative method for finding an approximate solution by the iteration rule

$$(6.22) \qquad x^{(\nu)} = x^{(\nu-1)} - \frac{Q(x^{(\nu-1)})}{\|\nabla Q(x^{(\nu-1)})\|_2^2} \nabla Q(x^{(\nu-1)}) := \varphi(x^{(\nu-1)}).$$

The iteration function is:

$$\varphi(x) = x - \frac{Q(x)}{\|\nabla Q(x)\|_2^2} \nabla Q(x).$$

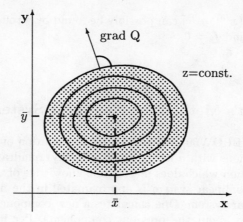

Figure 6.1: Method of steepest descent

Regarding convergence the same conditions as those for Newton's method hold. The vectors of the sequence $\{x^{(\nu)}\}$ converge towards the exact solution \bar{x} if the starting vector $x^{(0)}$ is close enough to \bar{x}. In general, it is feasible to use this *steepest descent method* for a coarser initial approximations $x^{(0)}$ than is possible for Newton's method. The steepest descent method, however, only converges linearly.

The usefulness of the steepest descent method is, however, limited if in a neighborhood of the exact solution \bar{x} there are also minima of the function $Q(x)$ that do not correspond to zeros of f. It might happen that the iteration sequence converges towards one of these non-zero minima of Q.

Generalized steepest descent methods and conditions for their convergence are given in [STOE91], 8.7; [GOLU89], 10.2; [STUM82], 9.2.2.

Steepest descent method for $n = 2$:

Write the iteration sequence as $\{(x^{(\nu)}, y^{(\nu)})\}$ for $x^{(\nu)}, y^{(\nu)} \in \mathbb{R}$. Then (6.22) becomes

$$x^{(\nu)} = x^{(\nu-1)} - \frac{Q(x^{(\nu-1)}, y^{(\nu-1)}) Q_x(x^{(\nu-1)}, y^{(\nu-1)})}{Q_x^2(x^{(\nu-1)}, y^{(\nu-1)}) + Q_y^2(x^{(\nu-1)}, y^{(\nu-1)})}$$

$$y^{(\nu)} = y^{(\nu-1)} - \frac{Q(x^{(\nu-1)}, y^{(\nu-1)}) Q_y(x^{(\nu-1)}, y^{(\nu-1)})}{Q_x^2(x^{(\nu-1)}, y^{(\nu-1)}) + Q_y^2(x^{(\nu-1)}, y^{(\nu-1)})}.$$

A starting vector $(x^{(0)}, y^{(0)})$ can possibly be found by a rough drawing of the graphs of $f_1 = 0$ and $f_2 = 0$.

6.2.4 Brown's Method for Nonlinear Systems

Brown's method [BROW69], [BROW71] for the solution of a system (6.1) of n nonlinear equations with n unknown is a (locally) quadratically convergent Newton-like iteration which does not assume knowledge of the partial derivatives of f. The nonlinear system is approximated in the neighborhood of a solution by a linear system. One calculates a new component from the most recent information about the previous components. Per iteration step, the procedure needs only about half as many function evaluations as Newton's method.

6.3 Choosing a Method

Among the indicated procedures, the damped Newton method or the damped simplified form of Newton's method are preferred to all other procedures. In practice, we recommend the modification where one uses an approximate Jacobi matrix in either procedure in order to avoid calculating and evaluating the partial derivatives needed for the Jacobi matrix in every iteration step, see especially the notes at the end of chapter 6.2.1.2 .

Bibliography for Chapter 6

[BJÖR74], 6.10; [BROW69]; [BROW71]; [CARN69], 5.9; [COLL73] I, 2.5; [CONT80], §5.2; [ENGE87], 4.2; [HÄMM91], 8. §2; [HENR72] vol.1, 5.4-5.5, 5.9; [IGAR85]; [ISAA66], 3.3; [McCA67], 3.3; [4] [NORT85]; [RICE83], p.239; [SCHW89], 5; [STIE63], 4.4; [STOE91], 5; [STUM82], 9.2-9.3; [TÖRN79] vol.1, 7.2 .

Chapter 7

Eigenvalues and Eigenvectors of Matrices

7.1 Basic Concepts

For a given (n, n) matrix $\boldsymbol{A} = (a_{ik})$ the eigenvalue problem consists of finding nonzero vectors \boldsymbol{x} so that \boldsymbol{Ax} is parallel to the vector \boldsymbol{x}. Such a vector \boldsymbol{x} is called an *eigenvector* of \boldsymbol{A}. It satisfies the *eigenvalue-eigenvector equation* for a scalar λ, called the *eigenvalue*:

$$(7.1) \qquad\qquad \boldsymbol{Ax} = \lambda \boldsymbol{x}.$$

With the (n, n) identity matrix \boldsymbol{I}_n, (7.1) can be rewritten as

$$(7.2) \qquad\qquad \boldsymbol{Ax} - \lambda \boldsymbol{x} = (\boldsymbol{A} - \lambda \boldsymbol{I})\boldsymbol{x} = \boldsymbol{0}.$$

Given λ, (7.2) is a homogeneous linear system in \boldsymbol{x} that has non-trivial solutions $\boldsymbol{x} \neq \boldsymbol{0}$, if and only if its system matrix $\boldsymbol{A} - \lambda \boldsymbol{I}$ is singular, i.e., if

$$(7.3) \qquad\qquad P(\lambda) := \det(\boldsymbol{A} - \lambda \boldsymbol{I}) = 0, \text{ or}$$

$$(7.4) \qquad P(\lambda) = det \begin{pmatrix} a_{11} - \lambda & a_{12} & a_{13} & \cdots & a_{1n} \\ a_{21} & a_{22} - \lambda & a_{23} & \cdots & a_{2n} \\ \vdots & & \ddots & & \vdots \\ a_{n1} & a_{n2} & a_{n3} & \cdots & a_{nn} - \lambda \end{pmatrix} = 0.$$

The equation $P(\lambda) = 0$ is called the *characteristic equation* of the matrix \boldsymbol{A}. $P(\lambda)$ is a polynomial of degree n in λ and is called the *characteristic polynomial* of \boldsymbol{A}. The real or complex zeros λ_i for $i = 1, \ldots, n$ of $P(\lambda)$ are called the *characteristic numbers* or *eigenvalues* of \boldsymbol{A}. (7.2) has non-trivial solutions $\boldsymbol{x}_i \neq \boldsymbol{0}$ only for the eigenvalues λ_i of \boldsymbol{A}. An eigenvector and eigenvalue pair \boldsymbol{x}_i and λ_i satisfies the eigenvalue-eigenvector equation

(7.5) $A x_i = \lambda_i x_i$ or $(A - \lambda_i I) x_i = 0$.

To determine the eigenvalues and the eigenvectors of a square matrix A is called the *eigenvalue problem*. It is a nonlinear problem since the unknown λ_i and x_i occur nonlinearly in (7.5).

We distinguish between the *complete* and a *partial* eigenvalue problem. The complete eigenvalue problem consists of finding all eigenvalues and eigenvectors of A; a partial eigenvalue problem asks only for one (or several) eigenvalues with or without the associated eigenvector(s).

There are essentially two kinds of *methods* for the solution of the matrix eigenvalue, eigenvector problem:

1) *Iterative methods*, which try to approximate some or all the eigenvalues and eigenvectors iteratively.

2) *Direct methods*: They generally form the characteristic polynomial $P(\lambda)$ of A. The eigenvalues are then determined as zeros of $P(\lambda)$. And the eigenvectors are found as solutions of the homogeneous linear systems (7.5). Direct methods can be used for the complete eigenvalue problem. Among them there are some that avoid the calculation of determinants, see e.g. the Krylov method in chapter 7.5.

Besides these methods, there are numerous techniques for estimating the eigenvalues of a given matrix, of which we shall quote only one:

THEOREM 7.1: (*Gershgorin's Theorem*) .
All eigenvalues of a real or complex (n, n) matrix A lie within the union of the *Gershgorin circles*

$$C_i = \{ z \in \mathbb{C} \mid |z - a_{ii}| \leq \sum_{k \neq i} |a_{ik}| \} \quad \text{for } i = 1, \dots, n.$$

If for some $1 \leq k < n$, k Gershgorin circles are disjoint from the remaining $n - k$ circles, then k eigenvalues of A lie inside the union of these k circles, the remaining $n - k$ eigenvalues of A lie in the union of the remaining $n - k$ circles.

The above theorem is one of the most useful tools to form a first idea of the location of the eigenvalues of a given matrix A. More involved estimates of eigenvalues can be found, e.g., in [ATKI89], ch. 9.1, [GOLU89], ch. 7.2, [HOUS75], ch. 3, or [STOE91], ch. 6.9.

If the matrix A is real, then all coefficients of $P(\lambda)$ in (7.4) will be real. If in this case $P(\nu) = 0$, then $P(\bar{\nu}) = \overline{P(\nu)} = \bar{0} = 0$, where the upper bar $^{-}$ denotes the complex conjugate $\bar{\nu} = a - bi$ of $\nu = a + bi \in \mathbb{C}$. Thus real matrices A can have real or complex eigenvalues, but their complex eigenvalues occur in complex conjugate pairs $\nu, \bar{\nu}$.

Computing the eigenvalues and eigenvectors of a complex (n, n) matrix A can be reduced to the corresponding problem for a real $(2n, 2n)$ matrix. Write $A = B + iC$ for real matrices B and C, and for any $x \in \mathbb{C}^n$ write $x = u + iv$ with real vectors u and v.

Then (7.1) splits into two real linear homogeneous systems:

$$Bu - Cv = \lambda u \text{ and}$$
$$Cu + Bv = \lambda v.$$

These equations can be written in block form with an $(2n, 2n)$ real matrix \tilde{A} and a $2n$-vector \tilde{x} thus:

$$\tilde{A}\tilde{x} = \tilde{\lambda}\tilde{x}, \text{ where } \tilde{A} = \begin{pmatrix} B & -C \\ C & B \end{pmatrix} \text{ and } \tilde{x} = \begin{pmatrix} u \\ v \end{pmatrix}.$$

Bibliography for Section 7.1

[ATKI89], 9; [BJÖR74], 5.92; [BOOR78], p.193; [COLL68], §8; [CONT80], 5.7; [ENGE87], 5.1; [FADD59], p.277/9; [HOUS75], 3, 6, 7; [ISAA66], 4.0; [KELL76]; [SPEL85], 2.1; [STIE63], 5.2; [STEW73], 6; [STOE91], 6; [TÖRN79] vol.2, part IV; [WERN80]; [WILK65]; [ZURM65], §9; [ZURM84], §§13.1-13.2, 15.

7.2 Diagonalizable Matrices and the Conditioning of the Eigenvalue Problem

If A were a truely random matrix, then its eigenvalues $\lambda_i \in \mathbb{R}$ would be random as well as roots of the polynomial equation $P(\lambda) = 0$. Hence it would be most unlikely that a random matrix would have two identical eigenvalues. In practice, however, matrices with multiple eigenvalues arise often from the structure of the physical problem or the mathematical model.

If A has n distinct eigenvalues $\lambda_1, \ldots, \lambda_n$, then the associated eigenvectors x_1, \ldots, x_n must be linearly independent. If we form a nonsingular matrix X with the eigenvectors x_i of A as its columns, then

$$(7.6)\, AX = A(x_1, \ldots, x_n) = (\lambda_1 x_1, \ldots, \lambda_n x_n) = X \begin{pmatrix} \lambda_1 & & 0 \\ & \ddots & \\ 0 & & \lambda_m \end{pmatrix}.$$

Thus $X^{-1}AX = D = \text{diag} (\lambda_1, \ldots, \lambda_n)$ is a diagonal matrix, and thus such an A is called *diagonalizable* .

If A has a repeated eigenvalue, A may or may not be diagonalizable in the form $X^{-1}AX = D$. A matrix A is diagonalizable precisely when there are as many linearly independent eigenvectors for each repeated eigenvalue λ_i as the multiplicity of λ_i as a root of $P(\lambda) = 0$ indicates.

For nondiagonalizable A, *many matrix normal forms* under *similarity transformations* $X^{-1}AX$ are known in theory.
The *Schur normal form* of a matrix A is an upper triangular matrix with the eigenvalues of A on its diagonal and the transforming matrix X unitary.
The *Jordan normal form* of A is a block diagonal matrix made up of *Jordan blocks* $J_i = \begin{pmatrix} \lambda_i & 1 & 0 \\ & \ddots & 1 \\ 0 & & \lambda_i \end{pmatrix}$ for the eigenvalues λ_i of A of various sizes determined in theory by A.
The *companion matrix normal form* of A is a block diagonal matrix made up of companion matrices $\begin{pmatrix} 0 & 1 & & 0 \\ & & \ddots & 1 \\ q_0 & \cdots & & q_\ell \end{pmatrix}$ for various factors $q(\lambda) = \lambda^{\ell+1} - q_\ell \lambda^\ell + \ldots + (-1)^\ell q_0$ of the characteristic polynomal $P(\lambda)$ of A. For an introduction into the theory of eigenvalues, eigenvectors and normal forms for matrices, see e.g. [LANC85].

Numerically, the eigenvalue problem $Ax = \lambda x$ behaves quite differently from the linear equations problem $Ax = b$ treated in chapters 4 and 6. The condition of the linear system $Ax = b$ is determined by the condition number *cond* A of A. The condition of the eigenvalue problem $Ax = \lambda x$, i.e., the intrinsic "goodness" of the problem of finding eigenvalues and eigenvectors of A reliably, is governed by the condition number *cond* X of the eigenvector matrix X (or a slightly more involved matrix X, if A is not diagonalizable). This is important to realize. For example, when solving linear equations $Ax = b$, we have the worst case of conditioning when A is the zero matrix, $A = 0_n$: any x will solve $0_n x = 0$, and *cond* $0_n = \infty$. But the eigenvalue problem $0_n x = \lambda x$

is extremely well conditioned: $\mathbf{0}_n$ is diagonal with each eigenvalue precisely equal to zero.

In this chapter, symmetric matrices \boldsymbol{A} with $\boldsymbol{A}^T = \boldsymbol{A}$ and hermitian matrices \boldsymbol{A} with $\boldsymbol{A}^* := \bar{\boldsymbol{A}}^T = \boldsymbol{A}$ will again play a special role. Such matrices have only real eigenvalues and can be diagonalized via orthogonal or unitary matrices:

$$(7.7) \quad \begin{cases} \boldsymbol{X}^T \boldsymbol{A} \boldsymbol{X} = \boldsymbol{D} = \operatorname{diag}(d_i) & \text{for} \quad \boldsymbol{X}^T \boldsymbol{X} = \boldsymbol{I} \text{ and } d_i \in \mathbb{R}, \text{ or} \\ \boldsymbol{U}^* \boldsymbol{A} \boldsymbol{U} = \boldsymbol{D} = \operatorname{diag}(d_i) & \text{for} \quad \boldsymbol{U}^* \boldsymbol{U} = \boldsymbol{I} \text{ and } d_i \in \mathbb{R}. \end{cases}$$

Thus from our previous remarks, the eigenvalue problem is well-conditioned for all symmetric or hermitian matrices \boldsymbol{A} since $cond_2(\boldsymbol{X}) = cond_2(\boldsymbol{U}) = 1$ for an orthogonal \boldsymbol{X} or a unitary \boldsymbol{U}.

Bibliography for Section 7.2

[ENGE87], 5.2; [PARL80]; [SCHW72], 4.3; [SPEL85], p.102-105; [ZURM65], §9.2; [ZURM84], §14.1.

7.3 Vector Iteration

7.3.1 The Dominant Eigenvalue and the Associated Eigenvector of a Matrix

Here we shall consider the eigenvalue problem (7.2) for a diagonalizable real matrix \boldsymbol{A}: Let \boldsymbol{A} have n linearly independent eigenvectors $\boldsymbol{x}_1, \boldsymbol{x}_2, \ldots, \boldsymbol{x}_n \in \mathbb{R}^n$. Take an arbitrary starting vector $\boldsymbol{z}^{(0)} \neq \boldsymbol{0}$ in \mathbb{R}^n and form the iteration sequence $\{\boldsymbol{z}^{(\nu)}\}$ according to the rule

$$(7.8) \quad \boldsymbol{z}^{(\nu)} := \boldsymbol{A}\boldsymbol{z}^{(\nu-1)}, \text{ where } \boldsymbol{z}^{(\nu-1)} = \begin{pmatrix} z_1^{(\nu-1)} \\ z_2^{(\nu-1)} \\ \vdots \\ z_n^{(\nu-1)} \end{pmatrix} \text{ for } \nu = 1, 2, \ldots.$$

Since the eigenvectors \boldsymbol{x}_i of \boldsymbol{A} form a basis for \mathbb{R}^n, $\boldsymbol{z}^{(0)}$ can be written uniquely as a linear combination of the \boldsymbol{x}_i:

$$\boldsymbol{z}^{(0)} = \sum_{i=1}^{n} c_i \boldsymbol{x}_i,$$

where $c_i \neq 0$ for at least one i. Thus as $\boldsymbol{A}\boldsymbol{x}_i = \lambda_i \boldsymbol{x}_i$ for each i, we obtain

$$z^{(\nu)} = c_1 \lambda_1^\nu \boldsymbol{x}_1 + c_2 \lambda_2^\nu \boldsymbol{x}_2 + \ldots + c_n \lambda_n^\nu \boldsymbol{x}_n \text{ for all } \nu = 0, 1, 2 \ldots.$$

Next we shall study the quotients $q_i^{(\nu)}$ of the i^{th} components of successive iterates $z^{(\nu)}$ and $z^{(\nu-1)}$:

$$q_i^\nu := \frac{z_i^{(\nu)}}{z_i^{(\nu-1)}} = \frac{c_1 \lambda_1^\nu x_{1,i} + c_2 \lambda_2^\nu x_{2,i} + \ldots + c_n \lambda_n^\nu x_{n,i}}{c_1 \lambda_1^{\nu-1} x_{1,i} + c_2 \lambda_2^{\nu-1} x_{2,i} + \ldots + c_n \lambda_n^{\nu-1} x_{n,i}}.$$

Case 1) \boldsymbol{A} has a single dominant eigenvalue, $|\lambda_1| > |\lambda_2| \geq |\lambda_3| \geq \ldots \geq |\lambda_n|$:

a) If $c_1 \neq 0$ and $x_{1,i} \neq 0$ for some i, then

$$q_i^{(\nu)} = \lambda_1 + O\left(\left|\frac{\lambda_2}{\lambda_1}\right|^\nu\right), \quad \text{i.e.,} \quad \lim_{\nu \to \infty} q_i^{(\nu)} = \lambda_1.$$

Clearly $x_{1,i} \neq 0$ for at least one i. Thus, at least one of the quotients $q_i^{(\nu)}$ tends towards λ_1. Regarding the other eigenvalues λ_i, see b) below.

For sufficiently large ν, $q_i^{(\nu)}$ will be a reasonable approximation for the eigenvalue λ_1 of largest magnitude of \boldsymbol{A}, called the *dominant eigenvalue* of \boldsymbol{A}. Thus

(7.9) $$q_i^{(\nu)} \approx \lambda_1.$$

In practice, one iterates until (7.9) is uniformly valid for all indices i with $x_{1,i} \neq 0$. For large ν, the vector $z^{(\nu)}$ is nearly parallel to \boldsymbol{x}_1. For $\nu \to \infty$, we have asymptotically

$$z^{(\nu)} \sim \lambda_1^\nu c_1 \boldsymbol{x}_1 \text{ and } z^{(\nu)} \sim \lambda_1 z^{(\nu-1)}.$$

If the eigenvectors \boldsymbol{x}_i have unit length, then asymptotically

$$\frac{z^{(\nu)}}{|z^{(\nu)}|} \approx \boldsymbol{x}_1.$$

b) If $c_1 = 0$ or $x_{1,i} = 0$, but $c_2 \neq 0$ and $x_{2,i} \neq 0$, and $|\lambda_2| > |\lambda_3| \geq \cdots \geq |\lambda_n|$:

If $c_1 = 0$ then the starting vector $z^{(0)}$ has no component in the direction of \boldsymbol{x}_1. If \boldsymbol{A} is symmetric and $c_1 = 0$, then $z^{(0)}$ is orthogonal to \boldsymbol{x}_1 as the eigenvectors of $\boldsymbol{A} = \boldsymbol{A}^T$ are mutually orthogonal by (7.7). In this case

$$q_i^{(\nu)} = \lambda_2 + O\left(\left|\frac{\lambda_3}{\lambda_2}\right|^\nu\right), \quad \text{i.e.,} \quad \lim_{\nu \to \infty} q_i^{(\nu)} = \lambda_2.$$

Thus

$$z^{(\nu)} \sim \begin{cases} c_1 \lambda_1^\nu x_1 & \text{if} \quad c_1 \neq 0, \\ c_2 \lambda_2^\nu x_2 & \text{if} \quad c_1 = 0. \end{cases}$$

For sufficiently large ν we have

$$\frac{z^{(\nu)}}{|z^{(\nu)}|} \approx x_1 \quad \text{for} \quad c_1 \neq 0, \quad \text{and} \quad \frac{z^{(\nu)}}{|z^{(\nu)}|} \approx x_2 \quad \text{for} \quad c_1 = 0, \quad \text{and}$$

$$q_i^{(\nu)} \approx \lambda_2 \begin{cases} \text{for all} \quad i = 1, \dots, n, \text{ if } c_1 = 0, c_2 \neq 0, \text{ and } x_{2,i} \neq 0, \\ \text{for all} \quad i \quad \text{with} \quad x_{1,i} = 0, \quad \text{if} \quad c_1 \neq 0. \end{cases}$$

Thus, it can theoretically happen that the $q_i^{(\nu)}$ tend towards an eigenvalue different from λ_i.

In computations, however, after a few iterations, $z^{(\nu)}$ will acquire a nonzero component in the direction of x_1 due to rounding errors. And hence the cases b) or c) do not really occur except for highly structured matrices A.

c) If $c_i = 0$ for $i = 1, \dots, j$, but $c_{j+1} \neq 0$, $x_{j+1,i} \neq 0$, and $|\lambda_{j+1}| > |\lambda_{j+2}| \geq \cdots \geq |\lambda_n|$, then asymptotically

$$q_i^{(\nu)} \approx \lambda_{j+1} \quad \text{and} \quad \frac{z^{(\nu)}}{|z^{(\nu)}|} \approx x_{j+1}.$$

If $x_{j+1,i} = 0$ for one i, the associated $q_i^{(\nu)}$ tends towards λ_{j+2}.

Before dealing with a multiple dominant eigenvalue of A, we shall give a

COMPUTATIONAL SCHEME 7.2 (*Vector iteration according to v. Mises*).

A				$z^{(0)}$	$z^{(1)}$	$z^{(2)}$	\ldots
a_{11}	a_{12}	\ldots	a_{1n}	$z_1^{(0)}$	$z_1^{(1)}$	$z_1^{(2)}$	
a_{21}	a_{22}	\ldots	a_{2n}	$z_2^{(0)}$	$z_2^{(1)}$	$z_2^{(2)}$	
\vdots	\vdots		\vdots	\vdots	\vdots	\vdots	
a_{n1}	a_{n2}	\ldots	a_{nn}	$z_n^{(0)}$	$z_n^{(1)}$	$z_n^{(2)}$	

During the computations, one calculates not only the vectors $z^{(\nu)} = Az^{(v-1)}$, but also normalizes each vector $z^{(\nu)}$ by dividing each of its components by the component of maximum modulus so that $\|z^{(\nu)}\|_1 = 1$. Denote the normalized vector by $z_n^{(\nu)}$, then $z^{(\nu+1)}$ is obtained by the rule $Az_n^{(\nu)} = z^{(\nu+1)}$. A different possibility would be to normalize each vector $z^{(\nu)}$ so that $\|z^{(v)}\|_2 = 1$. This however, would require more calculations. By normalizing one avoids over- or underflow of the components of $z_i^{(\nu)}$ in case $|\lambda_1| \neq 1$.

REMARK: Since the exact values of the eigenvalues are not known, one might want to carry out the calculations with several linearly independent starting vectors $z^{(0)}$ in order to be sure that the dominant eigenvalue has been obtained. In practice, this is, however, not a problem. For large n, the probability is almost zero that by chance one has chosen a starting vector $z^{(0)}$ that has no component in the direction of the dominant eigenvector x_1, as for example would be the case if $z^{(0)}$ were an eigenvector for another eigenvalue of A.

Case 2) A has a multiple dominant eigenvalue, $\lambda_1 = \lambda_2 = \ldots = \lambda_p$, $|\lambda_1| > |\lambda_{p+1}| \geq \ldots \geq |\lambda_n|$:

In this case there is a p dimensional *eigenspace* associated with λ_1:

$$(7.10) \qquad \dim E(\lambda_1) = \dim\{x \in \mathbb{C}^n \mid Ax = \lambda_1 x\} = p.$$

If x_1, \ldots, x_p form a basis for E_{λ_1} and if $z^{(0)}$ is a starting vector with $c_1 x_{1,i} + c_2 x_{2,i} + \ldots + c_p x_{p,i} \neq 0$, we obtain the relations

$$q_i^{(\nu)} = \lambda_1 + O\left(\left|\frac{\lambda_{p+1}}{\lambda_1}\right|^{\nu}\right), \quad \text{or} \quad \lim_{\nu\to\infty} q_i^{(\nu)} = \lambda_1.$$

Thus

$$z^{(\nu)} \sim \lambda_1^{\nu}(c_1^{(r)}x_1 + c_2^{(r)}x_2 + \ldots + c_p^{(r)}x_p) = y_r$$

$$\text{for } r = 1, \ldots, p, \ \nu = 0, 1, 2, \ldots.$$

There are p such linearly independent vectors y_r in the eigenspace $E(\lambda_1)$. As an approximation for λ_1, we again take $q_i^{(\nu)}$. And the computed y_r will be eigenvectors in $E(\lambda_1)$, but need not be linearly independent.

Case 3) The dominant eigenvalue of A occurs precisely twice with opposite sign, $\lambda_1 = -\lambda_2, |\lambda_1| > |\lambda_3| \geq \cdots \geq |\lambda_n|$:

In this case we form the quotients $\tilde{q}_i^{(\nu)}$ of the i^{th} components of the two vectors $z^{(\nu+2)}$ and $z^{(\nu)}$:

$$\tilde{q}_i^{(\nu)} := \frac{z_i^{(\nu+2)}}{z_i^{(\nu)}}.$$

If $c_1 x_{1,i} + (-1)^{\nu} c_2 x_{2,i} \neq 0$, then

$$(7.11) \qquad \tilde{q}_i^{(\nu)} = \lambda_1^2 + O\left(\left|\frac{\lambda_3}{\lambda_1}\right|^{\nu}\right), \quad \text{or} \quad \lim_{\nu\to\infty} \tilde{q}_i^{(\nu)} = \lambda_1^2.$$

Thus asymptotically

$$x_1 \ \sim \ z^{(\nu+1)} + \lambda_1 z^{(\nu)}, \text{ and}$$
$$x_2 \ \sim \ z^{(\nu+1)} - \lambda_1 z^{(\nu)}.$$

Thus, for sufficiently large v the computed $\tilde{q}_i^{(v)}$ can serve as approximations for λ_1 and λ_2:

$$\pm\sqrt{\tilde{q}_i^{(\nu)}} \approx \lambda_{1,2}.$$

With $\lambda_1^* = \sqrt{\tilde{q}_i^{(\nu)}}$, the approximate eigenvectors are

$$\frac{z^{(\nu+1)} + \lambda_1^* z^{(\nu)}}{|z^{(\nu+1)} + \lambda_1^* z^{(\nu)}|} \approx x_1, \text{ and}$$

$$\frac{z^{(\nu+1)} - \lambda_1^* z^{(\nu)}}{|z^{(\nu+1)} - \lambda_1^* z^{(\nu)}|} \approx x_2.$$

In practice, this last case can be noticed by the fact that convergence occurs only in iterates that are two steps apart in the computational scheme 7.2. Cases 2) and 3) may also occur if A has several eigenvalues very close to its dominant eigenvalue λ_1 or to $-\lambda_1$.

7.3.2 Determination of the Eigenvalue Closest to Zero

In (7.2) we can set $\lambda = 1/\kappa$ for $\lambda \neq 0$. Then the eigenvalue problem $Ax = \lambda x$ becomes

$$A^{-1}x = \kappa x.$$

Clearly, if A is nonsingular, then all of A's eigenvalues are nonzero. If $\lambda \neq 0$ is an eigenvalue of a nonsingular matrix A, then $1/\lambda$ is an eigenvalue for A^{-1} and vice versa. Thus we can use the vector iteration of chapter 7.3.1 to find the eigenvalue of A closest to zero as the dominant eigenvalue of A^{-1} by solving:

$$(7.12) \qquad\qquad z^{(\nu)} = A^{-1} z^{(\nu-1)}.$$

Here one should not compute A^{-1} explicitly, but rather use a Gauß algorithm from chapter 4 to calculate the vectors $z^{(\nu)}$ by solving $Az^{(\nu)} = z^{(\nu-1)}$. If A is singular, or $\det A = 0$, then at least one eigenvalue of A is zero.

7.3.3 Eigenvalues in Between

Let A be a symmetric matrix with orthonormal eigenvectors x_i, i.e., $x_i^T x_j = \delta_{ij}$ and $Ax_i = \lambda_i x_i$, where $\delta_{ij} = 1$ if $i = j$ and $\delta_{ij} = 0$ if $i \neq j$. If $z^{(0)} = \sum_{i=1}^{n} c_i x_i$, then $c_1 = z^{(0)T} x_1$. We shall use

$$y^{(0)} := z^{(0)} - c_1 x_1 = c_2 x_1 + c_3 x_3 + \ldots + c_n x_n$$

as a starting vector for the vector iteration with \boldsymbol{A}. Since

$$\boldsymbol{y}^{(0)T}\boldsymbol{x}_1 = \boldsymbol{z}^{(0)T}\boldsymbol{x}_1 - c_1 = 0,$$

$\boldsymbol{y}^{(0)}$ is orthogonal to \boldsymbol{x}_1, and case 1.b) of section 7.3.1 occurs, i.e., the quotients $q_i^{(\nu)}$ will tend towards λ_2. Since \boldsymbol{x}_1 is only known approximately, the iterates $\boldsymbol{y}^{(\nu)}$ will not be entirely free from components in the direction of \boldsymbol{x}_1. Thus one has to subtract from $\boldsymbol{y}^{(\nu)}$ its components in direction of \boldsymbol{x}_1 in each step of the procedure. For this purpose we form

$$\tilde{\boldsymbol{y}}^{(\nu)} = \boldsymbol{y}^{(\nu)} - (\boldsymbol{y}^{(\nu)T}\boldsymbol{x}_1)\boldsymbol{x}_1$$

and calculate $\boldsymbol{y}^{(\nu+1)} = \boldsymbol{A}\tilde{\boldsymbol{y}}^{(\nu)}$. Thus asymptotically we get the following approximations for λ_2 and for \boldsymbol{x}_2:

$$q_i^{(\nu)} = \frac{\tilde{\boldsymbol{y}}_i^{(\nu+1)}}{\tilde{\boldsymbol{y}}_i^{(\nu)}} \approx \lambda_2 \ \text{ and } \ \frac{\tilde{\boldsymbol{y}}^{(\nu)}}{|\tilde{\boldsymbol{y}}^{(\nu)}|} \approx \boldsymbol{x}_2.$$

To calculate other eigenvalues and eigenvectors one can proceed in an analogous way. If e.g. one should want to determine the third largest eigenvalue as well as the associated eigenvector of \boldsymbol{A}, one would take a vector void of components parallel to the known eigenvectors \boldsymbol{x}_1 and \boldsymbol{x}_2 as a starting vector:

$$\boldsymbol{y}^{(0)} := \boldsymbol{z}^{(0)} - c_1\boldsymbol{x}_1 - c_2\boldsymbol{x}_2 \ \text{ for an arbitrary } \boldsymbol{z}^{(0)} \in \mathbb{R}^n,$$

where $c_1 = \boldsymbol{z}^{(0)T}\boldsymbol{x}_1$ and $c_2 = \boldsymbol{z}^{(0)T}\boldsymbol{x}_2$ as before.

Since \boldsymbol{x}_1 and \boldsymbol{x}_2 are again given only approximately as \boldsymbol{x}_1^* and \boldsymbol{x}_2^*, the iterated $\boldsymbol{y}^{(\nu)}$ must always be "cleaned" of components in the directions of \boldsymbol{x}_1 and \boldsymbol{x}_2 .

These suggestions to obtain "in-between" eigenvalues of a symmetric matrix \boldsymbol{A} are unfortunately not all that practicable. The QR or QD algorithms below compute all eigenvalues, and, if desired, all eigenvectors of \boldsymbol{A} in a shorter time and more reliably than vector iteration and its variants, see chapters 7.6, 7.7 and 7.8.

Bibliography for Section 7.3

[COLL73] I, 2.2; [ENGE87], 5.3-5.4, 5.6; [FADD59], §53; [McCA67], 5.10; [NIEM87], 7.2; [NOBL65] I, 6.2-6.3; [SPEL85], 2.5; [ZURM65], §10.1-10.2; [ZURM84], §40.1.

7.4 The Rayleigh Quotient for Hermitian Matrices

Compared with (7.9), one can give an improved approximation for the dominant eigenvalue λ_1 of a Hermitian or symmetric matrix A. This method needs only slightly more calculation time and is based on the *Rayleigh quotient*.

DEFINITION 7.3 (*Rayleigh quotient*).
If A is an arbitrary (n, n) matrix and $x \neq 0$ is a vector in \mathbb{R}^n, then the Rayleigh quotient is defined as

$$R[x] = \frac{\bar{x}^T A x}{\bar{x}^T x}.$$

For an eigenvector x of A with $Ax = \lambda x$, clearly $R[x] = \lambda$, i.e., the Rayleigh quotient for an eigenvector x is equal to the corresponding eigenvalue λ. If A is Hermitian we have

THEOREM 7.4 For a Hermitian matrix A the Rayleigh quotient has local extrema at the eigenvectors of A. If the eigenvalues of $A = A^*$ are ordered such that $|\lambda_1| \geq |\lambda_2| \geq \cdots \geq |\lambda_n|$, then $|R[x]| \leq |\lambda_1|$ for all $x \in \mathbb{C}^n$.

The Rayleigh quotient for the vector $z^{(\nu)}$ derived from a vector iteration $z^{(\nu+1)} = A z^{(\nu)}$ is

$$R[z^{(\nu)}] = \frac{\bar{z}^{(\nu)^T} z^{(\nu+1)}}{\bar{z}^{(\nu)^T} z^{(\nu)}}.$$

By theorem 7.4, $|R[z^{(\nu)}]| \leq |\lambda_1|$ and thus $|R[z^{(\nu)}]|$ is a *lower bound* for $|\lambda_1|$.

The Rayleigh quotient for the vector $z^{(\nu)}$ of a vector iteration with a hermitian matrix A supplies a better approximate value for the dominant eigenvalue λ_1 than the quotients $q_i^{(\nu)}$ in chapter 7.3, namely

$$R[z^{(\nu)}] = \lambda_1 + O\left(\left|\frac{\lambda_2}{\lambda_1}\right|^{2\nu}\right).$$

Thus the order of convergence is quadratic in contrast to linear convergence $O(|\lambda_2/\lambda_1|^\nu)$ for the quotients $q_i^{(\nu)}$.

Bibliography for Section 7.4

[ENGE87], 5.5; [FADD59], §61; [HOUS75], 3.2; [ISAA66], p.149; [PARL80], 4.6; [SCHW72], 4.3; [STOE91], 6.9; [WILK65], 3.54; [ZURM65], §10.3; [ZURM84], §13.6.

7.5 The Krylov Method

We consider the eigenvalue problem (7.2) with a diagonalizable real matrix A (for the case of non-diagonalizable matrices see [ZURM84], p.175), and want to find all eigenvalues and eigenvectors of A.

7.5.1 Determining the Eigenvalues

1^{st} **case:** Assume all eigenvalues λ_i of A are distinct:

The characteristic polynomial $P(\lambda)$ of A has the form

$$(7.13) \qquad P(\lambda) = \sum_{j=0}^{n-1} a_j \lambda^j + \lambda^n.$$

And we can determine its coefficients a_j from the linear system:

$$(7.14) \qquad Za + z^{(n)} = 0,$$

where

$$\begin{aligned} Z &= (z^{(0)}, z^{(1)}, \dots, z^{(n-1)}), \\ z^{(\nu)} &= Az^{(\nu-1)} \text{ for } \nu = 1, \dots, n, \text{ and} \\ a^T &= (a_0, a_1, \dots, a_{n-1}). \end{aligned}$$

Here $z^{(0)} = \sum_{i=1}^{n} c_i x_i$ is an arbitrary starting vector. There are two cases involving the c_i:

 a) If $c_i \neq 0$ for all $i = 1, \dots, n$, then $\det Z \neq 0$ and the system (7.14) is uniquely solvable: there are n linearly independent vectors $z^{(\nu)}$, $\nu = 0, \dots, n-1$.

b) If $c_i = 0$ for $i = q + 1, q + 2, ..., n$, where the c_i have been reordered *w.l.o.g*, and $q < n$ is minimal, then

$$z^{(0)} = c_1 x_1 + c_2 x_2 + \ldots + c_q x_q \quad \text{with} \quad c_i \neq 0 \quad \text{for} \quad i = 1, \ldots, q.$$

Here the $q + 1$ vectors $z^{(0)}$ and $z^{(\nu+1)} = A z^{(\nu)}$ for $\nu = 0, 1, \ldots, q - 1$ are linearly dependent. Thus $\det Z = 0$. However, the (n, q) matrix

$$Z_q = (z^{(0)}, \ z^{(1)}, \ldots, z^{(q-1)})$$

has rank q, so that for

$$b = (b_0, b_1, \ldots, b_{q-1})^T,$$

the inhomogeneous system with n equations for q unknown b_j, $j = 0, \ldots, q - 1$,

$$(7.15) \qquad\qquad\qquad Z_q b + z^{(q)} = 0$$

has q free variables. The b_j for $j = 0, \ldots, q - 1$, with $b_q = 1$, are the coefficients of a factor $P_q(\lambda)$ of $P(\lambda)$:

$$(7.16) \qquad\qquad\qquad P_q(\lambda) = \sum_{j=0}^{q} b_j \lambda^j.$$

By solving $P_q(\lambda) = 0$, one can determine q of the n eigenvalues λ_i of A. In order to find all λ_i, the same method must be carried out repeatedly for several linearly independent $z^{(0)}$, until case a) occurs or all n eigenvalues of A have been found.

2^{nd} case: A has multiple eigenvalues:

Let A have s distinct eigenvalues $\lambda_j, j = 1, \ldots, s$ with $s < n$, each of multiplicity p_j, so that $p_1 + p_2 + \ldots + p_s = n$. In this case, proceed as follows: First determine how many linearly independent iterated vectors $z^{(\nu+1)} = A z^{(\nu)}$, $\nu = 0, 1, 2, \ldots$, can be found for an arbitrary starting vector

$$(7.17) \qquad\qquad\qquad z^{(0)} = c_1 x_1 + c_2 x_2 + \ldots + c_s x_s,$$

where each x_r is an eigenvector of A corresponding to λ_r for $1 \leq r \leq s$. If

$$z^{(0)}, z^{(1)}, \ldots, z^{(s)}$$

are linearly independent vectors, then the linear system of n equations with $s < n$ unknown \hat{b}_j:

(7.18) $\begin{cases} \hat{Z}\hat{b} + z^{(s)} = 0 & \text{with} \quad \hat{b} = (\hat{b}_0, \hat{b}_1, \dots, \hat{b}_{s-1})^T, \\ \text{and} \quad \hat{Z} = (z^{(0)}, z^{(1)}, \dots, z^{(s-1)}) \end{cases}$

determines the coefficients \hat{b}_j of the *minimum polynomial* $m(\lambda)$ of \boldsymbol{A}:

(7.19) $$m(\lambda) = \sum_{j=0}^{s-1} \hat{b}_j \lambda^j + \lambda^s = \prod_{k=1}^{s} (\lambda - \lambda_k).$$

$m(\lambda)$ has the s distinct eigenvalues of \boldsymbol{A} as simple zeros. If some of the c_i in 7.17 are zero, one needs to proceed analogously to case 1.b). To solve the polynomial equation (7.19) one can use the methods of chapter 3.3.

7.5.2 Determining the Eigenvectors

1^{st} **case**: Assume all eigenvalues λ_i of \boldsymbol{A} are distinct:

The corresponding eigenvectors can be found as linear combinations of the iterated vectors $z^{(\nu)}$. We have

$$\boldsymbol{x}_i = \sum_{j=0}^{n-1} \tilde{a}_{ij} z^{(j)},$$

where the \tilde{a}_{ij} are the coefficients of the polynomial

$$P_i(\lambda) = \frac{P(\lambda)}{\lambda - \lambda_i} = \sum_{j=0}^{n-1} \tilde{a}_{ij} \lambda^j.$$

The \tilde{a}_{ij} can be easily determined with the simple Horner scheme of chapter 3.2.1.

2^{nd} **case.** There are multiple eigenvalues for \boldsymbol{A}:

The above procedure is also applicable in this case. However, for the starting vector $z^{(0)}$ one will only obtain one eigenvector, i.e., the multiplicity p_j of the eigenvalue λ_j is not taken into account. For this reason, one has to choose exactly p_j linearly independent starting initial vectors $z^{(0)}$, and in doing so, one can obtain p_j linearly independent eigenvectors for λ_j.

The Krylov method should only be employed if the linear systems (7.14), (7.15) and (7.18) are well conditioned. For otherwise inaccuracies in determining the coefficients in (7.13), (7.16) and (7.19) may lead to gross errors when determining the λ_j.

Bibliography for Section 7.5

[FADD59] §§42, 43; [GOLU89], 9.1; [GOOS76]; [HOUS75], 6.1; [PARL80], 12; [WILK65], 6.20; [ZURM84] §14.

7.6 Eigenvalues of Positive Definite Tridiagonal Matrices, the qd Algorithm

For positive definite, tridiagonal real matrices A (see section 4.10.2) of the form (4.14) with diagonal entries d_i, $i = 1, ..., n$ and off–diagonal entries f_i, $i = 1, ..., n-1$ and $f_i \neq 0$ for every i, the eigenvalues of A can be determined by means of the qd algorithm of Rutishauser. Here the letters 'qd' stand for 'quotient difference'. The qd scheme is filled in row by row and has the form:

ν	$e_0^{(\nu)}$	$q_1^{(\nu)}$	$e_1^{(\nu)}$	$q_2^{(\nu)}$	$e_2^{(\nu)}$	$q_3^{(\nu)}$	$e_3^{(\nu)}$	\cdots	$q_n^{(\nu)}$	$e_n^{(\nu)}$
		$q_1^{(1)}$		0		0		\cdots	0	
1	0		$e_1^{(1)}$		$e_2^{(1)}$		$e_3^{(1)}$	\cdots		0
		$q_1^{(2)}$		$q_2^{(2)}$		$q_3^{(2)}$		\cdots	$q_n^{(2)}$	
2	0		$e_1^{(2)}$		$e_2^{(2)}$		$e_3^{(2)}$	\cdots		0
		$q_1^{(3)}$		$q_2^{(3)}$		$q_3^{(3)}$		\cdots	$q_n^{(3)}$	
3	0		$e_1^{(3)}$		$e_2^{(3)}$		$e_3^{(3)}$	\cdots		0
\vdots	\vdots	\vdots	\vdots	\vdots	\vdots	\vdots	\vdots	\cdots	\vdots	\vdots
\downarrow	\downarrow	\downarrow	\downarrow	\downarrow	\downarrow	\downarrow	\downarrow		\downarrow	\downarrow
∞	0	λ_1	0	λ_2	0	λ_3	0	\cdots	λ_n	0

One starts the qd scheme with the specific values

$$q_1^{(1)} = d_1, \quad q_{k+1}^{(1)} = d_{k+1} - e_k^{(1)}, \quad e_k^{(1)} = c_k^2 / q_k^{(1)} \quad \text{for } k = 1, \ldots, n - 1$$

in the first two rows and sets $e_0^{(\nu)} = e_n^{(\nu)} = 0$. Then the subsequent rows of the scheme are computed according to the rule:

$$e_k^{(\nu+1)} = q_{k+1}^{(\nu)} e_k^{(\nu)} / q_k^{(\nu+1)}, \quad q_k^{(\nu+1)} = e_k^{(\nu)} + q_k^{(\nu)} - e_{k-1}^{(\nu+1)}.$$

For a fixed ν, one calculates sequentially:

$$q_1^{(\nu)}, e_1^{(\nu)}, q_2^{(\nu)}, e_2^{(\nu)}, \ldots, e_{n-1}^{(\nu)}, q_n^{(\nu)}.$$

Then the eigenvalues of A will appear, arranged by their magnitudes, as $\lim_{\nu\to\infty} q_k^{(\nu)} = \lambda_k$. Also $\lim_{\nu\to\infty} e_k^{(\nu)} = 0$. The matrix A as stipulated above will only have distinct positive eigenvalues λ_i, see [STOE91], 6.6, or [SCHW72], p. 139 and 168. More intricate qd algorithms involving shifts that are extremely precise and fast are given in [PARL95].

NOTE. A direct method that is well applicable to data processing is the Jacobi method in the form given by Neumann ([RALS67] vol. I, chap. 7); see also [FADD59], §81; [SCHW72], 4.4; [COLL73] I, p.56 ff.

Bibliography for Section 7.6

[PARL95]; [RALS67], 10.3; [SCHW72], 4.6; [STOE91], 6.6; [WILK65], p.564.

7.7 Transformation to Hessenberg Form, the LR and QR Algorithms

The LR and QR factorization methods of chapter 4 can be used repeatedly to calculate all eigenvalues of an (n, n) matrix A simultaneously. An LR factorization may not always exist, but the QR factorization can always be realized. Thus the LR *eigenvalue algorithm* may not be feasible for a given A, while the QR *algorithm* always is. However, the latter uses about twice the amount of computational time.

The computating time decreases significantly for both algorithms if the matrix A is first transformed to upper Hessenberg form by a similarity. The Hessenberg matrix then has the same eigenvalues as A. And according to [WILK65], ch. 6, this transformation barely worsens the conditioning of the eigenvalue problem and does not change the eigenvalues of A by much.

7.7.1 Transformation of a Matrix to Upper Hessenberg Form

Each real (n,n) matrix $\boldsymbol{A} = (a_{ik})$ can be transformed via similarity using Householder matrices (see theorem 4.9) into *upper Hessenberg form* $\hat{\boldsymbol{A}}$:

$$\hat{\boldsymbol{A}} = \begin{pmatrix} * & * & * & \cdots & * \\ * & * & * & \cdots & * \\ 0 & * & * & \cdots & * \\ \vdots & \ddots & \ddots & \ddots & \vdots \\ 0 & \cdots & 0 & * & * \end{pmatrix} = (\tilde{a}_{ik}),$$

where $\tilde{a}_{ik} = 0$ for $i \geq k + 2$.

If \boldsymbol{A} is symmetric, i.e., if $\boldsymbol{A}^T = \boldsymbol{A}$, then the corresponding Hessenberg matrix is tridiagonal and symmetric.

Realization of a Hessenberg form for \boldsymbol{A}:

The process is iterative and takes $n-2$ steps. At the outset, set $\boldsymbol{A}_1 = \boldsymbol{A}$. For the first similarity we choose

$$\boldsymbol{H}_1 := \begin{pmatrix} 1 & 0 & \cdots & 0 \\ 0 & & & \\ \vdots & & \tilde{\boldsymbol{H}}_1 & \\ 0 & & & \end{pmatrix}$$

with an $(n-1, n-1)$ Householder matrix $\tilde{\boldsymbol{H}}_1$, which is formed according to theorem 4.9 as:

$$\tilde{\boldsymbol{H}}_1 = \boldsymbol{I}_{n-1} - \frac{2}{\|\boldsymbol{v}_1\|^2} \boldsymbol{v}_1 \boldsymbol{v}_1^T,$$

where

$$\boldsymbol{v}_1 = \begin{pmatrix} a_{21}^{(1)} + \operatorname{sign}(a_{21}^{(1)}) \| \boldsymbol{a}_1^{(1)} \| \\ a_{31}^{(1)} \\ \vdots \\ a_{n1}^{(1)} \end{pmatrix}, \quad \boldsymbol{a}_1^{(1)} = \begin{pmatrix} a_{21}^{(1)} \\ a_{31}^{(1)} \\ \vdots \\ a_{n1}^{(1)} \end{pmatrix}.$$

Here both \boldsymbol{v}_1 and $\boldsymbol{a}_1^{(1)} \in \mathbb{R}^{n-1}$ and \boldsymbol{a}_1 is the first column of the matrix \boldsymbol{A}_1 except for the entry $a_{11}^{(1)}$, which is omitted. If

$$\boldsymbol{A}_1 = \begin{pmatrix} a_{11}^{(1)} & a_{12}^{(1)} & \cdots & a_{1n}^{(1)} \\ a_{21}^{(1)} & & & \\ \vdots & & \tilde{\boldsymbol{A}}_1 & \\ a_{n1}^{(1)} & & & \end{pmatrix},$$

then $A_2 := H_1 A_1 H_1$ is the matrix

$$
\begin{pmatrix}
a_{11}^{(1)} & * & \cdots & * \\
* & & & \\
0 & & \tilde{H}_1 \tilde{A}_1 \tilde{H}_1 & \\
\vdots & & & \\
0 & & &
\end{pmatrix}.
$$

The process is repeated for $\tilde{H}_1 \tilde{A}_1 \tilde{H}_1$ of size $n-1$ by $n-1$, and after $n-2$ such similarities, an upper Hessenberg form is obtained for $A_1 = A$. The following algorithm describes the process in detail (see [STOE91], 6.5.4, [MART68], or [NIEM87], 8.5):

ALGORITHM 7.5. (*Transformation to Hessenberg form*).

Given: A real (n,n) matrix $A_1 = (a_{ik}^{(1)}) := A$, $i,k = 1, \ldots, n$.

Task: Find an upper Hessenberg matrix similar to A via Householder similarities.

For each $i = 1, \ldots, n-2$ do:

1) Calculate the $(n-i, n-i)$ Householder matrix \tilde{H}_i as

$$
\tilde{H}_i = I_{n-i} - \frac{2}{\|v_i\|^2} v_i v_i^T,
$$

where

$$
v_i = \begin{pmatrix}
a_{i+1,i}^{(i)} + \text{sign}(a_{i+1,i}^{(i)}) \|a_i^{(i)}\| \\
a_{i+2,i}^{(i)} \\
\vdots \\
a_{n,i}^{(i)}
\end{pmatrix}, \quad
a_i^{(i)} = \begin{pmatrix}
a_{i+1,i}^{(i)} \\
a_{i+2,i}^{(i)} \\
\vdots \\
a_{n,i}^{(i)}
\end{pmatrix} \in \mathbb{R}^{n-i-1}.
$$

2) Form the (n,n) orthogonal matrix

$$
H_i = \begin{pmatrix}
I_i & 0 \\
0 & \tilde{H}_i
\end{pmatrix}
\begin{matrix} \} & i \\ \} & n-i \end{matrix}
$$

and compute $A_{i+1} := H_i A_i H_i = (a_{jk}^{(i+1)})$.

Finally A_{n-1} will be in upper Hessenberg form. The operations count is $5/3n^3 + O(n^2)$ according to [WERN82], if in the computations the matrices \tilde{H}_i are never explicitly formed, but instead the updated matrix A_{i+1} is formed directly from the vector v_i. See [STOE91], 6.5.4, or [GOLU89], ch. 7.4.3 for example.

7.7.2 The LR Algorithm

The *LR algorithm* of Rutishauser is the earliest iterative eigenvalue algorithm that uses a factorization of the given matrix \boldsymbol{A}.

ALGORITHM 7.6 (*LR algorithm*).

Given: An (n,n) matrix \boldsymbol{A}.

 Task: Find all eigenvalues λ_i of \boldsymbol{A}.

1^{st} step: Set $\boldsymbol{A}_1 := \boldsymbol{A}$.

2^{nd} step: For each $i = 1, 2, 3, \ldots$ do:

> 2.1 Factor $\boldsymbol{A}_i = \boldsymbol{L}_i \boldsymbol{R}_i$ without pivot search, if possible, for a unit lower triangular matrix \boldsymbol{L}_i and an upper triangular matrix \boldsymbol{R}_i (see chapter 4.2).

> 2.2 Set $\boldsymbol{A}_{i+1} = \boldsymbol{R}_i \boldsymbol{L}_i$. (Each of the matrices \boldsymbol{A}_i is similar to \boldsymbol{A}).

3^{rd} step: Under certain additional assumptions, see theorem 7.7:

$$\lim_{i\to\infty} \boldsymbol{A}_i = \lim_{i\to\infty} \boldsymbol{R}_i = \begin{pmatrix} \lambda_1 & \cdots & * \\ & \ddots & \vdots \\ 0 & & \lambda_n \end{pmatrix}, \text{ while } \lim_{i\to\infty} \boldsymbol{L}_i = \boldsymbol{I}.$$

THEOREM 7.7 (*Convergence of the LR algorithm*).
Let \boldsymbol{A} be an (n,n) matrix with eigenvalues

$$|\lambda_1| > |\lambda_2| > \ldots > |\lambda_n| > 0.$$

If every *LR* factorization in step 2.1 of algorithm 7.6 can be computed, then all matrices \boldsymbol{A}_i of the *LR* algorithm are similar to \boldsymbol{A} and

$$\lim_{i\to\infty} \boldsymbol{A}_i = \lim_{i\to\infty} \boldsymbol{R}_i = \begin{pmatrix} \lambda_1 & \cdots & * \\ & \ddots & \vdots \\ 0 & & \lambda_n \end{pmatrix}, \quad \text{while} \quad \lim_{i\to\infty} \boldsymbol{L}_i = \boldsymbol{I}.$$

The assumptions of the above theorem are in fact very demanding. The existence of an LR factorization of A itself is not even assured for a nonsingular matrix A as no row interchanges can be allowed in the factorization.

It is recommended to transform the matrix A first to upper Hessenberg form and then to apply the LR algorithm. Note that all A_i will then have upper Hessenberg form as well. The operations count for each LR factorization of a Hessenberg matrix A_i is thus reduced to $O(n^2)$ operations. It decreases even more if A is symmetric, because in this case the Hessenberg matrix will also be symmetric and thus tridiagonal.

There are variants of the LR algorithm as well, that deal with multiple eigenvalues and complex eigenvalues. Some convergence accelerating techniques such as shifts are known. The main drawback of this method, however, is the uncertainty whether each A_i that is formed from the original matrix A will allow an LR factorization. This is remedied in the QR algorithm of Francis.

7.7.3 The Basic QR Algorithm

The QR factorization of a given square matrix A by means of Householder transformations has been described in section 4.14. It is the basis for the QR *algorithm*. The QR factorization of A is unique if the (n, n) matrix A is nonsingular, except for the signs of the diagonal elements of the upper triangular matrix R, see [WERN82]. If one wants to employ the QR algorithm for more general eigenvalue problems, such as for complex matrices A, then Q must be computed over the complex numbers and chosen unitary, i.e., $Q^{-1} = \bar{Q}^T$. The QR algorithm proceeds similarly to the LR algorithm except that its LR factorizations are replaced by QR factorizations. Again it is recommended to transform the (n, n) matrix A to upper Hessenberg form before factoring. This will lower the operations count considerably.

ALGORITHM 7.8 (*Basic QR algorithm according to Francis*).

Given: An (n, n) matrix A.

 Task: Find all eigenvalues of A.

1^{st} Step: Set $A_1 := A$

2^{nd} Step: For each $i = 1, 2, 3, \ldots$ do:

> 2.1 Factor $A_i = Q_i R_i$ with a unitary matrix Q_i (i.e. $Q_i^{-1} = \bar{Q}_i^T$) and an upper triangular matrix R_i.
>
> 2.2 Form $A_{i+1} = R_i Q_i$.

Under certain additional assumption, such as $|\lambda_1| > |\lambda_2| > ... > |\lambda_n| > 0$, it can be proved that

$$\lim_{i \to \infty} \boldsymbol{A}_i = \begin{pmatrix} \lambda_1 & \cdots & * \\ & \ddots & \vdots \\ 0 & & \lambda_n \end{pmatrix}.$$

For convergence theorems and proofs for the QR algorithm see [NIEM87], 8.; [WERN82], IV.; [WILK65], ch. 8. A practical version of the algorithm following Francis was developed by Peters and Wilkinson, see section 7.8.

Bibliography for Section 7.7

[GOLU89], 7.4; [MAES84], 4.4; [MART70]; [NIEM87], 8.4, 8.5; [PETE 70]; [RALS67], 10.5; [SCHM76], 6; [STOE91], 6.6; [TÖRN79] vol.2; [WERN82]; [WILK 65].

7.8 Eigenvalues and Eigenvectors of a Matrix via the QR Algorithm

If the matrix $\boldsymbol{A} = (a_{ik})$ has no specific structure, one should first precondition it to a form which allows for easy computations of the eigenvalues and the eigenvectors. From the papers [MART68], [PARL69], [PETE70], this preconditioning of \boldsymbol{A} can be done in the following two steps:

1^{st} step. Balance the matrix \boldsymbol{A} in order to improve its condition via the procedure given by B.N. Parlett and C. Reinsch [PARL69].

2^{nd} step. Transform \boldsymbol{A} to upper Hessenberg form \boldsymbol{B}:

$$(7.20) \qquad \boldsymbol{B} = (b_{ik}) = \begin{pmatrix} b_{11} & b_{12} & \cdots & b_{1n} \\ b_{21} & b_{22} & \cdots & b_{2n} \\ & \ddots & \ddots & \vdots \\ 0 & & b_{nn-1} & b_{nn} \end{pmatrix},$$

i.e., $b_{ik} = 0$ for $i > k + 1$, by using the procedure of R.S. Martin and J.H. Wilkinson [MART68].

In the second step, one determines a non-singular matrix C with

(7.21) $$B = C^{-1}AC.$$

Similarity by C transforms the eigenvalue problem $Ax = \lambda x$ into the equivalent problem

(7.22) $$By = \lambda y \quad \text{where} \quad y = C^{-1}x.$$

This corresponds to a Gauß algorithm applied to a system (4.2). The known right hand side b in (4.2) is replaced here by the unknown vector λx.
From (7.21)

$$\det(B - \lambda I) = \det(A - \lambda I),$$

i.e., B and A possess the same eigenvalues λ_i.
Thanks to the Hessenberg form (7.20) of B, its eigenvalues λ_i can be more easily determined:

3^{rd} step. The eigenvalues λ_i are then determined via Francis' QR algorithm according to G. Peters and J.H. Wilkinson [PETE70]. Starting with $B_1 := B$, one constructs a sequence $\{B_s\}$, $s = 1, 2, 3, ...$, of upper Hessenberg matrices which converge towards an upper triangular matrix $R = (r_{ik})$. For convergence conditions see [WERN82], p.255. Thus in the limit we have for all $i : r_{ii} \doteq \lambda_i$.

With $B_1 := B$ the *iteration rule* for $s = 1, 2, 3, ...$ is:

(i) $B_s - k_s I = Q_s R_s$,

(ii) $B_{s+1} = R_s Q_s + k_s I.$

The rule (i) asks for a QR factorization of the shifted Hessenberg matrix $B_s - k_s I$ into a product of an orthogonal matrix Q_s with $Q_s^T = Q_s^{-1}$ and an upper triangular matrix R_s. The next iterate B_{s+1} is then formed according to rule (ii). B_s is replaced by B_{s+1} and the steps are repeated.
By an appropriate selection of the so-called *shift parameter* k_s one can accelerate convergence. If $k_s = 0$ for all s, we have the basic QR algorithm of section 7.7.3 .

4^{th} step. Determine the eigenvectors according to [PETE70]. According to (7.22)

$$By_i = \lambda_i y_i \quad \text{with} \quad x_i = Cy_i.$$

For each λ_i one can calculate the components y_{ik} of y_i for $k = n - 1, n - 2, ..., 1$, with an arbitrary y_{in} recursively. Then $x_i = Cy_i$ gives the required eigenvectors x_i for $i = 1, ..., n$.

5^{th} step. Normalize the eigenvectors x_i.

Thus far we have only described the LR and QR algorithms for real matrices A with real eigenvalues. If a real matrix A has complex eigenvalues, the iterates A_i of either algorithm cannot converge to a triangular matrix, but they will rather converge to a block triangular matrix with $(2,2)$ diagonal blocks for each pair of complex conjugate eigenvalues of A. This convergence can be observed in the 3^{rd} step and a *real double shift* can be made in such a way as to accelerate convergence of the (2.2) diagonal blocks. This refinement, also due to Francis, is very intricate and is incorporated into the eigenvalue program EIGEN in the appendix. Details can be found in [STOE91], ch.6.6.6, [GOLU89], ch. 7.5, or [WILK65], ch. 8.

Bibliography for Section 7.8

[ANDE92]; [GOLU89]; [MART68]; [PARL69]; [PETE70]; [STOE91]; [WERN82]; [WILK65].

7.9 Decision Strategy

Vector iteration can be employed to solve the eigenvalue-eigenvector problem $Ax = \lambda x$ if, in the case of a diagonalizable matrix A, one wants to determine only the eigenvalue of largest or smallest modulus with the associated eigenvector. If one wants to calculate all eigenvalues and eigenvectors of the matrix A it is recommended to transform the matrix A to upper Hessenberg form and then to use the QR algorithm of chapter 7.8.

Bibliography for Chapter 7

[ANDE92]; [ATKI89], 9; [GOLU89]; [HÄMM91], 3.; [HOUS75]; [MAES84], 4.; [MART68]; [NIEM87]; [PARL69]; [PARL80]; [PETE70]; [SCHW72]; [SCHW89], 6.; [STEW73]; [STOE91]; [WERN82]; [WILK65].

Chapter 8

Linear and Nonlinear Approximation

Let f be a continuous function on the interval $[a, b] \subset \mathbb{R}$ which is to be approximated by an *approximation function* $\Phi \in C[a, b]$. Φ shall be dependent on $x \in [a, b]$ and on certain parameters $c_0, c_1, ..., c_n$:

$$\Phi(x) := \Phi(x, c_0, c_1, \ldots, c_n) = \Phi(x, \boldsymbol{c}) \text{ for } \boldsymbol{c} = (c_0, c_1, \ldots, c_n)^T.$$

The parameters are to be determined in such a way that the distance between f and Φ as functions on $[a, b]$ is minimized in a manner still to be specified.

We distinguish between two kinds of problems:

1) The given function f is to be approximated by Φ, whose behavior is more suited for numerical purposes. The approximating function Φ might be more easily differentiable or integrable, or it might be possible to compute its functional values more easily. In this case, we speak of a *continuous approximation* of f over the interval $[a, b]$.

2) An empirical function f which is known at finitely many points $x_i \in [a, b]$ by $y_i = f(x_i)$ is to be approximated by a function Φ which is defined by a "formula". If f is known by its graph one can obtain nodes $x_i \in [a, b]$ with functional values $y_i = f(x_i)$. In this case, we speak of a *discrete approximation* of f at the nodes $x_i \in [a, b]$.

Depending on the rules for forming the approximating function Φ, one distinguishes between *linear* and *nonlinear approximations*. An approximation problem is called *linear* if the approximation function Φ has the form

$$\Phi(x, \boldsymbol{c}) = c_0 \varphi_0(x) + c_1 \varphi_1(x) + \ldots + c_n \varphi_n(x) = \sum_{k=0}^{n} c_k \varphi_k(x)$$

for known linearly independent functions $\varphi_k \in C[a,b]$. Otherwise the problem is called *nonlinear*.

EXAMPLE:
The function

$$\Phi_1(x, c_0, c_1, c_2, c_3) = c_0 + c_1 e^{2x} + c_2 \ln x + c_3(x^2 + 1)$$

is a linear approximation function. On the other hand we cannot represent the function

$$\Phi_2(x, c_0, c_1, c_2, c_3, c_4) = c_0 \cosh(c_1 x) + c_2 e^{-c_3(x-2c_4)^3}$$

in linear form $\Sigma c_k \varphi_k(x)$. Hence for Φ_2 we are dealing with a nonlinear approximation function or a *nonlinear model*.

In the following sections, we will treat linear continuous approximations and linear discrete approximations in the root-mean-square sense, which is often called the discrete L_2 approximation or the discrete Gaussian least squares method. Moreover we outline the linear uniform approximation by Chebyshev polynomials and nonlinear discrete approximation in the root-mean-squares sense, also called nonlinear discrete L_2-approximation or the nonlinear discrete least squares method. Section 8.1.6 will deal with errors inherent in linear approximation and can be used as a guideline for improved node selection.

When working with the linear as well as the nonlinear least squares methods, we prefer to use Householder transformations, see chapter 4.14, in order to avoid a worsening of the condition number while solving normal equations. By doing so, we can hope for a better accuracy of the solution.

8.1 Linear Approximation

8.1.1 Statement of the Problem and Best Approximation

To each function $f \in C[a,b]$ we assign a real non-negative number $\|f\|$, called the *norm* of f, which satisfies the following *norm axioms*:

$$(8.1) \quad \begin{cases} 1) & \|f\| \geq 0, \\ 2) & \|f\| = 0 \quad \text{if and only if } f = 0 \text{ everywhere in } [a,b], \\ & \qquad\qquad\qquad \text{(positive definite)}, \\ 3) & \|\alpha f\| = |\alpha|\,\|f\| \quad \text{for all } \alpha \in R, \quad \text{(homogeneous)}, \\ 4) & \|f+g\| \leq \|f\| + \|g\| \quad \text{for all } f,g \in C[a,b], \\ & \qquad\qquad\qquad \text{(triangle inequality)}. \end{cases}$$

For two functions $f_1, f_2 \in C[a,b]$ we define their *distance* as

$$\varrho(f_1, f_2) := \|f_1 - f_2\|.$$

This satisfies the following *distance axioms*:

1) $\varrho(f_1, f_2) \geq 0,$

2) $\varrho(f_1, f_2) = 0$ if and only if $f_1(x) = f_2(x)$ for all $x \in [a,b],$

3) $\varrho(f_1, f_2) = \varrho(f_2, f_1),$ and

4) $\varrho(f_1, f_3) \leq \varrho(f_1, f_2) + \varrho(f_2, f_3)$ for all $f_1, f_2, f_3 \in C[a,b].$

In general we shall try to approximate a given function f using a set of $n+1$ linearly independent functions $\varphi_0, \varphi_1, \ldots, \varphi_n \in C[a,b]$. Functions $\varphi_0, \varphi_1, \ldots, \varphi_n \in C[a,b]$ are called *linearly dependent* if there are real numbers c_0, c_1, \ldots, c_n, not all zero, so that for every $x \in [a,b]$

$$c_0 \varphi_0(x) + c_1 \varphi_1(x) + \ldots + c_n \varphi_n(x) = 0.$$

Otherwise $\varphi_0, \varphi_1, \ldots, \varphi_n$ are called *linearly independent*. Given a set of linearly independent functions φ_k, $k = 0, \ldots, n$, then all linear combinations

$$(8.2) \quad \begin{cases} \Phi(x) := \Phi(x; c_0, c_1, \ldots, c_n) = \Phi(x, \boldsymbol{c}) = \sum_{k=0}^{n} c_k \varphi_k(x), \text{ defined for} \\ x \in [a,b], \text{ and } \boldsymbol{c} = (c_0, c_1, \ldots, c_n)^T \text{ with } c_k \in \mathbb{R} \end{cases}$$

are the admissable approximation functions. Any such Φ is called an *approximation function* or a *model function*; an approximation of the form (8.2) is called a *linear approximation*.

Let \bar{C} be the set of all Φ that can be formed via (8.2) from the given linearly independent set $\{\varphi_0, \ldots, \varphi_n\} \subseteq C[a,b]$. Each linear combination $\Phi \in \bar{C}$ is uniquely determined by the $(n+1)$ tuple $(c_0, c_1, ..., c_n)$ of its coefficients. The

distance between Φ and the given function $f \in C[a, b]$ depends only on Φ. It is

$$(8.3) \qquad \varrho(f, \Phi) = \|f - \Phi\| =: D(c_0, c_1, \ldots, c_n).$$

Specific sets of functions $\{\varphi_0, \varphi_1, \ldots, \varphi_n\}$ that are often used, are for example:

1) $\varphi_0 = 1$, $\varphi_1 = x$, $\varphi_2 = x^2, \ldots,$ $\varphi_n = x^n$. Here the approximating functions Φ are algebraic polynomials of degree n at most.

2) $\varphi_0 = 1$, $\varphi_1 = \cos x$, $\varphi_2 = \sin x$, $\varphi_3 = \cos 2x$, $\varphi_4 = \sin 2x, \ldots$. Here the approximation functions Φ are trigonometric polynomials.

3) $\varphi_0 = 1$, $\varphi_1 = e^{\alpha_1 x}$, $\varphi_2 = e^{\alpha_2 x}, \ldots,$ $\varphi_n = e^{\alpha_n x}$ for distinct real numbers α_i.

4) $\varphi_0 = 1$, $\varphi_1 = \frac{1}{(x-\alpha_1)^{p_1}}$, $\varphi_2 = \frac{1}{(x-\alpha_2)^{p_2}}, \ldots,$ $\varphi_n = \frac{1}{(x-\alpha_n)^{p_n}}$ for $\alpha_i \in \mathbb{R}$ and $p_i \in \mathbb{N}$.
 Several values of α_j or p_j can be equal here, but then the associated values of p_j or α_j must be different. Here the approximation functions Φ are specific rational functions, see also remark 8.2 below.

5) Orthogonal functions, see the particular cases in sections 8.1.2 and 8.1.3.

One criterion for linear independence of a set of functions $\varphi_0, \varphi_1, \ldots, \varphi_n \in C^n[a, b]$ is the *Wronskian determinant*: If

$$W(\varphi_0, \varphi_1, \ldots, \varphi_n) = \det \begin{pmatrix} \varphi_0 & \varphi_1 & \cdots & \varphi_n \\ \varphi_0' & \varphi_1' & \cdots & \varphi_n' \\ \vdots & \vdots & & \vdots \\ \varphi_0^{(n)} & \varphi_1^{(n)} & \cdots & \varphi_n^{(n)} \end{pmatrix} \not\equiv 0$$

does not vanish uniformly for all $x \in [a, b]$, then the family of functions $\{\varphi_0, \ldots \varphi_n\}$ is linearly independent, provided the φ_i are sufficiently often differentiable.

Statement of the approximation problem:

For a given function $f \in C[a, b]$ and a specified linearly independent set of functions $\varphi_0, \varphi_1, \ldots, \varphi_n \in C[a, b]$ find a function $\Phi^{(0)} \in \bar{C}$ of the form

$$(8.4) \qquad \Phi^{(0)}(x) := \Phi^{(0)}(x, c_0^{(0)}, c_1^{(0)}, \ldots, c_n^{(0)}) = \sum_{k=0}^{n} c_k^{(0)} \varphi_k(x) \in \bar{C},$$

so that

$$(8.5) \quad \begin{cases} D(c_0^{(0)}, c_1^{(0)}, \dots, c_n^{(0)}) = \|f - \Phi^{(0)}\| &= \min_{\Phi \in \bar{C}} \|f - \Phi\| \\ &= \min_{\Phi \in \bar{C}} D(c_0, c_1, \dots, c_n) \end{cases}$$

is minimal. Such a $\Phi^{(0)}$ is called a *best approximation* of f for the given set of functions $\varphi_0, \varphi_1, \dots, \varphi_n$ in the sense of the chosen norm $\|..\|$.

PROPOSITION 8.1 (*Existence theorem*).
For each system of linearly independent functions $\varphi_0, \varphi_1, \dots, \varphi_n \in C[a, b]$ and for each norm $\|..\|$, there is at least one best approximation $\Phi^{(0)}$ for the form (8.4) satisfying (8.5) for each function $f \in C[a, b]$.

The set of approximation functions $\varphi_0, \varphi_1, \dots, \varphi_n$ should be chosen with some consideration for the actual problem. It is useful, for instance, to employ trigonometric polynomials as model functions to determine the best approximation of a 2π-periodic function f.

REMARK 8.2 (*Rational approximation*).
For certain problems - for instance if it is known that f possesses poles outside of $[a, b]$ - a function of the form

$$(8.6) \qquad \Psi(x) = \frac{\displaystyle\sum_{k=0}^{m} a_k \varphi_k(x)}{\displaystyle\sum_{k=0}^{j} b_k \varphi_k(x)} \quad \text{with } \varphi_k \in C[a, b]$$

may be chosen as approximation function. If $\varphi_k(x) = x^k$ for all k, this defines a rational function whose numerator has degree m while the denominator has degree j. If w.l.o.g. $a_0 = 1$, then among all functions $\Psi \in \bar{\bar{C}}$ of the form (8.6), a best approximation $\Psi^{(0)}$ is to be determined with the property

$$D(a_1^{(0)}, a_2^{(0)}, \dots, a_m^{(0)}, b_0^{(0)}, b_1^{(0)}, \dots, b_j^{(0)}) = \|f - \Psi^{(0)}\| =$$

$$= \min_{\Psi \in \bar{\bar{C}}} \|f - \Psi\| = \min_{\Psi \in \bar{\bar{C}}} D(a_1, a_2, \dots, a_m, b_0, b_1, \dots, b_j),$$

see also sections 8.2 and 9.7.

8.1.2 Linear Continuous Root-Mean-Square Approximation

For approximation in the root-mean-square sense, we choose the L_2 *norm* for $C[a,b]$:

$$(8.7) \qquad \|g\|_2 := \left(\int\limits_a^b w(x) g^2(x) dx \right)^{\frac{1}{2}} \quad \text{for } g \in C[a,b].$$

Here $w(x) \in C[a,b]$, $w(x) > 0$ and assumed integrable on $[a,b]$. w is called a *weight function*. Set $g = f - \Phi$ and consider the square of the distance (8.3). Then condition (8.5) becomes

$$(8.8) \quad \|f - \Phi^{(0)}\|_2^2 = \min_{\Phi \in \bar{C}} \|f - \Phi\|_2^2 \;=\; \min_{\Phi \in \bar{C}} \int\limits_a^b w(x)(f(x) - \Phi(x))^2 dx$$

$$= \min_{\Phi \in \bar{C}} D^2(c_0, c_1, \ldots, c_n).$$

Thus the integral of the weighted least squares must be minimized. The necessary conditions $\partial(D^2)/\partial c_j = 0$ for $j = 0, \ldots, n$ supply $n+1$ linear equations that determine the $n+1$ coefficients $c_k^{(0)}$ of the best approximation (8.4):

$$(8.9) \; \sum_{k=0}^{n} c_k^{(0)} \int\limits_a^b w(x)\varphi_j(x)\varphi_k(x) dx = \int\limits_a^b w(x)f(x)\varphi_j(x) dx, \quad j = 0, \ldots, n.$$

Rewritten in matrix form this becomes

$$(8.10) \qquad\qquad G c^{(0)} = a \quad \text{with}$$

$$G := \begin{pmatrix} (\varphi_0, \varphi_0) & (\varphi_0, \varphi_1) & \cdots & (\varphi_0, \varphi_n) \\ (\varphi_1, \varphi_0) & (\varphi_1, \varphi_1) & \cdots & (\varphi_1, \varphi_n) \\ \vdots & \vdots & & \vdots \\ (\varphi_n, \varphi_0) & (\varphi_n, \varphi_1) & \cdots & (\varphi_n, \varphi_n) \end{pmatrix},$$

$$c^{(0)} = \begin{pmatrix} c_0^{(0)} \\ c_1^{(0)} \\ \vdots \\ c_n^{(0)} \end{pmatrix}, \text{ and } a = \begin{pmatrix} (f, \varphi_0) \\ (f, \varphi_1) \\ \vdots \\ (f, \varphi_n) \end{pmatrix}.$$

Here we have used the *weighted scalar products*:

$$(\varphi_k, \varphi_j) := \int\limits_a^b w(x)\varphi_k(x)\varphi_j(x)dx, \quad (f, \varphi_j) := \int\limits_a^b w(x)f(x)\varphi_j(x)dx.$$

The equations (8.9) are called *Gaussian normal equations*.

Since $(\varphi_j, \varphi_k) = (\varphi_k, \varphi_j)$, the matrix \boldsymbol{G} is symmetric, or $\boldsymbol{G} = \boldsymbol{G}^T$. The determinant $\det \boldsymbol{G}$ of the system matrix in (8.10) is called the *Gramian* of the family of functions $\varphi_0, \varphi_1, ..., \varphi_n$.

LEMMA 8.3 A set of function $\varphi_0, \varphi_1, ..., \varphi_n \in C[a, b]$ is linearly dependent if and only if its Gramian vanishes for all $x \in [a, b]$ (see [STUM82], p.133).

THEOREM 8.4 For every function $f \in C[a, b]$ and for each set of linearly independent functions $\varphi_0, \varphi_1, ..., \varphi_n \in C[a, b]$ there is exactly one best approximation $\Phi^{(0)}$ of the form (8.4) satisfying (8.8) for the norm (8.7). Its coefficients $c_k^{(0)}$ are determined by (8.9).

Special cases.
1) *Algebraic polynomials.*
Here the approximation functions Φ are algebraic polynomials of degree n at most:

$$(8.11) \qquad \Phi(x, \boldsymbol{c}) = \sum_{k=0}^n c_k x^k.$$

The normal equations (8.9) that determine $c_k^{(0)}$ are

$$(8.12) \qquad \sum_{k=0}^n c_k^{(0)} \int\limits_a^b w(x)x^{j+k}dx = \int\limits_a^b w(x)f(x)x^j dx \quad \text{for } j = 0, \dots, n.$$

With $w(x) \equiv 1$ (see remark 8.5), (8.10) has the form

$$(8.13) \qquad\qquad\qquad \boldsymbol{Gc} = \boldsymbol{a} \quad \text{with}$$

$$
G = \begin{pmatrix}
\int\limits_a^b dx & \int\limits_a^b x\,dx & \int\limits_a^b x^2\,dx & \cdots & \int\limits_a^b x^n\,dx \\
\int\limits_a^b x\,dx & \int\limits_a^b x^2\,dx & \int\limits_a^b x^3\,dx & \cdots & \int\limits_a^b x^{n+1}\,dx \\
\int\limits_a^b x^2\,dx & \int\limits_a^b x^3\,dx & \int\limits_a^b x^4\,dx & \cdots & \int\limits_a^b x^{n+2}\,dx \\
\vdots & \vdots & \vdots & & \vdots \\
\int\limits_a^b x^n\,dx & \int\limits_a^b x^{n+1}\,dx & \int\limits_a^b x^{n+2}\,dx & \cdots & \int\limits_a^b x^{2n}\,dx
\end{pmatrix},
$$

$$
c = \begin{pmatrix} c_0^{(0)} \\ c_1^{(0)} \\ c_2^{(0)} \\ \vdots \\ c_n^{(0)} \end{pmatrix}, \text{ and } a = \begin{pmatrix}
\int\limits_a^b f(x)\,dx \\
\int\limits_a^b f(x)x\,dx \\
\int\limits_a^b f(x)x^2\,dx \\
\vdots \\
\int\limits_a^b f(x)x^n\,dx
\end{pmatrix}.
$$

If $[a, b] = [0, 1]$, then the matrix G is the $(n + 1, n + 1)$ *Hilbert matrix* $H_n = (\frac{1}{i+j+1})_{i,j=0}^n$. This matrix is extremely ill conditioned, $cond H_n \approx e^{3.5n}$. This indicates that the linearly independent family of polynomials $\{x^k\}$ is very close to being linearly dependent. All functions x^k have very much the same shape on $[0, 1]$. This difficulty shows a fundamental problem with using the normal equations to find a best approximation numerically. It can be remedied by using orthogonal sets of approximating functions or by using Householder transformations, see chapter 8.1.3.4.

2) *Family of orthogonal functions.*

A set of functions φ_k forms an orthogonal system, if for the given weight function w

$$
(\varphi_j, \varphi_k) = \int\limits_a^b w(x)\varphi_j(x)\varphi_k(x)\,dx = 0 \quad \text{for} \quad j \neq k.
$$

In this case, (8.9) has the rather simple form:

(8.14) $\qquad\qquad (\varphi_j, \varphi_j)c_j^{(0)} = (f, \varphi_j) \text{ for } j = 0, \ldots, n.$

If the number of functions in \bar{C} is increased from n to $n + 1$, then the $c_j^{(0)}$ remain unchanged for $j = 0, \ldots, n$ in contrast to non-orthogonal systems of

function, and $c_{n+1}^{(0)}$ can be calculated from (8.14) by simply setting $j = n + 1$.

For discrete least squares approximation using orthogonal polynomials see section 8.1.3.2.

Examples of orthogonal families of functions.

a) $\varphi_k(x) = \cos kx$ for $x \in [0, 2\pi]$ and $k = 0, \ldots, n$ with $w(x) = 1$;

b) $\varphi_k(x) = \sin kx$ for $x \in [0, 2\pi]$ and $k = 1, \ldots, n$ with $w(x) = 1$;

c) *Legendre polynomials*

These are defined for $x \in [-1, +1]$ by

$$P_{k+1}(x) = \frac{1}{k+1}\left((2k+1)xP_k(x) - kP_{k-1}(x)\right) \text{ for } k = 1, 2, 3, \ldots$$

with $P_0(x) = 1$, $P_1(x) = x$, and $w(x) = 1$.

d) *Chebyshev polynomials*

These are defined for $x \in [-1, +1]$ by

$$T_{k+1}(x) = 2xT_k(x) - T_{k-1}(x) \text{ for } k = 1, 2, 3, \ldots$$

with $T_0(x) = 1$, $T_1(x) = x$, and $w(x) = 1/\sqrt{1 - x^2}$ (see section 8.1.4.2).

e) *Orthogonalization method of E. Schmidt:*

Let $\varphi_0, \varphi_1, \ldots, \varphi_n \in C[a, b]$ be $n + 1$ linearly independent functions. Then one can construct an orthogonal system of functions $\tilde{\varphi}_0, \tilde{\varphi}_1, \ldots, \tilde{\varphi}_n \in C[a, b]$ from the given $\varphi_0, \ldots, \varphi_n$. To achieve this, one forms linear combinations

$$\tilde{\varphi}_k = \varphi_k - a_{k0}\tilde{\varphi}_0 - a_{k1}\tilde{\varphi}_1 - \ldots - a_{k,k-1}\tilde{\varphi}_{k-1} \text{ for } k = 0, \ldots, n,$$

and determines the coefficients a_{kj} successively so that the orthogonality relations $(\tilde{\varphi}_j, \tilde{\varphi}_k) = 0$ are satisfied for $j \neq k$. Indeed setting

$$a_{kj} = (\varphi_k, \tilde{\varphi}_j)/(\tilde{\varphi}_j, \tilde{\varphi}_j) \text{ for } k = 0, \ldots, n \text{ and } j = 0, \ldots, k - 1$$

will achieve orthogonality. For $\varphi_k(x) = x^k$ and $x \in [-1, +1]$, this process generates the Legendre polynomials.

REMARK 8.5 In many cases, one can choose $w(x) = 1$ as a weight function for all $x \in [a, b]$. For some problems, however, other weight functions

are better suited. If for $w(x) = 1$ one obtains a best approximation $\Phi^{(0)}$, for which $(f(x) - \Phi^{(0)}(x))^2$ becomes large near an endpoint $x = a$ or $x = b$, one should choose a weight $\tilde{w}(x)$, which is large for $x \to a$ and $x \to b$. Then a best approximation $\tilde{\Phi}^{(0)}$, obtained with this weight function $\tilde{w}(x)$, will make $(f(x) - \tilde{\Phi}^{(0)}(x))^2$ small for $x \to a$ and $x \to b$. $\tilde{w}(x) = 1/\sqrt{1-x^2}$ is such a weight function for $a = -1$ and $b = +1$.

ALGORITHM 8.6 (*Continuous Gaussian least squares method*).

For a given function $f \in C[a, b]$ find a best approximation $\Phi^{(0)}$ by the continuous Gaussian least squares method.

1^{st} step: Choose a suitable system of linearly independent functions $\varphi_0, \varphi_1, \ldots, \varphi_n$.

2^{nd} step: Choose a suitable weight function $w(x) > 0$; see remark 8.5.

3^{rd} step: Solve the linear system (8.9) or (8.10) for the coefficients $c_k^{(0)}$ of the best approximation (8.4). If the chosen approximation functions are algebraic polynomials, then the system (8.12) or (8.13) must be solved. If the φ_k form an orthogonal system, solve (8.14).

8.1.3 Discrete Linear Root-Mean-Square Approximation

8.1.3.1 Normal Equations for Discrete Linear Least Squares

Here we are looking for a best approximation $\Phi^{(0)}$ of the form (8.4) for a given function $f \in C[a, b]$ which is known at $N + 1$ distinct points $x_i \in [a, b]$, $i = 0, \ldots, N$ with $N \geq n$ by its functional values $y_i = f(x_i)$. For any function $g \in C[a, b]$ we define the *semi-norm*

$$(8.15) \qquad \|g\|_2 := \left(\sum_{i=0}^{N} w_i g^2(x_i) \right)^{\frac{1}{2}}$$

with real numbers $w_i > 0$ as weights. Semi-norms fulfill all of the function norm axioms (8.1) with the exception of 2).

Set $g = f - \Phi$ and consider the square of the distance (8.3). Then condition (8.5) for a best approximation in terms of the semi-norm becomes

$$(8.16) \; \|f - \Phi^{(0)}\|_2^2 \;=\; \min_{\Phi \in \bar{C}} \|f - \Phi\|_2^2 = \min_{\Phi \in \bar{C}} \sum_{i=0}^{N} w_i \underbrace{(f(x_i) - \Phi(x_i, c))^2}_{\delta_i}$$

$$= \; \min_{\Phi \in \bar{C}} D^2(c_0, c_1, \ldots, c_n).$$

Thus the sum of the weighted least squares must be minimized.

Figure 8.1: Discrete least squares method

The conditions $\partial(D)^2/\partial c_j = 0$ for $j = 0, \ldots, n$ again supply $n + 1$ linear equations for the $n + 1$ coefficients $c_k^{(0)}$ of an optimal approximation. Since

$$\Phi(x_i, c) = \Phi(x_i, c_0, \ldots, c_n) = \sum_{k=0}^{n} c_k \varphi_k(x_i) \text{ and } \frac{\partial \Phi(x_i)}{\partial c_j} = \varphi_j(x_i),$$

we obtain the linear system or the *Gaussian normal equations*:

$$(8.17) \; \sum_{k=0}^{n} c_k^{(0)} \sum_{i=0}^{N} w_i \varphi_j(x_i) \varphi_k(x_i) = \sum_{i=0}^{N} w_i f(x_i) \varphi_j(x_i), \quad j = 0, \ldots, n, N \geq n.$$

Using the weighted scalar products

$$(8.18) \; \begin{cases} (\varphi_j, \varphi_k) & := \; \sum_{i=0}^{N} w_i \varphi_j(x_i) \varphi_k(x_i), \\ (f, \varphi_j) & := \; \sum_{i=0}^{N} w_i f(x_i) \varphi_j(x_i), \end{cases}$$

this is equivalent to the linear system $Gc = a$ with

$$(8.19) \; G = \begin{pmatrix} (\varphi_0, \varphi_0) & \cdots & (\varphi_0, \varphi_n) \\ \vdots & & \vdots \\ (\varphi_n, \varphi_0) & \cdots & (\varphi_n, \varphi_n) \end{pmatrix}, c = \begin{pmatrix} c_0^{(0)} \\ \vdots \\ c_n^{(0)} \end{pmatrix}, a = \begin{pmatrix} (f, \varphi_0) \\ \vdots \\ (f, \varphi_n) \end{pmatrix}.$$

Here again $G = G^T$ since $(\varphi_j, \varphi_k) = (\varphi_k, \varphi_j)$.

The normal equations (8.19) are, however, often very ill conditioned. Calculating their solution can unduly increase rounding errors and thereby lead to useless results. As an alternative we suggest using Householder transformations, see section 8.1.3.4.

DEFINITION 8.7 For given weights $w_i > 0$ we can assign to each approximating function $\varphi_k \in C[a, b]$ the vector

$$\boldsymbol{\varphi}_k := (\sqrt{w_0}\varphi_k(x_0), \sqrt{w_1}\varphi_k(x_1), \ldots, \sqrt{w_N}\varphi_k(x_N))^T \in \mathbb{R}^{n+1}.$$

With this definition the scalar products in (8.18) become

(8.18') $$(\varphi_j, \varphi_k) = \boldsymbol{\varphi}_j^T \boldsymbol{\varphi}_k, \text{ and } (f, \varphi_j) = \boldsymbol{f}^T \boldsymbol{\varphi}_j,$$

where $\boldsymbol{f} := (\sqrt{w_0}f(x_0), \ldots, \sqrt{w_N}f(x_N))^T$. Clearly if $w_i = 1$ for all i, then $\boldsymbol{\varphi}_k = (\varphi_k(x_0), ..., \varphi_k(x_N))^T$ and $\boldsymbol{f} = (f(x_0), ..., f(x_N))^T$.

THEOREM 8.8 The normal equations (8.17) are uniquely solvable if and only if the vectors $\boldsymbol{\varphi}_k$ are linearly independent. The $\boldsymbol{\varphi}_k, k = 0, \ldots, n$, cannot be linearly independent if $n \geq N$.

If $n > N$, the vectors $\boldsymbol{\varphi}_k$ must be linearly dependent. If $n = N$, we speak of an *interpolation problem*, see chapter 9.

REMARK 8.9 In most cases, the weights $w_i = 1$ are used. Different choices should be made if one knows that the functional values $f(x_i)$ have a different degree of accuracy for different x_i. If this is the case, one should generally assign smaller weights to the less precisely known functional values. If one normalizes the weights w_i so that $w_0 + w_1 + ... + w_N = 1$, one can interpret them as probabilities for the value $f(x_i)$ at the point x_i. Alternatively in (8.16) one can use $w_i = 1/f^2(x_i)$ as weights. This would correspond to minimizing the sum of the squares of the relative error terms:

$$\sum_{i=0}^N \left(\frac{f(x_i) - \Phi(x_i)}{f(x_i)}\right)^2 \stackrel{!}{=} \text{ Minimum.}$$

ALGORITHM 8.10 (*Discrete Gaussian least squares method*).

For a function $f \in C[a, b]$ let the values $f(x_i)$ be given at $N + 1$ distinct points $x_i \in [a, b]$ for $i = 0, \ldots, N$. Find the best approximation $\Phi^{(0)}$ of f in the discrete Gaussian least squares sense.

1^{st} step: Choose a suitable family of functions $\varphi_0, \varphi_1, \ldots, \varphi_n$ for forming the approximation functions (8.2) with $n \leq N$.

2^{nd} step: Determine suitable weights $w_i > 0$, see remark 8.9.

3^{rd} step: Form the normal equations (8.17) or (8.19) with the definitions (8.18) or (8.18′) and calculate the coefficients $c_k^{(0)}$ of the best approximation $\Phi^{(0)}$.

The requirement that all x_i are distinct can be ignored if at least $N' + 1$ points x_i are distinct, where $n \leq N' \leq N$.

REMARK 8.11 Only if the model functions used to form $\Phi \in \bar{C}$ are reasonably chosen for the problem, can we expect to find a good approximation by the least squares method. If one has no idea what functions would fit the problem, one should use least squares splines, see chapters 10 and 11. If there are several sets of suitable model functions then one should choose the best approximations $\Phi^{(0)}$ for which the least squares sum is minimized.

8.1.3.2 Discrete Least Squares via Algebraic Polynomials and Orthogonal Polynomials

If one uses algebraic polynomials (8.11) as approximation functions Φ, then with $\varphi_k(x_i) = x_i^k$ the normal equations (8.17) become

$$(8.20) \qquad \sum_{k=0}^{n} c_k^{(0)} \sum_{i=0}^{N} w_i x_i^{k+j} = \sum_{i=0}^{N} w_i f(x_i) x_i^j \text{ for } j = 0, \ldots, n.$$

For identical weights $w_i = 1$ this translates into the linear system

$$(8.21) \quad \begin{pmatrix} N+1 & \sum x_i & \sum x_i^2 & \cdots \sum x_i^n \\ \sum x_i & \sum x_i^2 & \sum x_i^3 & \cdots \sum x_i^{n+1} \\ \sum x_i^2 & \sum x_i^3 & \sum x_i^4 & \cdots \sum x_i^{n+2} \\ \vdots & \vdots & \vdots & \vdots \\ \sum x_i^n & \sum x_i^{n+1} & \sum x_i^{n+2} & \cdots \sum x_i^{2n} \end{pmatrix} \begin{pmatrix} c_0^{(0)} \\ c_1^{(0)} \\ c_2^{(0)} \\ \vdots \\ c_n^{(0)} \end{pmatrix} = \begin{pmatrix} \sum f(x_i) \\ \sum f(x_i)x_i \\ \sum f(x_i)x_i^2 \\ \vdots \\ \sum f(x_i)x_i^n \end{pmatrix},$$

where each sum extends over $i = 0, \ldots, N$. The system matrix in (8.20) or (8.21) is often very ill conditioned. In this case, the coefficients $c_i^{(0)}$ cannot be computed reliably. Thus one should not try to solve the system (8.21), but should work with discrete orthogonal polynomials $\varphi_k \equiv Q_k$ as approximation functions Φ instead:

$$\Phi : \quad \Phi(x) \quad = \quad \sum_{k=0}^{n} c_k Q_k(x)$$

$$= \quad c_0 Q_0(x) + c_1 Q_1(x) + \ldots + c_n Q_n(x),$$

where $(\varphi_j, \varphi_k) \equiv (Q_j, Q_k) = 0$ for $j \neq k$. Then the system matrix of the Gaussian normal equations (8.19) is diagonal, and one can read off the solution $c_j^{(0)}$ for $j = 0, \ldots, n$:

$$c_j^{(0)} = \frac{(f, Q_j)}{(Q_j, Q_j)} \quad \text{for } j = 0, \ldots, n.$$

Here the weighted scalar products are defined as

$$(f, Q_j) \quad := \quad \sum_{i=0}^{N} w_i f(x_i) Q_j(x_i), \text{ and}$$

$$(Q_j, Q_j) \quad := \quad \sum_{i=0}^{N} w_i Q_j^2(x_i).$$

The orthogonal polynomials Q_k can be calculated recursively as follows:

$$Q_0(x) \quad = \quad 1,$$
$$Q_1(x) \quad = \quad x - b_1,$$
$$Q_k(x) \quad = \quad (x - b_k) Q_{k-1}(x) - d_k Q_{k-2}(x), \text{ for } k \geq 2 \quad \text{with :}$$

$$b_k \quad = \quad \frac{(x Q_{k-1}, Q_{k-1})}{(Q_{k-1}, Q_{k-1})}, \quad \text{if } k \geq 1, \text{ and}$$

$$d_k \quad = \quad \frac{(Q_{k-1}, Q_{k-1})}{(Q_{k-2}, Q_{k-2})}, \quad \text{if } k \geq 2.$$

The scalar product in the numerator of b_k is computed as

$$(x Q_{k-1}, Q_{k-1}) := \sum_{i=0}^{N} w_i x_i Q_{k-1}^2(x_i).$$

The other scalar products are of the form (Q_j, Q_j). From b_k, d_k and c_k, one can calculate the least squares polynomial Φ at every point $x \in [a, b]$ by using a Horner scheme, see chapter 3.2, as follows:

1) $s_n = c_n$,
$s_{n-1} = c_{n-1} + s_n(x - b_n)$.

2) For each $k = n - 2, n - 3, ..., 0$ do:
$s_k = c_k + s_{k+1}(x - b_{k+1}) - s_{k+2}d_{k+2}$.
Then $\Phi(x) = s_0$.

One advantage of such an approach via orthogonal polynomials is that once one knows the polynomial $\Phi_n(x)$ of degree n, one also knows every least squares polynomial Φ_m of degree $m \leq n$, in fact

$$\Phi_m(x) = \sum_{k=0}^{m} c_k Q_k(x).$$

I.e., for every $m \leq n$, Φ_m has the same coefficients c_k for $k \leq m$ as Φ_n.

8.1.3.3 Linear Regression, the Least Squares Solution Using Linear Algebraic Polynomials

As an example for discrete least squares using linear algebraic polynomials we shall study *linear regression*:

Let $N + 1$ nodes $(x_i, y_i) \in \mathbb{R}^2$ be given for $i = 0, \dots, N$. To describe the correlation of the given points in the plane, we are looking for two regression lines, one for the dependence of y in terms of x, another for x in terms of y:

$$g_1: \quad y = \Phi^{(0)}(x) = c_0^{(0)} + c_1^{(0)}x \quad \text{(regression from y to x)},$$

$$g_2: \quad x = \tilde{\Phi}^{(0)}(y) = \tilde{c}_0^{(0)} + \tilde{c}_1^{(0)}y \quad \text{(regression from x to y)}.$$

The coefficients $c_0^{(0)}$ and $c_1^{(0)}$ for g_1 are determined by the normal equations (8.17) with $w_i = 1$, $\varphi_0(x) = 1$ and $\varphi_1(x) = x$, or by (8.21) for $n = 1$.

These normal equations give rise to a linear 2-by-2 system:

$$\begin{pmatrix} N+1 & \sum x_i \\ \sum x_i & \sum x_i^2 \end{pmatrix} \begin{pmatrix} c_0^{(0)} \\ c_1^{(0)} \end{pmatrix} = \begin{pmatrix} \sum y_i \\ \sum x_i y_i \end{pmatrix},$$

where all sums extend from $i = 0$ to $i = N$. Since the system matrix may already be ill-conditioned, one should preferably use the method described in

section 8.1.3.2 or use Householder transformations described below in order to solve for $c_i^{(0)}$.

If one considers y as an independent variable and x as the dependent variable, then one is looking for the coefficients $\tilde{c}_0^{(0)}, \tilde{c}_1^{(0)}$ defining g_2. The centroid (\bar{x}, \bar{y}) of the set of points $\{(x_i, y_i)\} \subset \mathbb{R}^2$ is defined as

$$\bar{x} = \frac{1}{N+1} \sum_{i=0}^{N} x_i, \quad \bar{y} = \frac{1}{N+1} \sum_{i=0}^{N} y_i.$$

It is the point of intersection of the two regression lines. The discrepancy between the two lines determines whether one can correctly say that there is a linear correlation between the given values for x_i and y_i.

8.1.3.4 Solving Linear Least Squares Problems using Householder Transformations

Assume that $N + 1$ nodes $(x_i, f(x_i)) \in \mathbb{R}^2$ are given for $i = 0, \ldots, N$ with $N + 1$ weights $w_i > 0$. We want to find a linear approximation function

$$\Phi(x, c) = c_0 \varphi_0(x) + c_1 \varphi_1(x) + \ldots + c_n \varphi_n(x)$$

in such a way that

$$\sum_{i=0}^{N} w_i \left(f(x_i) - \Phi(x_i, c_0, c_1, \ldots, c_n) \right)^2$$

is minimized.
To simplify the notation, assume for the moment that $w_i = 1$ for all i. With $c = (c_0, c_1, ..., c_n)^T$ and

$$\Phi(x_i, c) = c_0 \varphi_0(x_i) + c_1 \varphi_1(x_i) + \ldots + c_n \varphi_n(x_i)$$

for $i = 0, \ldots, N$ we have

$$(8.22) \quad \begin{pmatrix} \Phi(x_0, c) \\ \Phi(x_1, c) \\ \vdots \\ \Phi(x_N, c) \end{pmatrix} = \begin{pmatrix} \varphi_0(x_0) & \varphi_1(x_0) & \cdots & \varphi_n(x_0) \\ \varphi_0(x_1) & \varphi_1(x_1) & \cdots & \varphi_n(x_1) \\ \vdots & \vdots & & \vdots \\ \varphi_0(x_N) & \varphi_1(x_N) & \cdots & \varphi_n(x_N) \end{pmatrix} \begin{pmatrix} c_0 \\ c_1 \\ \vdots \\ c_n \end{pmatrix} =: Ac$$

with an $(N + 1, n + 1)$ matrix A whose columns are

$$\varphi_j := (\varphi_j(x_0), \varphi_j(x_1), \ldots, \varphi_j(x_N))^T \quad \text{for} \quad j = 0, \ldots, n.$$

With

$$(8.23) \qquad\qquad \boldsymbol{f} := (f(x_0), f(x_1), \ldots, f(x_N))^T,$$

and $w_i = 1$, the identities (8.18') and (8.19) give the following form to the normal equations

$$(8.24) \qquad\qquad \boldsymbol{A}^T \boldsymbol{A} \boldsymbol{c} = \boldsymbol{A}^T \boldsymbol{f},$$

for clearly $\boldsymbol{G} = \boldsymbol{A}^T \boldsymbol{A}$ and $\boldsymbol{a} = \boldsymbol{A}^T \boldsymbol{f}$. If $w_i \neq 1$, the vectors

$$\varphi_j := \begin{pmatrix} \sqrt{w_0}\, \varphi_j(x_0) \\ \vdots \\ \sqrt{w_N}\, \varphi_j(x_N) \end{pmatrix}, \quad \boldsymbol{f} := \begin{pmatrix} \sqrt{w_0}\, f(x_0) \\ \vdots \\ \sqrt{w_N}\, f(x_N) \end{pmatrix}$$

must be used in (8.22) instead, see definition 8.7 and theorem 8.8. Now

$$\sum_{i=0}^{N} (f(x_i) - \Phi(x_i))^2 \overset{!}{=} \quad \text{Minimum}$$

is equivalent to

$$(8.25) \qquad\qquad \|\boldsymbol{f} - \boldsymbol{A}\boldsymbol{c}\|_2^2 \overset{!}{=} \quad \text{Minimum}.$$

Hence \boldsymbol{c} is the solution of the normal equations (8.24), if \boldsymbol{c} is the optimal solution of (8.25), for a proof see [STOE91], 4.8.

Thus we are left with the problem of solving the overdetermined linear system

$$\boldsymbol{A}\boldsymbol{c} = \boldsymbol{f}.$$

If this is done via Householder transformations, the previously mentioned ill-conditioning problems will not occur. The condition of \boldsymbol{A} is not worsened, and besides, rounding errors introduced by forming $\boldsymbol{A}^T \boldsymbol{A}$ will not occur.

ALGORITHM 8.12 *(Linear least squares via Householder transformations).*

Given: An $(N + 1, n + 1)$ matrix \boldsymbol{A} according to (8.22) with $N \geq n$, $f \in \mathbb{R}^{N+1}$, $\text{rank}(\boldsymbol{A}) = n + 1$.

Task: Find the optimal solution c of the overdetermined system $Ac = f$ with respect to the root-mean-square error. Set $A^{(0)} := A$ and $f^{(0)} := f$ and calculate

(i) $A^{(i+1)} = H_i A^{(i)}$,

(ii) $f^{(i+1)} = H_i f^{(i)}$

for each $i = 0, 1, 2, ..., n$ and Householder transformations H_i that zero the lower trapezoid of A successively. Each H_i is an orthogonal $(N + 1, N + 1)$ matrix which must be constructed according to section 4.14 .

Then for the euclidean norm:

$$\|f - Ac\|_2 = \|f^{(n+1)} - A^{(n+1)} c\|_2.$$

If $f^{(n+1)} = \begin{pmatrix} b_1 \\ b_2 \end{pmatrix}$ and $A^{(n+1)} = \begin{pmatrix} B \\ 0 \end{pmatrix}$, then

$$f^{(n+1)} - A^{(n+1)} c = \begin{pmatrix} b_1 - Bc \\ b_2 \end{pmatrix}.$$

Here B is the $(n + 1, n + 1)$ upper triangular matrix obtained after the complete Householder reductions from A. 0 denotes the $(N - n, n + 1)$ zero matrix, $b_1 \in R^{n+1}, b_2 \in R^{N-n}$. And $\|f - Ac\|_2$ is minimized if c is chosen as the solution of $Bc = b_1$. This last system of equations is in row echelon form and can be solved by backsubstitution.

As $\mathrm{rank}(A) = n + 1$, we have $\det(B) \neq 0$, and hence there exists a unique solution c of the linear least squares problem.

8.1.4 Approximation of Polynomials by Chebyshev Polynomials

In order to compute functional values accurately, a function f should be approximated by a function Φ uniformly for all points $x \in [a, b]$ so that $|f(x) - \Phi(x)| \le \varepsilon$ for a given bound $\varepsilon > 0$. For the root-mean-square approximation such a bound for the absolute error cannot be given independently of x. In contrast, this is possible with the so-called *uniform* or *Chebyshev approximation* .

Here we shall only describe the uniform approximation of algebraic polynomials f by *Chebyshev polynomials*. The uniform approximation of a function f by means of an approximation function Φ can be reduced to this case in the following way:

The truncated Taylor series of a function f whose remainder is bounded in the interval $[a, b]$ is a polynomial which can be uniformly approximated by a linear combination of Chebyshev polynomials. Truncation and approximation errors should ideally have the same order of magnitude and add up to less than the given bound ε.

The degree of the approximating Chebyshev polynomial can moreover usually be kept smaller than that of the corresponding Taylor series expansion. For this reason, functional values can be computed faster via the approximating polynomial than via the truncated Taylor series. The approximating polynomial need only be computed once. This effort, however, can only be justified if, within a given problem, f has to be evaluated many times, as is usually the case in discretizations and when evaluating integrals, for example.

8.1.4.1 Best Uniform Approximation

As a norm of a function $g \in C[a, b]$ we choose the so-called *maximum norm* here:

$$\|g\|_\infty := \max_{x \in [a,b]} |g(x)| \, w(x)$$

for a continuous weight function $w(x) > 0$, see remark 8.9. For a given system of linearly independent functions $\varphi_0, \varphi_1, ..., \varphi_n \in C[a, b]$, we define \bar{C} to be the set of all linear combinations Φ as in (8.2). A best approximation $\Phi^{(0)} \in \bar{C}$ satisfies

$$
\begin{aligned}
\|f - \Phi^{(0)}\|_\infty &= \max_{x \in [a,b]} |f(x) - \Phi^{(0)}(x)| w(x) \\
&= \min_{\Phi \in \bar{C}} (\max_{x \in [a,b]} |f(x) - \Phi(x)| w(x)).
\end{aligned}
$$

(8.26)

for the maximum norm. Thus for the best uniform approximation $\Phi^{(0)}$ the maximum of the weighted absolute error $|f(x) - \Phi^{(0)}(x)|$ becomes minimal in the entire interval $[a, b]$. Consequently for the best approximation $\Phi^{(0)}$ we know that the absolute error $\|f - \Phi^{(0)}\|$ does not exceed a given $\varepsilon > 0$ for any $x \in [a, b]$. In other words, $|f(x) - \Phi^{(0)}(x)| \leq \varepsilon$ uniformly, i.e., f is approximated for all $x \in [a, b]$ by $\Phi^{(0)}$ with *accuracy* ε.

Such a best approximation $\Phi^{(0)}$ in the sense of the maximum norm is therefore called the *best uniform approximation*, or the *minimax approximation* for $f \in C[a, b]$.

For the uniform approximation of an arbitrary function $f \in C[a,b]$ – contrary to the discrete root-mean-square approximation – , there is no general procedure to determine the coefficients $c_k^{(0)}$ of $\Phi^{(0)}$. For approximate procedures see [ATKI89], 4.2, 4.9, [MEIN67], §7, or [WERN79], II, §§4-6. In the remainder, we shall only treat the important case of uniform approximation of polynomials by Chebyshev polynomials. This case is important in practice.

8.1.4.2 Approximation by Chebyshev Polynomials

A. *Introduction of Chebyshev Polynomials.*

The *Chebyshev polynomials* T_k are defined for $x \in [-1, +1]$ by

$$(8.27) \qquad T_k(x) = \cos(k \arccos x) \text{ for } k = 0, \dots, n.$$

Explicitly for small k:

$$(8.28) \qquad \begin{cases} T_0(x) = 1 & , \quad T_3(x) = 4x^3 - 3x, \\ T_1(x) = x & , \quad T_4(x) = 8x^4 - 8x^2 + 1, \\ T_2(x) = 2x^2 - 1 & , \quad T_5(x) = 16x^5 - 20x^3 + 5x. \end{cases}$$

In general, one can calculate Chebyshev polynomials from the triple recursion formula

$$(8.29) \quad T_{k+1} = 2x T_k - T_{k-1} \text{ with } T_0 = 1, \quad T_1 = x \text{ for } k = 1, \dots, .$$

Important *properties* of Chebyshev polynomials are:

1: Each T_k is a polynomial in x of degree k.

2: The coefficient of x^k in T_k is 2^{k-1}.

3: For all k and all $x \in [-1, +1]$, $|T_k(x)| \le 1$.

4: $T_k(x_j) = \pm 1$ at the $k+1$ points $x_j = \cos \frac{\pi j}{k}$, $j = 0, \dots, k$.

5: In $[-1, +1]$ T_k possesses exactly k real zeros $x_j = \cos \frac{2j+1}{k} \frac{\pi}{2}$, $j = 0, \dots, k-1$.

(See also [ATKI89], 4.)

B. *Representation of Polynomials as Linear Combinations of Chebyshev Polynomials.*

Powers of x can be easily represented as linear combinations of Chebyshev polynomials on account of (8.28) or (8.29). They are

$$
(8.30) \quad
\begin{cases}
1 = T_0 & , \quad x^3 = 2^{-2}(3T_1 + T_3), \\
x = T_1 & , \quad x^4 = 2^{-3}(3T_0 + 4T_2 + T_4), \\
x^2 = 2^{-1}(T_0 + T_2) & , \quad x^5 = 2^{-4}(10T_1 + 5T_3 + T_5),
\end{cases}
$$

and in general for $k = 0, 1, 2, \ldots$:

$$
(8.31) \qquad x^k = 2^{1-k}\left(T_k + \binom{k}{1}T_{k-2} + \binom{k}{2}T_{k-4} + \ldots + T^*\right).
$$

Here the last term T^* has the form

$$
T^* = \binom{k}{\frac{k-1}{2}} T_1
$$

for odd k, and

$$
T^* = \frac{1}{2}\binom{k}{\frac{k}{2}} T_0
$$

for even k. Each polynomial

$$
(8.32) \qquad P_m(x) = \sum_{i=0}^{m} a_i x^i, \qquad a_i \in \mathbb{R},
$$

of degree m can be expressed uniquely as a linear combination

$$
(8.33) \qquad P_m(x) = \sum_{j=0}^{m} b_j T_j(x)
$$

of Chebyshev polynomials by expressing the powers of x in (8.32) by Chebyshev polynomials using (8.30) and (8.31).

To determine the coefficients b_j of the Chebyshev expansion (8.33) from the coefficients a_i of (8.32), we can make use of the following calculation schemes for $i, j = 0, \ldots, 10$.

CALCULATION SCHEME 8.13 (*Even j*).

a_0	1					
$\dfrac{a_2}{2}$	1	1				
$\dfrac{a_4}{8}$	3	4	1			
$\dfrac{a_6}{32}$	10	15	6	1		
$\dfrac{a_8}{128}$	35	56	28	8	1	
$\dfrac{a_{10}}{512}$	126	210	120	45	10	1
	b_0	b_2	b_4	b_6	b_8	b_{10}

CALCULATION SCHEME 8.14 (*Odd j.*)

a_1	1				
$\dfrac{a_3}{4}$	3	1			
$\dfrac{a_5}{16}$	10	5	1		
$\dfrac{a_7}{64}$	35	21	7	1	
$\dfrac{a_9}{256}$	126	84	36	9	1
	b_1	b_3	b_5	b_7	b_9

With the exception of a_0, the coefficients a_i are divided by 2^{i-1} (left column) and multiplied with the number in the associated row which corresponds to the wanted coefficient b_j. The result is overwritten there. The column sums of the computed numbers supply the coefficient b_j of the Chebyshev expansion.

C. *Best Uniform Approximation.*

Besides the complete Chebyshev expansion (8.33) of degree n, it is useful to consider partial sums thereof:

$$S_n(x) = \sum_{j=0}^{n} b_j T_j(x) \quad \text{for } n \le m.$$

Clearly

$$P_m(x) = S_m(x) = S_{m-1}(x) + b_m T_m(x).$$

As approximation functions for P_m, we may choose linear combinations Φ

$$\Phi(x) = \sum_{k=0}^{n} c_k T_k(x) \text{ with } n < m.$$

The best uniform approximation $\Phi^{(0)}$ of P_n in the form

$$\Phi^{(0)}(x) = \sum_{k=0}^{n} c_k^{(0)} T_k(x) \text{ with } n < m,$$

is determined by

THEOREM 8.15

The best uniform approximation $\Phi^{(0)}$ of a polynomial P_m by a polynomial of $(m-1)^{st}$ degree in the interval $[-1, +1]$ is the uniquely determined partial sum

$$\Phi^{(0)}(x) = S_{m-1}(x) = \sum_{k=0}^{m-1} b_k T_k(x)$$

of its Chebyshev expansion S_m, where $c_k^{(0)} = b_k$ for $k = 0, \dots, m-1$. If $w(x) \equiv 1$, then

$$\|P_m - S_{m-1}\|_\infty = \max_{x \in [-1, +1]} |P_m(x) - S_{m-1}(x)| \le |b_m|.$$

(See [STIE63], p.202).

Thus in order to obtain the best approximation $\Phi^{(0)}$ of degree $m-1$, one must only delete the last term $b_m T_m$ of the Chebyshev expansion S_m.

D. *Uniform Approximation.*

Since the coefficients b_j of a Chebyshev expansion decrease rapidly with growing j in most cases, we obtain a rather good approximation of the polynomial P_m even if we omit more than one of the leading terms of its Chebyshev expansion. If

$$S_n(x) = \sum_{j=0}^{n} b_j T_j(x), \quad n \leq m - 1,$$

is a partial sum of the Chebyshev expansion S_m of P_m, then we have on account of property 3)

$$\|P_m - S_n\|_\infty = \max_{x \in [-1,+1]} |P_m(x) - S_n(x)| \leq \sum_{j=n+1}^{m} |b_j| =: \varepsilon_1.$$

Since ε_1 is independent of x, S_n is also a uniform approximation for P_m. For $n = m - 1$ it is the best uniform approximation. More details will follow in section 8.1.6 .

If one wants to find a suitable approximation function Φ for an f that is sufficiently often differentiable in the interval $[-1, +1]$, one can start with its Taylor expansion at $x = 0$:

$$f(x) = P_m(x) + R_{m+1}(x),$$

where P_m is the Taylor polynomial for f of degree m with remainder R_{m+1}. Assume that

$$|R_{m+1}(x)| \leq \varepsilon_2$$

uniformly for $x \in [-1, +1]$, i.e., that the truncation error ε_2 can be bounded independently of x. As an approximation for f, one can choose the partial sum S_n of the Chebyshev expansion S_m for P_m with $n \leq m - 1$. Then

$$
\begin{aligned}
\max_{x \in [-1,+1]} |f(x) - S_n(x)| &= \|f - S_n\|_\infty = \|P_m + R_{m+1} - S_n\|_\infty \\
&\leq \|P_m - S_n\|_\infty + \|R_{m+1}\|_\infty \leq \varepsilon_1 + \varepsilon_2.
\end{aligned}
$$

The maximal absolute error of the approximation S_n of f is therefore composed of the error ε_1 of the uniform approximation S_n of P_m and the truncation error ε_2. If for a given accuracy ε, the inequality $\varepsilon_1 + \varepsilon_2 \leq \varepsilon$ holds, then the function f is approximated by the polynomial S_n uniformly in the interval $[-1, +1]$.

ALGORITHM 8.16 *(Uniform Approximation by Chebyshev Polynomials).*

Let f be sufficiently often differentiable for $x \in [-1, +1]$. For a given $\varepsilon > 0$, find an approximating polynomial S_n with $|f(x) - S_n(x)| \leq \varepsilon$ for all $x \in [-1, +1]$.

1^{st} step: Form the Taylor series for f at $x = 0$:
$$f(x) = P_m(x) + R_{m+1}(x) = \sum_{i=0}^{m} a_i x^i + R_{m+1}(x), \ a_i = \frac{f^{(i)}(0)}{i!}.$$
Here, select m as the minimal index for which $|R_{m+1}(x)| \leq \varepsilon_2 < \varepsilon$ for all $x \in [-1, +1]$.

2^{nd} step: Calculate the Chebyshev expansion for P_m using the calculation schemes 8.13 and 8.14:
$$P_m(x) = \sum_{j=0}^{m} b_j T_j(x) \equiv S_m(x).$$

3^{rd} step: Choose the minimal $n \leq m - 1$, so that
$$|f(x) - S_n(x)| \leq \varepsilon_2 + |b_{n+1}| + |b_{n+2}| + \ldots + |b_m| \leq \varepsilon_2 + \varepsilon_1 \leq \varepsilon.$$
S_n is the desired approximating polynomial for f with an accuracy of ε uniformly in the interval $[-1, +1]$.

To calculate approximate values for the function f by means of S_n, S_n is converted via (8.27) or (8.28) into the standard form of a polynomial:

$$S_n(x) = \sum_{j=0}^{n} b_j T_j(x) \equiv \sum_{k=0}^{n} \tilde{a}_k x^k = \tilde{P}_n(x).$$

And \tilde{P}_n can then be evaluated via a Horner scheme.

An interval $[a, b] \neq [-1, +1]$ can be converted into the standard interval $[-1, +1]$ by means of a linear transformation: By setting $x = 2x'/(b-a) - (b+a)/(b-a)$, the x'-interval $[a, b]$ is converted into the x-interval $[-1, +1]$.

REMARK. According to the *approximation theorem of Weierstraß*, each function $f \in C[a, b]$ can be approximated uniformly by an algebraic polynomial of degree $n = n(\varepsilon)$ for a prescribed accuracy ε. And each 2π-periodic continuous function $f \in C(-\infty, +\infty)$ can be uniformly approximated by trigonometric polynomials (8.35) for $n = n(\varepsilon)$ for a prescribed accuracy ε ([ISAA66], 5.1; [WERN79] II, §1). The theorems of Weierstraß are only existence theorems, they do not indicate any constructive method for obtaining the approximation functions. In this chapter and in [MEIN67] §5, special methods are given for

the uniform approximation by algebraic or trigonometric polynomials; see also
[RALS67] vol. I,I; [BEZI72], §21. An algorithm for the uniform approximation
by means of rational functions is given in [WERN79], II, §5; [BEZI72], §22.
For tables of the coefficients for uniform approximations of important tran-
scendental functions see [ABRA72]; [HART68].

8.1.5 Approximation of Periodic Functions and the FFT

A 2π-periodic function can be represented under certain conditions (see for
instance [BJÖR74], 9.2, 9.4) by its *Fourier series*:

$$f(x) = \frac{\alpha_0}{2} + \sum_{k=1}^{\infty} (\alpha_k \cos\, kx\, + \beta_k \sin\, kx\,), \quad \text{where}$$

(8.34)
$$\begin{cases} \alpha_k = \frac{1}{\pi} \int_{-\pi}^{+\pi} f(x) \cos\, kx\ dx, & k = 0, \ldots, \text{ and} \\[2mm] \beta_k = \frac{1}{\pi} \int_{-\pi}^{+\pi} f(x) \sin\, kx\ dx, & k = 1, \ldots. \end{cases}$$

For *even functions* with $f(-x) = f(x)$ for all x we have $\beta_k \equiv 0$, while for
odd functions with $f(-x) = -f(x)$ we have $\alpha_k \equiv 0$ for all k. If the integrals
(8.34) can be evaluated, the truncated Fourier series can be used to obtain an
approximation for f. If, however, f is such that the integrals (8.37) cannot
be evaluated in closed form or if f is only given by a table of values, then we
have to seek a trigonometric polynomial Φ which approximates f:

(8.35)
$$\Phi(x) = \frac{a_0}{2} + \sum_{k=1}^{n} (a_k \cos\, kx\, + b_k \sin\, kx).$$

8.1.5.1 Root-Mean-Square Approximation of Periodic Functions

Given a 2π-periodic function f whose values are known at $x_j = \frac{j\pi}{N}$ for
$j = 0, \ldots, 2N - 1$, we want to find its best approximation $\Phi^{(0)}$ in the form
(8.35). In practice one always works with an even number of nodes. Using the
norm (8.15) with $w_i = 1$ we obtain a linear system analogous to (8.17) for the
$2n + 1$ coefficients $a_0^{(0)}, a_k^{(0)}, b_k^{(0)}, k = 1, \ldots, n$, where $2n + 1 < 2N$. These are
uniquely determined by theorem 8.8 for $2n + 1 < 2N$.

In this case, the approximation function (8.35) has the explicit form

$$\Phi^{(0)}(x) = \frac{a_0^{(0)}}{2} + \sum_{k=1}^{n}(a_k^{(0)}\cos kx + b_k^{(0)}\sin kx)$$

with

$$a_0^{(0)} = \frac{1}{N}\sum_{j=1}^{2N}y_j, \quad a_k^{(0)} = \frac{1}{N}\sum_{j=1}^{2N}y_j\cos kx_j, \text{ and } b_k^{(0)} = \frac{1}{N}\sum_{j=1}^{2N}y_j\sin kx_j.$$

For fixed N, the calculated coefficients $a_0^{(0)}, a_k^{(0)}, b_k^{(0)}$ do not change if n is increased.

For $n = N$ we have $b_N^{(0)} = 0$ so that there are only $2n$ coefficients to compute for $\Phi^{(0)}$. Thus the number of the coefficients equals the number $2N$ of nodes and we have to solve a *trigonometric interpolation* problem.

8.1.5.2 Trigonometric Interpolation

The trigonometric interpolation polynomial has the form

$$\Phi(x) = \frac{a_0}{2} + \sum_{k=1}^{N-1}(a_k\cos kx + b_k\sin kx) + \frac{a_N}{2}\cos Nx$$

with coefficients

$$(8.36) \begin{cases} a_0 = \frac{1}{N}\sum_{j=1}^{2N}y_j, \quad a_N = \frac{1}{N}\sum_{j=1}^{2N}(-1)^j y_j, \text{ and} \\ a_k = \frac{1}{N}\sum_{j=1}^{2N}y_j\cos kx_j, \quad b_k = \frac{1}{N}\sum_{j=1}^{2N}y_j\sin kx_j \text{ for } k=1,\dots,N-1. \end{cases}$$

Suitable algorithms for calculating the coefficients in (8.36) are found in [STOE91], 2.3.; [WERN79], p.50 ff.. Here we describe a method that can be used for hand computations as well as for computer applications. An algol program is given in [ZURM65], pp.368-370.

For the remainder of this chapter we assume that $2N = 12$, i.e., $x_j = j(2\pi/12) = j(\pi/6)$. Then the trigonometric interpolation polynomial becomes

$$\Phi(x) = \frac{a_0}{2} + \sum_{k=1}^{5}(a_k \cos kx + b_k \sin kx) + \frac{a_6}{2} \cos 6x$$

with coefficients ([ZURM65], p. 364):

$$a_k = \frac{1}{6}\sum_{j=1}^{12} y_j \cos kx_j, \quad k = 0,\ldots,6, \quad b_k = \frac{1}{6}\sum_{j=1}^{12} y_j \sin kx_j, \quad k = 1,\ldots,5.$$

The following computational scheme helps with the calculation of the coefficients a_k, b_k.

CALCULATION SCHEME 8.17 *(Numerical harmonic analysis according to Runge).*

1^{st} convolution (of the y_j):	-	y_1	y_2	y_3	y_4	y_5	y_6				
	y_{12}	y_{11}	y_{10}	y_9	y_8	y_7	-				
Sums s_j	s_0	s_1	s_2	s_3	s_4	s_5	s_6				
Differences d_j	-	d_1	d_2	d_3	d_4	d_5	-				
2^{nd} convolution (of the s_j):	s_0	s_1	s_2	s_3	2^{nd} convolution (of the d_j):	d_1	d_2	d_3			
	s_6	s_5	s_4	-		d_5	d_4	-			
Sums S_j	S_0	S_1	S_2	S_3	Sums \bar{S}_j	\bar{S}_1	\bar{S}_2	\bar{S}_3			
Differences D_j	D_0	D_1	D_2	-	Differences \bar{D}_j	\bar{D}_1	\bar{D}_2	-			

To calculate the coefficients of the cosine terms, we have to multiply S_j and D_j in every row with the cosine values in the first column of:

$\cos 0 = 1$	$+S_0$ $+S_2$	$+S_1$ $+S_3$	$+D_0$	-	$+S_0$	$-S_3$	$+D_0$	$-D_2$
$\cos\left(\frac{\pi}{6}\right) = \frac{\sqrt{3}}{2}$	-	-	-	$\frac{\sqrt{3}}{2}D_1$	-	-	-	-
$\cos\frac{\pi}{3} = \frac{1}{2}$	-	-	$\frac{1}{2}D_2$	-	$-\frac{1}{2}S_2$	$\frac{1}{2}S_1$	-	-
Sums	\sum_1	\sum_2	\sum_1	\sum_2	\sum_1	\sum_2	\sum_1	\sum_2
$\sum_1 + \sum_2$ $\sum_1 - \sum_2$	$6a_0$ $6a_6$		$6a_1$ $6a_5$		$6a_2$ $6a_4$		$6a_3$ -	

To calculate the coefficients of the sine terms, we have to multiply \bar{S}_j and \bar{D}_j in every row with the sine values in the first column of:

$\sin\frac{\pi}{6} = \frac{1}{2}$	$\frac{1}{2}\bar{S}_1$	-	-	-	-	-
$\sin\frac{\pi}{3} = \frac{\sqrt{3}}{2}$	-	$\frac{\sqrt{3}}{2}\bar{S}_2$	$\frac{\sqrt{3}}{2}\bar{D}_1$	$\frac{\sqrt{3}}{2}\bar{D}_2$	-	-
$\sin\frac{\pi}{2} = 1$	$+\bar{S}_3$	-	-	-	$+\bar{S}_1$	$-\bar{S}_3$
Sums	\sum_1	\sum_2	\sum_1	\sum_2	\sum_1	\sum_2
$\sum_1 + \sum_2$ $\sum_1 - \sum_2$	$6b_1$ $6b_5$		$6b_2$ $6b_4$		$6b_3$ -	

8.1.5.3 Complex Discrete Fourier Transformation (FFT)

If f is a real or complex valued periodic function with period $L = x_N - x_0$, i.e., if $f(x + L) = f(x)$ for all x, we shall assume that the functional values

$f_j := f(x_j)$ are given for $N = 2^\tau$ equidistant points

$$x_j = x_0 + j\frac{L}{N}, \qquad j = 0, \ldots, N-1,$$

where $\tau \in \mathbb{N}$. Then the function f can be approximated by its discrete L-periodic Fourier partial sum

$$\sum_{k=-N/2+1}^{N/2-1} c_k e^{i\left(k\frac{2\pi}{L}x\right)} + c_{N/2} \cos\left((N/2)\frac{2\pi}{L}x\right), \quad \text{where}$$

$i = \sqrt{-1} \in \mathbb{C}$ and $e^{i\xi} := \cos\xi + i\sin\xi$. This Fourier partial sum interpolates f at the nodes $(x_j, f(x_j))$.

Its N complex coefficients are the discrete Fourier coefficients

$$c_k = \frac{1}{N}\sum_{j=0}^{N-1} f_j e^{-i\left(k\frac{2\pi}{L}x_j\right)}.$$

They describe the frequency contents of the function f at the k^{th} harmonic of the fundamental frequency $2\pi/L$. For increasing k, this usually becomes smaller and smaller. An effective calculation of the discrete Fourier coefficients is available in the so-called *Fast Fourier Transform (FFT)* which is most effective in case $N = 2^\tau, \tau \in \mathbb{N}$. In this case the number of necessary complex multiplications is reduced from $2^\tau \cdot 2^\tau$ to $(\tau/2)2^\tau$. Algorithms with other radices and a conceptual overview of the FFT method are available in e.g. [LOAN92].

If the function f is real-valued, i.e., if all $f_j \in \mathbb{R}$, its discrete Fourier partial sum is also real and is given by

$$a_0 + 2\sum_{k=1}^{N/2-1}\left\{a_k\cos\left(k\frac{2\pi}{L}x\right) + b_k\sin\left(k\frac{2\pi}{L}x\right)\right\} + a_{N/2}\cos\left((N/2)\frac{2\pi}{L}x\right)$$

with the discrete Fourier coefficients

$$a_k = \frac{1}{N}\sum_{j=0}^{N-1} f_j\cos\left(k\frac{2\pi}{L}x_j\right) = Re(c_k),$$

$$b_k = \frac{1}{N}\sum_{j=0}^{N-1} f_j\sin\left(k\frac{2\pi}{L}x_j\right) = -Im(c_k).$$

The coefficients $2a_k$, $2b_k$ and $a_{N/2}$ describe the components of the periodic function f corresponding to the harmonic oscillations

$$\cos\left(k\frac{2\pi}{L}x\right), \quad \sin\left(k\frac{2\pi}{L}x\right) \quad \text{and} \quad \cos\left(\frac{N}{2}\frac{2\pi}{L}x\right).$$

The functional values f_j can be found from the discrete Fourier coefficients c_k via the same method as

$$f_j = \sum_{k=0}^{N/2} c_k e^{\left(j\frac{2\pi}{L}x_k\right)} + \sum_{k=N/2+1}^{N-1} c_{k-N} e^{i\left(j\frac{2\pi}{L}x_k\right)}.$$

This inverse transformation uses the same algorithm except for the division by N and replacing the N^{th} root

$$e^{-i\frac{2\pi}{N}} \quad \text{by} \quad e^{i\frac{2\pi}{N}}.$$

Bibliography for Section 8.1.4 and 8.1.5

[BJÖR74], 4; [BOOR79]; [COLL68], §25; [COLL73], 3.3, 3.4; [HÄMM91], Chap.1; [ISAA66], 5; [LOAN92]; [MEIN67]; [NIED84], 13; [SCHW89], 4.3; [SPÄ74/1]; [STIE63], 3.1, 7.2; [TÖRN79] vol. 2, 11.4-11.8; [WERN79], II; [ZURM65], §22-24 .

8.1.6 Error Estimates for Linear Approximation

In this section we shall list some methods for estimating the error of various procedures for approximations in linear spaces of functions. Most particularly, these estimates pertain to approximation by trigonometric and algebraic polynomials as previously treated. They are framed as estimates of the worst possible functions of the given type. Thus the particular numerical results achieved for a given problem may often be much better.

8.1.6.1 Estimates for the Error in Best Approximation

For a given function $f \in C[a,b]$, we shall denote by $E_n^*(f)$ the error in best approximation (in the uniform (maximum norm) or least square (2 norm) sense) by a trigonometric polynomial of degree at most n;

$$E_n^*(f) = \min_{T_n} \max_{x \in [a,b]} |f(x) - T_n(x)|, \text{ or}$$

(8.37) $$E_n^*(f) = \min_{T_n} \max_{x_i \in [a,b]} |f(x_i) - T_n(x_i)|$$

for given nodes $x_1, \ldots, x_N \in [a, b]$ in the discrete case .

Then the following theorem of Jackson holds:

THEOREM 8.18 (*Theorem of Jackson*):

Let $f \in C[a, b]$ be 2π-periodic.

a) If $f, f', \ldots, f^{(k-1)}$ are absolutely continuous, then

$$E_n^*(f) \leq \frac{\pi}{2(n+1)^k} \|f^{(k)}\|$$

for both the maximum and the 2-norm.

b) If $w(f; \delta) := \sup_{|\phi - \varphi| \leq \delta} |f(\theta) - f(\varphi)|$, then

$$E_n^*(f) \leq \omega(f; \frac{\pi}{n+1}).$$

c) $E_n(f) \leq \frac{\pi}{2(n+1)^k} E_n^*(f^{(k)})$.

An example which shows that the constant $\pi/2$ in a) is best possible is given by the antiderivative of sign $\sin nx$ or sign $\cos nx$.

Results similar to Jackson's can be obtained for algebraic polynomial approximations on $[-1, 1]$ by the transformation

$$x \leftrightarrow \cos \theta \qquad \text{for } -1 \leq x \leq 1 \text{ and } 0 \leq \theta \leq \pi.$$

This transformation maps every function defined on $[-1, 1]$ uniquely to a 2π-periodic even function and vice versa. Moreover this transformation maps algebraic polynomials of a given degree into trigonometric polynomials of the same degree and vice versa. In fact, we have already observed this in formula (8.27), and the Chebyshev expansion (8.33) of a function $f(x)$ is identical to the Fourier expansion (8.35) of $f(\cos \theta)$ under this transformation.

The result of Jackson may be restated for algebraic polynomial approximations with the error $E_n(f)$ defined as the error in best approximation of degree at most n in either the uniform or discrete sense:

THEOREM 8.19: *(Error in best polynomial approximation).*

Assume that for $f \in C[-1, 1]$ the functions $f, f', \ldots, f^{(k)}$ are absolutely continuous and $\|f^{(k)}\|$ exists. Then

a) for $n \geq k, E_n(f) \leq (\frac{\pi}{2})^k \frac{\|f^{(k)}\|}{(n+1)(n)\cdots(n-k+2)}$, where the norm used is either the maximum norm or the least squares norm (8.7) with weight $w(x) = \frac{1}{\sqrt{z-x^2}}$,

b) $E_n(f) \leq \omega(f; \frac{\pi}{n+1})$, where $\omega(R; \delta) := \sup_{|x-y| \leq \delta} |f(x) - f(y)|$,

c) in a) it is possible to replace $\|f^{(k)}\|$ by $E_{n-k}(f^{(k)})$.

Other estimates for algebraic polynomial approximation and refinements of theorem 8.19 have been proven more recently. For example, for a given function $f \in C[a, b]$ Timan has proved the existence of a polynomial P_n with

$$|f(x) - P_n(x)| \leq c_k \left(\frac{\sqrt{1 - x^2}}{n} + \frac{1}{n^2} \right)^k \omega(f^{(k)}; \frac{\sqrt{1 - x^2}}{n} + \frac{1}{n^2}),$$

while Sinwel and Runck [SINW82] give an estimate similar to theorem 8.19 a):

$$|f(x) - P_n(x)| \leq \frac{\pi}{2} \frac{1}{(n-2)(n-4)\cdots(n-2k)} \left(\frac{\sqrt{1-x^2}}{n} + \frac{2k}{n^2} \right)^k \|f^{(k)}\|.$$

For further discussions of these results, proofs and references consult [LORE86] or [CHEN82].

8.1.6.2 Error Estimates for Simultaneous Approximation of a Function and its Derivatives

There exist simultaneous estimates for both trigonometric and algebraic polynomial approximation of a function and its derivatives. For trigonometric approximation it is known that if $f(\theta)$ is 2π-periodic and $P_n(\theta)$ satisfies $\|f(\theta) - P_n(\theta)\| \leq C E_n^*(f)$ for some constant C, then there is a constant k_q such that

$$\|f^{(q)}(\theta) - P_n^{(q)}(\theta)\| \leq C k_q E_n^*(f^{(q)}),$$

see for example Czipszer and Freud [CZIP58]. Here $k_q = 5(2^q + 1)$ for the maximum norm, a very conservative estimate, and in the least square norm $k_q = 2 + \frac{3\pi}{2}$ for all q.

If in the least squares norm f is approximated by $S_n f$, the truncated Fourier series, then $S_n f$ is the best approximation of f, see section 8.1.4. And moreover $S_n(f^{(a)}) = (S_n f)^{(a)}$ is the best approximation of its derivatives $f^{(q)}$. Thus in this special case, the constant k_q can be taken as one. The results of Czipszer and Freud can be easily combined with theorem 8.18 to yield for example:

THEOREM 8.20:
Let $f \in C^q[0, 2\pi]$ be 2π-periodic. Then the best approximating polynomial $P_n(\theta)$ of f satisfies

$$\|f^{(k)}(\theta) - P_n^{(k)}(\theta)\| \leq \frac{\pi}{2(n+1)^{a-k}} k_k E_n^*(f^{(q)})$$

for $k = 0, \ldots, q$.

This is valid for both the uniform and least squares norm.

An algebraic polynomial analogue is the following result of Kilgore [KILG93]:

THEOREM 8.21:
Let $f \in C^q[-1, 1]$. Then there is a polynomial $P_n(x)$ such that for $k = 0, \ldots, q$ and every $|x| \leq 1$

$$|f^{(k)}(x) - P_n^{(k)}(x)| \leq M_{q,k} \left(\frac{\sqrt{1-x^2}}{n} \right)^{q-k} E_{n-q}(f^{(q)}).$$

We note that good estimates for the constants $M_{q,k}$ are known only for small values of q. For example, for $q = 1$ we may estimate $M_{1,0} = \frac{3\pi}{4}$ and $M_{1,1} = \frac{3\pi}{4} k_1$.

The utility of these results on best or near-best approximations lies in their depiction of what is theoretically feasible and what is not, as well as in their role as a benchmark for any technique of approximation which one may wish to consider. An important group of techniques will be examined in the next section.

8.1.6.3 Approximation Error Estimates using Linear Projection Operators

A linear projection is a linear operator that maps a linear space to a subspace and leaves the elements of its range space unchanged. One example of such is the Fourier series expansion (8.35) of a periodic function f up to degree n. Its operator S_n satisfies $S_n(\alpha f + \beta g) = \alpha S_n f + \beta S_n g$ for any two continuous or integrable functions f and g and any two scalars α, β. In this sense S_n is a linear operator. Since $S_n(P_n(\theta)) = P_n(\theta)$ if $P_n(\theta)$ is a trigonometric polynomial of degree at most n, S_n is a projection. In general, a linear projection operator on the space of functions does not provide the best approximation for a given function; but the approximation is usually good and easily computable. The Fourier series expansion operator S_n does provide the best approximation in the 2-norm. The fundamental relationship between linear projections and best approximations is given in the following:

THEOREM 8.22 (*Theorem of Lebesgue*):

Let L_n be a bounded linear projection from a normed linear space X to an n-dimensional subspace $Y_n \subset X$. Then for all $f \in X$

$$\|f - L_n f\| \le \|I - L_n\| E_n(f) \le (1 + \|L_n\|) E_n(f),$$

where I denotes the identity operator and $\|..\|$ is an operator norm with $\|I\| = 1$.

The results on simultaneous approximations of f and its derivatives may be similarly extended. The results of Cziszer and Freud for example would assert with $X = \{f \mid f \text{ is } 2\pi\text{-periodic}\}$ and $Y_n = \{$ trigonometric polynomial of degree at most $n\} \subseteq X$, that

$$\|f^{(k)}(\theta) - L_n^{(k)} f(\theta)\| \le k_p \|I - L_n\| E_n^*(f) \le k_p (1 + \|L_n\|) E_n^*(f).$$

With algebraic polynomial approximation one may adopt the simple expedient of subtracting from a function $f \in C^q[-1, 1]$ to be approximated the polynomial Q_{2q} of degree $2q + 1$ which agrees with f and its first q derivatives at the endpoints ± 1. Let us call the resulting function g. Then g has zeros of multiplicity q at 1 and -1. We will assume that L_n is a linear projection, mapping the 2π-periodic functions to trigonometric polynomials of the form $a_0 + \sum_{k=1}^n a_k \cos k\theta + b_k \sin k\theta$, in which the coefficients a_k and b_k are determined by application of L_n upon f and are not necessarily the Fourier coefficients of f. Moreover we assume that $L_n f$ is even if f is even and $L_n f$ is odd if f is odd. Then, identifying $x \leftrightarrow \cos \theta$, we see that the *conjugated operator*

$$\left(\sqrt{1-x}\right)^q L_n\left(\frac{g(t)}{(\sqrt{1-t^2})^q}\right) := \sin^q\theta\, L_n\left(\frac{g(\cos\phi)}{\sin^q\phi}\right)$$

is defined for a function $g \in C^q[-1,1]$ such that $g^{(k)}(\pm 1) = 0$ for $k = 0,\dots,q$, and furthermore always gives an algebraic polynomial in x.

A theorem of Balázs and Kilgore [BALÁ93] states:

THEOREM 8.23:

Let $g \in C^q[-1,1]$ be such that $g^{(k)}(\pm 1) = 0$ for $k = 0,\dots,q$. Let L_n be any bounded linear projection which maps even 2π-periodic functions to cosine polynomials of degree at most n, and odd 2π-periodic functions to sine polynomials of degree at most n. Then for $k = 0,\dots,q$ and for all $|x| \leq 1$

$$\left|g^{(k)}(x) - \left(\left(\sqrt{1-x^2}\right)^q L_n\left(\frac{g(t)}{(\sqrt{1-t^2})^q}\right)\right)^{(k)}(x)\right| \leq$$

$$\leq C_{q,k}\left(\frac{\sqrt{1-x^2}}{n} + \frac{1}{n^2}\right)^{q-k} E_{n-q}(g^{(q)})\|L_n\|.$$

The constants $C_{q,k}$ depend only upon q and k.

The function f that was to be approximated can simply be recovered by back-substitution and by noting that for $n \geq 2q + 1$ we have $E_{n-q}(g^{(q)}) = E_{n-p}(f^{(q)})$.

Again applications of this result can be made for small values of q giving good constants $C_{q,k}$. For example if L_n is T_n, the truncated Chebyshev expansion of f, and if $q = 1$, we obtain $\|T_n\| \approx \frac{2}{\pi}\log n$, and thus for $k = 0,1$:

$$\left|g^{(k)}(x) - \sqrt{1-x^2}\, T_n^{(k)}\left(\frac{g(t)}{\sqrt{1-t^2}}\right)(x)\right| \leq \frac{3\pi}{4} k_k(1 + \frac{2}{\pi}\log n) E_{n-1}(g').$$

We conclude by mentioning that most of the results described concerning linear projections will very much apply to the procedures for interpolation given in chapter 9.

Bibliography for Section 8.1.6

[ATKI89], 4.6; [BALÁ93]; [CHEN82]; [CZIP58]; [KILG93]; [KINC90], 6.8, 6.9; [LORE86]; [MEIN67]; [SINW82]

8.2 Nonlinear Approximation

If the approximation function Φ does not have the form (8.2)

$$\Phi(x, c_0, c_1, \ldots, c_n) = c_0 \varphi_0(x) + \ldots + c_n \varphi_n(x)$$

for linearly independent functions φ_k, but if instead the φ_k are dependent on free parameters of their own, such as in $\Phi(x, c) = c_0 + c_1 e^{c_2(x-c_n)^4} + c_4 \ln(c_5 x)$, then trying to minimize

$$(8.38) \quad D^2(c_0, c_1, \ldots, c_n) = \sum_{i=0}^{N} w_i (f(x_i) - \Phi(x_i, c))^2 \overset{!}{=} \quad \text{Minimum}$$

in general will lead to a nonlinear system of equations for the $c_k^{(0)}$:

$$(8.39) \qquad \frac{\partial(D^2)}{\partial c_j} = 0 \quad \text{for} \quad j = 0, \ldots, n.$$

In some special cases, the nonlinear model function Φ can be replaced by a linear model of the form (8.2) by an appropriate transformation (see section 8.2.1). Otherwise, the nonlinear system can be solved by the damped Newton method (see chapter 6.2.1.2) or by a combination of the Householder transformation method (see chapter 8.1.3.4) and the damped Newton method.

8.2.1 Transformation Method for Nonlinear Least Squares

If we use a nonlinear model $\Phi(x, c)$, for instance one of the forms

$$\Phi_1(x, c_0, c_1) = 1/(c_0 + c_1 \ln x)$$

or

$$\Phi_2(x, c) = e^{c_0 + c_1 x + c_2 x^2 + c_3 x^3},$$

then the minimization problem (8.38) leads to a nonlinear system for the coefficients c_k. If, however, we transform the above model function Φ_1 into $T(\Phi_1) = 1/\Phi_1 = \tilde{\Phi}_1$ then we obtain a linear model with

$$\tilde{\Phi}_1(x, c_0, c_1) = c_0 + c_1 \ln x.$$

For Φ_2 we can use $T(\Phi_2) = \ln \Phi_2 = \tilde{\Phi}_2$. Then the linear model

$$\tilde{\Phi}_2(x, \mathbf{c}) = c_0 + c_1 x + c_2 x^2 + c_3 x^3$$

arises. Instead of the sum of the squared errors

$$D^2(c_0, c_1, \ldots, c_n) = \sum_{i=0}^{N} w_i (f(x_i) - \Phi(x_i, \mathbf{c}))^2$$

for the nonlinear model function Φ, we now must minimize the transformed sum of the squared errors:

$$\begin{aligned}
\tilde{D}^2(c_0, c_1, \ldots, c_n) &= \sum_{i=0}^{N} \tilde{w}_i (T(f(x_i)) - T(\Phi(x_i)))^2 \\
&=: \sum_{i=0}^{N} \tilde{w}_i (\tilde{f}(x_i) - \tilde{\Phi}(x_i, \mathbf{c}))^2,
\end{aligned}$$

for the linear model function $\tilde{\Phi}$, so that the coefficients $c_k^{(0)}$ of the best approximation $\tilde{\Phi}^{(0)}$ are determined by a linear system. In order to achieve

$$D^2(c_0, c_1, \ldots, c_n) \overset{!}{=} \tilde{D}^2(c_0, c_1, \ldots, c_n),$$

at least in an approximate manner, one should best use weights \tilde{w}_i given by the following formula (see [SPÄT74/1]):

$$(8.40) \qquad\qquad \tilde{w}_i = \frac{w_i}{T'^2(\Phi)|_{\Phi = y_i}} \ .$$

ALGORITHM 8.24 (*Transformation method*).

Given: Points $(x_i, y_i = f(x_i))$, $i = 0, \ldots, N$, and weights $w_i > 0$ for
$i = 0, \ldots, N$ and a nonlinear model function Φ.

Task: Find the best approximation $\Phi^{(0)}$ so that

$$D^2(c_0, c_1, \ldots, c_n) = \sum_{i=0}^{N} w_i(f(x_i) - \Phi(x_i, \mathbf{c}))^2 \overset{!}{=} \quad \text{Minimum.}$$

1^{st} step: Choose a transformation T so that with $\tilde{\Phi} := T(\Phi)$,

$$\tilde{\Phi}(x) = \sum_{i=0}^{N} c_k \tilde{\varphi}_k(x) \text{ is a linear model.}$$

2^{nd} step: Calculate the transformed weights \tilde{w}_i from formula (8.40).

3^{rd} step: Calculate the coefficients $c_k^{(0)}$ of $\tilde{\Phi}^{(0)}$ from the normal equations

$$\begin{pmatrix} (\tilde{\varphi}_0, \tilde{\varphi}_0) & (\tilde{\varphi}_0, \tilde{\varphi}_1) & \cdots & (\tilde{\varphi}_0, \tilde{\varphi}_n) \\ (\tilde{\varphi}_1, \tilde{\varphi}_0) & (\tilde{\varphi}_1, \tilde{\varphi}_1) & \cdots & (\tilde{\varphi}_1, \tilde{\varphi}_n) \\ \vdots & \vdots & & \vdots \\ (\tilde{\varphi}_n, \tilde{\varphi}_0) & (\tilde{\varphi}_n, \tilde{\varphi}_1) & \cdots & (\tilde{\varphi}_n, \tilde{\varphi}_n) \end{pmatrix} \begin{pmatrix} c_0^{(0)} \\ c_1^{(0)} \\ \vdots \\ c_n^{(0)} \end{pmatrix} = \begin{pmatrix} (\tilde{f}, \tilde{\varphi}_0) \\ (\tilde{f}, \tilde{\varphi}_1) \\ \vdots \\ (\tilde{f}, \tilde{\varphi}_n) \end{pmatrix},$$

where $(\tilde{\varphi}_j, \tilde{\varphi}_k) := \sum_{i=0}^{N} \tilde{w}_i \tilde{\varphi}_j(x_i) \tilde{\varphi}_k(x_i)$ and

$(\tilde{f}, \tilde{\varphi}_j) := \sum_{i=0}^{N} \tilde{w}_i \tilde{f}_i \tilde{\varphi}_j(x_i)$ for $\tilde{f}_i = T(f(x_i))$.

4^{th} step: Use $c_k^{(0)}$ to form $\Phi^{(0)}$.

We again note that the solution of the normal equations in the third step should be found via Householder transformations (see section 8.1.3.3) in order to avoid the bad effects of its often ill-conditioned system matrix.

Special cases of this linearizing transformation can be found in [SPÄT74/1].

8.2.2 Nonlinear Root-Mean-Square Fitting

If one cannot find an appropriate transformation of a given approximation function Φ that depends nonlinearly on its coefficients c_k, then the minimization (8.38) will lead to a nonlinear system (8.39) for the optimal coefficients of the nonlinear model function Φ.

The resulting nonlinear system can for example be solved by the damped Newton method, see chapter 6.2.1. where one can solve the linear systems that arises either with a Gauß algorithm or via Householder transformations

(see section 4.14). The latter is to be preferred because Householder transformations do not worsen the conditioning of the system matrix.

For the damped Newton method, the user must prescribe $n + 1$ initial values for the coefficients c_k, $k = 0, \ldots, n$, as well as $N + 1$ weights w_i for the points (x_i, y_i), $i = 0, \ldots, N$, unless all $w_i = 1$. The partial derivatives which are necessary for forming the Jacobi matrix in Newton's method can be approximated by means of central difference quotients.

Bibliography for Section 8.2

[SPÄT74/1].

8.3 Decision Strategy

For the discrete linear and nonlinear approximation in the root-mean-square sense the crucial ingredient is to be able to select an appropriate model function. If a good model cannot be determined, one should use least square splines (see chapter 11). Algebraic polynomials will seldom be appropriate as model functions, unless the nodes are selected with great care, see section 8.1.6. If, however, algebraic polynomials are used, a least squares approximation can be recommended that uses orthogonal polynomials (see section 8.1.3.2).

For a nonlinear approximation, the necessity to solve a nonlinear system can be avoided if the model function can be transformed into a linear model. Otherwise one has to solve a nonlinear system - preferably by using Householder transformations, see sections 8.1.3.4 and 8.2.2.

For a real or complex valued periodic function with discrete and equidistant nodes, the discrete Fast Fourier transformation (FFT) is recommended (section 8.1.5.3).

Bibliography for Chapter 8

[ATKI89]; [CHEN82]; [CONT80], 6; [HÄMM91], 4; [KINC90]; [LORE86]; [MAES88], 6; [NIED87], 6; [SCHW89], 7.

Chapter 9

Polynomial and Rational Interpolation

9.1 The Problem

Assume that we know $n+1$ *data points* (x_i, y_i) with $x_i, y_i \in \mathbb{R}$, for $i = 0, \ldots, n$, in the form of a *table of values*

i	0	1	2	...	n
x_i	x_0	x_1	x_2	...	x_n
y_i	y_0	y_1	y_2	...	y_n

The *nodes* x_i are assumed distinct, but not necessarily equidistant or arranged in order. The data points (x_i, y_i) will be called *interpolation points*.

The task of polynomial interpolation is to find an algebraic polynomial $\Phi \equiv P_m$ of lowest possible degree m so that at the nodes x_i: $\Phi(x_i) = y_i$.

THEOREM 9.1 (*Existence and uniqueness theorem*).
For $n + 1$ interpolation points (x_i, y_i) with distinct nodes $x_i, i = 0, \ldots, n$, there exists a unique polynomial Φ:

$$\Phi(x) \equiv P_n(x) = \sum_{k=0}^{n} c_k x^k \qquad \text{with} \qquad c_k \in \mathbb{R},$$

so that

$$\Phi(x_i) \equiv P_n(x_i) = \sum_{k=0}^{n} c_k x_i^k = y_i \quad \text{for } i = 0, \ldots, n.$$

Φ is called the *interpolating polynomial* belonging for the given set of interpolation points. The equations $\Phi(x_i) = y_i$ for $i = 0, \ldots, n$ are called the *interpolation conditions*.

If the values $f(x_i)$ of a function $f \in C[a, b]$ are known at $n + 1$ nodes x_i and if $\Phi \in C[a, b]$ is the interpolating polynomial for the interpolation points $(x_i, y_i = f(x_i))$, then $\Phi(x_i) = f(x_i) = y_i$ and it is reasonable to assume that Φ approximates the function f in $[a, b]$. Determining a value $\Phi(\bar{x})$ for $\bar{x} \in [a, b]$ with $\bar{x} \neq x_i$ is called *interpolation*; if \bar{x} lies outside of $[a, b]$, one speaks of *extrapolation*.

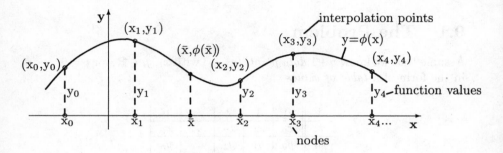

Figure 9.1: Interpolation

In the following we shall introduce different representations or *interpolation formulas* for the uniquely determined interpolating polynomial for $n + 1$ interpolation points.

REMARK (*Hermite interpolation*).
If for each $x_i \in [a, b]$ an $(m_i + 1)$ tuple of derivatives $(y_i, y_i', \ldots, y_i^{(m_i)})$ is given besides the value y_i, then the interpolating polynomial H with

$$H^{(k)}(x_i) = y_i^{(k)} \qquad \text{for} \quad k = 0, \ldots, m_i \text{ and } i = 0, \ldots, n$$

is called the *Hermite interpolation polynomial* (see [ATKI89], 3.6 or [WERN79], pp.7-16).

Bibliography for Section 9.1

[ATKI89], 3; [BJÖR74], 7.31; [BROS64], §21; [COLL73] I, 3.1; [ENGE87], 7.1; [HENR72] vol.2, 9.1; [McCA67], 6.0-1; [MEIN79], 3; [SCHW89], 3.1-3.3; [STOE91], 2.1.1; [STUM82], 3.1; [WERN79], I §1; [ZURM65], §11.1.

9.2 Lagrange Interpolation Formula

9.2.1 Lagrange Formula for Arbitrary Nodes

For as yet unspecified functions L_k that are independent of y_k, we define Φ in the form

$$(9.1) \qquad \Phi(x) \equiv L(x) = \sum_{k=0}^{n} L_k(x) y_k.$$

In order that the interpolation conditions $\Phi(x_i) = y_i$ are satisfied we must have that $L(x_i) = y_i$ at the nodes x_i for $i = 0, \ldots, n$.

At the nodes we thus should stipulate that

$$L_k(x_i) = \begin{cases} 1 & \text{for} \quad k = i, \\ 0 & \text{for} \quad k \neq i \end{cases}.$$

This can be achieved by setting

$$(9.2) \begin{cases} L_k(x) & = \dfrac{(x - x_0)(x - x_1) \ldots (x - x_{k-1})(x - x_{k+1}) \ldots (x - x_n)}{(x_k - x_0)(x_k - x_1) \ldots (x_k - x_{k-1})(x_k - x_{k+1}) \ldots (x_k - x_n)} \\ & = \displaystyle\prod_{\substack{i=0 \\ i \neq k}}^{n} \dfrac{x - x_i}{x_k - x_i}. \end{cases}$$

Here each L_k is a polynomial of degree n, so that $\Phi_n \equiv L$ is a polynomial of at most degree n. Formulas (9.1) and (9.2) form the *Lagrange interpolation* formula for arbitrary nodes.

ALGORITHM 9.2 (*Lagrange interpolation formula*).

Given: (x_i, y_i) for $i = 0, \ldots, n$, $y_i = f(x_i)$, and $x_i \neq x_k$ for $i \neq k$.

Task: Find the Lagrange interpolating polynomial.

1^{st} step: Determine L_k according to formula (9.2).

2^{nd} step: Find the interpolating polynomial L from formula (9.1).

Linear interpolation.

For two interpolation points $(x_0, y_0), (x_1, y_1)$, we can determine the Lagrange interpolating polynomial of degree $n = 1$. From (9.2)

$$L_0(x) = \frac{x - x_1}{x_0 - x_1}, \text{ and } L_1(x) = \frac{x - x_0}{x_1 - x_0},$$

so that the interpolation formula becomes:

$$L(x) = \sum_{k=0}^{1} L_k(x) y_k =$$

(9.3)

$$= \frac{x - x_1}{x_0 - x_1} y_0 + \frac{x - x_0}{x_1 - x_0} y_1 = \frac{\det \begin{pmatrix} y_0 & x_0 - x \\ y_1 & x_1 - x \end{pmatrix}}{x_1 - x_0}.$$

9.2.2 Lagrange Formula for Equidistant Nodes

Here we assume that the nodes x_i are equidistant with constant *step size*
$h = x_{i+1} - x_i$ for $i = 0, \ldots, n - 1$. Then $x_i = x_0 + hi$ for $i = 0, \ldots, n$ and

$$x = x_0 + ht, \text{ for } t \in [0, n].$$

Thus (9.2) becomes

$$L_k(x) = \prod_{\substack{i=0 \\ i \neq k}}^{n} \frac{t - i}{k - i} =: \tilde{L}_k(t) = \frac{t(t-1)\ldots(t-k+1)(t-k-1)\ldots(t-n)}{k!(-1)^{n-k}(n-k)!}.$$

The *Lagrange interpolation formula* for equidistant nodes thus reads:

$$\tilde{L}(t) = \sum_{k=0}^{n} \tilde{L}_k(t) y_k = \left(\prod_{i=0}^{n}(t - i) \right) \left(\sum_{k=0}^{n} \frac{(-1)^{n-k} y_k}{k!(n-k)!(t-k)} \right).$$

Bibliography for Section 9.2

[ATKI89], 3; [BJÖR74], 7.36; [CARN69], 1.7; [COLL73], p.89; [CONT80],
3.1-2; [ENGE87], 7.2; [HÄMM91], 6; [HENR72] vol.2, 10.2; [McCA67], 6.2;
[RICE83], chap.5; [STIE76], 7.11; [STOE91], 2.1.1; [STUM82], 3.1.1;
[TÖRN79] vol.2, 11.1; [ZURM65], §11.3.

9.3 The Aitken Interpolation Scheme for Arbitrary Nodes

Assume that $n+1$ interpolation points (x_i, y_i) are given, where the x_i are distinct but not necessarily equidistant. If one does not need the interpolating polynomial Φ itself but only wants to find its value $\Phi(\bar{x})$ at \bar{x}, then one can use the *Aitken interpolation scheme*.

One can find the value for $\Phi(\bar{x})$ of the interpolating polynomial by repeatedly using the linear interpolation (9.3). Let us call the linear interpolating polynomial associated with (x_0, y_0) and (x_1, y_1) by P_{01}. Then from (9.3):

$$P_{01}(x) = \frac{1}{x_1 - x_0} \quad \det \begin{pmatrix} y_0 & x_0 - x \\ y_1 & x_1 - x \end{pmatrix}.$$

If x_0 and x_i are two distinct nodes, then their associated linear interpolating polynomial shall be called P_{0i}, and:

$$(9.4)\ P_{0i}(x) = \frac{1}{x_i - x_0} \quad \det \begin{pmatrix} y_0 & x_0 - x \\ y_i & x_i - x \end{pmatrix} = P_{i0}(x) \text{ for each } i = 1, \ldots, n.$$

Here $P_{0i}(x_0) = y_0$ and $P_{0i}(x_i) = y_i$, i.e., P_{0i} solves the interpolation problem for the two interpolation points (x_0, y_0) and (x_i, y_i).

Using two linear polynomials P_{01} and P_{0i} for $i \geq 2$, we can produce polynomials P_{01i} of degree two at most by setting

$$(9.5)\ \ P_{01i}(x) = \frac{1}{x_i - x_1} \quad \det \begin{pmatrix} P_{01}(x) & x_1 - x \\ P_{0i}(x) & x_i - x \end{pmatrix} \text{ for each } i = 2, \ldots, n.$$

Each P_{01i} is the interpolating polynomial which solves the interpolation problem for the three interpolation points $(x_0, y_0), (x_1, y_1)$ and (x_i, y_i). The repeated application of such linear interpolations leads to interpolating polynomials of increasing degrees. The interpolating polynomial of degree n for the given $n + 1$ interpolation points is obtained by linear interpolation applied to two different interpolating polynomials of degree $n-1$, each of which has been formed from n of the given $n + 1$ nodes. In general, knowing two polynomials $P_{012\ldots(k-1)i}$ of degree $k - 1$, one can calculate a polynomial $P_{012\ldots ki}$ of degree k via the formula

$$(9.6)\qquad P_{012\ldots(k-1)ki}(x) = \frac{1}{x_i - x_k} \det \begin{pmatrix} P_{012\ldots(k-1)k}(x) & x_k - x \\ P_{012\ldots(k-1)i}(x) & x_i - x \end{pmatrix}$$

for $k = 0, \ldots, n - 1$ and $i = k + 1, \ldots, n$.

Hence the polynomial $P_{012\ldots ki}$ of degree k solves the interpolation problem for the $k+2$ interpolation points $(x_0, y_0), (x_1, y_1), \ldots, (x_k, y_k), (x_i, y_i)$.

CALCULATION SCHEME 9.3 (*Aitken interpolation scheme*).

i	x_i	y_i	$P_{0i}(\bar{x})$	$P_{01i}(\bar{x})$	$P_{012i}(\bar{x})$	\cdots	$P_{0123\ldots n}(\bar{x})$	$x_i - \bar{x}$
0	x_0	y_0						$x_0 - \bar{x}$
1	x_1	y_1	P_{01}					$x_1 - \bar{x}$
2	x_2	y_2	P_{02}	P_{012}				$x_2 - \bar{x}$
3	x_3	y_3	P_{03}	P_{013}	P_{0123}			$x_3 - \bar{x}$
\vdots	\vdots	\vdots	\vdots	\vdots	\vdots			\vdots
k	x_k	y_k	P_{0k}	P_{01k}	P_{012k}			$x_k - \bar{x}$
\vdots	\vdots	\vdots	\vdots	\vdots	\vdots	\ddots		\vdots
n	x_n	y_n	P_{0n}	P_{01n}	P_{012n}	\cdots	$P_{0123\ldots n}$	$x_n - \bar{x}$

Finally when $k = n - 1$, $P_{012\ldots n}$ will solve the interpolation problem for the given $n+1$ interpolation points (x_i, y_i), $i = 0, \ldots, n$. The above scheme evaluates $P_{012\ldots n}(\bar{x})$ effectively for one desired point \bar{x}.

ALGORITHM 9.4 (*Aitken interpolation scheme*).

Given: (x_i, y_i) for $i = 0, \ldots, n$, $y_i = f(x_i)$, and $x_i \neq x_k$ for $i \neq k$.

Task: Find the value of the interpolating polynomial
$\Phi(\bar{x}) = P_{012\ldots n}(\bar{x})$ at a point $\bar{x} \neq x_i$, which shall approximate $f(\bar{x})$.

1^{st} step: In the calculation scheme 9.3, fill in the three columns labelled x_i, y_i and $x_i - \bar{x}$ for $i = 0, \ldots, n$.

2^{nd} step: Calculate the column labelled $P_{0i}(\bar{x})$ according to formula (9.4) for $i = 1, \ldots, n$ and $x = \bar{x}$.

3^{rd} step: Calculate the column labelled $P_{01i}(\bar{x})$ according to formula (9.5) for $i = 2, \ldots, n$ and $x = \bar{x}$.

4^{th} step: Calculate the subsequent columns $P_{012\ldots ki}(\bar{x})$ according to formula (9.6) for $k = 3, \ldots, n-1$ and $i = k+1, \ldots, n$ until $P_{0123\ldots n}(\bar{x}) = \Phi(\bar{x})$.

An advantage of Aitken's scheme is that one need not decide in advance how many interpolation points (x_i, y_i) to use. It is possible to add new interpolation points at the end and thus augment the scheme *row by row*. The nodes themselves need not be ordered in any fashion.

Bibliography for Section 9.3

[CONT80], 3.3; [ENGE87], 7.3; [HÄMM91], 8a; [HENR72] vol.2, 10.4.5; [ISAA66], 6.2; [McCA67], 6.4; [NOBL65] II, 8.4; [WERN79], p.47.

9.4 Inverse Interpolation According to Aitken

If a table of functional values $(x_i, f(x_i))$ for a function $f \in C[a,b]$ does not contain an x-value for a given $\bar{y} = f(\bar{x})$, one can use the Aitken scheme by interchanging the roles of x and y in order to find \bar{x}. For this it is necessary that the inverse function $x = f^{-1}(y)$ exists. This for example is the case if f is strictly monotonic in $[a,b]$.

In the inverse Aitken scheme, one determines the value $\bar{x} = \Phi^*(\bar{y})$ for the interpolating polynomial Φ^* at the interpolation points $(y_i, x_i = f^{-1}(y_i))$.

CALCULATION SCHEME 9.5 (*Aitken inverse interpolation*).

i	y_i	x_i	x_{0i}	x_{01i}	\cdots	$x_{012\ldots n}$	$y_i - \bar{y}$
0	y_0	x_0					$y_0 - \bar{y}$
1	y_1	x_1	x_{01}				$y_1 - \bar{y}$
2	y_2	x_2	x_{02}	x_{012}			$y_2 - \bar{y}$
\vdots	\vdots	\vdots	\vdots	\vdots	\ddots		\vdots
n	y_n	x_n	x_{0n}	x_{01n}	\cdots	$x_{012\ldots n}$	$y_n - \bar{y}$

One proceeds according to algorithm 9.4 by interchanging the roles of x and y and by replacing P_{0i} by x_{0i}, P_{01i} by x_{01i} etc..

Bibliography for Section 9.4

[BJÖR74], 7.38; [CONT80], 3.4; [ENGE87], 7.4; [HENR72] vol.2, 10.6-7; [ISAA66], 6.2; [SCHW89], 3.5.

9.5 Newton Interpolation Formula

9.5.1 Newton Formula for Arbitrary Nodes

If $n + 1$ interpolation points $(x_i, y_i), i = 0, \ldots, n$, are given, the *Newton interpolating polynomial* N is given as:

$$(9.7) \qquad \Phi(x) \equiv N(x) = b_0 + b_1(x - x_0) + b_2(x - x_0)(x - x_1) + \ldots + \\ + b_n(x - x_0)(x - x_1)(x - x_2) \cdots (x - x_{n-1}).$$

The interpolating conditions $\Phi(x_i) \equiv N(x_i) = y_i$ for $i = 0, \ldots, n$ give rise to a system of $n + 1$ linear equations for the $n + 1$ coefficients b_k. *Divided differences* of first and higher order are defined by:

$$[x_i x_k] \quad := \quad \frac{y_i - y_k}{x_i - x_k},$$

$$[x_i x_k x_h] \quad := \quad \frac{[x_i x_k] - [x_k x_h]}{x_i - x_h},$$

$$[x_i x_k x_h x_m] \quad := \quad \frac{[x_i x_k x_h] - [x_k x_h x_m]}{x_i - x_m}, \text{ etc.}$$

These remain unchanged under permutations of the distinct nodes, see [ZURM65], §11.4. With them we can obtain the desired coefficients as:

$$(9.8) \begin{cases} b_0 &= y_0, \\ b_1 &= [x_1 x_0] = \dfrac{y_1 - y_0}{x_1 - x_0}, \\ b_2 &= [x_2 x_1 x_0] = \dfrac{[x_2 x_1] - [x_1 x_0]}{x_2 - x_0}, \\ b_3 &= [x_3 x_2 x_1 x_0] = \dfrac{[x_3 x_2 x_1] - [x_2 x_1 x_0]}{x_3 - x_0}, \\ \vdots \\ b_n &= [x_n x_{n-1} \ldots x_2 x_1 x_0] = \dfrac{[x_n x_{n-1} \ldots x_2 x_1] - [x_{n-1} x_{n-2} \ldots x_1 x_0]}{x_n - x_0}. \end{cases}$$

It is especially easy to determine the b_k in the following computational scheme in which the ordering of the nodes x_i is arbitrary.

CALCULATION SCHEME 9.6 (*Newton interpolation*).

i	x_i	y_i				
0	x_0	$y_0 = \underline{b_0}$				
			$[x_1 x_0] = \underline{b_1}$			
1	x_1	y_1		$[x_2 x_1 x_0] = \underline{b_2}$		
			$[x_2 x_1]$		$[x_3 x_2 x_1 x_0] = \underline{b_3}$	\cdots
2	x_2	y_2		$[x_3 x_2 x_1]$		
			$[x_3 x_2]$		\vdots	
3	x_3	y_3		\vdots		
			\vdots			
\vdots	\vdots	\vdots				

ALGORITHM 9.7 (*Newton interpolation formula*).

Given: (x_i, y_i) for $i = 0, \ldots, n$, $y_i = f(x_i)$, and $x_i \neq x_k$ for $i \neq k$.

Task: Find the interpolating polynomial in Newton's form (9.7).

1^{st} step: Calculate b_k via the calculation scheme 9.6 using (9.8).
2^{nd} step: Form $N(x)$ according to (9.7).

Bibliography for Section 9.5.1

[STUM82], p.52/53; [WERN79], p.35/36; [ZURM65], §11.4.

9.5.2 Newton Formula for Equidistant Nodes

Let the given nodes x_i be equidistant with a constant step size $h = x_{i+1} - x_i$ for $i = 0, \ldots, n - 1$. Then $x_i = x_0 + hi$ for $i = 0, \ldots, n$. As before we set

$$x = x_0 + ht \text{ for } t \in [0, n].$$

In order to calculate the coefficients b_i in the scheme 9.6, we introduce as abbreviations so-called *differences* Δ_i^k which will always refer to y-values. These differences are defined as follows:

$$(9.9) \quad \begin{cases} \Delta_i^0 & = \ y_i, \\ \Delta_{i+1/2}^{k+1} & = \ \Delta_{i+1}^k - \Delta_i^k, & k = 0, 2, 4, \ldots, \\ \Delta_i^{k+1} & = \ \Delta_{i+1/2}^k - \Delta_{i-1/2}^k, & k = 1, 3, 5, \ldots. \end{cases}$$

Thus for example:

$$\Delta_{i+1/2}^1 \ = \ y_{i+1} - y_i, \Delta_i^2 = \Delta_{i+1/2}^1 - \Delta_{i-1/2}^1 = y_{i+1} - 2y_i + y_{i-1},$$

$$\Delta_{i+1/2}^3 \ = \ \Delta_{i+1}^2 - \Delta_i^2 = y_{i+2} - 3y_{i+1} + 3y_i - y_{i-1}.$$

The differences Δ_i^k can be determined via the following scheme:

CALCULATION SCHEME 9.8 (*Difference scheme*).

i	y_i	$\Delta_{i+1/2}^1$	Δ_i^2	$\Delta_{i+1/2}^3$	\cdots
0	y_0				
		$y_1 - y_0 = \Delta_{1/2}^1$			
1	y_1		$\Delta_{3/2}^1 - \Delta_{1/2}^1 = \Delta_1^2$		
		$y_2 - y_1 = \Delta_{3/2}^1$		$\Delta_{3/2}^3$	
2	y_2		$\Delta_{5/2}^1 - \Delta_{3/2}^1 = \Delta_2^2$	\vdots	
		$y_3 - y_2 = \Delta_{5/2}^1$	\vdots		
3	y_3	\vdots			
\vdots	\vdots				

This scheme can also be augmented at will. If we set $h = x_{i+1} - x_i$, then the b_i are determined as

$$b_i = [x_i x_{i-1} \ldots x_1 x_0] = \frac{1}{i! h^i} \Delta_{i/2}^i \ \text{for } i = 1, \ldots, n.$$

And the Newton interpolation polynomial (9.7) is obtained as

$$N(x) = \tilde{N}(t) = y_0 + \binom{t}{1} \Delta_{1/2}^1 + \binom{t}{1} \Delta_1^2 + \ldots + \binom{t}{n} \Delta_{n/2}^n$$

where $\binom{t}{k} = \frac{t!}{(t-k)! k!}$ denotes the *binomial coefficient*. Here $N(x)$ or $\tilde{N}(t)$ is the *Newton interpolation formula for descending differences*; it is often denoted by $N_+(x)$ or $\tilde{N}_+(t)$.

Bibliography for Section 9.5

[BJÖR74], 7.33-34; [BOOR79] I; [CARN69], 1.6; [COLL73], 3.1; [CONT80], 3.6; [ENGE87], 7.5; [HÄMM91], 6.7; [ISAA66], 6.1; [McCA67], 6.3; [RICE83], chap.5; [SCHW89], 3.4; [STOE91], 2.1.3; [STUM82] 3.1.2; [TÖRN79] vol.2, 11.2; [WERN79], I §3; [ZURM65], §11.4-5.

9.6 Remainder of an Interpolation and Estimates of the Interpolation Error

The interpolating polynomial $\Phi \in C(I)$ formed for $n + 1$ interpolation points $(x_i, y_i = f(x_i))$ with $x_i \in I \subset \mathbb{R}$ for $i = 0, \ldots, n$ attains the values $f(x_i) = \Phi(x_i)$ at the nodes x_i, but in general it will differ at all other points $x \in I$ from $f(x)$. Define

$$R(x) = f(x) - \Phi(x) \text{ for } x \in I.$$

$R(x)$ is the true *interpolation error* and R is called the *remainder of the interpolation*. While the remainder R vanishes at the nodes x_i, in general one can make no statement about it in I for $x \neq x_i$. This is so, because one can for example arbitrarily change f at values $x \neq x_i$ without affecting Φ. If, however, f is $(n + 1)$ times continously differentiable in I the *Lagrange remainder formula* is valid for arbitrary nodes:

$$(9.10) \quad \begin{cases} R(x) = \dfrac{1}{(n + 1)!} f^{(n+1)}(\xi)\pi(x) \quad \text{with} \quad \pi(x) = \prod_{i=0}^{n}(x - x_i), \\ \text{and some } \xi = \xi(x) \in I. \end{cases}$$

In the case of equidistant nodes $x_i = x_0 + hi$ with $x = x_0 + ht$ for $t \in I_t := [0, n]$:

$$(9.11) \quad \begin{cases} R(x) = R(x_0 + ht) = h^{n+1} \dfrac{1}{(n + 1)!} f^{(n+1)}(\tilde{\xi})\pi^*(t) =: \tilde{R}(t) \quad \text{with} \\ \pi^*(t) = \prod_{i=0}^{n}(t - i) \text{ for some } \tilde{\xi} = \tilde{\xi}(t) \in I_t. \end{cases}$$

To examine $\pi(x)$ in (9.10) in order to estimate the interpolation error is very difficult if one chooses the nodes arbitrarily in I. In the case of equidistant nodes, there is the following result for $\pi^*(t)$ in (9.11):

The moduli of the extrema of $\pi^*(t)$ decrease up to the midpoint of the interval $[0, n]$ and afterwards they increase. Outside of this interval they grow very rapidly. Thus $\tilde{R}(t)$ can become rather large for values of t outside of the

interpolation interval, i.e., for *extrapolation*. The interpolation interval extends from the smallest to the largest node used for the interpolation. This statement is important for choosing appropriate interpolation points: One should choose the interpolation points (x_i, y_i) in such a way that the point \bar{x} lies near the center of the interpolation interval.

$$\pi^*(t) \text{ for } n = 5 \qquad\qquad \pi^*(t) \text{ for } n = 6$$

Figure 9.2: Interpolation error

An *estimation of the remainder* $R(x) = R(x_0 + th) = \tilde{R}(t)$ is possible for the case when one also knows interpolation points outside of the interpolation interval, so that the $(n+1)^{st}$ differences Δ^{n+1} can be formed. Then

$$\tilde{R}(t) \approx \frac{1}{(n+1)!}\Delta^{n+1}\pi^*(t),$$

provided that all the differences Δ^{n+1} do not differ too much. In this case it does not matter which of these $(n+1)$ differences is used, see [ZURM65], p.218.

As an estimate for the *Newton remainder for arbitrary nodes* we have for $\bar{x} \in I$:

$$(9.12) \qquad R(\bar{x}) \approx [x_{n+1}, x_n, x_{n-1}, \dots, x_1, x_0]\prod_{i=0}^{n}(\bar{x} - x_i),$$

where one further point (x_{n+1}, y_{n+1}) besides those used for constructing Φ is also assumed known.

Of course, it is not always the case that a function to be interpolated is $n+1$ times differentiable, or perhaps the magnitude of its $(n+1)^{st}$ derivative is not easily estimated. Interpolation, however, is a bounded linear projection if error is measured in the maximum norm, and thus the estimates in section 8.6 for linear projections do apply.

From this standpoint, the method chosen for computing the interpolating polynomial (Lagrange, Aitken, or Newton) is irrelevant, since the polynomial obtained is theoretically determined and unique regardless of the procedure. Given some nodes x_0, \ldots, x_n of interpolation, we let Lf be the polynomial interpolating f at the nodes x_0, \ldots, x_n with representation

$$Lf(x) = \sum_{k=0}^{n} f(x_k) L_k(x)$$

as described in Section 9.2.

It may be seen that

$$\|L\| = \| \sum_{k=0}^{n} |L_k(x)| \|,$$

and that furthermore

$$\|I - L\| = 1 + \|L\|.$$

Thus the quality of approximation by interpolation is related initimately (via the theorem of Lebesgue, Section 8.1.6.3) to $\|L\|$, which obviously depends upon the spacing of the nodes x_0, \ldots, x_n. From this point of view, equally spaced nodes are very bad indeed, with $\|L\|$ growing at an exponential rate with respect to n. Further reflection does show that the interval of interpolation may be standardized to $[-1, 1]$ in the ensuing discussion and that the nodes of interpolation may be shifted so that $x_0 = -1$ and $x_n = 1$ with no increase in $\|L\|$. If this is done, the shape of the function $\sum_{k=0}^{n} |L_k(x)|$ is scallop-shaped, with $n+1$ minima at x_0, \ldots, x_n where the value is 1 and n maxima in–between. Obviously $\|L\|$ is attained at at least one of these maxima, and it is a result of Kilgore [KILG77] that it must be attained at all of them if $\|L\|$ is to be minimized. Further results by Kilgore [KILG78] and de Boor and Pinkus [BOOR78] show that the configuration of nodes which gives the equality at all maxima is unique and that the best possible value of $\|L\|$ also enjoys another useful property: If the nodes x_0, \ldots, x_n are not the optimal set of nodes, then some of the resulting maxima will exceed the optimal value, but some others will fall short. Thus, the difference between

the greatest and least of these values provides a basis for evaluating the quality of a given set of nodes. Since inevitably $\|L\| \to \infty$ as $n \to \infty$ and since actual computation of the optimal nodes is a practical impossibility, this consideration becomes important. Interpolation based upon the zero set of the Chebyshev polynomial $T_{n+1}(x)$ gives excellent results, with the norm of the interpolation operator L of degree at most n satisfying

$$\|L\| \leq \frac{2}{\pi} \log n + \frac{1}{2},$$

for all n. Similar results may be obtained if the zero set of $T_{n+1}(x)$ is "extended" by dilating the zero x_0, \ldots, x_n until $x_0 = -1$ and $x_n = 1$.

Simultaneous approximation of derivatives may be affected by adding interpolation of the function f and its derivatives at ± 1, in the manner of Theorem 8.21. We may state the following theorem of Balázs and Kilgore [BALÁ91]:

THEOREM 9.9

Let $g \in C^q[-1, 1]$ be such that $g^{(k)}(\pm 1) = 0$ for $k = 0, \ldots, q$, and let L_n be a linear projection of functions in $C[-1, 1]$ into the algebraic polynomials of degree at most n. Then for $k = 0, \ldots, q$ and for all $|x| \leq 1$ we have

(i) for q even:

$$|g^{(k)}(x) - \left(\left(\sqrt{1-x^2} \right)^q L_n \left(\frac{g(t)}{(\sqrt{1-t^2})^q} \right) \right)^{(k)} (x)| \leq$$

$$\leq const. \left(\frac{\sqrt{1-x^2}}{n} + \frac{1}{n^2} \right)^{q-k} \|L_n\| E_n(g^{(q)}).$$

(ii) for q odd:

$$|g^{(k)}(x) - \left(\sqrt{1-x^2} \right)^{q+1} L_n \left(\frac{g(t)}{(\sqrt{1-x^2})^{q+1}} \right)^{(k)} (x)| \leq$$

$$\leq const. \left(\frac{\sqrt{1-x^2}}{n} + \frac{1}{n^2} \right)^{q-1-k} E_n(g^{(q)}) \| \sqrt{1-x^2} L_n \|.$$

(iii) also for q odd:

$$|g^{(k)}(x) - \left(\left(\sqrt{1-x^2} \right)^{q+1} L_n \left(\frac{q(t)}{(\sqrt{1-x^2})^{q+1}} \right) \right)^{(k)} (x)| \leq$$

$$\leq const. \left(\frac{\sqrt{1-x^2}}{n} + \frac{1}{n^2} \right)^{q-k} E_n(g^{(q)}) \|L_n^*\|,$$

where

$$L_n^* f(x) := \sqrt{1-x^2} L_n \left(\frac{f(t)}{\sqrt{1-t^2}} \right)(x),$$

and

$$\|L_n^*\| = \sup\{ \, \|L_n^* f\| \;\; | \;\; \|f\| \leq 1, \; \frac{f(x)}{\sqrt{1-x^2}} \in C[-1,1] \, \}.$$

Recall that the error function E_n was defined in section 8.1.6. The above theorem handles projections L_n more general than operators of interpolation and also more general than those in Theorem 8.21; the construction of L_n^* indeed substitutes for the hypothesis that L_n be defined for odd periodic functions as in Theorem 8.21. We have the estimate $\|L_n^*\| \leq const \, \log n$ valid for interpolation on the Chebyshev nodes, and therefore the following:

COROLLARY 9.10

Let $L_n f$ interpolate the function $f \in C^q[-1,1]$ on the zero set of the Chebyshev polynomial $T_{n+1}(x)$ and interpolate the derivatives $f, f', \ldots, f^{(r-1)}$ at ± 1 for $n = [\frac{q+1}{2}]$. Then for $k = 0, \ldots, q$ and for all $|x| \leq 1$

$$|f^{(k)}(x) - L_n^{(k)} f(x)| \leq C_{q,k} \left(\frac{\sqrt{1-x^2}}{n} + \frac{1}{n^2} \right)^{q-k} \log n \, E_n(f^{(q)}).$$

This result states in particular that the polynomial of interpolation on the zero set of $T_{n+1}(x)$ augmented merely by interpolation at ± 1 must converge uniformly to the function interpolated, and the derivatives of the polynomial must converge uniformly to the derivatives of the function. Reasonable values of the constants $C_{q,k}$ can be obtained for numerical work from the discussion in section 8.1.6.

Convergence of interpolation in the 2-norm.

We have stated that interpolation is not a bounded (continuous) linear operator if the domain is normed with the 2-norm. However, if the domain is $C[-1,1]$ with the usual maximum norm, it is still possible to discuss the least squares error in the interpolation procedure. One may obtain the following results:

THEOREM 9.11 *(Erdős–Turán theorem).*
Let $w(x) \geq 0$ be a weight function on $[-1,1]$, and let $f \in C[-1,1]$. Let Lf interpolate f on the zero set of $P_{n+1}(x)$, where P_0, P_1, \ldots is the sequence of orthogonal polynomials generated by $w(x)$ (see section 8.1.3.2). Then with $E_n(f)$ denoting the error in uniform approximation, we have as $n \to \infty$

$$\|(f - Lf)\|_{2,w} \leq 2E_n(f) \left(\int_{-1}^{1} w(x)dx \right)^{1/2}.$$

THEOREM 9.12
Let $f \in C^q[-1,1]$, and let Lf be as in the previous theorem. Let Pf interpolate on the nodes of Lf augmented by interpolation of $f, f', \ldots, f^{(r-1)}$ for $r = \lceil \frac{q+1}{2} \rceil$. Then, if the weight function $w(x)$ is positive and finite or is the product of a positive finite function with a Jacobi weight function, we have for $k = 0, \ldots, q$ and for all $|x| \leq 1$ as $n \to \infty$

$$\|f^{(k)}(x) - P^{(k)}f_{(x)}\|_{2,w} \leq \text{const } n^{k-q}E_n(f^{(q)}).$$

More striking is the next result, which underscores the great versatility of the Chebyshev nodes for interpolation.

THEOREM 9.13
Let $f \in C^q[-1,1]$, and let PF interpolate f on the zero set of the Chebyshev polynomial $T_{n+1}(x)$, augmented by interpolation of $f, f', \ldots, f^{(r-1)}$ for $r = \lceil \frac{q+1}{2} \rceil$ at ± 1. Then for any weight function $w(x)$ as described in the previous theorem we have for $k = 0, \ldots, q$ and for $|x| \leq 1$

$$\|f^{(k)}(x) - P^{(k)}f(x)\|_{2,w} \leq \text{const } n^{k-q}E_n(f^{(q)}).$$

Theorem 9.11 is the well known Erdös–Turán theorem, found for example in the book of Cheney [CHEN82]. Theorems 9.12 and 9.13 are discussed in [BALÁ91/2]

Bibliography for Section 9.6

[BALÁ91]; [BALÁ91/2]; [BOOR78]; [CHEN82]; [CONT80], 3.2; [ENGE87], 7.7; [ISAA66], 6.3.1-2; [KILG77]; [KILG78]; [NOBL65], II, 8.2; [ZURM65], §11.4-5.

9.7 Rational Interpolation

Due to their rapidly changing slopes and many extrema, the naive use of algebraic polynomials is not well suited to represent smooth curves. This can be remedied by using splines. If, however, one wants to use one single function for the interpolation, and not a spline function that is defined piecewise as a low degree polynomial, one can try to interpolate with rational functions because rational functions often give smoother approximations than polynomials.

We are given $n + 1$ interpolation points (x_i, y_i), $i = 0, \ldots, n$, with $y_i = f(x_i)$ and $x_i \neq x_k$ for $i \neq k$ and desire to find an interpolating rational function R given as

$$(9.13) \qquad R(x) = \frac{P_k(x)}{Q_j(x)},$$

where P_k and Q_j are algebraic polynomials of degree k and j, respectively, and $Q_j(x) \neq 0$ for all $x \in [a, b]$. R must fulfill the $n + 1$ interpolation conditions

$$(9.14) \qquad R(x_i) = \frac{P_k(x_i)}{Q_j(x_i)} = y_i \text{ for } i = 0, \ldots, n.$$

The two polynomials Q_j and P_k are dependent on $k + j + 2$ coefficients. If we normalize one of those to have leading coefficient 1, then R will depend only on $k + j + 1$ parameters. Thus we must have that $n = k + j$ in order to be able to calculate the $k + j + 1$ coefficients uniquely from (9.14). If R interpolates f at the points (x_i, y_i), then

$$(9.15) \qquad P_k(x_i) - Q_j(x_i)y_i = 0 \quad \text{for all } i = 0, \ldots, n.$$

Conversely, if all equations in (9.15) are satisfied, then (9.14) holds provided $Q_j(x_i) \neq 0$. If $Q_j(x_i) = 0$, then (9.15) implies that $P_k(x_i) = 0$.

In [WERN79] an easy method is given for calculating a specific rational function R as a continued fraction:

In this, *inverse difference quotients* for the points $(x_i, y_i), i = 0, \ldots, n$, are defined as:

$$(9.16)\begin{cases} \nabla^1(x_0, x_i)y = \dfrac{x_i - x_0}{y_i - y_0} \quad \text{for} \quad i = 1, \ldots, n, \\[2mm] \nabla^m(x_0, x_1, \ldots, x_{m-1}, x_i)y = \\[2mm] \quad = \dfrac{x_i - x_{m-1}}{\nabla^{m-1}(x_0, x_1, \ldots, x_{m-2}, x_i)y - \nabla^{m-1}(x_0, x_1, \ldots, x_{m-1})y} \\[2mm] \quad \text{for} \quad i = m, \ldots, n, \quad \text{and} \quad m \geq 2. \end{cases}$$

These quotients can be calculated easily via the following inverse difference scheme.

CALCULATION SCHEME 9.14 (*Inverse difference scheme*).

i	x_i	y_i	∇_y^1	∇_y^2	\cdots	∇_y^n
0	x_0	y_0				
			$\nabla^1(x_0, x_1)y = \alpha_1$			
1	x_1	y_1		$\nabla^2(x_0, x_1, x_2)y = \alpha_2$		
			$\nabla^1(x_0, x_2)y$		\ddots	
2	x_2	y_2		$\nabla^2(x_0, x_1, x_3)y$		
			$\nabla^1(x_0, x_3)y$			
3	x_3	y_3		\vdots		$\nabla^n(x_0, x_1, \ldots, x_n)y = \alpha_n$
			\vdots			
\vdots	\vdots	\vdots	$\nabla^1(x_0, x_{n-1})y$		\iddots	
$n-1$	x_{n-1}	y_{n-1}		$\nabla^2(x_0, x_1, x_n)y$		
			$\nabla^1(x_0, x_n)y$			
n	x_n	y_n				

From (9.16) we obtain

$$(9.17) \qquad y = f(x) = y_0 + \frac{x - x_0}{\nabla^1(x_0, x)y}$$

for the function $y = f(x)$. Furthermore (9.16) implies for $x_i := x$ that

$$
(9.18)
\begin{cases}
\nabla^1(x_0, x)y = \nabla^1(x_0, x_1)y + \frac{x - x_1}{\nabla^2(x_0, x_1, x)y}, \\[2mm]
\nabla^2(x_0, x_1, x)y = \nabla^2(x_0, x_1, x_2)y + \frac{x - x_2}{\nabla^3(x_0, x_1, x_2, x)y}, \\[2mm]
\quad \vdots \\[2mm]
\nabla^n(x_0, x_1, \ldots, x_{n-1}, x)y = \nabla^n(x_0, x_1, \ldots, x_n)y + \frac{x - x_n}{\nabla^{n+1}(x_0, x_1, \ldots, x_n, x)y}.
\end{cases}
$$

Using the formulas (9.18) repeatedly in (9.17), f has the *continued fraction* representation

$$
(9.19)
\begin{cases}
f(x) = y_0 + \cfrac{x - x_0}{\nabla^1(x_0, x_1)y + \cfrac{x - x_1}{\nabla^2(x_0, x_1, x_2)y + \cfrac{\ddots}{\quad + \cfrac{x - x_n}{\underbrace{\nabla^{n+1}(x_0, x_1, \ldots, x_n, x)y}_{(*)}}}}}.
\end{cases}
$$

If we omit the last term marked by $(*)$ in (9.19), we have found a rational function R for which the interpolation conditions $R(x_i) = y_i$ are satisfied for $i = 0, \ldots, n$. This can easily be seen by induction. Thus we have found a representation for a rational interpolating function R of f in the form of a continued fraction:

$$
(9.20)
\begin{cases}
R(x) = y_0 + \cfrac{x - x_0}{\nabla^1(x_0, x_1)y + \cfrac{x - x_1}{\nabla^2(x_0, x_1, x_2)y + \cfrac{\ddots}{\quad + \cfrac{x - x_{n-1}}{\nabla^n(x_0, x_1, \ldots, x_n,)y}}}} \\[6mm]
 = y_0 + \cfrac{x - x_0}{\alpha_1 + \cfrac{x - x_1}{\alpha_2 + \cfrac{x - x_2}{\cfrac{\ddots}{\alpha_{n-1} + \cfrac{x - x_{n-1}}{\alpha_n}}}}}, \\[6mm]
\text{where} \quad \alpha_m := \nabla^m(x_0, x_1, \ldots, x_m)y \quad \text{for} \quad m = 1, \ldots, n.
\end{cases}
$$

The quantities y_0 and α_m can be computed for $m = 1, \ldots, n$ via the inverse difference scheme 9.9.

To convert the continued fraction (9.20) into a rational fraction (9.13), one can compute the "polynomial vector" $r = (P_k, Q_j)^T$ as

$$(9.21) \quad \begin{cases} r(x) = \begin{pmatrix} P_k(x) \\ Q_j(x) \end{pmatrix} = \begin{pmatrix} y_0 & x - x_0 \\ 1 & 0 \end{pmatrix} \begin{pmatrix} \alpha_1 & x - x_1 \\ 1 & 0 \end{pmatrix} \\ \begin{pmatrix} \alpha_2 & x - x_2 \\ 1 & 0 \end{pmatrix} \cdots \begin{pmatrix} \alpha_{n-1} & x - x_{n-1} \\ 1 & 0 \end{pmatrix} \begin{pmatrix} \alpha_n \\ 1 \end{pmatrix}. \end{cases}$$

EXAMPLE: For $n = 4$, according to (9.20), the continued fraction rational function has the form

$$R(x) = y_0 + \cfrac{x - x_0}{\alpha_1 + \cfrac{x - x_1}{\alpha_2 + \cfrac{x - x_2}{\alpha_3 + \cfrac{x - x_3}{\alpha_4}}}}.$$

The quantities $\alpha_m = \nabla^m(x_0, x_1, ..., x_m)y$ for $m = 1, \ldots, 4$ arise from the inverse difference scheme (9.9). From (9.21) the equation

$$r(x) = \begin{pmatrix} P_k(x) \\ Q_j(x) \end{pmatrix} = \begin{pmatrix} y_0 & x - x_0 \\ 1 & 0 \end{pmatrix} \begin{pmatrix} \alpha_1 & x - x_1 \\ 1 & 0 \end{pmatrix} \begin{pmatrix} \alpha_2 & x - x_2 \\ 1 & 0 \end{pmatrix}$$
$$\cdot \begin{pmatrix} \alpha_3 & x - x_3 \\ 1 & 0 \end{pmatrix} \begin{pmatrix} \alpha_4 \\ 1 \end{pmatrix}$$

determines both P_k and Q_j.

ALGORITHM 9.15 *(Rational interpolation).*

Given: Interpolation points $(x_i, y_i), i = 0, \ldots, n, \; y_i = f(x_i), \; x_i \neq x_k$ for $i \neq k$.

Task: Find a rational interpolating function R as in (9.13) in the continued fraction expansion (9.20).

1) Calculate the inverse differences $\alpha_m = \nabla^m(x_0, x_1, ..., x_m)y$ for $m = 1, \ldots, n$ in scheme 9.9 using (9.16).

2) Calculate the polynomials P_k, Q_j according to (9.21). Their degrees k and j are determined in (9.21): $k = \left[\frac{n+1}{2}\right]$ and $j = \left[\frac{n}{2}\right]$, where $[x]$ denotes the largest integer $k \le x$.

In [WERN79], p.75, another algorithm is given by which one can construct a rational interpolating function (9.13) for which one can specify the degree k of the numerator polynomial P_k with $k \ge j$. The degree j of the denominator polynomial Q_j is determined by the requirement $k + j = n$.

Bibliography for Section 9.7

[MEIN79], 4.1; [STOE91], 2.2; [WERN79], I §5.

9.8 Interpolation for Functions in Several Variables

We shall consider here only the case of two independent variables x and y for functions $f : z = f(x,y)$ for $(x,y) \in B \subset \mathbb{R}^2$ with $(x,y,z) \in \mathbb{R}^3$. Given $N+1$ interpolation points $(x_j, y_j, z_j = f(x_j, y_j))$ for the distinct nodes (x_j, y_j), $j = 0, \ldots, N$, we want to find an interpolating function ϕ with $z = \phi(x,y)$ defined for $(x,y) \in B$ in such a way that for all $j = 0, \ldots, N$

$$\phi(x_j, y_j) = f(x_j, y_j).$$

9.8.1 Lagrange Interpolation Formula for Two Variables

We are looking for an algebraic polynomial P_r in two variables x and y of degree r as low as possible,

$$P_r(x,y) = \sum_{p,q} a_{pq} x^p y^q, \quad \text{so that} \quad P_r(x_j, y_j) = f(x_j, y_j) \text{ for } j = 0, \ldots, N.$$

Here the degree of P is defined as $r = \max_{a_{pq} \neq 0} \{p + q\}$.

In general, the existence and uniqueness of such a solution is not certain ([ISAA66], sector 6.6).

If specifically the nodes (x_i, y_k) are points of a rectangular grid for $i = 0, \ldots, m$, $k = 0, \ldots, n$, and if $f(x_i, y_k) =: f_{ik}$, then the interpolation points can be arranged in a rectangular array as follows:

	y_0	y_1	\cdots	y_n
x_0	f_{00}	f_{01}	\cdots	f_{0n}
x_1	f_{10}	f_{11}	\cdots	f_{1n}
\vdots	\vdots	\vdots		\vdots
x_m	f_{m0}	f_{m1}	\cdots	f_{mn}

This special interpolation problem is solvable uniquely in the form

$$\Phi(x, y) = \sum_{i=0}^{m} \sum_{k=0}^{n} a_{ik} x^i y^k.$$

The Lagrange interpolation formula for the above node distribution uses the definitions

$$(9.22) \quad \begin{cases} L_i^{(1)}(x) &= \dfrac{(x - x_0)\ldots(x - x_{i-1})(x - x_{i+1})\ldots(x - x_m)}{(x_i - x_0)\ldots(x_i - x_{i-1})(x_i - x_{i+1})\ldots(x_i - x_m)}, \\[2mm] L_k^{(2)}(y) &= \dfrac{(y - y_0)\ldots(y - y_{k-1})(y - y_{k+1})\ldots(y - y_n)}{(y_k - y_0)\ldots(y_k - y_{k-1})(y_k - y_{k+1})\ldots(y_k - y_n)} \end{cases}$$

to form

$$\Phi(x, y) \equiv L(x, y) = \sum_{i=0}^{m} \sum_{k=0}^{n} L_i^{(1)}(x) L_k^{(2)}(y) f_{ik}.$$

In (9.22), the nodes need not to equidistant, but we must have

$$\begin{aligned} x_{i+1} - x_i &= h_i^{(1)} = \quad \text{constant} \quad \text{for all} \quad y_k \quad \text{and each fixed} \quad i, \\ y_{k+1} - y_k &= h_k^{(2)} = \quad \text{constant} \quad \text{for all} \quad x_i \quad \text{and each fixed} \quad k. \end{aligned}$$

To approximate functions of several variables see e.g. [COLL68], §25. The methods are not as developed here as for functions of one variable. The use of multidimensional splines (see chapter 12) is recommended.

Bibliography for Section 9.8.1

[ENGE87], 7.11; [ISAA66], 6.6.

9.8.2 Shepard Interpolation

If the nodes $(x_j, y_j), j = 0, \ldots, N$, with $(x_j, y_j) \in B \subset \mathbb{R}^2$, do not form a rectangular grid but are arranged in a completely arbitrary and unordered way, we recommend to use the method of Shepard [SHEP68]. This method has proven well suited for the graphic representation of surfaces. Its interpolating function Φ is uniquely determined independently from the ordering of the nodes (x_j, y_j). The function $f : z = f(x, y)$ for $(x, y) \in B$, where B is an arbitrary region of the x, y plane, is interpolated for the given nodes (x_j, y_j) by the function

$$(9.23) \qquad \phi(x, y) = \sum_{j=0}^{N} w_j(x, y) f_j.$$

Here $\phi(x_j, y_j) = f(x_j, y_j)$ for $j = 0, \ldots, N$, where the f_j are the given functional values $f(x_j, y_j)$ at the nodes (x_j, y_j), $j = 0, \ldots, N$, and $w_j = w_j(x, y)$ is a weight function. Setting

$$(9.24) \qquad r_j(x, y) = \sqrt{(x - x_j)^2 + (y - y_j)^2}, \quad j = 0, \ldots, N,$$

we can define w_j as

$$(9.25) \qquad w_j(x, y) = \frac{\dfrac{1}{r_j^{\mu}}}{\displaystyle\sum_{i=0}^{N} \dfrac{1}{r_i^{\mu}}} \quad , \quad 0 < \mu < \infty.$$

With (9.25) the Shepard function ϕ has the representation

$$(9.26) \qquad \phi(x, y) = \frac{\displaystyle\sum_{j=0}^{N} \dfrac{1}{r_j^{\mu}} f_j}{\displaystyle\sum_{i=0}^{N} \dfrac{1}{r_i^{\mu}}}.$$

The exponent μ in (9.25) can be chosen arbitrarily. If $0 < \mu \leq 1$, the function ϕ has peaks at the nodes. If $\mu > 1$, the function is level at the nodes. The weights w_j satisfy

$$w_j(x,y) = \begin{cases} \geq & 0 \quad \text{for all } (x,y) \in B, \\ = & 1 \quad \text{for nodes } (x,y) = (x_j, y_j), \\ = & 0 \quad \text{for nodes } (x,y) = (x_k, y_k) \text{ with } k \neq j, \end{cases}$$

and $\sum_{j=0}^{N} w_j(x,y) = 1$. The interpolation conditions $\phi(x_j, y_j) = f_j$ are satisfied for all j.

The definition of ϕ in (9.23) means, that all $N + 1$ interpolation points (x_j, y_j, f_j), $j = 0, \ldots, N$, are used for calculating each new functional value $\phi(x, y)$, i.e., this is a global method. And hence it demands a high amount of computations. This can be alleviated by using a local variant. [SHEP68] has indicated a modification for calculating the $w_j(x,y)$ which uses only the nodes (x_j, y_j) within a circle of radius R around the point (x, y) when calculating a new functional value $\phi(x, y)$. The radius R of this circle must be chosen so that enough nodes (x_j, y_j) are situated within the local circle. In the *local Shepard method* the formula (9.25) is replaced by:

$$(9.27) \quad \begin{cases} w_j(x,y) = \dfrac{\dfrac{1}{\psi_j^\mu}}{\displaystyle\sum_{i=0}^{N} \dfrac{1}{\psi_i^\mu}} \quad , 0 < \mu < \infty, \text{ with} \\[2em] \psi_j(x,y) = \begin{cases} \dfrac{R}{r_j} - 1 & \text{for } 0 < r_j < R, \\ 0 & \text{for } r_j \geq R. \end{cases} \end{cases}$$

The r_j in (9.27) are calculated as in (9.24).

REMARK for formula (9.27): The ψ_j in (9.27) are in fact a modification of the weight functions $\tilde{\psi}_i$ suggested originally by Shepard:

$$(9.28) \quad \tilde{\psi}_j(x,y) = \begin{cases} \dfrac{1}{r_j(x,y)} & \text{for } 0 < r_j(x,y) < \frac{R}{3}, \\[1em] \dfrac{27}{4R} \cdot \left(\dfrac{r_j(x,y)}{R} - 1 \right)^2 & \text{for } \frac{R}{3} < r_j(x,y) \leq R, \text{ and} \\[1em] 0 & \text{for } R < r_j(x,y). \end{cases}$$

Tests in [TERW95] with both weight functions have shown that the local Shepard interpolation often fails to give good results, since it seems difficult to find near optimal values for the radius R and the smoothing factor μ.

If, however, one wants to use the faster local Shepard interpolation then one
ought to rely on the one with Franke–Little weights. This modified local Shep-
ard method generally produces good results and the choice of R does not seem
to be too critical, see Figures 9.3 (a) – (d) below.
The Shepard interpolation method thus has the form:

$$(9.29) \qquad \phi(x,y) = \frac{\sum_{j=0}^{N} \frac{1}{\tilde{\psi}_j^{\mu}} f_j}{\sum_{i=0}^{N} \frac{1}{\tilde{\psi}_i^{\mu}}} \quad .$$

Here we use $\tilde{\psi}_j$ from (9.28).

ALGORITHM 9.16 (*Global and local Shepard interpolation*).

Given: $N+1$ interpolation points $(x_j, y_j, z_j = f(x_j, y_j))$, $j = 0, \ldots, N$, for
$(x_j, y_j) \in B \subseteq \mathbb{R}^2$.

 Task: Find an interpolating function ϕ of the form (9.23) with
$\phi(x_j, y_j) = f(x_j, y_j)$ for $j = 0, \ldots, N$.

The following steps must be carried out for each point $(x, y) \in B$
with $(x, y) \neq (x_j, y_j)$:

1^{st} step: Choose an appropriate exponent μ with $0 < \mu < \infty$ and $R > 0$
in case of the local method.

2^{nd} step: Calculate $r_j(x, y)$ for $j = 0, \ldots, N$ using formula (9.24).

3^{rd} step: Calculate the weights $w_j(x, y)$ for $j = 0, \ldots, N$

- in case of the global method use (9.25),
- in case of the local method use (9.27).

4^{th} step: Calculate the functional value $\phi(x, y) \approx f(x, y)$ from formula
(9.23).

Recommendation: To avoid the long computations of the global Shepard
method, the local method is to be preferred. To avoid peaks at the nodes,
choose $2 \leq \mu \leq 6$. Our tests indicate that $0.1 < R < 0.5$ is the preferred
range, where we recommend to choose a small value for R in case of many
available nodes and a larger R for problems with few known nodes.
For the local method, however, any choice of R near the recommended upper
bound of 0.5 leads to unsatisfactory results. An improved version of the local
method is the following:

Local Shepard interpolation with Franke–Little weights

This very useful local variant uses the Franke-Little weights ([FRAN82], [HOSC89]):

$$(9.30) \qquad \xi_j(x,y) = \begin{cases} 1 - \dfrac{r_j(x,y)}{R} & \text{for} \quad 0 < r_j(x,y) < R, \\ \\ 0 & \text{for} \quad r_j(x,y) \geq R. \end{cases}$$

The weight function w_j then becomes

$$(9.31) \qquad w_j(x,y) = \frac{\xi_j^\mu(x,y)}{\displaystyle\sum_{i=0}^{N} \xi_i^\mu(x,y)} \quad .$$

With (9.29) and (9.30), the local Shepard function has the form

$$(9.32) \qquad \phi(x,y) = \sum_{j=0}^{N} w_j(x,y) f_j = \frac{\displaystyle\sum_{j=0}^{N} \left(1 - \frac{r_j(x,y)}{R}\right)^\mu f_j}{\displaystyle\sum_{i=0}^{N} \left(1 - \frac{r_i(x,y)}{R}\right)^\mu} \quad .$$

ALGORITHM 9.17 (*Local Shepard approximation with Franke-Little weights*).

Given: $N + 1$ points $(x_j, y_j, f_j = f(x_j, y_j)) \in \mathbb{R}^3$ for $j = 0, \dots, N$ with $(x_j, y_j) \in B \subseteq \mathbb{R}^2$.

Task: Find a Shepard function ϕ of the form (9.32).

The following steps have to be carried out for each point $(x, y) \in B$ with $(x, y) \neq (x_j, y_j)$:

1^{st} step: Choose suitable values for μ and R. Recommendation: $2 \leq \mu \leq 6$; $0.1 \leq R \leq 0.5$. See remarks following Algorithm 9.16 above.

2^{nd} step: Calculate $r_j(x,y)$ for $j = 0, \dots, N$ from (9.24).

3^{rd} step: Calculate the weights $w_j(x,y)$ for $j = 0, \dots, N$ according to (9.31) and (9.30).

4^{th} step: Calculate the functional value $\phi(x,y) \approx f(x,y)$ according to (9.32).

Example for Shepard interpolation:

We shall try to represent the surface defined by

$$T(x,y) \;=\; \frac{3}{4}e^{-\frac{1}{4}\left[(9x-2)^2+(9y-2)^2\right]} + \frac{3}{4}e^{-\left[\frac{1}{49}(9x+1)^2+\frac{1}{10}(9y+1)\right]}$$
$$-\frac{1}{5}e^{\left[(9x-4)^2+(9y-7)^2\right]} + \frac{1}{2}e^{-\frac{1}{4}\left[(9x-7)^2+(9y-3)^2\right]}$$

Figure 9.3 (a): Original function

To compute interpolating surfaces via the variants of the Shepard method, we select 100 random points $(x, y, T(x, y))$ with $0 \leq x, y \leq 1.1$. For $\mu = 2$ and $R = 0.2$ we obtain the most pleasing representation of the surface. Even smoother results could have been obtained by selecting additional nodes near the edges of our domain.

Figure 9.3 (b): Global Shepard interpolation with smoothing coefficient $\mu = 2$

Figure 9.3 (c): Local Shepard interpolation with Franke–Little weights,
radius $R = 0.4$, smoothing factor $\mu = 2$.

Figure 9.3 (d): Local Shepard interpolation with Franke–Little weights,
radius $R = 0.2$, smoothing factor $\mu = 2$.

Compared with the plot for the original function T given in Figure 9.3 (a), the
last plot using the local Shepard method with Franke–Little weights gives the
most accurate result. From these plots, the unmodified local Shepard method
and the global Shepard method (refer to plot (b)) cannot be recommended.
Note that the same surface shall be replotted using various interpolating sur-
face splines and B–splines in section 12.5 below.

Bibliography for Section 9.8.2

[FRAN82]; [FOLE84]; [HOSC89]; [SCHU76/1]; [SCHU76/2]; [SHEP68];
[TERW95].

9.9 Hints for Selecting an Appropriate Inter- polation Method

We must note that interpolation with polynomials has lost much of its im-
portance since the introduction of interpolating splines and subsplines which

do not have the detrimental ripple behavior of polynomials and which can be adapted more easily to the given interpolation points.

Polynomial interpolation can indeed give very good results if the user is free to choose the points of interpolation in a manner compatible with the method. It has the additional property that, if an analytic function is being approximated, then analyticity is preserved. If one has no freedom to choose nodes of interpolation, then one must usually avoid algebraic polynomials of degree greater than five because of the potential for wild oscillations in the approximating polynomial. This problem is of less acute nature if one is measuring error in the 2-norm. Often much better suited are rational functions for the interpolation of smooth curves (section 9.7). For them, interpolation is still possible in the vicinity of poles. However, rational interpolation demands more work to determine the coefficients when compared with polynomial interpolation. A special approach to rational interpolation is given in section 9.7 via continued fractions.

Remark for selecting the appropriate form of an algebraic interpolation polynomial: Since the interpolating polynomial of degree $\leq n$ is unique for $n + 1$ given interpolation points (x_i, y_i), $i = 0, \ldots, n$, it should not matter in which form it is represented. The Lagrange form is more important as a theoretical than as a practical representation, because each $L_k(x)$ must be newly calculated if one adds one new interpolation point. Because of this, the Newton interpolation formula is preferred over the Lagrange formula.

If the interpolating polynomial Φ is not needed in its general form, but only a functional value $\Phi(\bar{x})$ is wanted at one (or a few) point(s) \bar{x}, one should use the appropriate Aitken interpolation scheme (section 9.4). This method allows to add new interpolation points. Moreover, the nodes need not be ordered monotonically as is the case with some of the other formulas.

If there exists an extensive table of values (x_i, y_i), $i = 0, \ldots, n$, and if one wants to find $f(\bar{x}) = \bar{y}$ with an interpolation polynomial of m^{th} degree, one should always choose $m + 1$ neighboring nodes x_i in the table in such a way that \bar{x} is near the center of the chosen nodes, because then the interpolation error will be kept small (see section 9.6).

For multidimensional interpolation, spline methods (chapters 12, 13) are generally preferred. The Shepard interpolation (section 9.8.2) is, however, also well suited to represent surfaces graphically, if one uses the local Shepard interpolation with Franke–Little weights and $2 \leq \mu \leq 6$ as well as $0.1 \leq R \leq 0.5$. We refer to the plots (a) to (d) of section 9.8 above.

Bibliography for Chapter 9

[ATKI89], 3; [BERE71], vol. 1,2; [BJÖR79], 7.; [BROS76], §21; [CARN69], 1.7;
[CHEN82]; [COLL73]; [CONT80], 3.1.2; [ENGE87], 7; [FOLE84]; [FRAN82];
[HÄMM78], 6; [HÄMM 91], 5.; [HENR72] vol. 2, 9.1, 10.2; [HILD74];
[HOSC89]; [ISAA73], 6.6; [KINC90], 6; [KÜHN90]; [McCA67], 6.0 – 6.2;
[MEIN79], 3; [NIED87]; [POLO64] III §1; [RICE83], 5; [SAUE69], H §1–2.2;
[SCHU76]; [SCHUM76]; [SCHW86], 3.; [SHEP68]; [STIE76], 7,11; [STOE91],
2.; [STUM82], 3.1; [TERW95]; [TÖRN79], vol. 2, 11.1; [WERN], I §1;
[WILL71], §8.6–9, §13; [ZURM65], §11.1–11.3

Chapter 10

Interpolating Polynomial Splines for Constructing Smooth Curves

10.1 Cubic Polynomial Splines

Assume that $n+1$ values $y_i = f(x_i)$ of a function $f \in C[a, b]$ are known at the nodes $x_i \in [a, b]$, $i = 0, \ldots, n$. We desire to construct a "smooth" curve passing through the given points (x_i, y_i) with polynomial splines of third degree, also called *cubic splines*. If we assume that the nodes are ordered:

$$a = x_0 < x_1 < x_2 < \ldots < x_n = b,$$

then we can represent the desired curve by a spline function S with $S(x) \approx f(x)$, which is composed of cubic polynomials S_i for $x \in [x_i, x_{i+1}]$ and $i = 0, \ldots, n-1$. These S_i are defined differently for each subinterval and must satisfy certain *connecting conditions* at the nodes.

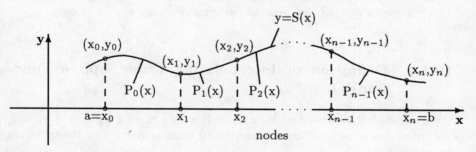

Figure 10.1: Spline interpolation

Depending on the conditions imposed at $x_0 = a$ and $x_n = b$, different types of spline functions S will result:

Types of Splines S:

(i) Natural cubic splines.

(ii) Periodic cubic splines.

(iii) Cubic splines with not-a-node condition.

(iv) Generalized natural cubic splines (cubic splines with specified second derivatives at the end points).

(v) Cubic splines with specified first derivatives at the end points.

(vi) Cubic splines with specified third derivatives at the end points.

Parametric cubic splines.

If the given nodes $x_i \in [a, b]$ are not monotonically ordered, then one has to employ parametric splines. Here the 2-dimensional or 3-dimensional parametrization $(x(t), y(t))$ or $(x(t), y(t), z(t))$ of the curve that passes through the points $\{(x_i, y_i)\}$ or $\{(x_i, y_i, z_i)\}$ is approximated by a vector valued spline function

$$\begin{pmatrix} S_x(t) \\ S_y(t) \end{pmatrix} \approx \begin{pmatrix} x(t) \\ y(t) \end{pmatrix}, \quad \text{or} \quad \begin{pmatrix} S_x(t) \\ S_y(t) \\ S_z(t) \end{pmatrix} \approx \begin{pmatrix} x(t) \\ y(t) \\ z(t) \end{pmatrix}.$$

t is the parameter of each of the component spline functions S_x, S_y or S_x, S_y, S_z: S_x is a spline function for the points (t_i, x_i), S_y a spline function for the points (t_i, y_i), and S_z for (t_i, z_i) for $i = 0, \ldots, n$ and $t_0 < t_1 < \ldots < t_n$. According to conditions imposed at the end points t_0 and t_n, the pair S_x, S_y or the triple S_x, S_y, S_z of spline functions is of one of the types specified above. The parameter nodes t_i are generally determined only approximately.

10.1.1 Definition of Interpolating Cubic Spline Functions

For monotonically ordered nodes $x_i \in [a, b]$ with $a = x_0 < x_1 < \ldots < x_n = b$, the spline functions mentioned in section 10.1 can be defined by the following set of conditions:

(1) $S \in C^2[a, b]$.

(2) S is a cubic polynomial S_i in each subinterval $[x_i, x_{i+1}], i = 0, \ldots, n-1$.

(3) S interpolates f, i.e., $S(x_i) = y_i$ for $i = 0, \ldots, n$.

(4) (i) For $x \in (-\infty, a]$ or $x \in [b, \infty)$, S is represented by the tangent of S at $a = x_0$ or $b = x_n$, thus $S''(x_0) = S''(x_n) = 0$. With these end point conditions, S is called a *natural cubic spline*.

(ii) If S satisfies the end point conditions $S(x_0) = S(x_n)$, $S'(x_0) = S'(x_n)$, and $S''(x_0) = S''(x_n)$, then S is called a *periodic cubic spline function* with period $x_n - x_0$.

(iii) If S satisfies the end point conditions $S_0(x_1) = S_1(x_1)$ and $S_{n-2}(x_{n-1}) = S_{n-1}(x_{n-1})$, then S is called a *cubic spline function* with *not-a-node-condition*. This condition means that the third derivative of the spline function is continuous at the nodes x_1 and x_{n-1}. And thus x_1 and x_{n-1} are not "genuine" nodes for the spline function.

(iv) For the end point conditions $S''(x_0) = \alpha$, $S''(x_n) = \beta$, S is called a *generalized natural cubic spline function*.

(v) For the end point conditions $S'(x_0) = \alpha$, $S'(x_n) = \beta$, S is called a *cubic spline function* with *specified first end point derivatives*.

(vi) For the end point conditions $S'''(x_0) = \alpha$, $S'''(x_n) = \beta$, S is called a *cubic spline function* with *specified third end point derivatives*.

To construct a function S that satisfies (1) and (2), we shall set

$$(10.1) \qquad S(x) \equiv S_i(x) := a_i + b_i(x - x_i) + c_i(x - x_i)^2 + d_i(x - x_i)^3$$
$$\text{for} \quad x \in [x_i, x_{i+1}], \quad i = 0, \ldots, n-1.$$

A cubic polynomial spline is twice continuously differentiable and depends on 4 parameters, hence it is a spline function of odd degree $3 = 2k - 1$ for $k = 2$ (k-times continuously differentiable, $2k$-parametric). In some cases, spline functions of even degree $2k$ are also used, e.g. in area preserving representation of histograms or of empirical distributions, see [SPÄT74/2].

The properties (1) and (3) for S in case of (i) lead to $4n$ conditions for the S_i:

(a) $S_i(x_i) = y_i$, $i = 0, \ldots, n$,

(b) $S_i(x_i) = S_{i-1}(x_i)$, $i = 1, \ldots, n$,

(c) $S_i'(x_i) = S_{i-1}'(x_i)$, $i = 1, \ldots, n-1$,

(d) $S_i''(x_i) = S_{i-1}''(x_i)$, $i = 1, \ldots, n$.

Here we have set $S_n(x_n) = a_n$ and $S_n''(x_n) = 2c_n$. Two additional end point conditions can be specified at will. Then we have to satisfy $4n + 2$ conditions for the $4n + 2$ coefficients a_i, b_i, c_i, d_i for $i = 0, \ldots, n-1$ and a_n, c_n. Other spline types lead to analogous conditions. Property (1), which gives rise to

conditions (c) and (d), is the strongest requirement for the spline function S. It effects a smooth transition between the graphs of the polynomials S_i and S_{i-1} at the node x_i for each $i = 1, \ldots, n-1$: The graphs of adjacent polynomials S_{i-1} and S_i have the same curvature at x_i. This property of spline functions is especially useful for approximating a function f which is known empirically, as through measurements, and which one can a draw using a flexible ruler, called a *spline* by shipbuilders. To compute such non-parametric spline functions see section 10.1.2 .

The *parametric cubic splines* mentioned in section 10.1 are used when one cannot sensibly order the nodes x_i, e.g. in case of interpolating closed curves, or curves with double points or other curves which cannot be described explicitly by a functional relation $y = S(x)$ and hence cannot be approximated by using splines of type (i),..., (vi). To compute such parametric spline functions, see section 10.1.3 .

10.1.2 Computation of Non-Parametric Cubic Splines

ALGORITHM 10.1 (*Natural cubic splines*).

Given: Points (x_i, y_i), $i = 0, \ldots, n$, $n \geq 2$ with
$$a = x_0 < x_1 < \ldots < x_n = b.$$

 Task: Find the coefficients a_i, b_i, c_i, d_i of the cubic polynomials S_i in (10.1).

1) Set $a_i = y_i$ for $i = 0, \ldots, n$.

2) Set $c_0 = c_n = 0$.

3) Solve $h_{i-1}c_{i-1} + 2c_i(h_{i-1} + h_i) + h_i c_{i+1} =$
$$= \frac{3}{h_i}(a_{i+1} - a_i) - \frac{3}{h_{i-1}}(a_i - a_{i-1}) \text{ for } i = 1, \ldots, n-1$$

with $h_i = x_{i+1} - x_i$ for $i = 0, \ldots, n-1$.

4) Set $b_i = \frac{1}{h_i}(a_{i+1} - a_i) - \frac{h_i}{3}(c_{i+1} + 2c_i)$ for $i = 0, \ldots, n-1$.

5) Set $d_i = \frac{1}{3h_i}(c_{i+1} - c_i)$ for $i = 0, \ldots, n-1$.

The equations in step 3) form a linear system with $n-1$ equations for the $n-1$ unknown $c_1, c_2, ..., c_{n-1}$. Using matrix notation, this becomes

$$Ac = g$$

with

$$A = \begin{pmatrix} 2(h_0+h_1) & h_1 & & & & \\ h_1 & 2(h_1+h_2) & h_2 & & & \\ & h_2 & 2(h_2+h_3) & h_3 & & \\ & & \ddots & \ddots & \ddots & \\ & & & h_{n-3} & 2(h_{n-3}+h_{n-2}) & h_{n-2} \\ & & & & h_{n-2} & 2(h_{n-2}+h_{n-1}) \end{pmatrix},$$

$$c = \begin{pmatrix} c_1 \\ c_2 \\ \vdots \\ c_{n-1} \end{pmatrix}, \quad g = \begin{pmatrix} \frac{3}{h_1}(a_2-a_1) - \frac{3}{h_0}(a_1-a_0) \\ \frac{3}{h_2}(a_3-a_2) - \frac{3}{h_1}(a_2-a_1) \\ \vdots \\ \frac{3}{h_{n-1}}(a_n-a_{n-1}) - \frac{3}{h_{n-2}}(a_{n-1}-a_{n-2}) \end{pmatrix}.$$

Properties of the matrix A:

The system matrix A is tridiagonal, symmetric, strongly diagonally dominant and elementwise positive. Hence it is strongly regular and positive definite. Thus the system $Ac = g$ has a unique solution. To find a solution numerically, one should employ a Gauß or Cholesky algorithm for tridiagonal matrices, see section 4.10. We note that the system matrix is well conditioned and does not require a pivot search or iterative refinement.

ALGORITHM 10.2 *(Periodic cubic splines).*

Given: Points $(x_i, y_i), i = 0, \ldots, n$ with $a = x_0 < x_1 < \ldots < x_n = b$.

Task: Find coefficients a_i, b_i, c_i, d_i for the cubic polynomials S_i in (10.1) of the periodic spline function S with $S(x) = S(x + kp)$ for $k \in \mathbb{Z}$ and the period $p = b - a$.

1) Set $a_i = y_i$ for $i = 0, \ldots, n-1$.

2) Set $a_0 = a_n,$ $b_0 = b_n,$ $c_0 = c_n.$

3) Solve the system $Ac = g$ with

$$A = \begin{pmatrix} 2(h_0+h_1) & h_1 & & & & & h_0 \\ h_1 & 2(h_1+h_2) & h_2 & & & & \\ & h_2 & 2(h_2+h_3) & h_3 & & & \\ & & \ddots & \ddots & \ddots & & \\ & & & h_{n-2} & 2(h_{n-2}+h_{n-1}) & h_{n-1} & \\ h_0 & & & & h_{n-1} & 2(h_{n-1}+h_n) \end{pmatrix}$$

$$c = \begin{pmatrix} c_1 \\ c_2 \\ \vdots \\ c_n \end{pmatrix}, \quad g = \begin{pmatrix} \dfrac{3}{h_1}(a_2 - a_1) - \dfrac{3}{h_0}(a_1 - a_0) \\ \dfrac{3}{h_2}(a_3 - a_2) - \dfrac{3}{h_1}(a_2 - a_1) \\ \vdots \\ \dfrac{3}{h_n}(a_{n+1} - a_n) - \dfrac{3}{h_{n-1}}(a_n - a_{n-1}) \end{pmatrix}$$

with $a_{n+1} = a_1, a_n = a_0, c_n = c_0, h_n = h_0,$ and $h_i = x_{i+1} - x_i$
if $n \geq 3$.

(For $n = 2$ the matrix A has the form

$$A = \begin{pmatrix} 2(h_0 + h_1) & h_0 + h_1 \\ h_0 + h_1 & 2(h_0 + h_1) \end{pmatrix} .$$

4) Set $b_i = \dfrac{1}{h_i}(a_{i+1} - a_i) - \dfrac{h_i}{3}(c_{i+1} + 2c_i)$ for $i = 0, \ldots, n - 1$.

5) Set $d_i = \dfrac{1}{3h_i}(c_{i+1} - c_i)$ for $i = 0, \ldots, n - 1$.

Properties of the matrix A.

Here the matrix A is cyclically tridiagonal, symmetric, diagonally dominant and elementwise positive. Hence A is strongly regular, positive definite and the system $Ac = g$ is well-conditioned. An algorithm for solving such systems with cyclically tridiagonal matrices is given in section 4.11.

Figure 10.2: Example of a periodic cubic spline

ALGORITHM 10.3 (*Cubic splines with not-a-node end point conditions*).

Given: Points (x_i, y_i), $i = 0, \ldots, n$ with $a = x_0 < x_1 < \ldots < x_n = b$.

Task: Find coefficients a_i, b_i, c_i, d_i for the cubic polynomials S_i of (10.1)

1) Set $a_i = y_i$ for $i = 0, \ldots, n$.

2) Solve $(h_0 + 2h_1)c_1 + (h_1 - h_0)c_2 = \dfrac{3}{h_1 + h_0}[(a_2 - a_1) - \dfrac{h_1}{h_0}(a_1 - a_0)]$
and
$$h_{i-1}c_{i-1} + 2c_i(h_{i-1} + h_i) + h_i c_{i+1} = \frac{3}{h_i}(a_{i+1} - a_i) - \frac{3}{h_{i-1}}(a_i - a_{i-1})$$

for $i = 2, \ldots, n-2$, and

$$(h_{n-2} - h_{n-1})c_{n-2} + (2h_{n-2} + h_{n-1})c_{n-1} =$$

$$= \frac{3}{h_{n-1} + h_{n-2}} \cdot \left[\frac{h_{n-2}}{h_{n-1}}(a_n - a_{n-1}) - (a_{n-1} - a_{n-2}) \right]$$

with $h_i = x_{i+1} - x_i$ for $i = 1, \ldots, n-1$.

3) Set $c_0 = c_1 + \dfrac{h_0}{h_1}(c_1 - c_2)$, $c_n = c_{n-1} + \dfrac{h_{n-1}}{h_{n-2}}(c_{n-1} - c_{n-2})$.

4) Set $b_i = \dfrac{1}{h_i}(a_{i+1} - a_i) - \dfrac{h_i}{3}(c_{i+1} + 2c_i)$ for $i = 0, \ldots, n-1$.

5) Set $d_i = \dfrac{1}{3h_i}(c_{i+1} - c_i)$ for $i = 0, \ldots, n-1$.

ALGORITHM 10.4 (*Generalized natural cubic splines*).

Given: Points $(x_i, y_i), i = 0, \ldots, n$, with $a = x_0 < x_1 < \ldots < x_n = b$ and
$S''(x_0) = \alpha, S''(x_n) = \beta$.

Task: Find coefficients a_i, b_i, c_i, d_i for the cubic polynomials S_i of (10.1).

Here Algorithm 10.1 is used, except that the steps 2) and 3) are replaced
by:

2') Set $c_0 = \frac{\alpha}{2}, \qquad c_n = \frac{\beta}{2}$.

3') Solve the system with the two altered equations:

$2(h_0 + h_1)c_1 + h_1 c_2 = \frac{3}{h_1}(a_2 - a_1) - \frac{3}{h_0}(a_1 - a_0) - h_0 \frac{\alpha}{2}$ for $i = 1$,

and $h_{n-2}c_{n-2} + 2(h_{n-2} + h_{n-1})c_{n-1} =$

$= \frac{3}{h_{n-1}}(a_n - a_{n-1}) - \frac{3}{h_{n-2}}(a_{n-1} - a_{n-2}) - h_{n-1}\frac{\beta}{2}$ for $i = n - 1$.

ALGORITHM 10.5 (*Cubic splines with specified first derivatives at
the end points*).

Given: Points $(x_i, y_i), i = 0, \ldots, n$, with $a = x_0 < x_1 < \ldots < x_n = b$ and
$S'(x_0) = \alpha, S'(x_n) = \beta$.

Task: Find coefficients a_i, b_i, c_i, d_i for the cubic polynomials S_i of (10.1).

Again algorithm 10.1 is used, except that steps 2) and 3) are replaced
by:

2'') Omit.

3'') Solve the system with two altered equations:

$\left(\frac{3}{2}h_0 + 2h_1\right) c_1 + h_1 c_2 = 3 \left[\frac{a_2 - a_1}{h_1} - \frac{1}{2}\left(3\frac{a_1 - a_0}{h_0} - \alpha\right)\right]$ for $i = 1$,

and $\left(2h_{n-2} + \frac{3}{2}h_{n-1}\right) c_{n-1} + h_{n-2}c_{n-2} =$

$= 3 \left[\frac{1}{2}\left(3\frac{a_n - a_{n-1}}{h_{n-1}} - \beta\right) - \frac{a_{n-1} - a_{n-2}}{h_{n-2}}\right]$ for $i = n - 1$.

Moreover set $c_0 = \frac{1}{2h_0}\left(\frac{3}{h_0}(a_1 - a_0) - 3\alpha - c_1 h_0\right)$ and

$$c_n = -\frac{1}{2h_{n-1}}\left(\frac{3}{h_{n-1}}(a_n - a_{n-1}) - 3\beta + c_{n-1}h_{n-1}\right).$$

ALGORITHM 10.6 (*Cubic splines with specified third derivatives at the end points*).

Given: Points $(x_i, y_i), i = 0, \ldots, n$, with $a = x_0 < x_1 < \ldots < x_n = b$,
$\quad S'''(x_0) = \alpha, S'''(x_n) = \beta$.

Task: Find coefficients a_i, b_i, c_i, d_i for the cubic polynomials S_i of (10.1).

Algorithm 10.1 is used, except that the steps 2) and 3) are replaced by:

((2)) Omit.

((3)) Solve the system with the two altered equations:

$(3h_0 + 2h_1)c_1 + h_1 c_2 = \frac{3}{h_1}(a_2 - a_1) - \frac{3}{h_0}(a_1 - a_0) + \frac{\alpha h_0^2}{2}$
for $i = 1$,

and $h_{n-2}c_{n-2} + (2h_{n-2} + 3h_{n-1})c_{n-1} =$

$= \frac{3}{h_{n-1}}(a_n - a_{n-1}) - \frac{3}{h_{n-2}}(a_{n-1} - a_{n-2}) - \frac{\beta h_{n-1}^2}{2}$
for $i = n - 1$.

Moreover set $c_0 = c_1 - \frac{\alpha h_0}{2}$, $c_n = c_{n-1} + \frac{\beta h_{n-1}}{2}$.

10.1.3 Computing Parametric Cubic Splines

If the nodes x_i cannot be ordered monotonically, then we have to compute the spline using a parameter.

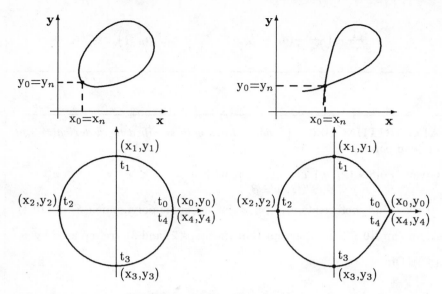

Figure 10.3: Types of curves for parametric splines, refer to
 table in section 10.3 also.

For $i = 0, ..., n-1$ the spline segments have the form

$$S_i(t) = a_i + b_i(t - t_i) + c_i(t - t_i)^2 + d_i(t - t_i)^3$$

for $a_i, b_i, c_i, d_i \in \mathbb{R}^m$ and $m = 2, 3$ with $t \in [t_i, t_{i+1}]$.

The interpolation nodes for the spline are the points

$$S_i(t_i) = a_i = P_i, \quad i = 0, ..., n-1$$

and P_n. Thus for each node P_i we must find a defining parameter value t_i with

$$t_0 < t_1 < \ldots < t_n.$$

Each spline segment $S_i(t)$ is defined on a parameter interval of length $h_i = t_{i+1} - t_i > 0$. Hence each spline segment can be represented in the form

$$S_i(t) = a_i + b_i t + c_i t^2 + d_i t^3, \quad t \in [0, h_i].$$

In the formulas for determining the spline coefficients in algorithms 10.1 to 10.6, only the interval lengths h_i have been used, but not the parameter nodes t_i themselves. Clearly if t_0 has been chosen, then

$$t_{i+1} = t_i + h_i, \quad i = 0, ..., n - 1 .$$

Thus we have to learn to judiciously assign a parameter interval of length h_i to each spline segment $\boldsymbol{S}_i(t)$, $i = 0, ..., n - 1$.

Every parametrization can be normalized by setting

$$\bar{h}_i = \frac{h_i}{\displaystyle\sum_{i=0}^{n-1} h_i} ,$$

so that $\displaystyle\sum_{i=0}^{n-1} \bar{h}_i = 1$. We note that such a normalization does not change the proportions of the parameter subintervals.

The simplest parametrization would involve setting $h_i = h = $ const uniformly with $h = 1$ for example. But such a parametrization will in general not lead to a satisfactory spline curve.

More fitting to the problem is a parametrization that tries to approximate the generally unknown arc length parametrization of the curve. The often used chordal parametrization uses the length of the line segment joining adjacent nodes as lengths for the parameter intervals:

$$h_i = |P_{i+1} - P_i|, \quad i = 0, ..., n - 1 .$$

This, however, is only a crude approximation of arc length.

For planar curves $(m = 2)$ the chordal parametrization is given by

$$h_i = \sqrt{(x_{i+1} - x_i)^2 + (y_{i+1} - y_i)^2}, \quad i = 0(1)n - 1 ,$$

while for space curves $(m = 3)$ it uses the interval lengths

$$h_i = \sqrt{(x_{i+1} - x_i)^2 + (y_{i+1} - y_i)^2 + (z_{i+1} - z_i)^2}, \quad i = 0, ..., n - 1 .$$

Figure 10.4: Determination of the parameter values t_i for the chordal parametrization

If three adjacent nodes are nearly collinear, then the chordal parametrization supplies parameter intervals whose lengths will differ only slightly from the arc length of each curve segment. If, however, the geometry of the given nodes indicates a highly curved segment, one ought to associate larger parameter intervals with those segments. R. Wodicka has modified the standard chordal parametrization to achieve this as follows:

Three adjacent non-collinear points define a circle. The length of the circular arc joining two of the nodes shall be taken into consideration when assigning a length to the respective parameter interval.

Let P, Q, R be three adjacent points on a circle with $|R - Q|$ denoting the distance between Q and R and γ the angle between PQ and PR, then the length of the circular arc from Q to R is

$$B = |R - Q|\frac{\gamma}{\sin\gamma}, \quad 0 \leq \gamma < \pi .$$

Since $\lim_{\gamma \to 0}(\gamma/\sin\gamma) = 1$ this is true even for collinear points P, Q, R.

For the points $P = P_{i-1}, Q = P_i, R = P_{i+1}$ and the angle γ_i formed by the line segments $P_{i-1}P_i$ and $P_{i-1}P_{i+1}$, the circular arc between P_i and P_{i+1} has the length

$$B_i = |P_{i+1} - P_i|\frac{\gamma_i}{\sin\gamma_i} .$$

For the points $P = P_{i+2}, Q = P_{i+1}, R = P_i$ and the angle γ_{i+1} between $P_{i+2}P_{i+1}$ and $P_{i+2}P_i$ we see analogously that the circular arc between P_i and P_{i+1} has the length

$$B_{i+1} = |P_{i+1} - P_i|\frac{\gamma_{i+1}}{\sin\gamma_{i+1}} .$$

We shall use the distance $|P_{i+1} - P_i|$ and the average of B_i and B_{i+1} to define the parameter interval length as

$$
\begin{aligned}
h_i &= (1-\sigma)|P_{i+1} - P_i| + \sigma\tfrac{1}{2}(B_i + B_{i+1}) \\
&= |P_{i+1} - P_i| \left(1 + \sigma\left(\tfrac{1}{2}\left(\tfrac{\gamma_i}{\sin\gamma_i} + \tfrac{\gamma_{i+1}}{\sin\gamma_{i+1}}\right) - 1\right)\right),
\end{aligned}
$$

where $\sigma \geq 0$ is a weight. Clearly $h_i \geq |P_{i+1} - P_i|$, the chordal distance.

If $\sigma = 0$, the above formula represents the standard chordal parametrization. If $\sigma > 0$, then by construction the term involving the length of the circular arcs B_i and B_{i+1} increases for strongly curved segments, thus enlarging the size of the corresponding parameter interval. By adjusting σ the shape of the resulting spline can be altered.

The sin term in the expression for B can easily be computed from the two vectors $s := Q - P$ and $t := R - P \in \mathbb{R}^m$ as

$$
\sin\angle(s,t) = \sqrt{1 - \cos^2\angle(s,t)} = \sqrt{1 - \frac{(s^T t)^2}{(s^T s)\,(t^T t)}}\ .
$$

For $m = 2$ this simplifies even further to:

$$
\sin\angle(s,t) = |\det(s,t)| / (\|s\|\,\|t\|),
$$

where $\|s\|$ stands for the euclidean norm $\|s\| = \|s\|_2 = \sqrt{s^T s}$ of $s \in \mathbb{R}^m$, see chapter 5.2.

In order to be able to use our weighted interval lengths h_i for the end intervals of an open curve as well, we need to introduce two additional points P_{-1} and P_{n+1}. P_{-1} shall be chosen by reflecting P_2 across the line (or plane) containing the midpoint of the line segment joining P_0 and P_1 that is perpendicular to $P_0 P_1$. Analogously we reflect P_{n-2} across the line (or plane) that passes through the midpoint of the segment $P_{n-1}P_n$ and is perpendicular to $P_{n-1}P_n$ to obtain P_{n+1}.

ALGORITHMUS 10.7 (*Computing the interval lengths and parameter values for parametric cubic splines*)

Given: Nodes $P_i = (x_i, y_i)$ or $P_i = (x_i, y_i, z_i)$ for $i = 0, ..., n$, $n \geq 3$, $P_i \neq P_{i+1} \neq P_{i+2} \neq P_i$ for $i = 0, ..., n-2$, a weight $\sigma \geq 0$.

Task: Find interval length $h_i > 0$, $i = 0, ..., n-1$, and parameter values t_i for $i = 0, ..., n$.

1. Construct chordal vectors for $i = 0, ..., n - 1$:
 $s_i := P_{i+1} - P_i \neq 0$.

2. If $\sigma = 0$: Set $h_i := \|s_i\|$, $i = 0, ..., n - 1$, continue with 3.
 If $\sigma > 0$:

 2.1 Construct additional chordal vectors:

 (a) If $P_n = P_0$ (closed curve)
 $s_{-1} := s_{n-1}$,
 $s_n := s_0$.

 (b) If $P_n \neq P_0$ (open curve)
 $$s_{-1} := 2 \frac{s_0^T s_1}{s_0^T s_0} s_0 - s_1,$$

 $$s_n := 2 \frac{s_{n-1}^T s_{n-2}}{s_{n-1}^T s_{n-1}} s_{n-1} - s_{n-2} .$$

 2.2 For $i = 0, ..., n-1$ do:

 (A) Compute C from 4. for
 $s := s_{i-1}$, $t := s_{i-1} + s_i$.
 Set $C_i := C$.

 (B) Compute C from 4. for
 $s := s_i + s_{i+1}$, $t := s_{i+1}$.
 Set $C_{i+1} := C$.

 Set $h_i := \|s_i\|(1 + \sigma(0.5(C_i + C_{i+1}) - 1))$.

3. Set $t_0 := 0$; for $i = 0, ..., n-1$ compute: $t_{i+1} := t_i + h_i$.

4. For $s \neq 0$ and $t \neq 0$ in 2.2 compute:
 $Z := s^T t$,
 $N := (s^T s)(t^T t)$,
 $S := \sqrt{1 - Z^2/N}$,
 If $S = 0$, set $C := 1$;
 If $S > 0$, set $\gamma := \arcsin(S)$,
 $\qquad\qquad$ If $Z < 0$, set $\gamma := \pi - \gamma$,
 $\qquad\qquad$ Set $C := \gamma/S$.

 Experience has shown that suitable spline curves can be constructed for weights with $0 \leq \sigma \leq 2$.

After we have found monotonically ordered values for the parameters t_i via the above algorithm, we proceed as follows:

Planar curve through the points $P_i = (x_i, y_i), i = 0, \ldots, n, n \geq 3$:

A spline function S_x must be calculated for the table of values (t_i, x_i), $i = 0, \ldots, n$, as well as a second spline function S_y for the values (t_i, y_i), $i = 0, \ldots, n$:

$$(10.2) \quad \begin{cases} S_x(t) & \equiv \quad S_{ix}(t) = a_{ix} + b_{ix}(t - t_i) + c_{ix}(t - t_i)^2 + d_{ix}(t - t_i)^3 \\ & \quad \text{for } t \in [t_i, t_{i+1}], \end{cases}$$

$$(10.3) \quad \begin{cases} S_y(t) & \equiv \quad S_{iy}(t) = a_{iy} + b_{iy}(t - t_i) + c_{iy}(t - t_i)^2 + d_{iy}(t - t_i)^3 \\ & \quad \text{for } t \in [t_i, t_{i+1}]. \end{cases}$$

According to the prescribed end point conditions, S_x and S_y can be calculated using one of the algorithms in section 10.1.2.

Space curve through the points $P_i = (x_i, y_i, z_i), i = 0, \ldots, n, n \geq 3$:

The spline functions S_x and S_y are calculated according to (10.2) and (10.3), while for the values $(t_i, z_i), i = 0, \ldots, n$, the spline function S_z is determined as

$$(10.4) \quad \begin{cases} S_z(t) & \equiv \quad S_{iz}(t) = a_{iz} + b_{iz}(t - t_i) + c_{iz}(t - t_i)^2 + d_{iz}(t - t_i)^3 \\ & \quad \text{for } t \in [t_i, t_{i+1}]. \end{cases}$$

In case of an open curve $(P_n \neq P_0)$, one of the types of non-periodic splines in section 10.1.2 may be used for calculating S_x, S_y or S_x, S_y, S_z, respectively. For a closed curve $(P_n = P_0)$, periodic splines should be employed if the curve is to be smooth, otherwise one of the other spline types can be used. This, however, may generate a discontinuous tangent at $P_0 = P_n$ with a corner there.

REMARK. The linear systems for calculating the coefficients c_{ix}, c_{iy} or c_{ix}, c_{iy}, c_{iz}, respectively, have the same system matrix for x and y, or x, y and z, respectively, but different right hand sides. Therefore the triangular factorization of the system matrix needs to be carried out only once. In the planar case the updating and backsubstiution is carried out twice, and for space curves thrice. This helps to cut down calculation time.

Computing the curvature of planar curves via splines.

Let C be a smooth planar curve that can be approximated only by parametric splines because in the given value table (x_i, y_i), the x_i cannot be ordered monotonically. Then

$$x(t) = \left(\begin{array}{c} x(t) \\ y(t) \end{array} \right) \approx \left(\begin{array}{c} S_x(t) \\ S_y(t) \end{array} \right) = S(t).$$

And the *curvature* of C is given by

$$\kappa = \frac{\dot{x}^T \ddot{x}}{\|\dot{x}\|^3} \approx \frac{\dot{S}^T \ddot{S}}{\|\dot{S}\|^3}.$$

Thus both functions S_x, S_y enter into the calculation. Since the parameter t generally is not precisely the arc length parameter, but only an approximation, large deviations may occur if one calculates the curvature κ of a curve from a parametric spline approximation.

To remedy this, one should calculate an angular value table (φ_i, r_i) from the points $(x_i, y_i), i = 0, \ldots, n$, by setting:

$$\varphi_i = \arctan \frac{y_i}{x_i}, \qquad r_i = \sqrt{x_i^2 + y_i^2}.$$

If the φ_i can be ordered monotonically and if $r = r(\varphi)$ is uniquely determined for all $\varphi \in [\varphi_0, \varphi_1]$, then r can be approximated by a spline $S(\varphi)$ for the value table $(\varphi_i, r_i), i = 0, \ldots, n$. The curvature can then be computed from the formula

$$\kappa(\varphi) = \frac{2\dot{r}^2 - r\ddot{r} + r^2}{(\dot{r}^2 + r^2)^{3/2}},$$

where we approximate $r(\varphi) \approx S(\varphi)$.

This *transformed parametric spline* $S(\varphi)$ has to be periodic if C is a closed, smooth curve. Then $[0, 2\pi]$ is the parameter interval for one full cycle, or it is $[\varphi_0, \varphi_1]$ if φ does not describe the angular parameter.

10.1.4 Joined Interpolating Polynomial Splines

A natural cubic polynomial spline $S \in C^2[x_0, x_n]$ is uniquely determined by $n + 1$ given interpolation points $(x_i, y_i), i = 0, \ldots, n$, and the end point conditions $S''(x_0) = S''(x_n) = 0$, see section 10.1.1. Thus $S'(x_0)$ and $S'(x_n)$ are uniquely determined.

If a spline function S is to be determined in the interval $[x_0, x_n] \subset [a, b]$ in such a way that for two functions $f \in C^2[a, x_0]$ and $g \in C^2[x_n, b]$ we obtain a twice continuously differentiable function in the whole interval $[a, b]$, then $f(x_0) = S(x_0)$, $S(x_n) = g(x_n)$, $f'(x_0) = S'(x_0)$, $f''(x_0) = S''(x_0)$, $S'(x_n) = g'(x_n)$, and $S''(x_n) = g''(x_n)$ must be satisfied. The given task might for example be to connect S to two linear functions f and g with a continuous second derivative at $x_0, x_n \in (a, b)$. During the production of fibre-reinforced plastic cylinders, rotation-symmetrical wound cores are often used whose meridians are composed of straight and curved sections that are joined with continuous slopes. If one defines the curved sections by interpolating splines, the tangents must be prescribed at the end-points.

In [WODI77], R. Wodicka has developed a joined spline method that satisfied the above end point conditions: Instead of cubic polynomials, polynomials of fourth degree are used in the intervals $[x_0, x_1]$ and $[x_{n-1}, x_n]$. This variant of cubic polynomial splines causes only slight changes in the calculation of the coefficients. It has the advantage that one can prescribe the direction and curvature at the end points.

Assume there are $n + 1$ nodes $x_i \in \mathbb{R}$ with

$$x_0 < x_1 < \ldots < x_i < x_{i+1} < \ldots < x_n$$

and the associated values $y_i \in \mathbb{R}$,

$$y_0, y_1, \ldots, y_i, y_{i+1}, \ldots, y_n,$$

and assume that $y_0'', y_n'' \in \mathbb{R}$ are given as well, where $n \geq 3$.

We want to find a spline function S on $[x_0, x_n]$ with the properties:

(1) $S \in C^2[x_0, x_n]$,

(2) $S(x_i) = y_i, i = 0, \ldots, n$,

(3) $S''(x_0) = y_0'', S''(x_n) = y_n''$,

(4) Additional conditions: There are four possible cases:

(α) no further conditions required for S, or

(β) $S'(x_0) = y_0'$, or

(γ) $S'(x_n) = y_n'$, or

(δ) $S'(x_0) = y_0', S'(x_n) = y_n'$ can be required.

For (α): No further end point conditions for S.

The spline function S with the properties (1), (2), (3) above is a generalized spline function (see section 10.1). It is defined as

$$S(x) \equiv S_i(x) = a_i + b_i(x - x_i) + c_i(x - x_i)^2 + d_i(x - x_i)^3$$
$$\text{for} \quad x \in [x_i, x_{i+1}], \quad i = 0, \ldots, n-1.$$

The equations used for determining the spline coefficients are given as follows:

1) $c_0 = \frac{y_0''}{2}, h_i = x_{i+1} - x_i$ for $i = 0, \ldots, n-1$.

2) The $c_i, i = 1, \ldots, n-1$, are the solutions of the $n-1$ linear equations

$$2(h_0 + h_1)c_1 + h_1 c_2 = \frac{3}{h_1}(y_2 - y_1) - \frac{3}{h_0}(y_1 - y_0) - h_0\frac{y_0''}{2},$$
$$h_i c_i + 2(h_i + h_{i+1})c_{i+1} + h_{i+1}c_{i+2} = \frac{3}{h_{i+1}}(y_{i+2} - y_{i+1}) - \frac{3}{h_i}(y_{i+1} - y_i),$$
$$\text{for } i = 1, \ldots, n-3 \quad \text{if} \quad n > 3, \text{ and}$$
$$h_{n-2}c_{n-2} + 2(h_{n-2} + h_{n-1})c_{n-1} = \frac{3}{h_{n-1}}(y_n - y_{n-1})$$
$$- \frac{3}{h_{n-2}}(y_{n-1} - y_{n-2}) - h_{n-2}\frac{y_n''}{2}.$$

(For n = 3, the system consists only of the first and last equation.)

3) $d_0 = \frac{1}{3h_0}\left(c_1 - \frac{y_0''}{2}\right),$

$d_i = \frac{1}{3h_i}(c_{i+1} - c_i), \quad i = 1, \ldots, n-2,$

$d_{n-1} = \frac{1}{3h_{n-1}}\left(\frac{y_n''}{2} - c_{n-1}\right) .$

4) $b_0 = \frac{1}{h_0}(y_1 - y_0) - \frac{h_0}{3}(c_1 + y_0'') ,$

$b_i = \frac{1}{h_i}(y_{i+1} - y_i) - \frac{h_i}{3}(c_{i+1} + 2c_i), \quad i = 1, \ldots, n-2,$

$b_{n-1} = \frac{1}{h_{n-1}}(y_n - y_{n-1}) - \frac{h_{n-1}}{3}\left(\frac{y_n''}{2} + 2c_{n-1}\right).$

5) $a_i = y_i, \quad i = 0, \ldots, n-1.$

If $y_0'' = y_n'' = 0$ we deal with natural cubic splines, see section 10.1.2.

For (β): Additional condition: $S'(x_0) = y_0'$.

Assume that the spline function S with the properties (1), (2), (3), and (4)(β) is defined as

$$S_0(x) = a_0 + b_0(x - x_0) + c_0(x - x_0)^2 + d_0(x - x_0)^3 + e_0(x - x_0)^4$$

for $x \in [x_0, x_1]$, and

$$S_i(x) = a_i + b_i(x - x_i) + c_i(x - x_i)^2 + d_i(x - x_i)^3$$

for $x \in [x_i, x_{i+1}]$ and $i = 1, \ldots, n-1$.

The polynomial coefficients are determined by the equations:

1) $c_0 = \frac{y_0''}{2}, h_i = x_{i+1} - x_i$ for $i = 0, \ldots, n-1$.

2) The $c_i, i = 1, \ldots, n-1$, are the solutions of the linear system

$$(h_0 + 2h_1)c_1 + h_1 c_2 = \frac{3}{h_1}(y_2 - y_1) - \frac{6}{h_0}(y_1 - y_0) + 3y_0' + h_0\frac{y_0''}{2},$$

$$h_i c_i + 2(h_i + h_{i+1})c_{i+1} + h_{i+1}c_{i+2} = \frac{3}{h_{i+1}}(y_{i+2} - y_{i+1}) - \frac{3}{h_i}(y_{i+1} - y_i),$$

for $i = 1, \ldots, n-3$ and $n > 3$, and

$$h_{n-2}c_{n-2} + 2(h_{n-2} + h_{n-1})c_{n-1} =$$
$$= \frac{3}{h_{n-1}}(y_n - y_{n-1}) - \frac{3}{h_{n-2}}(y_{n-1} - y_{n-2}) - h_{n-1}\frac{y_n''}{2}.$$

(For $n = 3$, the system consists only of the first and the last equation.)

3) $e_0 = \frac{1}{3h_0^2}(c_1 + y_0'') + \frac{1}{h_0^3}y_0' - \frac{1}{h_0^4}(y_1 - y_0)$.

4) $d_0 = \frac{1}{3h_0}(c_1 - \frac{y_0''}{2}) - 2h_0 e_0$,

$d_i = \frac{1}{3h_i}(c_{i+1} - c_i), \quad i = 1, \ldots, n-2$,

$d_{n-1} = \frac{1}{3h_{n-1}}\left(\frac{y_n''}{2} - c_{n-1}\right)$.

5) $b_0 = y_0'$,

$b_i = \frac{1}{h_i}(y_{i+1} - y_i) - \frac{h_i}{3}(c_{i+1} + 2c_i), \quad i = 1, \ldots, n-2,$

$b_{n-1} = \frac{1}{h_{n-1}}(y_n - y_{n-1}) - \frac{h_{n-1}}{3}\left(\frac{y_n''}{2} + 2c_{n-1}\right).$

6) $a_i = y_i, \quad i = 0, \ldots, n-1$.

For (γ): Additional condition: $S'(x_n) = y_n'$.

We assume that the spline function S with the properties (1), (2), (3), and $(4)(\gamma)$ is defined as

$$S_i(x) \;=\; a_i + b_i(x - x_i) + c_i(x - x_i)^2 + d_i(x - x_i)^3$$
$$\text{for } x \in [x_i, x_{i+1}], \ i = 0, \ldots, n - 2,$$

$$S_{n-1}(x) \;=\; a_{n-1} + b_{n-1}(x - x_{n-1}) + c_{n-1}(x - x_{n-1})^2 + d_{n-1}(x - x_{n-1})^3$$
$$+ e_{n-1}(x - x_{n-1})^4 \quad \text{for} \quad x \in [x_{n-1}, x_n].$$

The polynomial coefficients are determined by the equations

1) $c_0 = \frac{y_0''}{2}, h_i = x_{i+1} - x_i$ for $i = 0, \ldots, n - 1$.

2) The $c_i, i = 1, \ldots, n - 1$, are the solutions of the linear system

$$2(h_0 + h_1)c_1 + h_1 c_2 = \frac{3}{h_1}(y_2 - y_1) - \frac{3}{h_0}(y_1 - y_0) - h_0 \frac{y_0''}{2},$$

$$h_i c_i + 2(h_i + h_{i+1})c_{i+1} + h_{i+1}c_{i+2} = \frac{3}{h_{i+1}}(y_{i+2} - y_{i+1}) - \frac{3}{h_i}(y_{i+1} - y_i)$$

$$\text{for } i = 1, \ldots, n - 3 \quad \text{and} \quad n > 3, \text{ and}$$

$$h_{n-2}c_{n-2} + (2h_{n-2} + h_{n-1})c_{n-1} = \frac{6}{h_{n-1}}(y_n - y_{n-1})$$

$$- \frac{3}{h_{n-2}}(y_{n-1} - y_{n-2}) - 3y_n' + h_{n-1}\frac{y_n''}{2}.$$

(For n = 3, the system consists only of the first and the last equation.)

3) $e_{n-1} = \frac{1}{3h_{n-1}^2}(y_n'' + c_{n-1}) - \frac{1}{h_{n-1}^3}y_n' + \frac{1}{h_{n-1}^4}(y_n - y_{n-1})$.

4) $d_0 = \frac{1}{3h_0}\left(c_1 - \frac{y_0''}{2}\right)$,

$d_i = \frac{1}{3h_i}(c_{i+1} - c_i), \quad i = 1, \ldots, n - 2$,

$d_{n-1} = \frac{1}{3h_{n-1}}\left(\frac{y_n''}{2} - c_{n-1}\right) - 2h_{n-1}e_{n-1}$.

5) $b_0 = \frac{1}{h_0}(y_1 - y_0) - \frac{h_0}{3}(c_1 + y_0'')$,

$b_i = \frac{1}{h_i}(y_{i+1} - y_i) - \frac{h_i}{3}(c_{i+1} + 2c_i), \quad i = 1, \ldots, n - 2$,

$b_{n-1} = y_n' - h_{n-1}\left(\frac{y_n''}{2} + c_{n-1}\right) + 2h_{n-1}^3 e_{n-1}$.

6) $a_i = y_i, \quad i = 0, \ldots, n-1$.

For (δ): Additional conditions: $S'(x_0) = y_0', S'(x_n) = y_n'$.

Let the spline function S with properties (1), (2), (3), and (4)(δ) be defined as

$$S_0(x) = \quad a_0 + b_0(x - x_0) + c_0(x - x_0)^2 + d_0(x - x_0)^3 + e_0(x - x_0)^4$$
$$\text{for} \quad x \in [x_0, x_1],$$

$$S_i(x) = \quad a_i + b_i(x - x_i) + c_i(x - x_i)^2 + d_i(x - x_i)^3$$
$$\text{for} \quad x \in [x_i, x_{i+1}], \quad i = 1, \ldots, n-2,$$

$$S_{n-1}(x) = \quad a_{n-1} + b_{n-1}(x - x_{n-1}) + c_{n-1}(x - x_{n-1})^2 + d_{n-1}(x - x_{n-1})^3$$
$$+ e_{n-1}(x - x_{n-1})^4$$
$$\text{for} \quad x \in [x_{n-1}, x_n].$$

The polynomial coefficients are determined by the equations:

1) $c_0 = \frac{y_0''}{2}, h_i = x_{i+1} - x_i$ for $i = 0, \ldots, n-1$.

2) The $c_i, i = 1, \ldots, n-1$, are the solutions of the linear system

$$(h_0 + 2h_1)c_1 + h_1 c_2 = \frac{3}{h_1}(y_2 - y_1) - \frac{6}{h_0}(y_1 - y_0) + 3y_0' + h_0 \frac{y_0''}{2},$$

$$h_i c_i + 2(h_i + h_{i+1})c_{i+1} + h_{i+1}c_{i+2} = \frac{3}{h_{i+1}}(y_{i+2} - y_{i+1}) - \frac{3}{h_i}(y_{i+1} - y_i),$$

$$\text{for } i = 1, \ldots, n-3 \quad \text{and} \quad n > 3, \text{ and}$$

$$h_{n-2}c_{n-2} + (2h_{n-2} + h_{n-1})c_{n-1} =$$

$$\frac{6}{h_{n-1}}(y_n - y_{n-1}) - \frac{3}{h_{n-2}}(y_{n-1} - y_{n-2}) - 3y_n' + h_{n-1}\frac{y_n''}{2}.$$

(For $n = 3$, the system consists only of the first and last equation.)

3) $e_0 = \frac{1}{3h_0^2}(c_1 + y_0'') + \frac{1}{h_0^3}y_0' - \frac{1}{h_0^4}(y_1 - y_0),$

$$e_{n-1} = \frac{1}{3h_{n-1}^2}(y_n'' + c_{n-1}) - \frac{1}{h_{n-1}^3}y_n' + \frac{1}{h_{n-1}^4}(y_n - y_{n-1}).$$

4) $d_0 = \frac{1}{3h_0}\left(c_1 - \frac{y_0''}{2}\right) - 2h_0 e_0,$

$$d_i = \frac{1}{3h_i}(c_{i+1} - c_i), \quad i = 1, \ldots, n-2,$$

$$d_{n-1} = \frac{1}{3h_{n-1}} \left(\frac{y_n''}{2} - c_{n-1} \right) - 2h_{n-1}e_{n-1}.$$

5) $b_0 = y_0'$,

$$b_i = \frac{1}{h_i}(y_{i+1} - y_i) - \frac{h_i}{3}(c_{i+1} + 2c_i), \quad i = 1, \ldots, n-2,$$

$$b_{n-1} = y_n' - h_{n-1} \left(\frac{y_n''}{2} + c_{n-1} \right) + 2h_{n-1}^3 e_{n-1}.$$

6) $a_i = y_i, \quad i = 0, \ldots, n-1.$

Bibliography for Section 10.1.4

[WODI77].

10.1.5 Convergence and Error Estimates for Interpolating Cubic Splines

In contrast to the uncertain convergence of interpolation, the convergence of interpolating splines and their derivatives towards the function and its derivatives is always guaranteed if the nodes x_k form a partition of $[a, b]$ whose diameter decreases.
Let

$$P: \quad a = x_0 < x_1 < \ldots < x_n = b$$

be a partition of the interval $[a, b]$ and define its *diameter* by

$$h = \max_{0 \le i < n} (x_{i+1} - x_i).$$

Then the following theorem holds for interpolating cubic splines (see [STOE91], 2.4.3; [NIED87], p.176 ff.).

THEOREM 10.8. Let $f \in C^4(a, b), |f^{(4)}(x)| \le M$ and S be an interpolating cubic spline for the partition P of $[a, b]$. Then:

1) If S is a spline with prescribed end point derivatives $S'(a) = f'(a)$ and $S'(b) = f'(b)$, or with prescribed second end point derivatives

$S''(a) = f''(a)$ and $S''(b) = f''(b)$, then

$$|f^{(k)}(x) - S^{(k)}(x)| \le C_k M h^{4-k}$$

for $x \in [a, b], k = 0, 1, 2$, with $C_0 = 5/384$, $C_1 = 1/24$, $C_2 = 3/8$.

2) If f is periodic with period $b - a$ and S is the associated periodic spline, then

$$|f^{(k)}(x) - S^{(k)}(x)| \le 3/8 M h^{4-k}$$

for $x \in [a, b]$ and $k = 0, 1, 2$.

REMARK: For interpolating natural splines, the only global approximation estimate is $|f(x) - S(x)| = O(h^2)$ in contrast to $|f(x) - S(x)| = O(h^4)$ for splines with prescribed first or second end point derivatives and splines with not-a-node condition.

Example: Convergence of polynomial interpolation and polynomial splines:

We shall work with the function

$$f(x) = \frac{1}{1 + x^2} \quad x \in [-5, 5] .$$

For f we select an odd number of equidistant nodes and interpolate: The first plot uses just three nodes -5, 0 and 5. Thereafter we half each of the subintervals used in the previous interpolation. When increasing the number of nodes thus, it is obvious from the plots below that the cubic interpolating spline converges towards f, while the interpolating polynomial represents f well only near the center of the interval with divergence at its boundaries.

Interpolation
polynomial

Cubic
spline

Exact function
plot

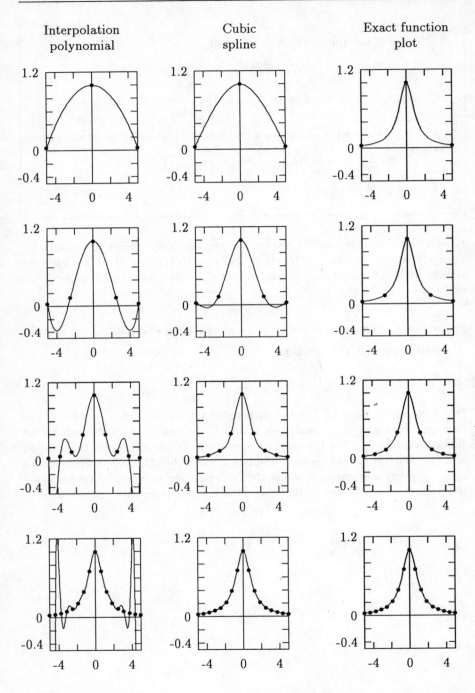

Figure 10.5: Comparison: Interpolating polynomial – cubic spline – exact plot of
$f(x) = 1/(1 + x^2)$ for $x \in [-5, 5]$ and increasing number of nodes

Bibliography for Section 10.1

[AHLB67]; [BEZI72]; [BÖHM74]; [BÖHM76]; [BOOR79]; [STOE91], 2.4; [ENGE72]; [GREV69]; [HÄMM91], 9; [NIED87], 6.3; [RALS67] vol.II, 8; [SAUE69], H §4; [SPÄT73], 1.7; [SPÄT74/2]; [STUM82], III; [WERN79], III; [WODI77].

10.2 Hermite Splines of Fifth Degree

10.2.1 Definition of Hermite Splines

Let us assume here that we know $n + 1$ functional values $y_i = f(x_i)$ and the slopes $y_i' = f'(x_i)$ at $n + 1$ nodes x_i for a function f. For these $n + 1$ value triples (x_i, y_i, y_i'), $i = 0, \ldots, n$, we can obtain a particularly good fitting by using *Hermite splines*. We aim to construct a curve as "smooth" as possible through the given points (x_i, y_i) with slopes y_i' by using polynomial splines of fifth degree, so called Hermite splines. If the x_i are ordered

$$a = x_0 < x_1 < \ldots < x_n = b,$$

the given curve can be represented by a spline function S with $S(x) \approx f(x)$, which is composed of polynomials S_i of fifth degree for $x \in [x_i, x_{i+1}]$ and $i = 0, \ldots, n - 1$.

As each S_i must satisfy certain end point conditions, we shall obtain various Hermite spline functions S. If the nodes x_i cannot be ordered, then we must use parametric Hermite splines; we refer back to section 10.1. Parametric Hermite splines for other end point conditions are treated in section 10.2.3.

Types of Hermite splines.

We aim to find a spline function S on $[a, b] = [x_0, x_n]$ with the properties:

(1) $S \in C^3[a, b]$.

(2) In each interval $[x_i, x_{i+1}]$ for $i = 0, \ldots, n - 1$, S is a polynomial S_i of fifth degree.

(3) S fulfills the interpolation conditions $S(x_i) = y_i, S'(x_i) = y_i', i = 0, \ldots, n$.

(4) S satisfies one of the following end point conditions (i) through (v):

(i) $S''(x_0) = S''(x_n) = 0$. Such an S is called a *natural Hermite spline*.

(ii) If $S(x_0) = S(x_n), S'(x_0) = S'(x_n), S''(x_0) = S''(x_n)$,
$S'''(x_0) = S'''(x_n)$, then S is called a *periodic Hermite spline*.

(iii) $S''(x_0) = y_0'', S''(x_n) = y_n''$ where $y''(x_0) = y_0'', y''(x_n) = y_n''$.

(iv) S has curvature radii r_0 and r_n at x_0 and x_n.

(v) $S'''(x_0) = y'''(x_0) = y_0''', S'''(x_n) = y'''(x_n) = y_n'''$.

10.2.2 Computation of Non-Parametric Hermite Splines

The coefficients of the various Hermite splines described in section 10.2.1 shall be computed now. In order to construct S with property (2) in section 10.2.1, we set

$$(10.5) \quad \begin{cases} S(x) & \equiv & S_i(x) := a_i + b_i(x - x_i) + c_i(x - x_i)^2 + \\ & & + d_i(x - x_i)^3 + e_i(x - x_i)^4 + f_i(x - x_i)^5, \\ & & \text{for } x \in [x_i, x_{i+1}] \text{ and } i = 0, \ldots, n-1. \end{cases}$$

The properties (1) and (3) on S lead to the following connecting conditions:

(a) $\qquad S_i(x_i) = y_i,$ $\qquad\qquad\qquad$ for $i = 0, \ldots, n,$

(b) $\qquad S_i'(x_i) = y_i',$ $\qquad\qquad\qquad$ for $i = 0, \ldots, n,$

(c) $\qquad S_i(x_i) = S_{i-1}(x_i),$ $\qquad\qquad$ for $i = 1, \ldots, n,$

(d) $\qquad S_i'(x_i) = S_{i-1}'(x_i),$ $\qquad\qquad$ for $i = 1, \ldots, n-1,$

(e) $\qquad S_i''(x_i) = S_{i-1}''(x_i),$ $\qquad\qquad$ for $i = 1, \ldots, n-1,$

(f) $\qquad S_i'''(x_i) = S_{i-1}'''(x_i),$ $\qquad\qquad$ for $i = 1, \ldots, n-1,$

\qquad with $S_n(x_n) = a_n, \ S_n'(x_n) = b_n.$

ALGORITHM 10.9 *(Non-periodic Hermite splines).*

Given: n triples (x_i, y_i, y_i'), $i = 0, \ldots, n$, where the x_i are ordered monotonically.

Task: Find coefficients $a_i, b_i, c_i, d_i, e_i, f_i$ of the polynomials (10.6).

1) Set $a_i = y_i$, $b_i = y_i'$ for $i = 0, \ldots, n$.

2) (i) $c_0 = c_n = 0$,

 (iii) $c_0 = \frac{1}{2} y_0''$, $c_n = \frac{1}{2} y_n''$,

 (iv) $c_0 = (1 + b_0^2)^{3/2}/2r_0$, $c_n = (1 + b_n^2)^{3/2}/2r_n$,

 (v) $c_0 = \frac{1}{3} \left[\dfrac{10(a_1 - a_0)}{h_0^2} - \dfrac{2(2b_1 + 3b_0)}{h_0} - \dfrac{y_0''' h_0}{6} + c_1 \right]$,

 $c_n = \frac{1}{3} \left[\dfrac{10(a_{n-1} - a_n)}{h_{n-1}^2} + \dfrac{2(2b_{n-1} + 3b_n)}{h_{n-1}} + \dfrac{y_n''' h_{n-1}}{6} + c_{n-1} \right]$.

3) Solve the following linear system for $c_1, c_2, \ldots, c_{n-1}$ with $h_i = x_{i+1} - x_i$:

$$3 \left(\frac{\alpha}{h_0} + \frac{1}{h_1} \right) c_1 - \frac{1}{h_1} c_2 = 10 \left[\frac{a_2 - a_1}{h_1^3} - \frac{a_1 - a_0}{h_0^3} \right] +$$

$$+ 4 \left[\frac{b_0}{h_0^2} - \frac{3}{2} \left(\frac{1}{h_1^2} - \frac{1}{h_0^2} \right) b_1 - \frac{b_2}{h_1^2} \right] + \beta_1,$$

$$- \frac{c_{i-1}}{h_{i-1}} + 3 \left(\frac{1}{h_{i-1}} + \frac{1}{h_i} \right) c_i - \frac{1}{h_i} c_{i+1} = 10 \left[\frac{a_{i+1} - a_i}{h_i^3} - \frac{a_i - a_{i-1}}{h_{i-1}^3} \right] +$$

$$+ 4 \left[\frac{b_{i-1}}{h_{i-1}^2} - \frac{3}{2} \left(\frac{1}{h_i^2} - \frac{1}{h_{i-1}^2} \right) b_i - \frac{b_{i+1}}{h_i^2} \right] \text{ for } i = 2, \ldots, n-2,$$

$$- \frac{c_{n-2}}{h_{n-2}} + 3 \left(\frac{1}{h_{n-2}} + \frac{\alpha}{h_{n-1}} \right) c_{n-1} = 10 \left[\frac{a_n - a_{n-1}}{h_{n-1}^3} - \frac{a_{n-1} - a_{n-2}}{h_{n-2}^3} \right] +$$

$$+ 4 \left[\frac{b_{n-2}}{h_{n-2}^2} - \frac{3}{2} \left(\frac{1}{h_{n-1}^2} - \frac{1}{h_{n-2}^2} \right) b_{n-1} - \frac{b_n}{h_{n-1}^2} \right] + \beta_2.$$

Here the quantities α, β_1 and β_2 must be set according to the end point conditions (i), (iii), (iv), (v) as:

$$\alpha = \begin{cases} 1 & \text{in case of (i),(iii),(iv),} \\ 8/9 & \text{in case of (v).} \end{cases}$$

$$\beta_1 = \begin{cases} 0 & \text{in case of (i),} \\ y_0''/2h_0 & \text{in case of (iii),} \\ (1 + b_0^2)^{3/2}/(2h_0 r_0) & \text{in case of (iv),} \\ \dfrac{10}{3h_0^3}(a_1 - a_0) - \dfrac{2}{3h_0^2}(2b_1 + 3b_0) - \dfrac{y_0'''}{18} & \text{in case of (v).} \end{cases}$$

$$\beta_2 = \begin{cases} 0 & \text{in case of (i),} \\ y_n''/2h_{n-1} & \text{in case of (iii),} \\ (1+b_n^2)^{3/2}/(2h_{n-1}r_n) & \text{in case of (iv),} \\ -\dfrac{10}{3h_{n-1}^3}(a_n - a_{n-1}) + \dfrac{2}{3h_{n-1}^2}(3b_n + 2b_{n-1}) + \dfrac{y_n'''}{18} & \text{in case of (v).} \end{cases}$$

4) $d_i = \dfrac{10}{h_i^3}(a_{i+1} - a_i) - \dfrac{2}{h_i^2}(2b_{i+1} + 3b_i) + \dfrac{1}{h_i}(c_{i+1} - 3c_i)$, $i = 0, \ldots, n-1$,

$d_n = d_{n-1} - \dfrac{2}{h_{n-1}^2}(b_n - b_{n-1}) + \dfrac{2}{h_{n-1}}(c_n + c_{n-1})$.

5) $e_i = \dfrac{1}{2h_i^3}(b_{i+1} - b_i) - \dfrac{1}{h_i^2}c_i - \dfrac{1}{4h_i}(d_{i+1} + 5d_i)$, $i = 0, \ldots, n-1$.

6) $f_i = \dfrac{1}{10h_i^3}(c_{i+1} - c_i - 3d_i h_i - 6e_i h_i^2)$, $i = 0, \ldots, n-1$.

The linear system in step 3) of the above algorithm 10.9 has the form $Ac = g$ with

$$A = \begin{pmatrix} 3\left(\frac{\alpha}{h_0} + \frac{1}{h_1}\right) & -\frac{1}{h_1} & & & \\ -\frac{1}{h_1} & 3\left(\frac{1}{h_1} + \frac{1}{h_2}\right) & -\frac{1}{h_2} & & \\ \ddots & \ddots & \ddots & & \\ & -\frac{1}{h_{n-3}} & 3\left(\frac{1}{h_{n-3}} + \frac{1}{h_{n-2}}\right) & -\frac{1}{h_{n-2}} \\ & & -\frac{1}{h_{n-2}} & 3\left(\frac{1}{h_{n-2}} + \frac{\alpha}{h_{n-1}}\right) \end{pmatrix},$$

$$g = \begin{pmatrix} 10\left[\frac{a_2 - a_1}{h_1^3} - \frac{a_1 - a_0}{h_0^3}\right] + 4\left[\frac{b_0}{h_0^2} - \frac{3}{2}\left(\frac{1}{h_1^2} - \frac{1}{h_0^2}\right)b_1 - \frac{b_2}{h_1^2}\right] + \beta_1 \\ 10\left[\frac{a_3 - a_2}{h_2^3} - \frac{a_2 - a_1}{h_1^3}\right] + 4\left[\frac{b_1}{h_1^2} - \frac{3}{2}\left(\frac{1}{h_2^2} - \frac{1}{h_1^2}\right)b_2 - \frac{b_3}{h_2^2}\right] \\ \vdots \\ 10\left[\frac{a_n - a_{n-1}}{h_{n-1}^3} - \frac{a_{n-1} - a_{n-2}}{h_{n-2}^3}\right] + 4\left[\frac{b_{n-2}}{h_{n-2}^2} - \frac{3}{2}\left(\frac{1}{h_{n-1}^2} - \frac{1}{h_{n-2}^2}\right)b_{n-1} - \frac{b_n}{h_{n-1}^2}\right] + \beta_2 \end{pmatrix}$$

and $c = (c_1, c_2, \ldots, c_{n-1})^T$.

Properties of the matrix A.

A is tridiagonal, symmetric and strongly diagonally dominant, it has positive diagonal elements and negative co-diagonal elements. Hence A is positive definite and strongly nonsingular. As a consequence, the system $Ac = g$ is

uniquely solvable by the methods described in section 4.10. Pivoting and iterative refinement are not necessary.

ALGORITHM 10.10 (*Periodic Hermite splines*).

Given: n triples (x_i, y_i, y_i'), $i = 0, \ldots, n$, with monotonically ordered x_i.

Task: Find coefficients $a_i, b_i, c_i, d_i, e_i, f_i$ of the polynomials (10.6) with the end point conditions $S^{(k)}(x_0) = S^{(k)}(x_n)$ for $k = 0, \ldots, 3$.

1) Set $a_i = y_i$, $b_i = y_i'$ for $i = 0, \ldots, n$.

2) Set $c_0 = c_n$, $c_1 = c_{n+1}$, $a_1 = a_{n+1}$, $b_1 = b_{n+1}$, $h_0 = h_n$.

3) Solve the linear system for c_1, c_2, \ldots, c_n with $h_i = x_{i+1} - x_i$:

$$3\left(\frac{1}{h_0} + \frac{1}{h_1}\right)c_1 - \frac{1}{h_1}c_2 - \frac{1}{h_0}c_n == 10\left[\frac{a_2 - a_1}{h_1^3} - \frac{a_1 - a_0}{h_0^3}\right] +$$
$$+ 4\left[\frac{b_0}{h_0^3} - \frac{3}{2}\left(\frac{1}{h_1^2} - \frac{1}{h_0^2}\right)b_1 - \frac{b_2}{h_1^2}\right],$$

$$-\frac{c_{i-1}}{h_{i-1}} + 3\left(\frac{1}{h_{i-1}} + \frac{1}{h_i}\right)c_i - \frac{1}{h_i}c_{i+1} = 10\left[\frac{a_{i+1} - a_i}{h_i^3} - \frac{a_i - a_{i-1}}{h_{i-1}^3}\right] +$$
$$+ 4\left[\frac{b_{i-1}}{h_{i-1}^3} - \frac{3}{2}\left(\frac{1}{h_i^2} - \frac{1}{h_{i-1}^2}\right)b_i - \frac{b_{i+1}}{h_i^2}\right], \text{ for } i = 2, \ldots, n-1, \text{ and}$$

$$-\frac{c_1}{h_0} - \frac{1}{h_{n-1}}c_{n-1} + 3\left(\frac{1}{h_{n-1}} + \frac{1}{h_0}\right)c_n = 10\left[\frac{a_1 - a_n}{h_0^3} - \frac{a_n - a_{n-1}}{h_{n-1}^3}\right] +$$
$$+ 4\left[\frac{b_{n-1}}{h_{n-1}^3} - \frac{3}{2}\left(\frac{1}{h_0^2} - \frac{1}{h_{n-1}^2}\right)b_n - \frac{b_1}{h_0^2}\right].$$

4) Set $d_i = \frac{10}{h_i^3}(a_{i+1} - a_i) - \frac{2}{h_i^2}(2b_{i+1} + 3b_i) + \frac{1}{h_i}(c_{i+1} - 3c_i)$,
 for $i = 0, \ldots, n-1$ and
 $$d_n = d_{n-1} - \frac{2}{h_{n-1}^2}(b_n - b_{n-1}) + \frac{2}{h_{n-1}}(c_n + c_{n-1}).$$

5) Set $e_i = \frac{1}{2h_i^3}(b_{i+1} - b_i) - \frac{1}{h_i^2}c_i - \frac{1}{4h_i}(d_{i+1} + 5d_i)$ for $i = 0, \ldots, n-1$.

6) Set $f_i = \frac{1}{10h_i^3}(c_{i+1} - c_i - 3d_i h_i - 6e_i h_i^2)$ for $i = 0, \ldots, n-1$.

The linear system in 3) of algorithm 10.10 has n equations for the n unknown c_1, c_2, \ldots, c_n with a cyclically tridiagonal, symmetric, and strongly diagonally

dominant system matrix \boldsymbol{A} which has positive main diagonal elements and negative elements off its main diagonal. Thus this matrix is positive definite and strongly nonsingular. The system should be solved using a Cholesky method for cyclically tridiagonal matrices from section 4.11.

10.2.3 Computation of Parametric Hermite Splines

If we are given a set of triples $(x_i, y_i, y_i'), i = 0, \ldots, n$, on a curve C, where the x_i cannot be ordered monotonically, we have to find an approximate parametrization $\{x(t), y(t)\} \approx \{S_x(t), S_y(t)\}$ of C. Then the Hermite spline functions S_x and S_y for the triples (t_i, x_i, \dot{x}_i) and (t_i, y_i, \dot{y}_i) must be calculated, respectively, with parameter values t_i that are ordered monotonically. The calculations can be carried out as in section 10.1.3. The values for \dot{x}_i, \dot{y}_i can be found from the given y_i', $i = 0, \ldots, n$, by using $\dot{x}_i^2 + \dot{y}_i^2 = 1$ and $y_i' = \dot{y}_i / \dot{x}_i$ as follows:

$$\dot{x}_i = \frac{\sigma_i}{+\sqrt{1 + y_i'^2}}, \text{ where } \begin{cases} \sigma_i = \operatorname{sgn}(\boldsymbol{p}_{i+1} - \boldsymbol{p}_i)^T \dot{\boldsymbol{p}}_{i+} & \text{for } i = 0, \ldots, n-1, \\ \sigma_i = \operatorname{sgn}(\boldsymbol{p}_n - \boldsymbol{p}_{n-1})^T \dot{\boldsymbol{p}}_{n+} & \text{for } i = n, \end{cases}$$

and $\dot{y}_i = \dot{x}_i y_i'$ for $i = 0, \ldots, n$,

where we define

$$\boldsymbol{p}_i := \begin{pmatrix} x_i \\ y_i \end{pmatrix}, \quad \dot{\boldsymbol{p}}_i := \begin{pmatrix} \dot{x}_i \\ \dot{y}_i \end{pmatrix}, \text{ and } \dot{\boldsymbol{p}}_{i+} := \begin{pmatrix} |\dot{x}_i| \\ |\dot{x}_i| y_i' \end{pmatrix}.$$

The sign of \dot{x}_i in the definition of σ_i is determined so that the angle between the vectors $(\boldsymbol{p}_{i+1} - \boldsymbol{p}_i)$ and $\dot{\boldsymbol{p}}_i$ is acute, i.e., $(\boldsymbol{p}_{i+1} - \boldsymbol{p}_i)^T \dot{\boldsymbol{p}}_i > 0$. If the scalar product $(\boldsymbol{p}_{i+1} - \boldsymbol{p}_i)^T \dot{\boldsymbol{p}}_{i+}$ vanishes for an index i, we choose $\sigma_i = \operatorname{sgn}(\boldsymbol{p}_i - \boldsymbol{p}_{i-1})^T \dot{\boldsymbol{p}}_{i+}$, provided $(\boldsymbol{p}_i - \boldsymbol{p}_{i-1})^T \dot{\boldsymbol{p}}_{i+} \neq 0$; otherwise the problem cannot be solved uniquely. Similarly, the quantities $\boldsymbol{p}_0, \boldsymbol{p}_1$ and y_0' are not unique if $(\boldsymbol{p}_1 - \boldsymbol{p}_0)^T \dot{\boldsymbol{p}}_{0+} = 0$.

For a vertical tangent we set:

$$\dot{\boldsymbol{p}}_{i+} = \begin{pmatrix} 0 \\ 1 \end{pmatrix} \quad \text{and} \quad \dot{\boldsymbol{p}}_i = \sigma_i \dot{\boldsymbol{p}}_{i+}.$$

The spline functions S_x and S_y are computed analogously to section 10.2.1. Thus S_x is calculated for the triples (t_i, x_i, \dot{x}_i) and S_y for the triples (t_i, y_i, \dot{y}_i). The \dot{x}_i, \dot{y}_i are computed from the y_i' as described above. The following end point boundary conditions can be specified here:

(1) natural end point conditions,

(2) periodic end point conditions,

(3) specified values for y_0'', y_n'',

(4) specified values for $(\ddot{x}_0, \ddot{y}_0), (\ddot{x}_n, \ddot{y}_n)$,

(5) radii of curvature for r_0 and r_n,

(6) specified values for $(\dddot{x}_0, \dddot{y}_0), (\dddot{x}_n, \dddot{y}_n)$.

The spline functions S_x and S_y with end point conditions (1) through (6) can be calculated as follows:

For (1): S_x, S_y are natural splines. S_x is computed for the triples (t_i, x_i, \dot{x}_i) according to algorithm 10.9 with end point condition (i) where x_i is replaced by t_i, y_i by x_i, and y_i' by \dot{x}_i. S_y is computed for the triples (t_i, y_i, \dot{y}_i) according to algorithm 10.9 with end point condition (i), replacing x_i by t_i, y_i' by \dot{y}_i, while y_i remains unchanged.

For (2): S_x, S_y are periodic splines. S_x is computed according to algorithm 10.10. Here we replace x_i by t_i, y_i by x_i and y_i' by \dot{x}_i. S_y is calculated according to algorithm 10.10 by replacing x_i by t_i, y_i' by \dot{y}_i, and y_i remains unchanged.

For (3): Calculate S_x according to algorithm 10.9 with end point condition (iii), where we use t_i instead of x_i, x_i instead of y_i, and \dot{x}_i instead of y_i'. The end point conditions on y_0'', y_n'' are replaced by $\ddot{x}_0 = 1, \ddot{x}_n = 1$, and $\dot{x}_0, \dot{x}_n \neq 0$. Calculate S_y according to algorithm 10.9 with end point condition (iii) with t_i instead of x_i, \dot{y}_i instead of y_i', and y_i unchanged and end point conditions on \ddot{y}_0, \ddot{y}_n instead of y_0'', y_n''. The values for \ddot{y}_0 and \ddot{y}_n are calculated as follows:

$$\ddot{y}_0 = \frac{1}{\dot{x}_0}(\dot{x}_0^3 y_0'' + \dot{y}_0), \quad \ddot{y}_n = \frac{1}{\dot{x}_n}(\dot{x}_n^3 y_n'' + \dot{y}_n).$$

For (4): Calculate S_x according to algorithm 10.9 with end point condition (iii) with t_i instead of x_i, x_i instead of y_i, \dot{x}_i instead of y_i', \ddot{x}_0 instead of y_0'', and \ddot{x}_n instead of y_n''. Calculate S_y according to algorithm 10.9 with end point condition (iii) with t_i instead of x_i, \dot{y}_i instead of y_i', \ddot{y}_0 instead of y_0'', and \ddot{y}_n instead of y_n''.

For (5): Calculate S_x according to algorithm 10.9 with end point condition (iii) with t_i instead of x_i, x_i instead of y_i, \dot{x}_i instead of y_i', \ddot{x}_0 instead of y_0'' and \ddot{x}_n instead of y_n''. The values of \ddot{x}_0 and \ddot{x}_n are calculated as follows:

$$\ddot{x}_0 = \begin{cases} -1/(r_0 \dot{y}_0) & \text{if } \dot{x}_0 = 0, \\ 1 & \text{otherwise.} \end{cases}$$

$$\ddot{x}_n = \begin{cases} -1/(r_n \dot{y}_n) & \text{if } \dot{x}_n = 0, \\ 1 & \text{otherwise.} \end{cases}$$

We can calculate S_y according to algorithm 10.9 with end point condition (iii)
with t_i instead of x_i, \dot{y}_i instead of y_i', \ddot{y}_0 instead of y_0'', \ddot{y}_n instead of y_n'', while
y_i remains unchanged. The values of \ddot{y}_0 and \ddot{y}_n are calculated according to
the following formulas:

$$\ddot{y}_0 = \begin{cases} 1 & \text{if } \dot{x}_0 = 0, \\ \left(\frac{1}{r_0} + \dot{y}_0\right)/\dot{x}_0 & \text{else,} \end{cases}$$

$$\ddot{y}_n = \begin{cases} 1 & \text{if } \dot{x}_n = 0, \\ \left(\frac{1}{r_n} + \dot{y}_n\right)/\dot{x}_n & \text{else.} \end{cases}$$

For (6): We calculate of S_x according to algorithm 10.9 with end point con-
dition (v) with t_i instead of x_i, x_i instead of y_i, \dot{x}_i instead of y_i', \dddot{x}_0 instead
of y_0''', and \dddot{x}_n instead of y_n'''.
The calculation of S_y is carried out according to algorithm 10.9 with end point
condition (v) with t_i instead of x_i, y_i remains, \dot{y}_i instead of y_i', \dddot{y}_0 instead of
y_0''', and \dddot{y}_n instead of y_n''';

If other end points conditions are given, the formulas must be altered ac-
cordingly. Formulas for prescribed quadruples (x_i, y_i, y_i', y_i'') can be found in
[SPÄT74/2], pp. 55.

Bibliography for Section 10.2

[AHLB67], IV; [BÖHM74], 5.2; [SPÄT74/2], p. 52 ff.

10.3 Hints for Selecting Appropriate Inter-
polating or Approximating Splines

In general, cubic splines are superior to Hermite splines for the analytic rep-
resentation of smooth curves. This is due to the fact that cubic splines es-
sentially describe mechanical sheer in an analytical way and minimize the
bending energy, whereas Hermite splines, being locally polynomials of fifth
degree, exhibit stronger ripple effects. If, however, one needs to achieve given
derivatives at the interpolating points for the spline or if one needs a higher
order of differentiability, then one must use Hermite splines.
For approximating cubic splines (fitting splines), which will be discussed in
chapter 11, the same assessment holds as to their applicability. One uses fit-
ting splines if the given functional values contain errors, and the associated

data should not be linked by a smooth curve, but instead should be approximated by a fitting curve.

If there are clues for a suitable model function for the curve, one should employ the root-mean-square approximation of chapter 8.1.3. If, however, there is no known model, it is advisable to employ fitting splines.

Besides chosing a suitable spline type, one must also determine a partition $P : x_0 < x_1 < x_2 < ... < x_n$ of the interval $[a, b]$. For steep slopes of the function $y = f(x)$ the nodes x_i must be chosen closely together. In areas of little y variation, they can be chosen more distant. In general, equidistant partitions will be inappropriate.

Monotone and specifically concave functions often create problems with overshoot and ripple effects when approximated by cubic splines. Besides the recommendations given in the table below, we can also recommend the method of [FRIT80] to alleviate such problems.

One could say that the prescribed second end point derivatives $S''(x_0) = S''(x_n) = 0$ of natural splines often have an ill effect, and that splines with free first or second end point derivatives exhibit better approximation properties. If one has no information about the end point derivatives, they can be determined by interpolation: For instance one can interpolate the first four and the last four (or three) given interpolation points (x_i, y_i) by a polynomial of third (or second) degree and then determine the derivatives of the interpolating polynomial at the end points x_0 and x_n. Then one can continue with the chosen spline method. If one wants to preserve line segments as lines and peaks as peaks, one should use Akima subsplines or Renner subsplines, see chapter 13.

The formula for calculating the parameter values of the Renner subsplines in chapter 13 is also well suited for an approximate calculation of the arc length of a curve.

Parametric splines.

If the nodes x_i are not and cannot be arranged in a monotonic order, the spline calculations must be carried out by means of a *parameter*. To calculate the parameter values, the methods of chapter 10.1.3 should be employed.

For further literature concerning splines of high order, rational splines, generalized cubic splines (4-parameter splines with non-polynomial model functions) see e.g. [AHLB67]; [GREV69]; [RALS67], II.; [BÖHM74]; [BÖHM76]; [SPÄT74/2].

No.	Type of curve (of the form...)	n. c. s.	p. c. s.	cub. sp. l.e.d.	par. nat. c.s.	par. per. c.s.	par. c.s. l.e.d.	cub. Bez. sp.	A. ss.	R. ss.	B-s. c.
1	concave up, open, monontone order	+	- -	++	+	- -	+	+	+	+	++
2	concave up, open, non monotone order	- -	- -	- -	+	- -	++	+	- -	+	++
3	concave up and down, open, monotone order	+	- -	++	+	- -	+	++	++	+	++
4	concave up and down, open, non monotone order	- -	- -	- -	+	- -	+	+ +	- -	++	++
5	concave up and down, periodic in $f(x)$ and $f'(x)$, monotone order	- -	+	- -	- -	- -	- -	- -	++	+	- -

No.	Type of curve, (of the form...)	n. c. s.	p. c. s.	cub. sp. l.e.d.	par. nat. c.s.	par. per. c.s.	par. c.s. l.e.d.	cub. Bez. sp.	A. ss.	R. ss.	B-s. c.
6	concave up or concave up and down, smooth, closed	--	--	--	--	++	--	+	--	++	++
7	concave up or concave up and down, not smooth, closed	--	--	--	+	--	++	+	--	--	++
8	concave up, open, non monotone order	--	--	--	+	--	++	+	--	++	++
9	concave up, open, monotone order, steep slopes	-	--	++	-	--	+	++	++	+	+
10		--	--	--	--	--	--	-	++	+	-
11		--	--	--	--	--	--	-	--	++	-

In the above table, we have used the following abbreviations:

abbreviation	meaning
++	well suited
+	suited
-	use is possible, results may not be useful
- -	not suitable
n. c. s.	natural cubic spline
p. c. s.	periodic cubic spline
cub. sp. 1. e. d.	cubic spline with prescribed first end point derivatives
par. nat. c. s.	parametric natural cubic spline
par. per. c. s.	parametric periodic cubic spline
par. c. s. 1. e. d.	parametric cubic spline with prescribed first end point derivative
cub. Bez. sp.	cubic Bézier spline
A. ss.	Akima subspline
R. ss.	Renner subspline
B-s. c.	B-spline curve

The large general table above is designed to help the user decide which of the interpolating cubic splines may be most suitable for a given type of curve. In this table we have included the cubic Bézier splines of section 12.3 as well as the Akima and Renner subsplines presented in chapter 13. The table also includes remarks for B–splines of chapter 12 which are extremely well suited for graphing all smooth curves. Besides the information given in this table, one should also consult the mentioned chapters 12 and 13.

For the *construction of smooth surfaces* one can employ one of the methods discussed in chapter 12, such as bicubic splines in section 12.1, interpolating surface splines in section 12.2, Bézier splines in section 12.3 and B-splines in 12.4. Detailed hints on their use for generating surfaces and curves are given in section 12.5.

Bibliography for Chapter 10

[BÖHM74]; [BÖHM76]; [STOE91], 2.4; [ENGE87]; [FRIT80]; [GREV69]; [HÄMM91], 6.; [HOSC89]; [MAES88], 6.2; [NIED87], 6.3; [RALS67], IV. 8; [SCHW89], 3.7; [SHAM73], II. 1.3; [SPÄT74/2]; [WODI77].

Chapter 11

Cubic Fitting Splines for Constructing Smooth Curves

11.1 The Problem

For a given function $f \in C[a, b]$, we assume that the ordinates $u_i \approx f(x_i)$ are known approximately for $n + 1$ nodes $x_i \in [a, b]$. If the u_i are obtained by measurements which are in general imprecise, the deviations in the data u_i render a straightforward approximation using interpolating methods meaningless. Instead, one needs to find an error compensating function S whose graph passes near the interpolation points (x_i, u_i) and is as smooth as possible.

If we assume that the nodes are arranged monotonically:

(11.1)
$$a = x_0 < x_1 < \ldots < x_n = b,$$

then the desired curve can be found as a spline function of the form $S(x) \approx f(x)$, which is composed of piecewise cubic polynomials S_i for $x \in [x_i, x_{i+1}]$, $i = 0, \ldots, n - 1$, where the S_i must satisfy certain *connecting conditions*. Depending on the given end point conditions, one can obtain several different types of *fitting splines S*.

Figure 11.1: Polynomial cubic fitting spline

Types of non-parametric cubic fitting splines S on $[a, b]$:

(i) Cubic fitting spline with prescribed first end point derivatives.

(ii) Cubic fitting spline with prescribed second end point derivatives.
 (If the second end point derivatives are zero, this is a
 natural cubic fitting spline).

(iii) Cubic fitting spline with prescribed third end point derivatives.

(iv) Periodic cubic fitting spline.

If the montonicity condition (11.1) cannot be fulfilled, one should use para-
metric splines instead, see section 11.4.

11.2 Definition of Fitting Spline Functions

For the measurement points (x_i, u_i), $i = 0, \ldots, n$, $n \geq 2$, with nodes x_i
arranged monotonically, the non-parametric fitting spline functions S men-
tioned under (i),..., (iv) in section 11.1 are defined using weights $w_i > 0$ by
the following conditions:

1) $S \in C^2[a, b]$.

2) In each interval $[x_i, x_{i+1}]$ for $i = 0, \ldots, n-1$, S is a cubic polynomial S_i:

$$S(x) \equiv S_i(x) = a_i + b_i(x - x_i) + c_i(x - x_i)^2 + d_i(x - x_i)^3$$
$$\text{for} \quad x \in [x_i, x_{i+1}], \quad i = 0, \ldots, n-1.$$

3) $S(x_i) = y_i$ for $u_i \approx y_i$ and $i = 0, \ldots, n$.

4) With $w_i(u_i - y_i) = r_i$, $i = 0, \ldots, n$, for weights $w_i > 0$, we have

$$
\begin{aligned}
r_0 &= S_0'''(x_0), \\
r_i &= S_i'''(x_i) - S_{i-1}'''(x_i), \qquad i = 1, \ldots, n-1, \\
r_n &= -S_{n-1}'''(x_n).
\end{aligned}
$$

5) End point conditions for the different types are:

(i) Cubic fitting spline with prescribed first end point derivatives:
$$S'(x_0) = \alpha, \qquad S'(x_n) = \beta.$$

(ii) Cubic fitting spline with prescribed second end point derivatives:

$$S''(x_0) = \alpha, \qquad S''(x_n) = \beta.$$

(If $\alpha = \beta = 0$ this is a natural fitting spline.)

(iii) Cubic fitting spline with prescribed third end point derivatives:

$$S'''(x_0) = \alpha, \qquad S'''(x_n) = \beta.$$

(iv) Periodic cubic fitting spline:

period interval : $[x_0, x_n] = [a, b]$,

periodic conditions : $u_0 = u_n$, $w_0 = w_n$, and

$$S^{(k)}(x_0) = S^{(k)}(x_n) \text{ for } k = 0, 1, 2.$$

Remark concerning the choice of weights:
It can be shown that if the weights w_i in 4) are chosen to be large, then the resulting curve will be the interpolating cubic spline for the points (x_i, u_i), while for $w_i \approx 0$ the least squares regression line is computed. Hence, by selecting the weights suitably one can guarantee that the ensuing spline function passes closely near the measured values u_i (in case of large w_i), or that it approximates them in a least squares sense (for small w_i). In practice, this can be monitored interactively on a computer display. If the test conditions that give rise to the measured values u_i do not change, then one can use the previously obtained values for the weights repeatedly.

11.3 Non-Parametric Cubic Fitting Splines

In the following, we shall indicate algorithms for calculating the fitting splines of section 11.1.

(i) Fitting spline with prescribed first end point derivatives.

In order to calculate the coefficients a_i, b_i, c_i, d_i of the cubic polynomials S_i one should evaluate the following equations in the indicated order:

1) *Linear system for* $c_1, c_2, ..., c_{n-1}$:

Set $h_i = x_{i+1} - x_i$ for $i = 0, \ldots, n-1$, $W_i = \frac{6}{w_i}$ for $i = 0, \ldots, n$,

$H_i = \frac{1}{h_i} + \frac{1}{h_{i+1}}$ for $i = 0, \ldots, n-2$ and abbreviate

$$F_1 = \frac{h_0 - \dfrac{W_0}{h_0^2} - \dfrac{W_1}{h_0}H_0}{2h_0^2 + \dfrac{1}{h_0}(W_0 + W_1)}, \quad F_2 = \frac{\dfrac{W_1}{h_0h_1}}{2h_0^2 + \dfrac{1}{h_0}(W_0 + W_1)},$$

$$F_3 = \frac{\dfrac{W_{n-1}}{h_{n-2}h_{n-1}}}{2h_{n-1}^2 + \dfrac{1}{h_{n-1}}(W_{n-1} + W_n)}, \quad F_4 = \frac{h_{n-1} - \dfrac{W_{n-1}}{h_{n-1}}H_{n-2} - \dfrac{W_n}{h_{n-1}^2}}{2h_{n-1}^2 + \dfrac{1}{h_{n-1}}(W_{n-1} + W_n)},$$

$$G_1 = W_1 H_0 + \frac{W_0}{h_0} - h_0^2, \quad G_2 = \frac{W_1}{h_1}, \quad G_3 = 3\left((u_1 - u_0) - \alpha h_0\right)$$

$$G_4 = \frac{W_{n-1}}{h_{n-2}}, \quad G_5 = \frac{W_n}{h_{n-1}} + W_{n-1}H_{n-2} - h_{n-1}^2,$$

$$G_6 = 3\left(\beta h_{n-1} - (u_n - u_{n-1})\right).$$

Then the c_i are determined by the following five-diagonal system:

$$\left(F_1 G_1 + 2(h_0 + h_1) + \frac{W_0}{h_0^2} + W_1 H_0^2 + \frac{W_2}{h_1^2}\right)c_1 +$$

$$+ \left(h_1 - \frac{W_1}{h_1}H_0 - \frac{W_2}{h_1}H_1 - F_1 G_2\right)c_2 + \left(\frac{W_2}{h_1 h_2}\right)c_3 =$$

$$= 3\left(\frac{u_2 - u_1}{h_1} - \frac{u_1 - u_0}{h_0}\right) - F_1 G_3,$$

$$\left(h_1 - \frac{W_1}{h_1}H_0 - \frac{W_2}{h_1}H_1 - F_1 G_2\right)c_1 +$$

$$+ \left(2(h_1 + h_2) + \frac{W_1}{h_1^2} + W_2 H_1^2 + \frac{W_3}{h_2^2} - F_2 G_2\right)c_2 +$$

$$+ \left(h_2 - \frac{W_2}{h_2}H_1 - \frac{W_3}{h_2}H_2\right)c_3 + \left(\frac{W_3}{h_2 h_3}\right)c_4$$

$$= 3\left(\frac{u_3 - u_2}{h_2} - \frac{u_2 - u_1}{h_1}\right) - F_2 G_3,$$

$$\left(\frac{W_{i-1}}{h_{i-2}h_{i-1}}\right)c_{i-2} + \left(h_{i-1} - \frac{W_{i-1}}{h_{i-1}}H_{i-2} - \frac{W_i}{h_{i-1}}H_{i-1}\right)c_{i-1} +$$

$$+ \left(2(h_{i-1} + h_i) + \frac{W_{i-1}}{h_{i-1}^2} + W_i H_{i-1}^2 + \frac{W_{i+1}}{h_i^2}\right)c_i$$

$$+ \left(h_i - \frac{W_i}{h_i}H_{i-1} - \frac{W_{i+1}}{h_i}H_i\right)c_{i+1} + \left(\frac{W_{i+1}}{h_i h_{i+1}}\right)c_{i+2} =$$

$$= 3\left(\frac{u_{i+1} - u_i}{h_i} - \frac{u_i - u_{i-1}}{h_{i-1}}\right) \text{ for } i = 3, \dots, n-3,$$

$$\left(\frac{W_{n-3}}{h_{n-4}h_{n-3}}\right)c_{n-4} + \left(h_{n-3} - \frac{W_{n-3}}{h_{n-3}}H_{n-4} - \frac{W_{n-2}}{h_{n-3}}H_{n-3}\right)c_{n-3}+$$

$$+ \left(2(h_{n-3}+h_{n-2}) + \frac{W_{n-3}}{h_{n-3}^2} + W_{n-2}H_{n-3}^2 + \frac{W_{n-1}}{h_{n-2}^2} - F_3 G_4\right)c_{n-2}+$$

$$+ \left(h_{n-2} - \frac{W_{n-2}}{h_{n-2}}H_{n-3} - \frac{W_{n-1}}{h_{n-2}}H_{n-2} + F_3 G_5\right)c_{n-1} =$$

$$= 3\left(\frac{u_{n-1}-u_{n-2}}{h_{n-2}} - \frac{u_{n-2}-u_{n-3}}{h_{n-3}}\right) - F_3 G_6, \quad \text{and}$$

$$\left(\frac{W_{n-2}}{h_{n-3}h_{n-2}}\right)c_{n-3} + \left(h_{n-2} - \frac{W_{n-2}}{h_{n-2}}H_{n-3} - \frac{W_{n-1}}{h_{n-2}}H_{n-2} + F_3 G_5\right)c_{n-2}+$$

$$+ \left(2(h_{n-2}h_{n-1}) + \frac{W_{n-2}}{h_{n-2}^2} + W_{n-1}H_{n-2}^2 + \frac{W_n}{h_{n-1}^2} + F_4 G_5\right)c_{n-1} =$$

$$= 3\left(\frac{u_n-u_{n-1}}{h_{n-1}} - \frac{u_{n-1}-u_{n-2}}{h_{n-2}}\right) - F_4 G_6 .$$

2) *Calculation of c_0 and c_n:*

$$c_0 = \frac{3\left((u_1-u_0) - \alpha h_0\right) + \left(W_1 H_0 + \frac{W_0}{h_0} - h_0^2\right)c_1 - \frac{W_1}{h_1}c_2}{2h_0^2 + \frac{1}{h_0}(W_0 + W_1)},$$

$$c_n = \frac{3(\beta h_{n-1} - (u_n - u_{n-1})) + \left(W_{n-1}H_{n-2} + \frac{W_n}{h_{n-1}} - h_{n-1}^2\right)c_{n-1} - \frac{W_{n-1}}{h_{n-2}}c_{n-2}}{2h_{n-1}^2 + \frac{1}{h_{n-1}}(W_{n-1} + W_n)}.$$

3) *Calculation of a_i for $i = 0, \ldots, n$:*

$$a_0 = u_0 + \frac{2}{w_0 h_0}(c_0 - c_1),$$

$$a_i = u_i - \frac{2}{w_i}\left[\frac{1}{h_{i-1}}c_{i-1} - \left(\frac{1}{h_{i-1}} + \frac{1}{h_i}\right)c_i + \frac{1}{h_i}c_{i+1}\right] \text{ for } i = 1, \ldots, n-1,$$

$$a_n = u_n - \frac{2}{w_n h_{n-1}}(c_{n-1} - c_n).$$

4) *Calculation of b_i:*

$$b_i = \frac{1}{h_i}(a_{i+1} - a_i) - \frac{h_i}{3}(c_{i+1} + 2c_i) \text{ for } i = 0, \ldots, n-1.$$

5) *Calculation of d_i:*

$$d_i = \frac{1}{3h_i}(c_{i+1} - c_i) \text{ for } i = 0, \ldots, n-1.$$

(ii) Fitting spline with prescribed second end point derivatives.

To calculate the coefficients a_i, b_i, c_i, d_i of the cubic polynomials S_i we have to evaluate the following equations in the indicated order:

1) *Linear system for* $c_1, c_2, ..., c_{n-1}$:

With $h_i = x_{i+1} - x_i$ for $i = 0, \dots, n-1$, $W_i = \frac{6}{w_i}$ for $i = 0, \dots, n$, $H_i = \frac{1}{h_i} + \frac{1}{h_{i+1}}$ for $i = 0, \dots, n-2$, the five-diagonal linear system for the c_i is:

$$\left(2(h_0 + h_1) + \frac{W_0}{h_0^2} + W_1 H_0^2 + \frac{W_2}{h_1^2} \right) c_1 + \left(h_1 - \frac{W_1}{h_1} H_0 - \frac{W_2}{h_1} H_1 \right) c_2 +$$

$$+ \left(\frac{W_2}{h_1 h_2} \right) c_3 = 3 \left(\frac{u_2 - u_1}{h_1} - \frac{u_1 - u_0}{h_0} \right) - \frac{\alpha}{2} \left(h_0 - \frac{W_0}{h_0^2} - \frac{W_1}{h_0} H_0 \right) ,$$

$$\left(h_1 - \frac{W_1}{h_1} H_0 - \frac{W_2}{h_1} H_1 \right) c_1 + \left(2(h_1 + h_2) + \frac{W_1}{h_1^2} + W_2 H_1^2 + \frac{W_3}{h_2^2} \right) c_2 +$$

$$+ \left(h_2 - \frac{W_2}{h_2} H_1 - \frac{W_3}{h_2} H_2 \right) c_3 + \left(\frac{W_3}{h_2 h_3} \right) c_4 =$$

$$= 3 \left(\frac{u_3 - u_2}{h_2} - \frac{u_2 - u_1}{h_1} \right) - \frac{\alpha}{2} \left(\frac{W_1}{h_0 h_1} \right) ,$$

$$\left(\frac{W_{i-1}}{h_{i-2} h_{i-1}} \right) c_{i-2} + \left(h_{i-1} - \frac{W_{i-1}}{h_{i-1}} H_{i-2} - \frac{W_i}{h_{i-1}} H_{i-1} \right) c_{i-1} +$$

$$+ \left(2(h_{i-1} + h_i) + \frac{W_{i-1}}{h_{i-1}^2} + W_i H_{i-1}^2 + \frac{W_{i+1}}{h_i^2} \right) c_i +$$

$$+ \left(h_i - \frac{W_i}{h_i} H_{i-1} - \frac{W_{i+1}}{h_i} H_i \right) c_{i+1} + \left(\frac{W_{i+1}}{h_i h_{i+1}} \right) c_{i+2} =$$

$$= 3 \left(\frac{u_{i+1} - u_i}{h_i} - \frac{u_i - u_{i-1}}{h_{i-1}} \right) , \quad i = 3, \dots, n-3 ,$$

$$\left(\frac{W_{n-3}}{h_{n-4} h_{n-3}} \right) c_{n-4} + \left(h_{n-3} - \frac{W_{n-3}}{h_{n-3}} H_{n-4} - \frac{W_{n-2}}{h_{n-3}} H_{n-3} \right) c_{n-3} +$$

$$+ \left(2(h_{n-3} + h_{n-2}) + \frac{W_{n-3}}{h_{n-3}^2} + W_{n-2} H_{n-3}^2 + \frac{W_{n-1}}{h_{n-2}^2} \right) c_{n-2} +$$

$$+ \left(h_{n-2} - \frac{W_{n-2}}{h_{n-2}} H_{n-3} - \frac{W_{n-1}}{h_{n-2}} H_{n-2} \right) c_{n-1} =$$

$$= 3 \left(\frac{u_{n-1} - u_{n-2}}{h_{n-2}} - \frac{u_{n-2} - u_{n-3}}{h_{n-3}} \right) - \frac{\beta}{2} \left(\frac{W_{n-1}}{h_{n-2} h_{n-1}} \right), \text{ and}$$

$$\left(\frac{W_{n-2}}{h_{n-3} h_{n-2}} \right) c_{n-3} + \left(h_{n-2} - \frac{W_{n-2}}{h_{n-2}} H_{n-3} - \frac{W_{n-1}}{h_{n-2}} H_{n-2} \right) c_{n-2} +$$

$$+ \left(2(h_{n-2} + h_{n-1}) + \frac{W_{n-2}}{h_{n-2}^2} + W_{n-1} H_{n-2}^2 + \frac{W_n}{h_{n-1}^2} \right) c_{n-1} =$$

$$= 3 \left(\frac{u_n - u_{n-1}}{h_{n-1}} - \frac{u_{n-1} - u_{n-2}}{h_{n-2}} \right) - \frac{\beta}{2} \left(h_{n-1} - \frac{W_{n-1}}{h_{n-1}} H_{n-2} - \frac{W_n}{h_{n-1}^2} \right).$$

2) $c_0 = \frac{\alpha}{2}, \qquad c_n = \frac{\beta}{2}.$

3) *Calculation of the* $a_i, i = 0, \ldots, n$:

$$a_0 = u_0 + \frac{2}{w_0 h_0} (c_0 - c_1),$$

$$a_i = u_i - \frac{2}{w_i} \left[\frac{1}{h_{i-1}} c_{i-1} - \left(\frac{1}{h_{i-1}} + \frac{1}{h_i} \right) c_i + \frac{1}{h_i} c_{i+1} \right], \quad i = 1, \ldots, n-1,$$

$$a_n = u_n - \frac{2}{w_n h_{n-1}} (c_{n-1} - c_n).$$

4) *Calculation of the* b_i:

$$b_i = \frac{1}{h_i} (a_{i+1} - a_i) - \frac{h_i}{3} (c_{i+1} + 2c_i), \quad i = 0, \ldots, n-1.$$

5) *Calculation of the* d_i:

$$d_i = \frac{1}{3 h_i} (c_{i+1} - c_i), \quad i = 0, \ldots, n-1.$$

(iii) **Fitting spline with prescribed third end point derivatives.**

To calculate the coefficients a_i, b_i, c_i, d_i of the cubic polynomials S_i, one has to evaluate the following equations in the indicated order:

1) *Linear system for* $c_1, c_2, \ldots, c_{n-1}$:

With $h_i = x_{i+1} - x_i$ for $i = 0, \ldots, n-1$, $W_i = \frac{6}{w_i}$ for $i = 0, \ldots, n$, $H_i = \frac{1}{h_i} + \frac{1}{h_{i+1}}$ for $i = 0, \ldots, n-2$, the five-diagonal linear system for the c_i is:

$$\left(3h_0 + 2h_1 + \frac{W_1}{h_1}H_0 + \frac{W_2}{h_1^2}\right)c_1 + \left(h_1 - \frac{W_1}{h_1}H_0 - \frac{W_2}{h_1}H_1\right)c_2 +$$

$$+ \left(\frac{W_2}{h_1 h_2}\right)c_3 = 3\left(\frac{u_2 - u_1}{h_1} - \frac{u_1 - u_0}{h_0}\right) + \frac{\alpha}{2}\left(h_0^2 - \frac{W_0}{h_0} - W_1 H_0\right),$$

$$\left(h_1 - \frac{W_1}{h_1^2} - \frac{W_2}{h_1}H_1\right)c_1 + \left(2(h_1 + h_2) + \frac{W_1}{h_1^2} + W_2 H_1^2 + \frac{W_3}{h_2^2}\right)c_2 +$$

$$+ \left(h_2 - \frac{W_2}{h_2}H_1 - \frac{W_3}{h_2}H_2\right)c_3 + \left(\frac{W_3}{h_2 h_3}\right)c_4 =$$

$$= 3\left(\frac{u_3 - u_2}{h_2} - \frac{u_2 - u_1}{h_1}\right) + \frac{\alpha}{2}\left(\frac{W_1}{h_1}\right),$$

$$\left(\frac{W_{i-1}}{h_{i-2}h_{i-1}}\right)c_{i-2} + \left(h_{i-1} - \frac{W_{i-1}}{h_{i-1}}H_{i-2} - \frac{W_i}{h_{i-1}}H_{i-1}\right)c_{i-1} +$$

$$+ \left(2(h_{i-1} + h_i) + \frac{W_{i-1}}{h_{i-1}^2} + W_i H_{i-1}^2 + \frac{W_{i+1}}{h_i^2}\right)c_i +$$

$$+ \left(h_i - \frac{W_i}{h_i}H_{i-1} - \frac{W_{i+1}}{h_i}H_i\right)c_{i+1} + \left(\frac{W_{i+1}}{h_i h_{i+1}}\right)c_{i+2} =$$

$$= 3\left(\frac{u_{i+1} - u_i}{h_i} - \frac{u_i - u_{i-1}}{h_{i-1}}\right), \qquad i = 3, \ldots, n - 3,$$

$$\left(\frac{W_{n-3}}{h_{n-4}h_{n-3}}\right)c_{n-4} + \left(h_{n-3} - \frac{W_{n-3}}{h_{n-3}}H_{n-4} - \frac{W_{n-2}}{h_{n-3}}H_{n-3}\right)c_{n-3} +$$

$$+ \left(2(h_{n-3} + h_{n-2}) + \frac{W_{n-3}}{h_{n-3}^2} + W_{n-2}H_{n-3}^2 + \frac{W_{n-1}}{h_{n-2}^2}\right)c_{n-2} +$$

$$+ \left(h_{n-2} - \frac{W_{n-2}}{h_{n-2}}H_{n-3} - \frac{W_{n-1}}{h_{n-2}^2}\right)c_{n-1} =$$

$$= 3\left(\frac{u_{n-1} - u_{n-2}}{h_{n-2}} - \frac{u_{n-2} - u_{n-3}}{h_{n-3}}\right) - \frac{\beta}{2}\left(\frac{W_{n-1}}{h_{n-2}}\right), \text{ and}$$

$$\left(\frac{W_{n-2}}{h_{n-3}h_{n-2}}\right)c_{n-3} + \left(h_{n-2} - \frac{W_{n-2}}{h_{n-2}}H_{n-3} - \frac{W_{n-1}}{h_{n-2}}H_{n-2}\right)c_{n-2} +$$

$$+ \left(2h_{n-2} + 3h_{n-1} + \frac{W_{n-2}}{h_{n-2}^2} + \frac{W_{n-1}}{h_{n-2}}H_{n-2}\right)c_{n-1} +$$

$$= 3\left(\frac{u_n - u_{n-1}}{h_{n-1}} - \frac{u_{n-1} - u_{n-2}}{h_{n-2}}\right) - \frac{\beta}{2}\left(h_{n-1}^2 - W_{n-1}H_{n-2} - \frac{W_n}{h_{n-1}}\right).$$

2) $c_0 = c_1 - \frac{\alpha}{2} h_0, \quad c_n = c_{n-1} + \frac{\beta}{2} h_{n-1}.$

3) *Calculation of the* a_i, $i = 0, \ldots, n$:

$$a_0 = u_0 + \frac{2}{w_0 h_0}(c_0 - c_1),$$

$$a_i = u_i - \frac{2}{w_i}\left[\frac{1}{h_{i-1}} c_{i-1} - \left(\frac{1}{h_{i-1}} + \frac{1}{h_i}\right) c_i + \frac{1}{h_i} c_{i+1}\right], \quad i = 1, \ldots, n-1,$$

$$a_n = u_n - \frac{2}{w_n h_{n-1}}(c_{n-1} - c_n).$$

4) *Calculation of the* b_i:

$$b_i = \frac{1}{h_i}(a_{i+1} - a_i) - \frac{h_i}{3}(c_{i+1} + 2c_i), \quad i = 0, \ldots, n-1.$$

5) *Calculation of the* d_i:

$$d_i = \frac{1}{3h_i}(c_{i+1} - c_i), \quad i = 0, \ldots, n-1.$$

(iv) Periodic fitting spline.

To calculate the coefficients a_i, b_i, c_i, d_i of the cubic polynomials S_i, one has to evaluate the following equations in the indicated order:

1) *Linear system for* c_1, c_2, \ldots, c_n:

With $h_i = x_{i+1} - x_i$ for $i = 0, \ldots, n-1$, $W_i = \frac{6}{w_i}$ for $i = 0, \ldots, n$, $H_i = \frac{1}{h_i} + \frac{1}{h_{i+1}}$ for $i = 0, \ldots, n-2$, one obtains an almost cyclically five-diagonal system of the form:

$$\left(\frac{W_{i-1}}{h_{i-2} h_{i-1}}\right) c_{i-2} + \left(h_{i-1} - \frac{W_{i-1}}{h_{i-1}} H_{i-2} - \frac{W_i}{h_{i-1}} H_{i-1}\right) c_{i-1} +$$

$$+ \left(2(h_{i-1} + h_i) + \frac{W_{i-1}}{h_{i-1}^2} + W_i H_{i-1}^2 + \frac{W_{i+1}}{h_i^2}\right) c_i +$$

$$+ \left(h_i - \frac{W_i}{h_i} H_{i-1} - \frac{W_{i+1}}{h_i} H_i\right) c_{i+1} + \left(\frac{W_{i+1}}{h_i h_{i+1}}\right) c_{i+2} =$$

$$= 3\left(\frac{u_{i+1} - u_i}{h_i} - \frac{u_i - u_{i-1}}{h_{i-1}}\right), \qquad i = 1, \ldots, n.$$

Because of periodicity, all negative indices $(-k)$ in the above equations should be replaced by $(n-k)$ and all indices $(n+k)$ larger than n by (k) for $k \in \mathbb{N}$.

2) $c_0 = c_n$.

3) *Calculation of the a_i, $i = 0, \ldots, n$:*

$$
a_0 = u_0 - \frac{2}{w_0} \left[\frac{1}{h_0}(c_1 - c_0) + \frac{1}{h_{n-1}}(c_n - c_{n-1}) \right], \quad a_n = a_0,
$$

$$
a_i = u_i - \frac{2}{w_i} \left[\frac{1}{h_{i-1}} c_{i-1} - \left(\frac{1}{h_{i-1}} + \frac{1}{h_i} \right) c_i + \frac{1}{h_i} c_{i+1} \right], \quad i = 1, \ldots, n-1,
$$

4) *Calculation of the b_i:*

$$
b_i = \frac{1}{h_i}(a_{i+1} - a_i) - \frac{h_i}{3}(c_{i+1} + 2c_i), \quad i = 0, \ldots, n-1.
$$

5) *Calculation of the d_i:*

$$
d_i = \frac{1}{3h_i}(c_{i+1} - c_i), \quad i = 0, \ldots, n-1.
$$

11.4 Parametric Cubic Fitting Splines

A planar curve C with a parametrization $(x(t), y(t))$ which is given approximately by measurement points (x_i, u_i), can be approximated by a vectorial fitting spline with t as the curve parameter for $i = 0, \ldots, n-1$:

$$
\begin{pmatrix} S_x(t) \\ S_y(t) \end{pmatrix} \equiv \begin{pmatrix} S_{ix}(t) \\ S_{iy}(t) \end{pmatrix} \equiv \begin{pmatrix} a_{ix} + b_{ix}(t - t_i) + c_{ix}(t - t_i)^2 + d_{ix}(t - t_i)^3 \\ a_{iy} + b_{iy}(t - t_i) + c_{iy}(t - t_i)^2 + d_{iy}(t - t_i)^3 \end{pmatrix}.
$$

Here S_x is the fitting spline for the points (t_i, x_i), $i = 0, \ldots, n$, and S_y is the fitting spline for the points (t_i, u_i), $i = 0, \ldots, n$. According to the type of spline, one must employ the corresponding algorithms of section 11.3 to calculate S_x and S_y. To calculate S_x one replaces x_i by t_i and u_i by x_i within the algorithm, while when calculating S_y one only replaces x_i by t_i.

To use a parameter t that approximates the arc length has proven most useful. One such approximate parameter is the chordal parametrization:

$$
t_0 = 0, \quad t_{i+1} = t_i + \sqrt{(x_{i+1} - x_i)^2 + (u_{i+1} - u_i)^2} \text{ for } i = 0, \ldots, n-1,
$$

another more suited parametrization is given by the method described in algorithm 10.7.

By using a parametrization $(x(t), y(t), z(t))$, one can proceed analogously for space curves C^* which are given approximately by measurement points $(x_i, u_i, v_i) \in \mathbb{R}^3$. C^* can be approximated by a vectorial fitting spline

$$\begin{pmatrix} S_x(t) \\ S_y(t) \\ S_z(t) \end{pmatrix} \equiv \begin{pmatrix} S_{ix}(t) \\ S_{iy}(t) \\ S_{iz}(t) \end{pmatrix} \approx \begin{pmatrix} x(t) \\ y(t) \\ z(t) \end{pmatrix} \text{ for } i = 0, \dots, n-1,$$

which contains the points (x_i, y_i, z_i) that are close to the measured points (x_i, u_i, v_i). Here S_x is the spline formed for the points $(t_i, x_i), S_y$ the spline for the points (t_i, u_i) and S_z the one for (t_i, v_i).

For every curve, the specific algorithm of section 11.3 has to be solved three times in order to calculate S_x, S_y and S_z. The LR factorization of the system matrix in step 1), however, has to be computed only once and can be used to solve the three linear systems for each set of c_i. The parameter values could be taken as:

$$t_0 = 0, \quad t_{i+1} = t_i + \sqrt{(x_{i+1} - x_i)^2 + (u_{i+1} - u_i)^2 + (v_{i+1} - v_i)^2}.$$

Better results, however, will arise from using algorithm 10.7 which finds a parametrization of the curve given by the data points that more closely approximates the arc length parametrization.

11.5 Decision Strategy

All hints on how to proceed with fitting splines can be found in section 10.3.

Bibliography for Chapter 11

[BÖHM74], 6; [ENGE87], p.235 ff.; [REIN71]; [SPÄT73]; [SPÄT74/1]; [SPÄT74/2], p.74-85.

Chapter 12

Two-Dimensional Splines, Surface Splines, Bézier Splines, B-Splines

12.1 Interpolating Two-Dimensional Cubic Splines for Constructing Smooth Surfaces

Assume that we are given a rectangular grid G in the x, y-plane

$$G = \left\{ \ (x_i, y_j) \ \middle| \ \begin{array}{l} a = x_0 < x_1 < \ldots < x_n = b \\ c = y_0 < y_1 < \ldots < y_m = d \end{array} \right\}$$

with heights u_{ij} defined at each point $(x_i, y_j) \in G$

$$u_{ij} := u(x_i, y_j), \quad i = 0, \ldots, n, \quad j = 0, \ldots, m.$$

Figure 12.1: Rectangular grid

We desire to find an interpolating smooth surface for the ordinates u_{ij}, which is described by a two-dimensional spline function $S = S(x, y)$ for (x, y) in the rectangle $R = \{(x, y) \mid a \leq x \leq b,\ c \leq y \leq d\}$.

The bicubic spline function $S = S(x, y)$ is defined for $(x, y) \in R$ by the following properties:

(1) S fulfills the interpolation conditions

$$S(x_i, y_j) = u_{ij}, \quad i = 0, \ldots, n, \quad j = 0, \ldots, m.$$

(2) S is continuously differentiable on R and $\partial^2 S / \partial x \partial y$ is continuous on R.

(3) In each subrectangle R_{ij}, where

$$R_{ij} := \left\{ (x, y) \ \middle| \ \begin{array}{ll} x_i \leq x \leq x_{i+1}, & i = 0, \ldots, n-1 \\ y_j \leq y \leq y_{j+1}, & j = 0, \ldots, m-1 \end{array} \right\},$$

S is defined as a bicubic polynomial $f_{ij} = f_{ij}(x, y)$.

(4) S fulfills certain boundary conditions, to be specified later.

Due to property (3), the bicubic spline function S has the form

(12.1)
$$S(x, y) \equiv f_{ij}(x, y) = \sum_{k=0}^{3} \sum_{s=0}^{3} a_{ijks}(x - x_i)^k (y - y_j)^s$$
$$\text{for } (x, y) \in R_{ij}, \quad i = 0, \ldots, n-1, \quad j = 0, \ldots, m-1.$$

We shall write out equation (12.1) explicitly:

$$f_{ij}(x, y) = \sum_{k=0}^{3} \sum_{s=0}^{3} a_{ijks}(x - x_i)^k (y - y_j)^s =$$

a_{ij00} $+a_{ij10}(x - x_i)$ $+a_{ij01}(y - y_j)+$

$a_{ij20}(x - x_i)^2$ $+a_{ij11}(x - x_i)(y - y_j)$ $+a_{ij02}(y - y_j)^2+$

$a_{ij30}(x - x_i)^3$ $+a_{ij21}(x - x_i)^2(y - y_j)$ $+a_{ij12}(x - x_i)(y - y_j)^2+$

$a_{ij03}(y - y_j)^3$ $+a_{ij31}(x - x_i)^3(y - y_j)$ $+a_{ij22}(x - x_i)^2(y - y_j)^2+$

$a_{ij13}(x - x_i)(y - y_j)^3$ $+a_{ij32}(x - x_i)^3(y - y_j)^2$ $+a_{ij23}(x - x_i)^2(y - y_j)^3+$

$a_{ij33}(x - x_i)^3(y - y_j)^3$

The 16 $m \cdot n$ coefficients a_{ijks} of (12.1) must be determined so that S fulfills the properties (1) and (2). If the a_{ijks} are to be uniquely determined, certain boundary conditions must be specified for R as in the one-dimensional case. One possibility is to specify the following partial derivatives for S:

$$
(12.2) \begin{cases}
\dfrac{\partial}{\partial x} S(x_i, y_j) & =: \quad p_{ij} = a_{ij10}, \quad i = 0, n, \quad j = 0, \dots, m, \\[2ex]
\dfrac{\partial}{\partial y} S(x_i, y_j) & =: \quad q_{ij} = a_{ij01}, \quad i = 0, \dots, n, \quad j = 0, m, \\[2ex]
\dfrac{\partial^2}{\partial x \partial y} S(x_i, y_j) & =: \quad s_{ij} = a_{ij11}, \quad i = 0, n, \quad j = 0, m.
\end{cases}
$$

These coefficients can be calculated approximately by means of one-dimensional cubic splines or other interpolation methods, see chapter 10. Depending on the given boundary conditions, one of the following algorithms should be chosen. We note that the problem can always be solved in theory, for [BOOR62] has proved that there exists a bicubic spline function (12.1) for given u_{ij} and given boundary derivatives (12.2) which interpolates the u_{ij}.

Calculation of the bicubic spline function S.

We shall give three algorithms for calculating S:

ALGORITHM 12.1 (*Bicubic spline function with prescribed boundary*).

Given: (i) $u_{ij} = u(x_i, y_j) = a_{ij00}$ for $i = 0, \dots, n, \ j = 0, \dots, m,$
(ii) boundary values for the partial derivatives in (12.2).

Task: Find the bicubic spline function S of the form (12.1).

1^{st} step: Calculate $a_{ij10} = p_{ij}$ for $i = 1, \dots, n-1, \ j = 0, \dots, m$ from

$$
(12.3) \begin{cases}
a_{i-1,j10}\, \dfrac{1}{h_{i-1}} + 2a_{ij10}\left(\dfrac{1}{h_{i-1}} + \dfrac{1}{h_i}\right) + a_{i+1,j10}\, \dfrac{1}{h_i} = \\[2ex]
\quad = \dfrac{3}{h_{i-1}^2}(a_{ij00} - a_{i-1,j00}) + \dfrac{3}{h_i^2}(a_{i+1,j00} - a_{ij00}) \\[2ex]
\text{for} \quad i = 1, \dots, n-1, \quad j = 0, \dots, m, \\[2ex]
\text{with} \quad h_i = x_{i+1} - x_i \quad \text{for} \quad i = 0, \dots, n-1.
\end{cases}
$$

These equations give $(m+1)$ linear systems with $(n-1)$ equations each for $(n+1)$ unknowns. By prescribing $2(m+1)$ quantities $a_{ij10}, i = 0, n$ for

$j = 0, \ldots, m$, each of these systems can be solved uniquely.

2^{nd} step: Determine $a_{ij01} = q_{ij}$ for $i = 0, \ldots, n$, $\quad j = 1, \ldots, m-1$ with

(12.4)
$$
\begin{cases}
a_{i,j-1,01}\dfrac{1}{k_{j-1}} + 2a_{ij01}\left(\dfrac{1}{k_{j-1}} + \dfrac{1}{k_j}\right) + a_{i,j+1,01}\dfrac{1}{k_j} = \\[2mm]
\qquad = \dfrac{3}{k_{j-1}^2}(a_{ij00} - a_{i,j-1,00}) + \dfrac{3}{k_j^2}(a_{i,j+1,00} - a_{ij00}) \\[2mm]
\text{for} \quad i = 0, \ldots, n, \quad j = 1, \ldots, m-1, \\[2mm]
\text{with} \quad k_j = y_{j+1} - y_j \text{ for } i = 0, \ldots, m-1.
\end{cases}
$$

With the given $2(n+1)$ boundary values a_{ij01} for $i = 0, \ldots, n$, $\quad j = 0, m$, each of these $n+1$ linear systems is uniquely solvable.

3^{rd} step: Calculate $a_{ij11} = s_{ij}$ for $i = 1, \ldots, n-1$, $\quad j = 0, m$ from the two linear systems

(12.5)
$$
\begin{cases}
\dfrac{1}{h_{i-1}}a_{i-1,j11} + 2a_{ij11}\left(\dfrac{1}{h_{i-1}} + \dfrac{1}{h_i}\right) + \dfrac{1}{h_i}a_{i+1,j11} = \\[2mm]
\qquad = \dfrac{3}{h_{i-1}^2}(a_{ij01} - a_{i-1,j01}) + \dfrac{3}{h_i^2}(a_{i+1,j01} - a_{ij01}) \\[2mm]
\text{with} \quad h_i = x_{i+1} - x_i \text{ for } j = 0 \text{ and } m.
\end{cases}
$$

The four values at the corners $a_{0011}, a_{n011}, a_{0m11}$ and a_{nm11} are given.

4^{th} step: Calculate the derivatives $s_{ij} = a_{ij11}$, $\quad i = 0, \ldots, n$, $j = 1, \ldots, m-1$ from

(12.6)
$$
\begin{cases}
\dfrac{1}{k_{j-1}}a_{i,j-1,11} + 2a_{ij11}\left(\dfrac{1}{k_{j-1}} + \dfrac{1}{k_j}\right) + a_{i,j+1,11}\dfrac{1}{k_i} = \\[2mm]
\qquad = \dfrac{3}{k_{j-1}^2}(a_{ij10} - a_{i,j-1,10}) + \dfrac{3}{k_j^2}(a_{i,j+1,10} - a_{ij10}) \\[2mm]
\text{with} \quad k_j = y_{j+1} - y_j \quad \text{for} \quad j = 0, \ldots, m-1.
\end{cases}
$$

The required boundary values a_{ij11} for $i = 1, \ldots, n-1$, $\quad j = 0, m$ have been determined in the third step; the a_{ij11} for $i = 0, n$ and $j = 0, m$ have been prescribed.

5^{th} step: Determine the matrices $(\boldsymbol{G}(x_i))^{-1}$. Since

(12.7)
$$\boldsymbol{G}(x_i) = \begin{pmatrix} 1 & 0 & 0 & 0 \\ 0 & 1 & 0 & 0 \\ 1 & h_i & h_i^2 & h_i^3 \\ 0 & 1 & 2h_i & 3h_i^2 \end{pmatrix}$$

with $\det \boldsymbol{G}(x_i) = h_i^4 \neq 0$ and $h_i = x_{i+1} - x_i$
for $i = 0, \ldots, n-1$, the inverse $(\boldsymbol{G}(x_i))^{-1}$ exists. In fact

$$(\boldsymbol{G}(x_i))^{-1} = \begin{pmatrix} 1 & 0 & 0 & 0 \\ 0 & 1 & 0 & 0 \\ -\frac{3}{h_i^2} & -\frac{2}{h_i} & \frac{3}{h_i^2} & -\frac{1}{h_i} \\ \frac{2}{h_i^3} & \frac{1}{h_i^2} & -\frac{2}{h_i^3} & \frac{1}{h_i^2} \end{pmatrix}.$$

6^{th} step: Determine the matrices $(\boldsymbol{G}(y_j)^T)^{-1}$. Since

(12.8)
$$\boldsymbol{G}(y_j) = \begin{pmatrix} 1 & 0 & 0 & 0 \\ 0 & 1 & 0 & 0 \\ 1 & k_j & k_j^2 & k_j^3 \\ 0 & 1 & 2k_j & 3k_j^2 \end{pmatrix}$$

with $\det \boldsymbol{G}(y_j) = k_j^4 \neq 0$ for $k_j = y_{j+1} - y_j$ and

$j = 0, \ldots, m-1$, the inverse $(\boldsymbol{G}(y_j)^T)^{-1}$ exists. In fact

$$(\boldsymbol{G}(y_j)^T)^{-1} = \begin{pmatrix} 1 & 0 & -\frac{3}{k_j^2} & \frac{2}{k_j^3} \\ 0 & 1 & -\frac{2}{k_j} & \frac{1}{k_j^2} \\ 0 & 0 & \frac{3}{k_j^2} & -\frac{2}{k_j^3} \\ 0 & 0 & -\frac{1}{k_j} & \frac{1}{k_j^2} \end{pmatrix}.$$

7^{th} step: Determine the matrices \boldsymbol{M}_{ij} as

(12.9)
$$\boldsymbol{M}_{ij} = \begin{pmatrix} a_{ij00} & a_{ij01} & a_{i,j+1,00} & a_{i,j+1,01} \\ a_{ij10} & a_{ij11} & a_{i,j+1,10} & a_{i,j+1,11} \\ a_{i+1,j00} & a_{i+1,j01} & a_{i+1,j+1,00} & a_{i+1,j+1,01} \\ a_{i+1,j10} & a_{i+1,j11} & a_{i+1,j+1,10} & a_{i+1,j+1,11} \end{pmatrix}$$
for $i = 0, \ldots, n-1$, $j = 0, \ldots, m-1$.

8^{th} step: Calculate the coefficient matrix \boldsymbol{A}_{ij} for f_{ij} from the equation

(12.10)
$$\left\{ \begin{array}{l} \boldsymbol{A}_{ij} = \boldsymbol{G}(x_i)^{-1} \boldsymbol{M}_{ij} (\boldsymbol{G}(y_j)^T)^{-1} = \{a_{ijks}\} \\[2mm] \text{for } k = 0, \ldots, 3 \ \text{ and } s = 0, \ldots, 3. \end{array} \right.$$

9^{th} step: Form the bicubic spline function $S(x,y) \equiv f_{ij}(x,y)$ for each rectangle R_{ij} according to (12.1).

ALGORITHM 12.2 (*Bicubic spline function without prescribed boundary*).

Given: Functional values $u_{ij} = u(x_i, y_j)$ for $i = 0, \ldots, n$, $j = 0, \ldots, m$ at the grid points $(x_i, y_j) \in G$.

Task: Find the interpolating spline function S of the form (12.1).

The values of the partial derivatives $p_{ij} = a_{ij10}$, $q_{ij} = a_{ij01}$, $s_{ij} = a_{ij11}$ at the boundary points of S in (12.2) can be obtained here by one-dimensional natural cubic splines which can be determined each time using three points and their slopes. We can form splines that pass through the points (x_i, u_{ij}) $i = 0, 1, 2$ and $i = n - 2, n - 1, n$ for $j = 0, \ldots, m$. Their derivatives supply the p_{ij} at the boundary points. Similarly we can form splines for the points (y_j, u_{ij}) for $j = 0, 1, 2$ and $j = m - 2, m - 1, m$ for $i = 0, \ldots, n$. Their derivatives determine the values of q_{ij} at the boundary points. In order to be able to recognize this inside the algorithm, we have not simplified its formulation, but have rather stated the formulas (12.12) outside of (12.11), those of (12.14) outside of (12.13) and those of (12.16) outside of (12.15).

To calculate $s_{ij} = a_{ij11}$ for $i = 0, n$ and $j = 0, m$, one-dimensional natural splines are formed for the points (x_i, q_{ij}) for $i = 0, 1, 2$ and $i = n-2, n-1, n$ and $j = 0, m$, and their derivatives are computed. The a_{ij00} are given via the u_{ij}. Thus the bicubic spline function S can be calculated according to (12.1) as follows:

1^{st} step: Calculate the boundary values a_{ij10} for $i = 0, n$ and $j = 0, \ldots, m$ as

(12.11)
$$
\begin{cases}
a_{0j10} &= S_x(x_0, y_j) = b_{oj} = \\
&= \dfrac{1}{h_0}(a_{1j00} - a_{0j00}) - \dfrac{h_0}{3}c_{1j} \\
&\quad \text{for} \quad j = 0, \ldots, m, \quad \text{and} \\
a_{nj10} &= S_x(x_n, y_n) = b_{n-1,j} + 2c_{n-1,j}h_{n-1} + \\
&\quad + 3d_{n-1,j}h_{n-1}^2 \\
&\quad \text{for} \quad j = 0, \ldots, m \text{ and } h_i = x_{i+1} - x_i.
\end{cases}
$$

Here the values for the coefficients b_{ij}, c_{ij} and d_{ij} are to be found in (12.12):

(12.12)
$$
\begin{cases}
1. \ u_{ij} &= a_{ij00} \text{ for } i = 0, 1, 2, n-2, n-1, n, \\
&\quad \text{and } j = 0, \ldots, m, \\[4pt]
2. \ c_{0j} &= c_{2j} = c_{n-2,j} = c_{nj} = 0 \text{ for } j = 0, \ldots, m, \\
&\quad \text{(natural splines)} \\[4pt]
3. \ c_{ij} &= \dfrac{3}{2(h_i + h_{i-1})}\left[\dfrac{1}{h_i}(a_{i+1,j00} - a_{ij00}) + \right. \\
&\quad \left. - \dfrac{1}{h_{i-1}}(a_{ij00} - a_{i-1,j00})\right] \\
&\quad \text{for } i = 1, n-1 \text{ and } j = 0, \ldots, m, \\[4pt]
4. \ b_{n-1,j} &= \dfrac{1}{k_{n-1}}(a_{nj00} - a_{n-1,j00}) - \dfrac{2k_{n-1}}{3}c_{n-1,j} \\
&\quad \text{for } j = 0, \ldots, m, \\[4pt]
5. \ d_{n-1,j} &= \dfrac{1}{3k_{n-1}}c_{n-1,j} \text{ for } j = 0, \ldots, m.
\end{cases}
$$

2^{nd} step: Calculate the boundary values a_{ij01} for $i = 0, \ldots, n$ and $j = 0, m$ as

(12.13)
$$
\begin{cases}
a_{i001} &= S_y(x_i, y_0) = \beta_{i0} = \dfrac{1}{k_0}(a_{i100} - a_{i000}) - \dfrac{k_0}{3}\gamma_{i1}, \\
&\quad \text{for } i = 0, \ldots, n, \quad \text{and} \\
a_{im01} &= S_y(x_i, y_m) = \beta_{im-1} + 2\gamma_{im-1}k_{m-1} + 3\delta_{im-1}k_{m-1}^2, \\
&\quad \text{for } i = 0, \ldots, n \text{ with } \quad k_j = y_{j+1} - y_j.
\end{cases}
$$

Here the needed coefficients can be found from (12.14):

$$(12.14) \begin{cases}
1.\ u_{ij} &= \alpha_{ij} = a_{ij00} \ \text{ for } i = 0, \ldots, n, \\
&\quad j = 0, 1, 2, m-2, m-1, m, \\[2mm]
2.\ \gamma_{i0} &= \gamma_{i2} = \gamma_{i,m-2} = \gamma_{im} = 0, \quad i = 0, \ldots, n, \\[2mm]
3.\ \gamma_{ij} &= \dfrac{3}{2(k_{j-1}+k_j)} \left[\dfrac{1}{k_j}(a_{ij+1,00} - a_{ij00}) - \right. \\
&\quad \left. - \dfrac{1}{k_{j-1}}(a_{ij00} - a_{ij-1,00}) \right] \\
&\quad \text{for } i = 0, \ldots, n, \ j = 1, m-1, \\[2mm]
4.\ \beta_{i,m-1} &= \dfrac{1}{h_{m-1}}(a_{im00} - a_{i,m-1,00}) - \dfrac{2h_{m-1}}{3}\gamma_{i,m-1} \\
&\quad \text{for } i = 0, \ldots, n, \\[2mm]
5.\ \delta_{im-1} &= -\dfrac{1}{3h_{m-1}}\gamma_{im-1}, \quad i = 0, \ldots, n.
\end{cases}$$

3^{rd} step: Calculate the boundary values a_{ij11} for $i = 0, n$ and $j = 0, m$ from

$$(12.15) \begin{cases}
a_{0j11} &= S_{yx}(x_0, y_j) = s_{0j} = \tilde{b}_{0j} = \dfrac{1}{k_0}(a_{1j01} - a_{0j01}) - \dfrac{k_0}{3}\tilde{c}_{1j}, \\
&\quad \text{for } j = 0, m, \quad \text{and} \\[2mm]
a_{nj11} &= S_{yx}(x_n, y_j) = s_{nj} = \\
&= \tilde{b}_{n-1,j} + 2\tilde{c}_{n-1,j}h_{n-1} + 3\tilde{d}_{n-1,j}h_{n-1}^2, \\
&\quad \text{for } j = 0, m \quad \text{with} \quad h_i = x_{i+1} - x_i,
\end{cases}$$

where the coefficients on the right are given by (12.16):

$$(12.16) \begin{cases}
\text{1. } q_{ij} = a_{ij01} \quad \text{for } i = 0,1,2,n-2,n-1,n \text{ and} \\
\qquad\qquad j = 0,m, \\[2mm]
\text{2. } \tilde{c}_{0j} = \tilde{c}_{2j} = \tilde{c}_{n-2,j} = \tilde{c}_{n,j} = 0, \quad j = 0,m, \\[2mm]
\text{3. } \tilde{c}_{ij} = \frac{3}{2}\frac{1}{(h_{i-1}+h_i)}\left[\frac{1}{h_i}(a_{i+1,j01}-a_{ij01})-\right. \\
\qquad\qquad \left. -\frac{1}{h_{i-1}}(a_{ij01}-a_{i-1,j01})\right] \\
\qquad\qquad \text{for } i = 1, n-1, \, j = 0,m, \\[2mm]
\text{4. } \tilde{b}_{n-1,j} = \frac{1}{h_{n-1}}(a_{nj01}-a_{n-1,j01})-\frac{2h_{n-1}}{3}\tilde{c}_{n-1,j} \\
\qquad\qquad \text{for } j = 0,m, \\[2mm]
\text{5. } \tilde{d}_{n-1,j} = -\frac{1}{3h_{n-1}}\tilde{c}_{n-1,j} \quad \text{for } j = 0,m.
\end{cases}$$

4^{th} step: Calculate of the partial derivatives a_{ij10} for $i = 1,\ldots,n-1$, $j = 0,\ldots,m$ satisfying (12.3).

5^{th} step: Solve the linear systems (12.4) in order to determine a_{ij01}, $i = 0,\ldots,n$, $j = 1,\ldots,m-1$.

6^{th} step: Determine the values a_{ij11}, $i = 1,\ldots,n-1$, $j = 0,m$ from (12.5).

7^{th} step: Calculate the partial derivatives a_{ij11}, $i = 0,\ldots,n$, $j = 1,\ldots,m-1$ using (12.6).

8^{th} step: Determine the matrices $\boldsymbol{G}(x_i)^{-1}$ in (12.7).

9^{th} step: Determine the matrices $(\boldsymbol{G}(y_j)^T)^{-1}$ in (12.8).

10^{th} step: Determine the matrices \boldsymbol{M}_{ij} according to (12.9).

11^{th} step: Calculate the coefficient matrices \boldsymbol{A}_{ij} according to (12.10) for $i = 0,\ldots,n-1$, $j = 0,\ldots,m-1$.

12^{th} step: Form the bicubic spline $S(x,y) \equiv f_{ij}(x,y)$ for $(x,y) \in R_{ij}$ according to (12.1).

One-dimensional splines are used in this algorithm 12.2 using three points each. Alternatively one could determine one-dimensional splines and their derivatives through all given points (x_i, u_{ij}), $i = 0,\ldots,n$, for fixed j, or through (y_j, u_{ij}), $j = 0,\ldots,m$ for fixed i.

ALGORITHM 12.3 (*Bicubic spline function with prescribed normals*).

Given: (i) Functional values $u_{ij} = u(x_i, y_j)$ for $i = 0, \ldots, n$, $j = 0, \ldots, m$
at the grid points (x_i, y_j);

(ii) Surface normals \boldsymbol{n}_{ij} for each grid point (x_i, y_j) with
$\boldsymbol{n}_{ij}^T = (n_{ij1}, n_{ij2}, n_{ij3})$ with $n_{ij3} \neq 0$ for $i = 0, \ldots, n$,
$j = 0, \ldots, m$.

Task: Find the spline function S of the form (12.1), which assumes the
values u_{ij} in (i) and has the normals (ii) at the grid points (x_i, y_j).

From the normal vectors \boldsymbol{n}_{ij}, the p_{ij} and q_{ij} can be determined everywhere
on the grid, so that we do not need to form one-dimensional splines here.

A point on the surface is represented by

$$\boldsymbol{g}(x, y) = \begin{pmatrix} x \\ y \\ u(x, y) \end{pmatrix}.$$

With the tangent vectors

$$\boldsymbol{g}_x = \frac{\partial}{\partial x}\boldsymbol{g}(x, y) = \begin{pmatrix} 1 \\ 0 \\ u_x(x, y) \end{pmatrix} \quad \text{and} \quad \boldsymbol{g}_y = \frac{\partial}{\partial y}\boldsymbol{g}(x, y) = \begin{pmatrix} 0 \\ 1 \\ u_y(x, y) \end{pmatrix},$$

the surface normal \boldsymbol{n} can be described using the cross product

$$\boldsymbol{n} = \boldsymbol{g}_x \times \boldsymbol{g}_y = \begin{pmatrix} -u_x \\ -u_y \\ 1 \end{pmatrix}.$$

If we normalize the third component of the given normal vectors \boldsymbol{n}_{ij} to
have the value one, we get

$$\frac{\boldsymbol{n}_{ij}}{n_{ij3}} = \begin{pmatrix} \dfrac{n_{ij1}}{n_{ij3}} \\ \dfrac{n_{ij2}}{n_{ij3}} \\ 1 \end{pmatrix} \overset{!}{=} \boldsymbol{n}(x_i, y_j) = \begin{pmatrix} -u_x(x_i, y_j) \\ -u_y(x_i, y_j) \\ 1 \end{pmatrix}.$$

Thus we have the relations

$$\left. \begin{array}{rcl} p_{ij} = a_{ij10} &=& -\dfrac{n_{ij1}}{n_{ij3}}, \\[2mm] q_{ij} = a_{ij01} &=& -\dfrac{n_{ij2}}{n_{ij3}}, \end{array} \right\} \quad \text{for} \quad \begin{array}{rcl} i &=& 0, \ldots, n, \quad \text{and} \\ j &=& 0, \ldots, m. \end{array}$$

The only derivatives left to be calculated are the $s_{ij} = a_{ij11}$ which we can obtain by using one-dimensional splines.

1^{st} step: Calculate the partial derivatives $p_{ij} = a_{ij10} = -n_{ij1}/n_{ij3}$ for $i = 0, \ldots, n,\ j = 0, \ldots, m$.

2^{nd} step: Calculate the partial derivatives $q_{ij} = a_{ij01} = -n_{ij2}/n_{ij3}$ for $i = 0, \ldots, n,\ j = 0, \ldots, m$.

3^{rd} step: Calculate the four boundary values of the mixed partials $s_{ij} = a_{ij11}$ for $i = 0, n$ and $j = 0, m$ according to (12.15) and (12.16).

4^{th} step: Determine a_{ij11} for $i = 1, \ldots, n-1,\quad j = 0, m$ according to (12.5).

5^{th} step: Determine a_{ij11} for $i = 0, \ldots, n,\ j = 1, \ldots, m-1$ according to (12.6).

6^{th} step: Determine the matrices $\boldsymbol{G}(x_i)^{-1}$ according to (12.7).

7^{th} step: Determine the matrices $(\boldsymbol{G}(y_j)^T)^{-1}$ according to (12.8).

8^{th} step: Determine the matrices \boldsymbol{M}_{ij} according to (12.9).

9^{th} step: Determine the coefficient matrices \boldsymbol{A}_{ij} for $i = 0, \ldots, n-1,\ j = 0, \ldots, m-1$ according to (12.10).

10^{th} step: Form the bicubic spline function $S \equiv f_{ij}$ for each rectangle R_{ij} in the form (12.1).

Bibliography for Section 12.1

[BEZI72]; [BOOR62]; [GREV69], VII; [REIN71]; [RITT69]; [SHAH70]; [SPÄT71]; [SPÄT73]; [SPÄT74/2], 8.

12.2 Two-Dimensional Interpolating Surface Splines

Here we want to compute a smooth two-dimensional interpolating surface for a set of prescribed points $(x_i, y_i, f_i) \in \mathbb{R}^3, i = 1, \ldots, NX$, with $f_i = f(x_i, y_i)$. We shall not use one-dimensional splines as was the case with bicubic and

bicubic Bézier splines. Instead the surface splines shall be interpreted in a physical way as described by the bending of a thin plate of infinite dimensions that is deflected at several independent points vertically in such a way as to minimize the bending energy. This approach generalizes natural splines from one dimension to two dimensions.

We will describe how to construct such surface splines. In order to understand the algorithm fully, one ought to consult the literature, and specifically the paper of J. Meinguet, [MEING79]. Unfortunately the linear systems that arise in this context are generally ill conditioned.

Statement of the Problem.

Assume that NX distinct points (x_i, y_i), $i = 1, \ldots, NX$, and NX functional values f_i with

$$f_i = f(x_i, y_i), \; i = 1, \ldots, NX \; ,$$

are known. Moreover we want to prescribe a derivative order $M > 1$ which determines the "smoothness" of the interpolating surface. For this let $p_i^{(M)}$ denote the two-dimensional monomials of degree less than or equal to M. Clearly there are $(M + 1)(M + 2)/2$ of them.

For example if $M = 1$, then there are $(M + 1)(M + 2)/2 = 3$ one–dimensional monomials of degree at most 1, namely:

$$p_1^{(1)} = 1, \; p_2^{(1)} = x, \; \text{and} \; p_3^{(1)} = y;$$

while for $M = 2$ we have $(M + 1)(M + 2)/2 = 6$ two–dimensional monomials:

$$\begin{aligned} p_1^{(2)} &= 1, & p_2^{(2)} &= x, & p_3^{(2)} &= y, & p_4^{(2)} &= x^2, \\ p_5^{(2)} &= xy, & p_6^{(2)} &= y^2. \end{aligned}$$

Next one selects a set of points from the given points (x_i, y_i) so that for $MM := M(M + 1)/2$ and $M > 1$, the following condition is fulfilled:

There is a subset $\{(x_{i_j}, y_{i_j}) \mid i_j \in \{1, 2, \ldots, NX\}, j = 1, \ldots, MM\}$ of the set $\{(x_i, y_i), i = 1, \ldots, NX\}$ so that the matrix $\boldsymbol{P}_j^{(M-1)} := (p_j^{(M-1)}(x_{i_j}, y_{i_j}))$ is nonsingular.

EXAMPLE: (for the p_j from above)

1) If $M = 2$ then $MM = 3$. If one chooses the subset

$$\{(x_1, y_1) = (0,0), \quad (x_2, y_2) = (0,1), \quad (x_3, y_3) = (1,0)\} \subseteq \{(x_i, y_i)\},$$

then the determinant of the matrix $P_j^{(M-1)}$ is:

$$\det\left(P_j^{(M-1)}\right) = \det(p_j^{(1)}(x_{i_j}, y_{i_j})) =$$

$$= \begin{vmatrix} 1 & x_1 & y_1 \\ 1 & x_2 & y_2 \\ 1 & x_3 & y_3 \end{vmatrix} = \begin{vmatrix} 1 & 0 & 0 \\ 1 & 0 & 1 \\ 1 & 1 & 0 \end{vmatrix} = -1 \neq 0,$$

hence the matrix is nonsingular.

2) If, however, one chooses the subset

$$\{(x_1, y_1) = (0,0), \quad (x_2, y_2) = (0.5, 0.5), \quad (x_3, y_3) = (1,1)\} \subseteq \{(x_i, y_i)\},$$

instead, then

$$\det\left(P_j^{(M-1)}\right) = \begin{vmatrix} 1 & 0 & 0 \\ 1 & 0.5 & 0.5 \\ 1 & 1 & 1 \end{vmatrix} = 0,$$

or in this case the subset does not fulfill the required condition for the matrix $P_j^{(M-1)}$.

We want to find a function Φ of the form

(12.17) $$\Phi(x,y) := \sum_{j=1}^{n} c_j \varphi_j(x, y),$$

so that the interpolating conditions

(12.18) $$\Phi(x_i, y_i) = \sum_{j=1}^{n} c_j \varphi_j(x_i, y_i) \stackrel{!}{=} f_i$$

are fulfilled for $i = 1, \ldots, NX$, as well as certain minimal properties (see [MEING79]). We define the functions φ_j as follows:

$$(12.19) \begin{cases} \textbf{1)} \quad \varphi_{j+NX}(x,y) = p_j^{(M-1)}(x,y) \text{ for } j = 1,\ldots,MM, \text{ monomials} \\ \qquad \text{up to degree M-1,} \\ \\ \textbf{2)} \quad \varphi_j(x,y) = E(x-x_j, y-y_j) \text{ for } j = 1,\ldots,NX \text{ with} \\ \qquad E(x,y) := (r^2)^{M-1}\ln r^2 \text{ for } r^2 = x^2+y^2 \quad \text{(kernel function).} \end{cases}$$

These conditions lead to a linear system with $NX + MM$ rows and columns
for an (NX,NX) matrix \boldsymbol{G} and an (NX,MM) matrix \boldsymbol{P} of the form

$$(12.20) \qquad \begin{pmatrix} \boldsymbol{G} & \boldsymbol{P} \\ \boldsymbol{P}^T & \boldsymbol{0} \end{pmatrix} \begin{pmatrix} \boldsymbol{c}^{(1)} \\ \boldsymbol{c}^{(2)} \end{pmatrix} = \begin{pmatrix} \boldsymbol{f} \\ \boldsymbol{0} \end{pmatrix}, \text{ where}$$

$$(12.21) \qquad \boldsymbol{G} = (E(x_i-x_j, y_i-y_j)), \quad i,j=1,\ldots,NX, \text{ and}$$

$$(12.22) \quad \boldsymbol{P} = \left(p_j^{(M-1)}(x_i,y_i)\right), \quad i=1,\ldots,NX, \quad j=1,\ldots,MM,$$

and the vectors

$$\boldsymbol{c}^{(1)} = \begin{pmatrix} c_1 \\ c_2 \\ \vdots \\ c_{NX} \end{pmatrix}, \boldsymbol{c}^{(2)} = \begin{pmatrix} c_{NX+1} \\ c_{NX+2} \\ \vdots \\ c_{NX+MM} \end{pmatrix}, \boldsymbol{f} = \begin{pmatrix} f_1 \\ f_2 \\ \vdots \\ f_{NX} \end{pmatrix}.$$

Here \boldsymbol{P}^T is the transpose of \boldsymbol{P}. For large NX and M, the condition of the
system matrix in (12.20) unfortunately worsens.

We speak of interpolating surface splines if the diagonal elements g_{ii} of \boldsymbol{G} are
zero, i.e., if $E(0,0) = 0$.

REMARK: We recommend to transform the nodes $(x_i,y_i), i = 1,\ldots,NX$,
first onto the unit circle:

$$(12.23) \begin{cases} (x_i,y_i) \to \frac{1}{r}\left[(x_i,y_i)-(\bar{x},\bar{y})\right], \\ \\ \text{where } \bar{x} = \frac{1}{NX}\sum_{i=1}^{NX} x_i, \quad \bar{y} = \frac{1}{NX}\sum_{i=1}^{NX} y_i, \text{ and} \\ \\ r = \max_{1\le i\le NX}\sqrt{(\bar{x}-x_i)^2+(\bar{y}-y_i)^2}. \end{cases}$$

Similar algorithms can be formulated in three dimensions as well. Two pro-
grams for such three-dimensional surface splines are available in the program
part of this book.

Bibliography for Section 12.2

[DONG79]; [GORD69]; [GREV69]; [MEING79].

12.3 Bézier Splines

12.3.1 Bézier Spline Curves

A Bézier spline curve is composed of local Bézier curves that are joined smoothly.

A Bézier curve of order n for $n \geq 1$ can be represented via *Bernstein polynomials* of degree n

$$(12.24) \qquad B_j^n(v) = \binom{n}{j}(1-v)^{n-j}v^j, \qquad j = 0, 1, \ldots, n,$$

and $n+1$ *Bézier points* b_j, $j = 0, \ldots, n$, $b_j \in \mathbb{R}^2$ or $b_j \in \mathbb{R}^3$ as:

$$(12.25) \qquad P(v) = \sum_{j=0}^{n} B_j^n(v)b_j \quad \text{for } v \in [0,1].$$

In the following we shall only consider Bézier curves of order $n = 3$. Such curves can be described via (12.24) and (12.25) as

$$(12.26) \qquad P(v) = (1-v)^3 b_0 + 3(1-v)^2 v b_1 + 3(1-v)v^2 b_2 + v^3 b_3, \ v \in [0,1].$$

$P(0) = b_0$ is the starting point, $P(1) = b_3$ is the endpoint of the Bézier curve. The points b_1 and b_2 do not lie on the curve. The slopes at the two endpoints are

$$(12.27) \qquad P'(0) = 3(b_1 - b_0), \quad P'(1) = 3(b_3 - b_2).$$

Thus the chord joining b_0 and b_1, provided $b_0 \neq b_1$, is the tangent to the curve at b_0. Similarly the tangent at b_3 is the line joining b_2 and b_3 if $b_2 \neq b_3$. Furthermore we note that the Bézier curve is always contained in the smallest convex region containing the Bézier points b_0, b_1, b_2, b_3. From these properties one can get a rough idea how a Bézier curve will look for a given set of Bézier points b_i in case $n = 3$.

The second derivative can be computed at the endpoints from (12.26) as

$$(12.28) \qquad \begin{cases} P''(0) &= 6(b_0 - 2b_1 + b_2), \\ P''(1) &= 6(b_1 - 2b_2 + b_3). \end{cases}$$

A cubic Bézier spline curve is composed of m local Bézier curves of third order. The spline segments can be described as

$$(12.29)\begin{cases} P_k(v) & = (1-v)^3 b_{3k} + 3(1-v)^2 v b_{3k+1} + 3(1-v)v^2 b_{3k+2} + \\ & \quad +v^3 b_{3k+3}, \ v \in [0,1], \ \text{for } k = 0,1,\ldots,m-1, \ m \geq 2. \end{cases}$$

The Bézier points are numbered in such a way that

$$P_{k-1}(1) = b_{3(k-1)+3} = b_{3k} = P_k(0), \quad k = 1,\ldots,m-1.$$

Since the spline curve passes through each of the endpoints b_0, b_3, \ldots, b_{3m} of the m curve segments, we shall call these points the *interpolation points* of the cubic Bézier spline curve in what follows.

At the points $b_3, b_6, \ldots, b_{3m-3}$ two curve segments will join. We shall demand that adjoining spline segments have identical slopes and second derivatives at these Bézier points. Hence we must have

$$(12.30) \ P'_{k-1}(1) = P'_k(0), \quad P''_{k-1}(1) = P''_k(0) \quad \text{for } k = 1,\ldots,m-1.$$

By using (12.27) the first condition becomes

$$(12.31) \qquad\qquad b_{3k} - b_{3k-1} = b_{3k+1} - b_{3k}.$$

Hence the points $b_{3k-1}, b_{3k}, b_{3k+1}$ must be collinear and the interpolation point b_{3k} must be the center of the line segment $b_{3k-1} b_{3k+1}$, see Figure 12.2. The second condition in (12.30) implies with (12.28) that

$$(12.32) \quad d_k := b_{3k-1} + (b_{3k-1} - b_{3k-2}) = b_{3k+1} - (b_{3k+2} - b_{3k+1}).$$

The points d_k, $k = 1,\ldots,m-1$, are called *weight points* or weights for short. Their actual location is best illustrated below :

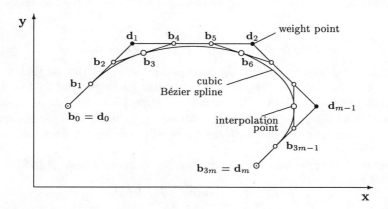

Figure 12.2: • Given nodes, weight points, o interpolation points

By using (12.32) and (12.31), the Bézier points $b_1, b_2, \ldots, b_{3m-2}, b_{3m-1}$ can be represented via the $(m+1)$ weights $d_0, d_1, \ldots, d_{m-1}, d_m$ as

$$(12.33) \quad \begin{cases} b_{3k-2} = \frac{1}{3}(2d_{k-1} + d_k), & k = 1, \ldots, m, \\[2mm] b_{3k-1} = \frac{1}{3}(d_{k-1} + 2d_k), & k = 1, \ldots, m, \\[2mm] b_{3k} = \frac{1}{6}(d_{k-1} + 4d_k + d_{k+1}), & k = 1, \ldots, m-1. \end{cases}$$

Hence in order to produce a Bézier spline, precisely $m+1$ weight points d_0, \ldots, d_m need to be specified. Thus one can do without the interpolation points.

Besides the weight points, the end points b_0 und b_{3m} need to be specified as well. Choosing

$$(12.34) \qquad\qquad b_0 = d_0, \quad b_{3m} = d_m$$

results in

$$P_0''(0) = 0, \quad P_{m-1}''(1) = 0,$$

i.e., this prescribes a natural cubic spline.

ALGORITHM 12.4 (*Cubic Bézier spline curve*).

Given: $m+1$ weights d_k, $k = 0, \ldots, m, m \geq 2$ with $d_k \in \mathbb{R}^2$ or $d_k \in \mathbb{R}^3$.

Task: Find m cubic polynomials
$$P_k(v) = (1-v)^3 b_{3k} + 3(1-v)^2 v b_{3k+1} + \\ +3(1-v)v^2 b_{3k+2} + v^3 b_{3k}, \quad v \in [0,1],$$
for $k = 0, \ldots, m-1$.

1^{st} step: Determine the $3m-1$ Bézier points $b_1, b_2, \ldots, b_{3m-1}$ from (12.33).

2^{nd} step: Choose the endpoints b_0, b_{3m} as in (12.34).

One drawback of this method is that one cannot prescribe the interpolation points b_{3k} for $k = 1, \ldots, m-1$, but instead one has to give the weight points d_k which have no directly obvious relation to the shape of the curve. The interpolation points are calculated only during the process of the algorithm. A modification can be found in section 12.3.3. There we describe how to obtain modified interpolating Bézier splines, for another representation see [HOSC89].

An advantage of Bézier splines is that one can add new weight points to the data and this will only affect a local change of the course of the

curve, i.e., a different weight point d_k will only affect the Bézier points $b_{3k-3}, b_{3k-2}, \ldots, b_{3k+2}, b_{3k+3}$. For this reason, cubic Bézier splines can be employed with good control in modelling. Moreover cubic Bézier splines can be formed without having to solve any system of linear equations.

Special features of cubic Bézier splines.

If one wants to create a corner with cubic Bézier splines, one simply needs to make three successive weight points identical. If for instance $d_{i-1} = d_i = d_{i+1}$, then the interpolation point b_{3i} will become identical to these weight points, and thus the Bézier curve will not be differentiable at the point $b_{3i} = d_i$.

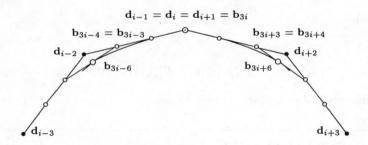

Figure 12.3: Forming a non–differentiable corner with Bézier splines

The curve segments between b_{3i-4} and $d_i = b_{3i}$, and between $b_{3i} = d_i$ and b_{3i+4} are line segments.

Bibliography for Section 12.3.1

[BEZI72].

12.3.2 Bézier Spline Surfaces

A Bézier spline surface is composed locally of Bézier surfaces, which are called *patches*.

Figure 12.4: Bézier surface

In the following we shall restrict our attention to bicubic Bézier spline surfaces and bicubic patches. A bicubic Bézier surface can be represented as

$$(12.35) \quad \boldsymbol{P}(v,w) = \sum_{j=0}^{3} \sum_{s=0}^{3} B_j^3(v) B_s^3(w) \boldsymbol{b}_{js} \quad \text{for } v \in [0,1] \text{ and } w \in [0,1].$$

Here $B_j^3(v)$ and $B_s^3(w)$ are Bernstein polynomials of third degree, see (12.24), and $\boldsymbol{b}_{js} \in \mathbb{R}^3$, $j = 0,1,2,3$, $s = 0,1,2,3$, are the 16 Bézier points of the surface.

This representation can be derived using the following method to generate a Bézier surface: Let

$$\boldsymbol{P}(v) = \sum_{j=0}^{3} B_j^3(v) \boldsymbol{b}_j, \quad v \in [0,1],$$

be a spacial Bézier curve in \mathbb{R}^3, see (12.25), (12.26). Assume that each Bézier point moves along a cubic Bézier curve of its own :

$$\boldsymbol{b}_j(w) = \sum_{s=0}^{3} B_s^3(w) \boldsymbol{b}_{js}, \quad w \in [0,1], \quad j = 0,1,2,3.$$

Thus the curve $\boldsymbol{P}(v)$ will move in space while changing its shape and it will cover the surface described by

$$\begin{aligned} \boldsymbol{P}(v,w) &= \sum_{j=0}^{3} B_j^3(v) \boldsymbol{b}_j(w) \\ &= \sum_{j=0}^{3} B_j^3(v) \sum_{s=0}^{3} B_s^3(w) \boldsymbol{b}_{js} \end{aligned}$$

as stipulated in (12.35).

For $w = const$ we obtain the v-curves while $v = const$ defines the w-curves of the Bézier surface. The properties of Bernstein polynomials imply that the Bézier points $\boldsymbol{P}(0,0) = \boldsymbol{b}_{00}$, $\boldsymbol{P}(1,0) = \boldsymbol{b}_{30}$, $\boldsymbol{P}(0,1) = \boldsymbol{b}_{03}$, $\boldsymbol{P}(1,1) = \boldsymbol{b}_{33}$ lie on the Bézier surface as boundary points, see Figure 12.5.

The points $b_{j0}, b_{j3}, j = 0, 1, 2, 3$, and $b_{0s}, b_{3s}, s = 0, 1, 2, 3$, are Bézier points of the boundary curves for the Bézier surface, see Figure 12.5.

Due to (12.27), applied to the boundary curves, the tangent planes to the Bézier surface at the boundary points are determined by the point triples (b_{00}, b_{10}, b_{01}), (b_{30}, b_{31}, b_{20}), (b_{03}, b_{02}, b_{13}), and (b_{33}, b_{23}, b_{32}), each of which is non-collinear.

The 16 Bézier points should be chosen in \mathbb{R}^3 so as to generate 9 generally non-planar quadrilaterals, see Figure 12.5.

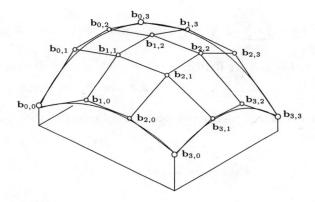

Figure 12.5: Bicubic Bézier surface, including patches for $i = 0$ and $k = 0$.

A bicubic Bézier spline surface is composed of $m \cdot n$ bicubic patches, see Figure 12.5:

$$(12.36) \qquad P_{ik}(v, w) = \sum_{j=0}^{3} \sum_{s=0}^{3} B_j^3(v) B_s^3(w) \boldsymbol{b}_{3i+j, 3k+s},$$
$$i = 0, \ldots, m-1, \ k = 0, \ldots, n-1.$$

Explicitly the parametrization of one patch (i, k) can be described as

$$P_{ik}(v,w) = \sum_{j=0}^{3}\left(\sum_{s=0}^{3}B_s^3(w)\boldsymbol{b}_{3i+j,3k+s}\right)B_j^3(v)$$

$$= [\boldsymbol{b}_{3i,3k}(1-w)^3+$$
$$+3\boldsymbol{b}_{3i,3k+1}(1-w)^2w+3\boldsymbol{b}_{3i,3k+2}(1-w)w^2+$$
$$+\boldsymbol{b}_{3i,3k+3}w^3](1-v)^3+3[\boldsymbol{b}_{3i+1,3k}(1-w)^3+$$
(12.37)
$$+3\boldsymbol{b}_{3i+1,3k+1}(1-w)^2w+3\boldsymbol{b}_{3i+1,3k+2}(1-w)w^2+$$
$$+\boldsymbol{b}_{3i+1,3k+3}w^3](1-v)^2v+3[\boldsymbol{b}_{3i+2,3k}(1-w)^3+$$
$$+3\boldsymbol{b}_{3i+2,3k+1}(1-w)^2w+3\boldsymbol{b}_{3i+2,3k+2}(1-w)w^2+$$
$$+\boldsymbol{b}_{3i+2,3k+3}w^3](1-v)v^2+[\boldsymbol{b}_{3i+3,3k}(1-w)^3+$$
$$+3\boldsymbol{b}_{3i+3,3k+1}(1-w)^2w+3\boldsymbol{b}_{3i+3,3k+2}(1-w)w^2+$$
$$+\boldsymbol{b}_{3i+3,3k+3}w^3]v^3.$$

The numbering of the Bézier points has been chosen in such a way that the
common boundary curve of two adjacent patches is generated by the identical
set of Bézier points. Thus adjacent patches are insured to join smoothly.
(12.37) implies for example for the adjacent patches $(i-1,k)$ and (i,k) :

$$P_{i-1,k}(1,w) = \sum_{s=0}^{3}B_s^3(w)\boldsymbol{b}_{3(i-1)+3,3k+s} =$$
$$= \sum_{s=0}^{3}B_s^3(w)\boldsymbol{b}_{3i,3k+s} = P_{ik}(0,w).$$

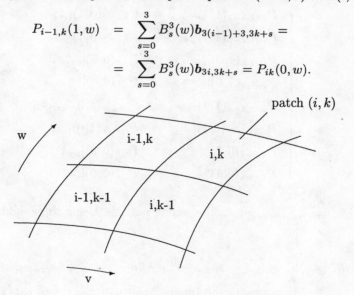

Figure 12.6: Adjacent patches of a bicubic Bézier spline surface

By using bicubic Bézier splines one can also represent surfaces whose nodes
are not given on a rectangular grid G. Moreover surfaces $z = f(x,y)$ with
more than one z-value for one (x,y)-value can also be represented as bicubic
Bézier spline surfaces.

Computing bicubic Bézier splines

When constructing a Bézier spline surface we shall assume that the partial derivatives of first and second order with respect to v and w exist for $\boldsymbol{P}_{ik}(v,w)$ in (12.37) and satisfy the following smoothness conditions for adjacent patches $(i-1,k-1), (i-1,k), (i,k-1)$ and (i,k) :

$$(12.38) \quad \begin{cases} \boldsymbol{P}_{i-1,k-1}(1,1)|_v &= \boldsymbol{P}_{i,k-1}(0,1)|_v, \\ \boldsymbol{P}_{i-1,k}(1,0)|_v &= \boldsymbol{P}_{i,k}(0,0)|_v, \\ \boldsymbol{P}_{i-1,k-1}(1,1)|_w &= \boldsymbol{P}_{i-1,k}(1,0)|_w, \\ \boldsymbol{P}_{i,k-1}(0,1)|_w &= \boldsymbol{P}_{i,k}(0,0)|_w. \end{cases}$$

Here the symbols $|_v$, and $|_w$ denote the first partial derivatives with respect to v and w, respectively. Moreover we must have:

$$(12.39) \quad \begin{cases} \boldsymbol{P}_{i-1,k-1}(1,1)|_{vv} &= \boldsymbol{P}_{i,k-1}(0,1)|_{vv}, \\ \boldsymbol{P}_{i-1,k-1}(1,1)|_{vw} &= \boldsymbol{P}_{i,k-1}(0,1)|_{vw}, \\ \boldsymbol{P}_{i-1,k}(1,0)|_{vv} &= \boldsymbol{P}_{i,k}(0,0)|_{vv}, \\ \boldsymbol{P}_{i-1,k}(1,0)|_{vw} &= \boldsymbol{P}_{i,k}(0,0)|_{vw}, \\ \boldsymbol{P}_{i-1,k-1}(1,1)|_{ww} &= \boldsymbol{P}_{i-1,k}(1,0)|_{ww}, \\ \boldsymbol{P}_{i-1,k-1}(1,1)|_{wv} &= \boldsymbol{P}_{i-1,k}(1,0)|_{wv}, \\ \boldsymbol{P}_{i,k-1}(0,1)|_{ww} &= \boldsymbol{P}_{i,k}(0,0)|_{ww}, \\ \boldsymbol{P}_{i,k-1}(0,1)|_{wv} &= \boldsymbol{P}_{i,k}(0,0)|_{wv}, \end{cases}$$

where $|_{vv}, |_{vw}, |_{ww}, |_{wv}$ denote the second partials. Using the weight points

$$\begin{aligned} \boldsymbol{d}_{ik} :=\ & \boldsymbol{b}_{3i-2,3k-2} - 2\boldsymbol{b}_{3i-2,3k-1} - 2\boldsymbol{b}_{3i-1,3k-2} + 4\boldsymbol{b}_{3i-1,3k-1} \\ =\ & 4\boldsymbol{b}_{3i+1,3k+1} - 3\boldsymbol{b}_{3i+1,3k+2} - 2\boldsymbol{b}_{3i+2,3k+1} + \boldsymbol{b}_{3i+2,3k+2} \\ =\ & -2\boldsymbol{b}_{3i-2,3k+1} + 4\boldsymbol{b}_{3i-1,3k+1} - 2\boldsymbol{b}_{3i-1,3k+2} + \boldsymbol{b}_{3i-2,3k+2} \\ =\ & -2\boldsymbol{b}_{3i+1,3k-2} + 4\boldsymbol{b}_{3i+1,3k-1} + \boldsymbol{b}_{3i+2,3k-2} - 2\boldsymbol{b}_{3i+2,3k-1}, \end{aligned}$$

the following equations can be derived from (12.38) and (12.39):

$$(12.40) \begin{cases} \boldsymbol{b}_{3i-2,3k-2} = \frac{4}{9}\boldsymbol{d}_{i-1,k-1} + \frac{2}{9}\boldsymbol{d}_{i-1,k} + \frac{2}{9}\boldsymbol{d}_{i,k-1} + \frac{1}{9}\boldsymbol{d}_{i,k} \\ \qquad\qquad \text{for} \quad i = 1,\ldots,m,\ k = 1,\ldots,n, \\[4pt] \boldsymbol{b}_{3i-2,3k+2} = \frac{4}{9}\boldsymbol{d}_{i-1,k+1} + \frac{2}{9}\boldsymbol{d}_{i-1,k} + \frac{2}{9}\boldsymbol{d}_{i,k+1} + \frac{1}{9}\boldsymbol{d}_{i,k} \\ \qquad\qquad \text{for} \quad i = 1,\ldots,m,\ k = 0,\ldots,n-1, \\[4pt] \boldsymbol{b}_{3i+2,3k-2} = \frac{4}{9}\boldsymbol{d}_{i+1,k-1} + \frac{2}{9}\boldsymbol{d}_{i,k-1} + \frac{2}{9}\boldsymbol{d}_{i+1,k} + \frac{1}{9}\boldsymbol{d}_{i,k} \\ \qquad\qquad \text{for} \quad i = 0,\ldots,m-1,\ k = 1,\ldots,n, \\[4pt] \boldsymbol{b}_{3i+2,3k+2} = \frac{4}{9}\boldsymbol{d}_{i+1,k+1} + \frac{2}{9}\boldsymbol{d}_{i,k+1} + \frac{2}{9}\boldsymbol{d}_{i+1,k} + \frac{1}{9}\boldsymbol{d}_{i,k} \\ \qquad\qquad \text{for} \quad i = 0,\ldots,m-1,\ k = 0,\ldots,n-1, \\[4pt] \boldsymbol{b}_{3i-2,3k} = \frac{1}{9}\boldsymbol{d}_{i-1,k-1} + \frac{4}{9}\boldsymbol{d}_{i-1,k} + \frac{1}{9}\boldsymbol{d}_{i-1,k+1} + \\ \qquad\qquad + \frac{1}{18}\boldsymbol{d}_{i,k-1} + \frac{2}{9}\boldsymbol{d}_{i,k} + \frac{1}{18}\boldsymbol{d}_{i,k+1} \\ \qquad\qquad \text{for} \quad i = 1,\ldots,m,\ k = 1,\ldots,n-1, \\[4pt] \boldsymbol{b}_{3i,3k-2} = \frac{1}{9}\boldsymbol{d}_{i-1,k-1} + \frac{4}{9}\boldsymbol{d}_{i,k-1} + \frac{1}{9}\boldsymbol{d}_{i+1,k-1} + \\ \qquad\qquad + \frac{1}{18}\boldsymbol{d}_{i-1,k} + \frac{2}{9}\boldsymbol{d}_{i,k} + \frac{1}{18}\boldsymbol{d}_{i+1,k} \\ \qquad\qquad \text{for} \quad i = 1,\ldots,m-1,\ k = 1,\ldots,n, \\[4pt] \boldsymbol{b}_{3i,3k+2} = \frac{1}{9}\boldsymbol{d}_{i-1,k+1} + \frac{4}{9}\boldsymbol{d}_{i,k+1} + \frac{1}{9}\boldsymbol{d}_{i+1,k+1} + \\ \qquad\qquad + \frac{1}{18}\boldsymbol{d}_{i-1,k} + \frac{2}{9}\boldsymbol{d}_{i,k} + \frac{1}{18}\boldsymbol{d}_{i+1,k} \\ \qquad\qquad \text{for} \quad i = 1,\ldots,m-1,\ k = 0,\ldots,n-1, \\[4pt] \boldsymbol{b}_{3i+2,3k} = \frac{1}{9}\boldsymbol{d}_{i+1,k-1} + \frac{4}{9}\boldsymbol{d}_{i+1,k} + \frac{1}{9}\boldsymbol{d}_{i+1,k+1} + \\ \qquad\qquad + \frac{1}{18}\boldsymbol{d}_{i,k-1} + \frac{2}{9}\boldsymbol{d}_{i,k} + \frac{1}{18}\boldsymbol{d}_{i,k+1} \\ \qquad\qquad \text{for} \quad i = 0,\ldots,m-1,\ k = 1,\ldots,n-1, \\[4pt] \boldsymbol{b}_{3i,3k} = \frac{1}{36}\boldsymbol{d}_{i-1,k-1} + \frac{1}{9}\boldsymbol{d}_{i,k-1} + \frac{1}{36}\boldsymbol{d}_{i+1,k-1} + \\ \qquad\qquad + \frac{1}{9}\boldsymbol{d}_{i-1,k} + \frac{4}{9}\boldsymbol{d}_{i,k} + \frac{1}{9}\boldsymbol{d}_{i+1,k} + \\ \qquad\qquad + \frac{1}{36}\boldsymbol{d}_{i-1,k+1} + \frac{1}{9}\boldsymbol{d}_{i,k+1} + \frac{1}{36}\boldsymbol{d}_{i+1,k+1} \\ \qquad\qquad \text{for} \quad i = 1,\ldots,m-1,\ k = 1,\ldots,n-1. \end{cases}$$

If the \boldsymbol{d}_{ik} in (12.40) are known, then from (12.40) one can determine the Bézier points that lie in the interior of the surface: If the $(m+1)(n+1)$ weight points \boldsymbol{d}_{ik} are given, then the interior Bézier points are known. To compute the bicubic polynomial (12.37) we only miss the $6(m+n)$ Bézier points $\boldsymbol{b}_{0,s}, \boldsymbol{b}_{3m,s}, \boldsymbol{b}_{j,0}$ and $\boldsymbol{b}_{j,3n}$ for $s = 0,\ldots,3n,\ j = 0,\ldots,3m$, which lie on the boundary. These can also be prescribed independently.

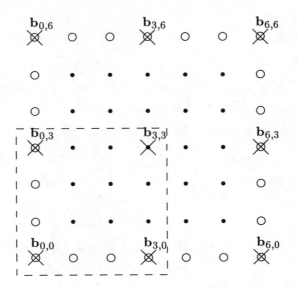

Figure 12.7: ○ prescribed Bézier points, • computed Bézier points,
× interpolation points in the parameter plane.

The rectangular frame in Figure 12.7 gives only information about the order of the Bézier points, but not about their relative position in space.

Corrections to bicubic Bézier splines.
If one moves a weight point d_{ik} of a bicubic Bézier spline by $36d$, the Bézier points b_{js} in the adjacent patches will be changed as follows: From the equations (12.40), the new Bézier points b_{js}^* become

$$
\begin{aligned}
b_{3i-2,3k-2}^* &= [4d_{i-1,k-1} + 2d_{i-1,k} + 2d_{i,k-1} + d_{i,k} + 36d]/9 \\
&= b_{3i-2,3k-2} + 4d, \\
b_{3i-2,3k+2}^* &= [4d_{i-1,k+1} + 2d_{i-1,k} + 2d_{i,k+1} + d_{i,k} + 36d]/9 \\
&= b_{3i-2,3k+2} + 4d, \\
b_{3i+2,3k-2}^* &= [4d_{i+1,k-1} + 2d_{i,k-1} + 2d_{i+1,k} + d_{i,k} + 36d]/9 \\
&= b_{3i+2,3k-2} + 4d, \\
b_{3i+2,3k+2}^* &= [4d_{i+1,k+1} + 2d_{i,k+1} + 2d_{i+1,k} + d_{i,k} + 36d]/9 \\
&= b_{3i+2,3k+2} + 4d, \\
b_{3i-2,3k}^* &= [2d_{i-1,k-1} + 8d_{i-1,k} + 2d_{i-1,k+1} + d_{i,k-1} + \\
&\quad + 4(d_{i,k} + 36d) + d_{i,k+1}]/18 = b_{3i-2,3k} + 8d
\end{aligned}
$$

$$b^*_{3i,3k-2} = [2d_{i-1,k-1} + 8d_{i,k-1} + 2d_{i+1,k-1} + d_{i-1,k} +$$
$$+4(d_{i,k} + 36d) + d_{i+1,k}]/18 = b_{3i,3k-2} + 8d,$$

$$b^*_{3i,3k+2} = [2d_{i-1,k+1} + 8d_{i,k+1} + 2d_{i+1,k+1} + d_{i-1,k} +$$
$$+4(d_{i,k} + 36d) + d_{i+1,k}]/18 = b_{3i,3k+2} + 8d,$$

$$b^*_{3i+2,3k} = [2d_{i+1,k-1} + 8d_{i+1,k} + 2d_{i+1,k+1} + d_{i,k-1} +$$
$$+4(d_{i,k} + 36d) + d_{i,k+1}]/18 = b_{3i+2,3k} + 8d,$$

$$b^*_{3i,3k} = [d_{i-1,k-1} + 4d_{i,k-1} + d_{i+1,k-1} + 4d_{i-1,k} +$$
$$+16(d_{i,k} + 36d) + 4d_{i+1,k} + d_{i-1,k+1} + 4d_{i,k+1} +$$
$$+d_{i+1,k+1}]/36 = b_{3i,3k} + 16d.$$

The changes of the remaining Bézier points, that are given in the following table, are calculated in an analogous way. They are the only points affected by the original change of the weight.

$$(12.41) \begin{cases} \begin{array}{c|ccccccc} 3i+3 & +d & +2d & +4d & +4d & +4d & +2d & +d \\ 3i+2 & +2d & +4d & +8d & +8d & +8d & +4d & +2d \\ 3i+1 & +4d & +8d & +16d & +16d & +16d & +8d & +4d \\ 3i & +4d & +8d & +16d & +16d & +16d & +8d & +4d \\ 3i-1 & +4d & +8d & +16d & +16d & +16d & +8d & +4d \\ 3i-2 & +2d & +4d & +8d & +8d & +8d & +4d & +2d \\ 3i-3 & +d & +2d & +4d & +4d & +4d & +2d & +d \\ \hline & 3k-3 & 3k-2 & 3k-1 & 3k & 3k+1 & 3k+2 & 3k+3 \end{array} \end{cases}$$

Note that a change of some or all Bézier points $b_{j,s}, j \in \{0,\dots,3m\}$, $s \in \{0,\dots,3n\}$, on the boundary does not change any but the above mentioned Bézier points.

If instead of the weights d_{ik}, one changes the Bézier point $b_{3i,3k}$ by $16d$, then the weight point d_{ik} changes by $36d$ and the corresponding Bézier points b_{js} need to be altered by the values indicated in the table (12.41). A significant consequence of our detailed study of the "changing" behavior of bicubic Beziér splines is to realize that bicubic Bézier spline surfaces can be altered locally in contrast to the bicubic splines of section 12.1, which cannot. Thus the bicubic Bézier splines can offer important advantages over bicubic splines in practical applications.

When using cubic or bicubic Bézier splines one will soon realize that one can determine the course of the spline surface or curve at the boundary accurately by prescribing the boundary Bézier points. However, one can influence the course of the spline curves in the interior of the surface only indirectly by altering the "unintuitive" weight points. Hence one would want to modify this method so as to use the interpolation points of the cubic or bicubic Bézier

splines instead of the weight points. This leads to a genuine interpolation problem which is more useful in practice, see section 12.3.3.

ALGORITHM 12.5 (*Bicubic Bézier splines*).

Given: 1) $b_{0,s}, b_{3m,s}, b_{j,0}, b_{j,3n}$ for $s = 0, \ldots, 3n$, $j = 0, \ldots, 3m$, the Bézier points on the boundary.

2) d_{ik}, $i = 0, \ldots, m$, $k = 0, \ldots, n$, the weight points.

Task: Find the bicubic Bézier spline (12.36).

1^{st} step: Determine the $9(m-1)(n-1)$ interior Bézier points by solving the equations (12.40). The parameters v and w vary on each of the $m \cdot n$ patches from 0 to 1.

2^{nd} step: Construct the $m \cdot n$ bicubic polynomials (12.37) for $i = 0, \ldots, m-1$, $k = 0, \ldots, n-1$.

12.3.3 Modified Interpolating Cubic Bézier Splines

To avoid the limitations imposed by the weight points of the Beziér splines mentioned in 12.3.2, a modification is given in [STEU79] and [HOSC89]. In [STEU79] the bicubic Bézier procedure is extended to a "genuine" interpolation procedure. Instead of the weight points d_{ik}, the $6(m+n)$ boundary Bézier points $b_{0,s}, b_{3m,s}, b_{j,0}, b_{j,3n}$, $j = 1, \ldots, 3m$, $s = 0, \ldots, 3n$, and the $(m-1)(n-1)$ interior Bézier points $b_{3i,3k}$, $i = 1, \ldots, m-1$, $k = 1, \ldots, n-1$, are prescribed as interpolation points for the bicubic Bézier spline.

At first, these interpolation points are regarded as weights for which one computes so-called pseudo-interpolation points by averaging. Then, using the correction behavior of bicubic Bézier splines (see (12.42)), these are shifted until they differ by less than a given bound $\varepsilon > 0$ from the prescribed interpolation points.

Bibliography for Section 12.3

[BEZI72], [MICH69], [STEU79], [HOSC89] .

12.4 B-Splines

12.4.1 B-Spline-Curves

For a given set of $n+1$ *de Boor points* (control points) \boldsymbol{d}_i, $i = 0, \ldots, n$, $n \geq 1$ with $\boldsymbol{d}_i \in \mathbb{R}^2$ or $\boldsymbol{d}_i \in \mathbb{R}^3$ and $1 \leq k \leq n+1$ normalized *B-Spline functions* $N_{i,k}$, $k = 1, \ldots, n+1$, a B-spline curve of order k can be constructed as

$$(12.42) \qquad \boldsymbol{P}(t) = \sum_{i=0}^{n} N_{i,k}(t)\boldsymbol{d}_i, \qquad t \in I.$$

We remark that the de Boor points are designated by \boldsymbol{d}_i as were the weight points in section 12.3. This has historic reasons and the two kinds of points are quite different.

We begin with the definitions and some properties of B-spline functions $N_{i,k}$.

We assume that $k+1$ real ordered parameter values t_j, $j = i, \ldots, i + k$ for $k \geq 1$ are given:

$$t_i < t_{i+1} < \ldots < t_{i+k}.$$

The t_i shall be called *nodes*. The normalized B-spline function $N_{i,k}$ of order k shall be defined on \mathbb{R} as

$$(12.43) \quad \begin{cases} \text{If } k = 1: \quad N_{i,1}(t) = \begin{cases} 1 \text{ for } t_i \leq t < t_{i+1} \\ 0 \text{ for } t < t_i \quad \text{and} \quad t \geq t_{i+1}; \end{cases} \\ \text{If } k \geq 2: \quad N_{i,k}(t) = \dfrac{t - t_i}{t_{i+k-1} - t_i} N_{i,k-1}(t) + \dfrac{t_{i+k} - t}{t_{i+k} - t_{i+1}} N_{i+1,k-1}(t). \end{cases}$$

The interval $[t_i, t_{i+k}]$ is the *support* of $N_{i,k}$. The order k of $N_{i,k}$ indicates how many subintervals compose the support of $N_{i,k}$. The B-spline function $N_{i,k}$ assumes nonzero values precisely on its support. For $k \geq 2$ for example we have:

$$(12.44) \quad \begin{aligned} N_{i,k}(t) &> 0 \quad \text{for} \quad t_i < t < t_{i+k}, \quad \text{while} \\ N_{i,k}(t) &= 0 \quad \text{for} \quad t \leq t_i \text{ and } t \geq t_{i+k}. \end{aligned}$$

The recursive definition (12.43) of $N_{i,k}$ implies for $k \geq 2$ that $N_{i,k}(t)$ is composed of k polynomials of degree $k-1$, each defined over one subinterval $[t_j, t_{j+1}]$ of its support. These polynomials joint $k-2$ times continuously differentiable at the inner nodes t_j with $t_i < t_j < t_{i+k}$. For this smooth joining of polynomial segments we must require $k-2 \geq 1$ or $k \geq 3$.

Due to (12.42), a B-spline curve with $n+1$ control points \boldsymbol{d}_i, $i = 0,\ldots,n$, involves $n+1$ B-spline functions $N_{i,k}$ for the $n+1+k$ nodes

$$t_0, t_1, \ldots, t_n, t_{n+1}, \ldots, t_{n+k}.$$

The nodes can be normalized by setting $t_j := j$ for $j = 0,\ldots,n+k$, i.e., taking the nodes to be

$$0, 1, \ldots, n, \ldots, n+k.$$

Then all subintervals have the same length one and (12.43) becomes

$$(12.45)\begin{cases} \text{If } k = 1: \quad N_{i,1}(t) = \begin{cases} 1 & \text{for} \quad i \le t < i+1 \\ 0 & \text{for} \quad t < i \quad \text{and} \quad t \ge i+1; \end{cases} \\ \text{If } k \ge 2: \quad N_{i,k}(t) = \dfrac{t-i}{k-1}N_{i,k-1}(t) + \dfrac{i+k-t}{k-1}N_{i+1,k-1}(t). \end{cases}$$

A B-spline curve with normalized nodes is called a *uniform B-spline curve*.

Let us attempt to compute the value of $\boldsymbol{P}(t)$ for $t \in [t_r, t_{r+1})$ for the B-spline curve (12.42). In the sum (12.42) we only have to use those $N_{i,k}$ whose support contains the interval $[t_r, t_{r+1})$. These are the B-spline functions numbered $N_{r-k+1,k}$ with support $[t_{r-k+1}, t_{r+1}]$ until $N_{r,k}$ with support $[t_r, t_{r+k}]$. Thus $\boldsymbol{P}(t)$ can be computed as

$$(12.46) \qquad \boldsymbol{P}(t) = \sum_{i=r-k+1}^{r} N_{i,k}(t)\boldsymbol{d}_i, \quad t \in [t_r, t_{r+1}).$$

By construction $\sum\limits_{i=0}^{n} N_{i,k}(t) = 1$, and thus or $t \in [t_r, t_{r+1})$ we have $\sum\limits_{i=r-k+1}^{r} N_{i,k}(t) = 1$, i.e., for every $t \in [t_r, t_{r+1}]$ a total weight of one is distributed among the k de Boor points $\boldsymbol{d}_{r-k+1}, \ldots, \boldsymbol{d}_r$.

To compute $\boldsymbol{P}(t)$ it is advisable to use *de Boor's algorithm* which uses the auxiliary values

$$(12.47) \qquad \alpha_{i,k}^j := \frac{t - t_i}{t_{i+k-j} - t_i}$$

and

$$(12.48) \qquad \boldsymbol{D}_i^j := \alpha_{i,k}^j \boldsymbol{D}_i^{j-1} + (1 - \alpha_{i,k}^j)\boldsymbol{D}_{i-1}^{j-1} \quad \text{with} \quad \boldsymbol{D}_i^0 := \boldsymbol{d}_i$$

for $k \ge 2$, $j \ge 1$. For $k \ge 2$, (12.43) becomes in light of (12.47)

(12.49) $N_{i,k}(t) = \alpha^1_{i,k} N_{i,k-1}(t) + (1 - \alpha^1_{i+1,k}) N_{i+1,k-1}(t).$

Thus from (12.46)

$$P(t) = \sum_{i=r-k+1}^{r} [N_{i,k-1}(t)\alpha^1_{i,k} d_i + N_{i+1,k-1}(t)(1 - \alpha^1_{i+1,k}) d_i].$$

If we replace $i+1$ by i in the second term of each summand above and note that for $t \in [t_r, t_{r+1})$ it is sufficient to sum the $N_{i,k-1}$ terms from $i = r-(k-1)+1$ on, then

$$P(t) = \sum_{i=r-(k-1)+1}^{r} N_{i,k-1}(t)[\alpha^1_{i,k} d_i + (1 - \alpha^1_{i,k}) d_{i-1}].$$

Thus using (12.48) with $j = 1$ we obtain

$$P(t) = \sum_{i=r-(k-1)+1}^{r} N_{i,k-1}(t) D^1_i.$$

This process can easily be repeated. For $k-j+1$ instead of k, (12.49) becomes

$$N_{i,k-j+1}(t) = \alpha^1_{i,k-j+1} N_{i,k-j}(t) + (1 - \alpha^1_{i+1,k-j+1}) N_{i+1,k-j}(t).$$

Using $\alpha^1_{i,k-j+1} = \alpha^j_{i,k}$, from (12.47) we obtain

$$N_{i,k-j+1}(t) = \alpha^j_{i,k} N_{i,k-j}(t) + (1 - \alpha^j_{i+1,k}) N_{i+1,k-j}(t).$$

Thus as in the first step above:

$$\begin{aligned}
P(t) &= \sum_{i=r-(k-j)}^{r} N_{i,k-(j-1)}(t) D^{j-1}_i \\
&= \sum_{i=r-(k-j)+1}^{r} N_{i,k-j}(t)[\alpha^j_{i,k} D^{j-1}_i + (1 - \alpha^j_{i,k}) D^{j-1}_{i-1}] \\
&= \sum_{i=r-(k-j)+1}^{r} N_{i,k-j}(t) D^j_i.
\end{aligned}$$

This will stop when the sum contains only one term, i.e., when $r-(k-j)+1 = r$ or $j = k-1$. Then from (12.43)

$$P(t) = N_{r,k-j}(t) D^j_r = N_{r,1}(t) D^{k-1}_r = D^{k-1}_r.$$

De Boor's algorithm can be described in the following scheme:

	$j = 1$	$j = 2$	\ldots	$j = k - 2$	$j = k - 1$
$d_{r-k+1} = D_{r-k+1}^0$					
$d_{r-k+2} = D_{r-k+2}^0$	D_{r-k+2}^1				
$d_{r-k+3} = D_{r-k+3}^0$	D_{r-k+3}^1	D_{r-k+3}^2			
\vdots					
$d_{r-1} = D_{r-1}^0$	D_{r-1}^1	D_{r-1}^2	\ldots	D_{r-1}^{k-2}	
$d_r = D_r^0$	D_r^1	D_r^2	\ldots	D_r^{k-2}	$D_r^{k-1} = P(t)$ $t \in [t_r, t_{r+1})$

An entry D_i^j in column j is derived from the entry D_i^{j-1} to the left and the entry D_{i-1}^{j-1} one row up on the left according to (12.48) and (12.47). When progressing horizontally we multiply with $\alpha_{i,k}^j$, while going down is achieved by multiplying with $(1 - \alpha_{i,k}^j)$. Since the final formulas use only $\alpha_{i,k}^j$ for the same index k, one can omit this index k.

For the first de Boor point d_o we have $r - k + 1 = 0$ or $r = k - 1$; for d_n we have $r = n$. Thus de Boor's algorithm allows us to compute $P(t)$ for $t \in [t_r, t_{r+1})$ for all $k - 1 \leq r \leq n$, i.e., for all $t_{k-1} \leq t < t_{n+1}$. Since we always will need k de Boor points for a B-spline of order k, we must have $n + 1 \geq k$.
For $j = 1$ de Boor's algorithm uses $\alpha_{r-k+2,k}^1$ until $\alpha_{r,k}^1$, or by (12.47) the nodes

$$t_{r-k+2}, \ldots, t_r, t_{r+1}, \ldots, t_{r+k-1}.$$

Thus for $r = k - 1$ until $r = n$ this list encompasses $t_1, \ldots, t_{k-1}, t_k, \ldots, t_n, t_{n+1}, \ldots, t_{n+k-1}$. And the end nodes t_0 and t_{n+k} will never be used.

We can thus generate open as well as closed B-spline curves. To ensure smoothness we must stipulate $k \geq 3$ and thus the local curves will be generated from at least 3 de Boor points.

Open B-Spline-Curves

As with Bézier curves, an open B-spline curve shall pass through the first and last de Boor point d_0 and d_n and the line segments $d_0 d_1$ and $d_{n-1} d_n$ shall be tangent to the curve. This can be achieved by taking the first (and last) k nodes identical. Thus for an open B-spline curve, the nodes are

$$t_0 = t_1 = \ldots = t_{k-1} < t_k < \ldots < t_n < t_{n+1} = t_{n+2} = \ldots = t_{n+k}.$$

The interval (t_{k-1}, t_{n+1}) is then divided into $n - k + 2$ nontrivial subintervals. A B-spline curve of maximal order $k = n+1$ is a Bézier curve in $[t_n, t_{n+1}]$. For a uniform and open B-spline curve of order $k \geq 3$ we can generate the normalized nodes by setting

$$(12.50) \quad \begin{cases} t_j = k - 1 & \text{for} \quad j = 0, \ldots, k-1, \\ t_j = j & \text{for} \quad j = k, \ldots, n, \quad \text{and} \\ t_j = n + 1 & \text{for} \quad j = n+1, \ldots, n+k. \end{cases}$$

In case that one of the denominators of $\alpha_{i,k}^j$ in (12.47) is zero, i.e., if some nodes coincide, we shall set $\alpha_{i,k}^j = 0$ in de Boor's algorithm.

Closed B-Spline Curves

A closed B-spline curve will occur if

$$\boldsymbol{d}_{n+1} := \boldsymbol{d}_0, \quad \boldsymbol{d}_{n+2} := \boldsymbol{d}_1, \ldots,$$

i.e., if the de Boor points are cyclically repeated. De Boor's algorithm can then be performed for $(n+1)$ k-tuples of de Boor points $(\boldsymbol{d}_{r-k+1}, \ldots, \boldsymbol{d}_r)$ for $r = k - 1, \ldots, n + k - 1$:

$$(\boldsymbol{d}_0, \boldsymbol{d}_1, \ldots, \boldsymbol{d}_{k-1}),$$
$$\vdots$$
$$(\boldsymbol{d}_{n-k+1}, \ldots, \boldsymbol{d}_{n-1}, \boldsymbol{d}_n),$$
$$(\boldsymbol{d}_{n-k+2}, \ldots, \boldsymbol{d}_n, \boldsymbol{d}_{n+1}),$$
$$\vdots$$
$$(\boldsymbol{d}_n, \boldsymbol{d}_{n+1}, \ldots, \boldsymbol{d}_{n+k-1}).$$

For this we need to know the nodes $t_1, \ldots, t_{n+k-1}, t_{n+k}, \ldots, t_{n+2k-2}$ and these nodes must all be distinct.

For a closed B-spline curve we hence must use $k - 1$ additional points $\boldsymbol{d}_{n+1} = \boldsymbol{d}_0, \quad \boldsymbol{d}_{n+2} = \boldsymbol{d}_1, \quad \ldots, \quad \boldsymbol{d}_{n+k-1} = \boldsymbol{d}_{k-2}$ besides the $(n + 1)$ de Boor points $\boldsymbol{d}_0, \ldots, \boldsymbol{d}_n$.

The points $\boldsymbol{P}(t)$ with $t \in [t_{k-1}, t_{n+k}]$ generate the closed B-spline curve. The number of subintervals used is $n+1$ independent of the order k.

A uniform closed B-spline curve of order k has the normalized nodes

$$(12.51) \qquad t_j = j \quad \text{for} \quad j = 0, \ldots, n+2k-1.$$

The first and last node are again never used.

ALGORITHM 12.6 (*Uniform B-Spline-Curves*).

Given: $n+1$ de Boor points \boldsymbol{d}_i, $i = 0,\ldots,n$, with $n \geq 2$, $\boldsymbol{d}_i \in \mathbb{R}^2$ or $\boldsymbol{d}_i \in \mathbb{R}^3$. An order k, with $3 \leq k \leq n+1$.
The desired curve may be open or closed.

Task: Find $\boldsymbol{P}(t)$ on the uniform B-spline curve.

1^{st} step: (In case of an open curve)

Generate the nodes in (12.50). The interval $I = [k-1, n+1]$ is subdivided into $n-k+2$ subintervals.

(In case of a closed curve)

Generate the nodes in (12.51). The interval $I = [k-1, n+k]$ is subdivided into $n+1$ subintervals. Generate the remaining $k-1$ de Boor points $\boldsymbol{d}_{n+1} = \boldsymbol{d}_0$, $\boldsymbol{d}_{n+2} = \boldsymbol{d}_1$, \ldots, $\boldsymbol{d}_{n+k-1} = \boldsymbol{d}_{k-2}$.

2^{nd} step: For $t \in I$ determine the subinterval $[r, r+1]$ with $r \leq t < r+1$.

3^{rd} step: For $j = 1,\ldots,k-1$ do:
For $i = (r-k+j+1),\ldots,r$ do:

$$\alpha_i^j = \frac{t-t_i}{t_{i+k-j}-t_i}, \qquad (12.47)$$

$$\boldsymbol{D}_i^j = \alpha_i^j \boldsymbol{D}_i^{j-1} + (1-\alpha_i^j)\boldsymbol{D}_{i-1}^{j-1}, \quad (12.48)$$

$$\boldsymbol{D}_i^0 = \boldsymbol{d}_i.$$

4^{th} step: Set $\boldsymbol{P}(t) = \boldsymbol{D}_r^{k-1}$.

In de Boor's algorithm, the shape of the curve segment for the subinterval $[t_r, t_{r+1}]$ is only affected by k de Boor points. Conversely one de Boor point \boldsymbol{d}_i influences $\boldsymbol{P}(t)$ only in the k subintervals $[t_i, t_{i+1}]$ to $[t_{i+k-1}, t_{i+k}]$.
Thus changing one control point only affects the local behavior of the B-spline curve. This effect is less widely spread the smaller the order k has been chosen. For low orders k, the B-spline curve nearly follows the polygon of the control points, so that its shape can be judged easily from the \boldsymbol{d}_i. To use cubic B-spline curves of order 4 which are twice continuously differentiable seems to be most advantageous here.
In contrast to B-spline curves, the number $n+1$ of Bézier points \boldsymbol{b}_i is linked to the degree of the Bernstein polynomials B_i^n used. Moreover cubic Bézier spline curves depend on the given Bézier weights \boldsymbol{d}_i whose influence on the resulting curve is not easily assessed.
Another strong point of B-spline curves is their ease in generating closed curves.

12.4.2 B-Spline-Surfaces

Similar to Bézier surfaces in section 12.3.2, one can use B-spline functions $N_{i,k}(v)$ and $N_{j,k}(w)$, both of the same order k, together with $(m+1) \cdot (n+1)$ control points $\boldsymbol{d}_{ij} \in \mathbb{R}^3$ to generate B-spline surfaces:

$$(12.52) \quad \boldsymbol{P}(v,w) = \sum_{i=0}^{m} \sum_{j=0}^{n} N_{i,k}(v) N_{j,k}(w) \boldsymbol{d}_{ij} \qquad \text{for } v \in I_v, w \in I_w.$$

Here the de Boor points \boldsymbol{d}_{ij} form the vertices of the *de Boor polyhedron*, see the Bézier polyhedron in Figure 12.5.

For $w = w^* = \text{const}$ we can obtain a v-curve of the B-spline surface

$$(12.53) \quad \begin{aligned} \boldsymbol{P}(v,w^*) &= \sum_{i=0}^{m} N_{i,k}(v) \left(\sum_{j=0}^{n} N_{j,k}(w^*) \boldsymbol{d}_{ij} \right) \\ &= \sum_{i=0}^{m} N_{i,k}(v) \boldsymbol{d}_i(w^*), \quad v \in I_v. \end{aligned}$$

This is a B-spline curve of order k for the $m+1$ de Boor points

$$(12.54) \quad \boldsymbol{d}_i(w^*) = \sum_{j=0}^{n} N_{j,k}(w^*) \boldsymbol{d}_{ij}, \quad i = 0, \ldots, m.$$

Similarly if $v = v^* = \text{const}$ we obtain a w-curve

$$(12.55) \quad \begin{aligned} \boldsymbol{P}(v^*,w) &= \sum_{j=0}^{n} N_{j,k}(w) \left(\sum_{i=0}^{m} N_{i,k}(v^*) \boldsymbol{d}_{ij} \right) \\ &= \sum_{j=0}^{n} N_{j,k}(w) \boldsymbol{d}_j(v^*), \quad w \in I_w. \end{aligned}$$

This is a B-spline curve of order k for the $n+1$ de Boor points

$$(12.56) \quad \boldsymbol{d}_j(v^*) = \sum_{i=0}^{m} N_{i,k}(v^*) \boldsymbol{d}_{ij}, \quad j = 0, \ldots, n.$$

As with B-spline curves, refer to Algorithm 12.6, we assume that $3 \le k \le \min(m+1, n+1)$. To achieve the minimal order of $k = 3$, we must start with a de Boor polyhedron with at least 9 vertices.

From now on we shall study uniform open B-spline surfaces whose v- and w-curves form uniform and open B-spline curves of order k.

According to (12.50), the normalized nodes for $N_{i,k}(v)$ are

$$(12.57) \quad \begin{cases} v_\ell = k - 1 & \text{for} \quad \ell = 0, \ldots, k - 1, \\ v_\ell = \ell & \text{for} \quad \ell = k, \ldots, m, \quad \text{and} \\ v_\ell = m + 1 & \text{for} \quad \ell = m + 1, \ldots, m + k; \end{cases}$$

while those for $N_{j,k}(w)$ are

$$(12.58) \quad \begin{cases} w_\ell = k - 1 & \text{for} \quad \ell = 0, \ldots, k - 1, \\ w_\ell = \ell & \text{for} \quad \ell = k, \ldots, n, \quad \text{and} \\ w_\ell = n + 1 & \text{for} \quad \ell = n + 1, \ldots, n + k. \end{cases}$$

The parameter intervals for v and w are $I_v = [k-1, m+1]$ and $I_w = [k-1, n+1]$, respectively.

Next we shall describe how to compute $\boldsymbol{P}(v^*, w^*)$ for $v^* \in I_v$ and $w^* \in I_w$: These are integers $r, s \in \mathbb{N}$ so that

$$k - 1 \leq r \leq v^* < r + 1 \leq m+1,$$

$$k - 1 \leq s \leq w^* < s + 1 \leq n+1.$$

From (12.46) $v^* \in [r, r-1)$, thus using the v-curve description of (12.53) we have

$$(12.59) \qquad \boldsymbol{P}(v^*, w^*) = \sum_{i=r-k+1}^{r} N_{i,k}(v^*) \boldsymbol{d}_i(w^*) \,.$$

Hence we need k de Boor points (12.54), which can be found as

$$(12.60) \qquad \boldsymbol{d}_i(w^*) = \sum_{j=s-k+1}^{s} N_{j,k}(w^*) \boldsymbol{d}_{ij}, \quad i = r - k + 1, \ldots, r \,,$$

since $w^* \in [s, s + 1]$. De Boor's algorithm 12.6 can be used to find these k de Boor points as well as the value of $\boldsymbol{P}(v^*, w^*)$. Alternately, $\boldsymbol{P}(v^*, w^*)$ can also be found as a point of the w-curve:

$$(12.61) \qquad \boldsymbol{P}(v^*, w^*) = \sum_{j=s-k+1}^{s} N_{j,k}(w^*) \boldsymbol{d}_j(v^*) \,, \quad \text{where}$$

$$(12.62) \qquad \boldsymbol{d}(v^*) = \sum_{i=r-k+1}^{r} N_{i,k}(v^*) \boldsymbol{d}_{ij} \quad \text{for } j = s - k + 1, \ldots, s.$$

In either approach to find $\boldsymbol{P}(v^*, w^*)$, one needs to apply de Boor's algorithm $k + 1$ times.

ALGORITHM 12.7 (*Uniform open B-spline surfaces*).

Given: $(m+1) \cdot (n+1)$ de Boor points $d_{ij} \in \mathbb{R}^3$, $i = 0, \ldots, m$, $m \geq 2$, $j = 0, \ldots, n$, $n \geq 2$. An order k with $3 \leq k \leq \min(m+1, n+1)$.

Task: Find $\boldsymbol{P}(v^*, w^*)$ of the uniform open B-spline surface.

1^{st} step: Compute the nodes (12.57) and (12.58). Set
$I_v = [k-1, m+1]$, $I_w = [k-1, n+1]$.

2^{nd} step: For the given $v^* \in I_v$ determine the subinterval with $v^* \in [r, r+1]$, similarly for $w^* \in [s, s+1]$.

Choose between the equivalent third steps: $3(v)$ or $3(w)$:

3^{rd} step (v): Compute $\boldsymbol{P}(v^*, w^*)$ as a point of the v-curve $w = w^*$ according to (12.60) and (12.59):

$3(v).1$: Form the k de Boor points $\boldsymbol{d}_i(w^*)$ for
$i = (r - k + 1), \ldots, r$:
For $q = 1, \ldots, k-1$ do:
For $p = (s - k + q + 1), \ldots, s$ do:

$$\alpha_p^q = \frac{w^* - w_p}{w_{p+k-q} - w_p},$$

$$\boldsymbol{D}_p^q = \alpha_p^q \boldsymbol{D}_p^{q-1} + (1 - \alpha_p^q) \boldsymbol{D}_{p-1}^{q-1},$$

$$\boldsymbol{D}_p^0 = \boldsymbol{d}_{ip}.$$

$$\boldsymbol{d}_i(w^*) = \boldsymbol{D}_s^{k-1}.$$

$3(v).2$: Compute $\boldsymbol{P}(v^*, w^*)$:
For $q = 1, \ldots, k-1$ do:
For $p = (r - k + q + 1), \ldots, r$ do:

$$\alpha_p^q = \frac{v^* - v_p}{v_{p+k-q} - v_p},$$

$$\boldsymbol{D}_p^q = \alpha_p^q \boldsymbol{D}_p^{q-1} + (1 - \alpha_p^q) \boldsymbol{D}_{p-1}^{q-1},$$

$$\boldsymbol{D}_p^0 = \boldsymbol{d}_p(w^*).$$

$$\boldsymbol{P}(v^*, w^*) = \boldsymbol{D}_r^{k-1}.$$

3^{rd} step (w): Compute $\boldsymbol{P}(v^*, w^*)$ as a point of the w-curve $v = v^*$ according to (12.62) and (12.61).

3(w).1: Form the k de Boor points for $j = s - k + 1, \ldots, s$:
For $q = 1, \ldots, k - 1$ do:
For $p = (r - k + q + 1), \ldots, r$ do:

$$\alpha_p^q = \frac{v^* - v_p}{v_{p+k-q} - v_p},$$

$$D_p^q = \alpha_p^q D_p^{q-1} + (1 - \alpha_p^q) D_{p-1}^{q-1},$$

$$D_p^0 = d_{pj}.$$

$$d_j(v^*) = D_r^{k-1}.$$

3(w).2: Compute $P(v^*, w^*)$:
For $q = 1, \ldots, k - 1$ do:
For $p = (s - k + q + 1), \ldots, s$ do:

$$\alpha_p^q = \frac{w^* - w_p}{w_{p+k-q} - w_p},$$

$$D_p^q = \alpha_p^q D_p^{q-1} + (1 - \alpha_p^q) D_{p-1}^{q-1}.$$

$$D_p^0 = d_p(v^*).$$

$$P(v^*, w^*) = D_s^{k-1}.$$

If one wants to find a number of values $P(v^*, w^*)$ on the same v-curve, then we advise to first determine all the de Boor points in (12.54) in order to avoid recomputing them in algorithm 12.7. If $v^* \in [r, r + 1)$ we only need to compute the de Boor points $d_{r-k+1}(w^*), \ldots, d_r(w^*)$ in 3(v).1 which are needed in 3(v).2. One should proceed analogously with v and w interchanged if a large number of points is to be computed on the same w-curve.

As for B-spline curves, changes in the de Boor points d_{ij} only affect the local shape of a uniform B-spline surface. A value $P(v^*, w^*)$ for $v^* \in [r, r + 1)$, $w^* \in [s, s+1)$ can be computed from the de Boor points d_{ij}, $i = r-k+1, \ldots, r$, $j = s-k+1, \ldots, s$ using formulas (12.59) to (12.62). Thus the values of $P(v^*, w^*)$ for all $r \le v^* < r+1$, $s \le w^* < s+1$ are only affected by the just mentioned $k \cdot k$ de Boor points. Conversely one de Boor point d_{ij} influences the value of $P(v, w)$ only on the v-intervals $[i, i + 1]$ to $[i + k - 1, i + k]$ and on the w-intervals $[j, j + 1]$ to $[j+k-1, j+k]$. Thus one point d_{ij} has a bearing exactly on the associated $k \cdot k$ surface segments.

A tube-like uniform B-spline surface can be generated by for example choosing open uniform B-spline curves as v-curves while making the w-curves closed uniform B-spline curves. The de Boor points of the closed w-curves then become

$$d_{i,n+1} := d_{i,0}, \;\; d_{i,n+2} := d_{i,1}, \;\; \ldots, \;\; d_{i,n+k-1} := d_{i,k-2}.$$

for $i = 0, \ldots, m$. Then analogous to (12.51) the B-spline functions $N_{j,k}(w)$ must be defined for the nodes $w_\ell = \ell$ for $\ell = 0, \ldots, n+2k-1$.

Bibliography for Section 12.4

[BART87]; [BÖHM84]; [HOSC89].

12.5 Hints

Some hints for this chapter are combined with those for chapters 10, 11 and 13 in section 10.3. For Bézier splines, the comments at the end of sections 12.3.1 and 12.3.2 should also be taken into consideration.
When dealing with multidimensional interpolation or with multidimensional approximation, we suggest to consult section 9.8 and [NIEM91].

Bézier and B-Spline curves

Both types of curves are parametric and do not interpolate. For Bézier splines we have described an interpolating modification in section 12.3.3 (modified interpolating cubic Bézier splines). The table in section 10.3 contains hints for using cubic Bézier splines.
Uniform B-spline curves can be handled more easily than Bézier spline curves, since one only has to prescribe the de Boor polygon which determines the path of the curve. Closed curves should always be generated via B-spline curves.

Surface representations via splines

When constructing smooth surfaces, the bicubic splines of section 12.1 can only be used over a rectangular grid G with monotonically ordered nodes in both the x- and y-direction. This limits their use considerably.

The triples $(x_i, y_i, f_i = f(x_i, y_i))$ can be prescribed in any order for the interpolating surface splines of section 12.2. This is a great advantage in practice. However, we must assume that for each point (x_i, y_i) there is only one functional value f_i. The order of the derivatives used for surface splines can be chosen freely. Computational tests have shown that in general the derivatives involved should be of order 3, 4 or 5. With higher derivatives and a larger

number of interpolation points, the condition number of the system matrix will worsen for the linear system that has to be solved for the spline coefficients.

As Bézier splines are given parametrically, they require no ordering of the interpolation points, see section 12.3. They can therefore also be used for representing closed surfaces with double points. Besides, they can produce curves with a corner and surfaces with a crease. And local changes can be incorporated with ease.

The advantages of B-spline curves also hold for B-spline surfaces; they are generally preferred over Bézier spline surfaces, since they are globally continously differentiable, provided the order $k \geq 3$.

EXAMPLE:

We shall again try to represent the function

$$
\begin{aligned}
T(x,y) \;=\; & \frac{3}{4}e^{-\frac{1}{4}\left[(9x-2)^2 + (9y-2)^2\right]} + \frac{3}{4}e^{-\left[\frac{1}{49}(9x+1)^2 + \frac{1}{10}(9y+1)\right]} \\
& -\frac{1}{5}e^{\left[(9x-4)^2 + (9y-7)^2\right]} + \frac{1}{2}e^{-\frac{1}{4}\left[(9x-7)^2 + (9y-3)^2\right]}
\end{aligned}
$$

from section 9.8 approximately via interpolating surface splines and B–splines for $0 \leq x, y \leq 1.1$. To compute the various spline surfaces we select one set of 100 random points $(x, y, T(x, y))$, with the corners $(0,0,T(0,0))$, $(0,1.1,T(0,1.1))$, $(1.1,0,T(1.1,0))$, $(1.1,1.1,T(1.1,1.1))$ amongst them. All spline surfaces will be constructed from the same 100 points. The equivalent B-spline surface will be constructed using the equidistant grid generated by $x = 0$, 0.11, 0.22, ..., 1.1, and $y = 0$, 0.11, 0.22, ..., 1.1. A plot of the original surface determined by T was shown in Figure 9.3 (a) in section 9.8. There various interpolating surfaces were generated from the same set of nodes using Shepard interpolation.

Figure 12.8: Two–dimensional interpolating surface spline with derivative order $m = 2$.

Figure 12.9: Two–dimensional interpolating surface spline with derivative order $m = 3$.

Figure 12.10: Two–dimensional interpolating surface spline with derivative
order $m = 4$.

The surface spline of derivative order $m = 5$ does not show any visual differ-
ences from the one with derivative order 4 or from the plot for the original
function T in Figure 9.3 (a) of section 9.8.
Our last plot shows the B–spline surface of T for an equidistant set of nodes
$x = 0,\ 0.11,\ 0.22, ...,\ 1.1$, and $y = 0,\ 0.11,\ 0.22, ...,\ 1.1$.

Figure 12.11: B–spline surface of derivative order $m = 3$.

Remark: If one uses the same set of random nodes that was used for generating the spline surfaces above and for Shepard interpolation earlier in section 9.8, the B–spline surface will turn out more jagged. Since B–splines are approximating functions, one often can obtain better results with them by cutting down on the number of nodes while adding a few extra nodes near and on the boundary of the domain.

B–splines can not only be used to approximate surfaces given by analytic functions $z = F(x,y)$, but also surfaces that are given by analytic parametrizations. Specifically, surfaces with one family parameter curves such as cylinders, tori and combinations thereof can easily be represented as B–splines. B–spline surfaces on the other hand do not need the defining function F to be represented analytically; the only information needed for B–spline surfaces are the defining de Boor polytopes.

Further details for non–uniform rational B–splines can be found in [PIEGL95].

Bibliography for Chapter 12

[BART87]; [BEZI72]; [BÖHM84]; [BOXB79]; [DONG79]; [GORD69];
[GREV69]; [HOSC89]; [MEING79]; [MICH79]; [REIN71]; [RITT69];
[SHAH70]; [SPÄT71]; [SPÄT73]; [SPÄT86]; [STEU79].

Chapter 13

Akima and Renner Subsplines

As in the case of interpolating cubic splines, the interpolating *Akima subsplines* and *Renner subsplines* are composed of cubic polynomials on subintervals. The main difference, however, consists in the fact that these subsplines need only be once continuously differentiable in contrast to the cubic splines of chapter 10, which were constructed to be twice continuously differentiable everywhere. Deviating slightly from the original papers of Akima [AKIM70] and Renner [RENN81], [RENN82], we present useful modifications that follow the suggestions of R. Wodicka [WODI91]. One of them is to represent corners as corners, i.e., a continuous derivative is not required everywhere. If corners are not desired for the curve, they can be avoided by inserting additional nodes. An essential advantage of the Akima and Renner subsplines lies in the fact that no linear system of equations must be solved to form them.

13.1 Akima Subsplines

For an explicitly known or an unknown function $f \in C^1[a, b]$ assume that the functional values $y_i = f(x_i)$ are known at the $n + 1$ monotonically ordered nodes

$$a = x_0 < x_1 < \ldots < x_n = b.$$

Assume $n \geq 4$. A smooth curve is to be laid through the $n + 1$ points $P_i = (x_i, y_i) \in \mathbb{R}^2$. This shall be represented by a spline function S, defined for each subinterval:

$$S(x) \equiv S_i(x) = a_i + b_i(x - x_i) + c_i(x - x_i)^2 + d_i(x - x_i)^3,$$
$$\text{for } x \in [x_i, x_{i+1}], \text{ and } i = 0, \ldots, n - 1.$$

Thus S is given by a polynomial of at most third degree in each subinterval. The coefficients a_i, b_i, c_i, d_i can be calculated for $i = 0, \ldots, n - 1$, provided that at the nodes x_i and x_{i+1} we know the functional values

$$S_i(x_i) = y_i \quad , \qquad S_i(x_{i+1}) = y_{i+1},$$

and the derivatives (or the slope of the tangents)

$$S_i'(x_i) = t_i \quad , \qquad S_i'(x_{i+1}) = t_{i+1}.$$

Since, in general, the function f is not explicitly known, it is necessary to determine the required slopes numerically first. In order to find the slope t_i at P_i, Akima uses only the adjacent points $P_{i-2}, P_{i-1}, P_{i+1}, P_{i+2}$ as one might if one drew the curve free-hand through the set of given points. Since one would connect three (or more) collinear points by a straight line, the spline S ought to do likewise.
At the end of a straight segment the slope of the tangent is determined for the adjoining bent curve. By not demanding continuity for the second derivative of S we can avoid the overshoot or "ripple" effect of cubic splines when joining a curved piece of the curve with a straight line piece.

If a point P_i lies between two straight linear segments of the curve with different slopes then the curve ought to have a corner at P_i. This is inconsistent with demanding continuous differentiability. Akima excluded drawing such corners and favored continuously differentiable curves. We, however, want to be able to trace corners. Therefore it is necessary to assign each node a left and a right hand slope.

Using the chordal slopes

$$m_i = \frac{y_{i+1} - y_i}{x_{i+1} - x_i},$$

the Akima slope formula is

$$t_i = \frac{|m_{i+1} - m_i| m_{i-1} + |m_{i-1} - m_{i-2}| m_i}{|m_{i+1} - m_i| + |m_{i-1} - m_{i-2}|} .$$

If the denominator does not vanish, then

$$t_i = m_{i-1} \quad \text{for} \quad m_{i-2} = m_{i-1}, \quad \text{and}$$
$$t_i = m_i \quad \text{for} \quad m_i = m_{i+1} .$$

If $m_{i-1} \neq m_i$ and the denominator vanishes, then there is a corner at P_i.

For t_0, t_1, t_{n-1}, and t_n the slope formula is only usable if the four additional slopes m_{-2}, m_{-1}, m_n, and m_{n+1} are supplied.

ALGORITHM 13.1 (*Akima Subsplines*).

Given: $(x_i, y_i), i = 0, \ldots, n, \quad n \geq 4$, with $x_i < x_{i+1}$ for $i = 0, \ldots, n-1$, non-periodic or periodic $(y_n = y_0)$.

Task: Find the Akima subspline coefficients a_i, b_i, c_i, d_i for $i = 0, \ldots, n-1$.

1) Calculate the slopes $m_i = (y_{i+1} - y_i)/(x_{i+1} - x_i)$ for each $i = 0, \ldots, n-1$.

2) Prescribe additional slopes in the

 non-periodic case:
 $$\begin{aligned}
 m_{-2} &:= 3m_0 - 2m_1, \\
 m_{-1} &:= 2m_0 - m_1, \\
 m_n &:= 2m_{n-1} - m_{n-2}, \\
 m_{n+1} &:= 3m_{n-1} - 2m_{n-2}, \text{ and in the}
 \end{aligned}$$

 periodic case:
 $$\begin{aligned}
 m_{-2} &:= m_{n-2}, \\
 m_{-1} &:= m_{n-1}, \\
 m_n &:= m_0, \\
 m_{n+1} &:= m_1 \;.
 \end{aligned}$$

3) Compute the left and the right hand slopes t_i^L and t_i^R for each $i = 0, \ldots, n$.

 (The values of t_0^L, t_n^R are calculated, but not used in 4).

 Set $NE := |m_{i+1} - m_i| + |m_{i-1} - m_{i-2}|$.

 If $NE > 0$: *Set*
 $$\begin{aligned}
 \alpha_i &:= |m_{i-1} - m_{i-2}|/NE, \\
 t_i^L &:= (1 - \alpha_i)m_{i-1} + \alpha_i m_i \\
 &= m_{i-1} + \alpha_i(m_i - m_{i-1}), \\
 t_i^R &:= t_i^L.
 \end{aligned}$$

 If $NE = 0$: *Set*
 $$\begin{aligned}
 t_i^L &:= m_{i-1}, \\
 t_i^R &:= m_i.
 \end{aligned}$$

4) For $i = 0, \ldots, n-1$ calculate the coefficients a_i, b_i, c_i, d_i:
 $$\begin{aligned}
 a_i &:= y_i, \\
 b_i &:= t_i^R, \\
 h_i &:= x_{i+1} - x_i, \\
 c_i &:= \frac{1}{h_i}\left(3m_i - 2t_i^R - t_{i+1}^L\right), \\
 d_i &:= \frac{1}{h_i^2}\left(t_i^R + t_{i+1}^L - 2m_i\right).
 \end{aligned}$$

If the nodes cannot be arranged in a monotonic order then one must proceed parametrically analogously as with splines, see section 10.1.3. There is, however, a disadvantage: The approximation of arc length by the chordal length is often not very good. Renner subsplines generally will approximate arc length better.

13.2 Renner Subsplines

For $n + 1$ points $P_i = (x_i, y_i) \in \mathbb{R}^2$ or $P_i = (x_i, y_i, z_i) \in \mathbb{R}^3$ with $n \geq 4$ and $P_{i+1} \neq P_i$ for all $i = 0, \ldots, n-1$, we want to find a smooth curve that connects the points in their given order. The curve is closed if $P_n = P_0$. The curve is to be composed of piecewise cubic vectorial polynomials:

$$(13.1) \qquad \boldsymbol{S}_i(t) = \boldsymbol{a}_i + t\boldsymbol{b}_i + t^2\boldsymbol{c}_i + t^3\boldsymbol{d}_i, \quad t \in [0, T_i], \; i = 0, \ldots, n - 1.$$

Here $\boldsymbol{a}_i, \boldsymbol{b}_i, \boldsymbol{c}_i, \boldsymbol{d}_i \in \mathbb{R}^m$ for $m = 2$ or $m = 3$. The segment of the curve joining P_i and P_{i+1} is consequently given by a polynomial of at most third degree. The coefficients $\boldsymbol{a}_i, \boldsymbol{b}_i, \boldsymbol{c}_i, \boldsymbol{d}_i \in \mathbb{R}^m$ must be calculated for $i = 0, \ldots, n-1$, as well as the lengths T_i of the parameter intervals. For this purpose, we must know the points

$$\boldsymbol{S}_i(0) = P_i \quad , \qquad \boldsymbol{S}_i(T_i) = P_{i+1}$$

and the unit tangent vectors

$$\boldsymbol{S}'_i(0) = \boldsymbol{t}_i \quad , \qquad \boldsymbol{S}'_i(T_i) = \boldsymbol{t}_{i+1}.$$

If we demand in addition that $\|\boldsymbol{S}'(T_i/2)\| = 1$, then the T_i can be easily computed:

According to Simpson's rule, see chapter 15.3.2, the length of the curve segment between P_i and P_{i+1} can be approximated due to the fact that all vectors involved are unit vectors as

$$\int_0^{T_i} \|\boldsymbol{S}'_i(t)\| dt \approx \frac{T_i}{6} \left(\|\boldsymbol{S}'_i(0)\| + 4\|\boldsymbol{S}'_i(T_i/2)\| + \|\boldsymbol{S}'_i(T_i)\| \right) = T_i \; .$$

For this reason, t is an approximation of the arc length parametrization .

If the tangents to the curve at P_i are not known, the unit tangent vectors t_i must first be determined. As for Akima curves, we again want to be able to reproduce straight line segments and corners. For the chordal vectors

$$s_i = P_{i+1} - P_i \neq 0, \text{ normalized to } s_i^0 = s_i/\|s_i\|,$$

the formula of Renner describes the non-normalized tangent vector as

$$
\begin{aligned}
t_i &= (1 - \alpha_i)s_{i-1} + \alpha_i s_i = s_{i-1} + \alpha_i(s_i - s_{i-1}) \quad \text{with} \\
\alpha_i &= \frac{A(s_{i-2}^0, s_{i-1}^0)}{A(s_{i-2}^0, s_{i-1}^0) + A(s_i^0, s_{i+1}^0)}.
\end{aligned}
$$

Here $A(s, t)$ measures the area of the parallelogram spanned by the vectors s and $t \in \mathbb{R}^m$. This area can be most easily computed for unit vectors $s, t \in \mathbb{R}^3$ using the identity

$$A^2(s,t) = \|s \times t\|^2 = \|s\|^2\|t\|^2 - (s^T t)^2 = 1 - (s^T t)^2,$$

where $s \times t$ and $\|s\|$ denote the cross product and the euclidean vector norm respectively. This formula further simplifies for $m = 2$ and unit vectors s and $t \in \mathbb{R}^2$ to

$$A(s,t) = \sqrt{1 - (s^T t)^2} = |\det(s,t)|.$$

These two formulas are meaningful for all points P_i, which possess two adjacent points to the "left" and to the "right". This will be the case for all points on a closed curve.

If the denominator in α_i does not vanish, we have

$$
\begin{aligned}
t_i &= s_{i-1} \quad, \quad \text{if} \quad P_{i-2}, P_{i-1}, P_i \quad \text{are collinear, and} \\
t_i &= s_i \quad\quad, \quad \text{if} \quad P_i, P_{i+1}, P_{i+2} \quad \text{are collinear.}
\end{aligned}
$$

If the denominator is zero, we have a corner if $A(s_{i-1}^0, s_i^0) > 0$. In order to represent corners with a Renner subspline, we shall assign each point P_i a left and a right unit tangent vector t_i^L and t_i^R, respectively. If we desire to have a corner at P_i we shall set $t_i^L = s_{i-1}^0$ and $t_i^R = s_i^0$. Otherwise we set $t_i^L = t_i^R = t_i$

ALGORITHM 13.2 (*Renner subsplines*).

Given: $n + 1$ points $P_i = (x_i, y_i)$ or $P_i = (x_i, y_i, z_i)$ for $i = 0, \ldots, n$ with
$\qquad n \geq 4$, and $P_i \neq P_{i+1}$ for $i = 0, \ldots, n - 1$.

Task: Find the Renner subspline in the form (13.1).

1) For $i = 0, \ldots, n - 1$ find the chordal vectors:
 $s_i := P_{i+1} - P_i \neq \mathbf{0}$.

2) Determine additional chordal vectors:

 2.1 For closed curves with $(P_n = P_0)$:
 $$
 \begin{aligned}
 s_{-2} &:= s_{n-2}, \\
 s_{-1} &:= s_{n-1}, \\
 s_n &:= s_0, \\
 s_{n+1} &:= s_1.
 \end{aligned}
 $$

 2.2 For not closed curves with $(P_n \neq P_0)$:
 $$
 \begin{aligned}
 s_{-2} &:= 3s_0 - 2s_1 \\
 s_{-1} &:= 2s_0 - s_1 \\
 s_n &:= 2s_{n-1} - s_{n-2} \\
 s_{n+1} &:= 3s_{n-1} - 2s_{n-2}
 \end{aligned}
 $$

3) Normalize the chordal vectors s_i for $i = -2, -1, \ldots . n + 1$:
 If $\|s_i\| > 0$, set $s_i^0 := s_i / \|s_i\|$;

 if $\|s_i\| = 0$, set $s_i^0 := \mathbf{0}$.

 (Note if the points P_0, P_1, P_2 or P_{n-2}, P_{n-1}, P_n are collinear, then in
 2.2 the vectors s_{-2}, s_{-1} or s_n, s_{n+1} might be zero vectors.)

4) Determine the left and right unit tangent vectors for $i = 0, \ldots, n$:
 Set $\mathrm{NE} := \sqrt{1 - (s_{i-2}^{0T} s_{i-1}^0)^2} + \sqrt{1 - (s_i^{0T} s_{i+1}^0)^2}$
 If $\mathrm{NE} > 0$:
 $\alpha_i := \sqrt{1 - (s_{i-2}^{0T} s_{i-1}^0)^2} / \mathrm{NE}$
 $t_i^L := s_{i-1} + \alpha_i(s_i - s_{i-1})$
 $t_i^L := t_i^L / \|t_i^L\|$
 $t_i^R := t_i^L$
 If $\mathrm{NE} = 0$:
 $t_i^L := s_{i-1}^0$
 $t_i^R := s_i^0$
 (t_0^L and t_n^R will not be used.)

5) For each $i = 0, \ldots, n-1$ compute the length T_i of the parameter interval:

$A := 16 - ||t_i^R + t_{i+1}^L||^2$

$B := 6s_i^T \cdot (t_i^R + t_{i+1}^L)$

$C := 36||s_i||^2$

$T_i := (-B + \sqrt{B^2 + AC})/A$

6) Compute the coefficients for the Renner subspline for $i = 0, \ldots, n-1$:

$a_i := P_i$

$b_i := t_i^R$

$c_i := \frac{3}{T_i^2} s_i - \frac{1}{T_i}(2t_i^R + t_{i+1}^L)$

$d_i := \frac{1}{T_i^2}(t_i^R + t_{i+1}^L) - \frac{2}{T_i^3} s_i.$

13.3 Rounding of Corners with Akima and Renner Splines

For Akima and Renner curves, a corner is produced at P_i if the set of points P_{i-2}, P_{i-1}, P_i and the set of points P_i, P_{i+1}, P_{i+2} are each collinear, but if the three points P_{i-1}, P_i, P_{i+1} are not.

If such corners are undesirable in the graph, they can be eliminated by adding a point on each side of P_i, so that the curve is rounded out at P_i:

If there is a corner at P_i, then P_i should be replaced by two points, one shifted in direction $P_{i-1} - P_i$ by $\lambda \cdot ||P_{i-1} - P_i||$ to the "left" of P_i, and the other shifted by $\mu \cdot ||P_{i+1} - P_i||$ to the "right" of P_i, where $0 < \lambda, \mu < 1$. The first such point is called P_i. The second is called P_{i+1}, and the old points P_{i+1} to P_n are renamed as P_{i+2} to P_{n+1}. Now the three points P_{i-1}, P_i, P_{i+1}, as well as P_i, P_{i+1}, P_{i+2} are not collinear, and the corner will be eliminated.

The constants λ, μ can be chosen so that the curve between P_i and P_{i+1} is in part made up of a parabola. This rounded segment can be further adjusted by a constant $0 < \beta < 1$.

Since for $n + 1$ given points $P_i, i = 0, \ldots, n$, on a curve there can be at most $k = \left[\frac{n+1}{2}\right]$ corners, there will be at most $n + 1 + k$ points $P_i, i = 0, \ldots, n + k$, with $0 \leq k \leq \left[\frac{n+1}{2}\right]$ after the complete curve has been "rounded" out. Here $[x]$ denotes the greatest integer $n \leq x$ for $x \in \mathbb{R}$.

ALGORITHM 13.3 (*Rounded Akima and Renner splines*).

Given: $n + 1$ points $\boldsymbol{P}_i = (x_i, y_i)$ or $\boldsymbol{P}_i = (x_i, y_i, z_i)$ for $i = 0, \ldots, n$ with $n \geq 4$, and $\boldsymbol{P}_i \neq \boldsymbol{P}_{i+1}$ for $i = 0, \ldots, n-1$, and a constant β, $0 < \beta < 1$.

Task: Find at most $k = \left[\frac{n+1}{2}\right]$ new points so that the Akima or Renner subspline calculated for the $n + 1 + k$ points $\boldsymbol{P}_i = (x_i, y_i)$ or $\boldsymbol{P}_i = (x_i, y_i, z_i)$ does not contain a corner.

1) Set $i := 2$.

2) If, when using Akima splines: $|m_{i+1} - m_i| + |m_{i-1} - m_{i-2}| = 0$ and $|m_i - m_{i-1}| > 0$, or if for Renner splines:
$$\sqrt{1 - (s_{i-2}^{0T} s_{i-1}^0)^2} + \sqrt{1 - (s_i^{0T} s_{i+1}^0)^2} = 0 \text{ and } \sqrt{1 - (s_{i-1}^{0T} s_i^0)^2} > 0,$$
then there is a corner at \boldsymbol{P}_i.

 In this case set SWITCH := 1, otherwise SWITCH := 0.

3) If SWITCH = 0, continue with 4), otherwise

 3.1 Relabel the \boldsymbol{P}_i to \boldsymbol{P}_n:
 For $j = n, n-1, \ldots, i+1, i$ do:
 $$\boldsymbol{P}_{j+1} := \boldsymbol{P}_j$$
 (Now $\boldsymbol{P}_{i+1} = \boldsymbol{P}_i$).

 3.2 Shift \boldsymbol{P}_i and \boldsymbol{P}_{i+1} (the old \boldsymbol{P}_i):

 For Akima: $L := 2(x_i - x_{i-1})$
 $R := 2(x_{i+2} - x_{i+1})$

 For Renner: $L := 2\|\boldsymbol{P}_i - \boldsymbol{P}_{i-1}\|$
 $R := 2\|\boldsymbol{P}_{i+2} - \boldsymbol{P}_{i+1}\|$

 $B := \beta \min(L, R)$

 $\lambda := B/L$

 $\mu := B/R$

 $\boldsymbol{P}_i := \boldsymbol{P}_i - \lambda(\boldsymbol{P}_i - \boldsymbol{P}_{i-1})$

 $\boldsymbol{P}_{i+1} := \boldsymbol{P}_{i+1} + \mu(\boldsymbol{P}_{i+2} - \boldsymbol{P}_{i+1})$

 3.3 $n := n + 1$.

4) $i := i + 1$
 If $i \leq n - 2$, go to 2), otherwise stop.

13.4 Approximate Computation of Arc Length

Since the length T_i of the parameter interval of a Renner curve approximates the arc length of the curve segment between the points P_i and P_{i+1}, the entire arc length of a curve can be determined approximately. For this the steps 1) to 5) of algorithm 13.2 can be used. The arc length is approximately

$$L \approx \sum_{i=0}^{n-1} T_i \ .$$

If a smooth planar or space curve is parametrized by $x = x(t), y = y(t)$ with $(x'(t), y'(t)) \neq (0,0)$ or $x = x(t), y = y(t), z = z(t)$ with $(x'(t), y'(t), z'(t)) \neq (0,0,0)$ for all $t \in [a,b]$, then a partition

$$a = t_0 < t_1 < \ldots < t_n = b$$

produces $n + 1$ points

$$P_i = (x_i, y_i) = (x(t_i), y(t_i))$$

or

$$P_i = (x_i, y_i, z_i) = (x(t_i), y(t_i), z(t_i))$$

for $i = 0, \ldots, n$. And the steps 2) to 4) of algorithm 13.2 can be replaced by

$$t_i^R = t_i^L = \frac{1}{\sqrt{x'^2(t_i) + y'^2(t_i)}} (x'(t_i), y'(t_i))^T, \quad i = 0, \ldots, n \ ,$$

or

$$t_i^R = t_i^L = \frac{1}{\sqrt{x'^2(t_i) + y'^2(t_i) + z'^2(t_i)}} (x'(t_i), y'(t_i), z'(t_i))^T, \quad i = 0, \ldots, n \ .$$

If a curve is given explicitly by a function $y = f(x)$ for $x \in [a,b]$, one can find its arc length with the following algorithm:

ALGORITHMUS 13.4. (*Computation of arc length*).

Given: Graph $y = f(x)$ for $x \in [a,b]$ with $f \in C^1[a,b]$.

Task: Find the arc length of the curve.

1) Generate $n + 1$ points P_i as $P_i = \begin{pmatrix} x_i \\ y_i = f(x_i) \end{pmatrix}$ for $i = 0, \ldots, n$ with $a = x_0 < x_1 < \ldots < x_n = b$.

2) Calculate the tangent vectors
$$t_i = \frac{1}{\sqrt{1 + f'^2(x_i)}} \begin{pmatrix} 1 \\ f'(x_i) \end{pmatrix}, \quad i = 0, \ldots, n.$$

3) Compute approximate values T_i for the arc length of the curve segment between P_i and P_{i+1} for $i = 0, \ldots, n-1$ from
$A := 16 - \|t_i + t_{i+1}\|^2,$

$B := 6(P_{i+1} - P_i)^T(t_i + t_{i+1}),$

$C := 36\|P_{i+1} - P_i\|^2,$

$T_i := \left(-B + \sqrt{B^2 + AC}\right)/A.$

4) Calculate the total arc length L of the curve for $x \in [a, b] = [x_0, x_n]$:
$$L \approx \sum_{i=0}^{n-1} T_i.$$

REMARK: If the derivative of f is infinite at a point x_i, then t_i should be chosen as the vector $(0, 1)^T$ in 2).

13.5 Selection Hints

The decision to use Akima or Renner subsplines depends first of all on whether a once continuously differentiable curve is sufficient for the problem. If so, these subsplines are preferred because the "overshoot ripple" of the standard splines is avoided to a large degree. Furthermore, collinear nodes are connected by a straight line, and adjacent straight line segments can be joined with non differentiable "corners", or alternatively they can be joined in a rounded way, see algorithm 13.3. If for spacial or planar curves, the nodes x_i are arranged in a monotonic order, one uses Akima subsplines; if the nodes are not ordered, one uses Renner subsplines.

The large table at the end in section 10.3 contains further hints about the suitability of Akima and Renner subsplines for specific types of curves. To illustrate our points, we include a number of examples below.

Akima spline (non periodic, not rounded) Akima spline (non periodic, with rounding of corners)

Akima spline (periodic, no rounding) Akima spline (periodic, with rounding)

Figure 13.1: Examples for Akima splines

REMARK:

In the following figure we define a curve to be "closed" if its starting and end points coincide and the curve has a unique tangent at this point, such as in (3) or (6). Otherwise the curve is called "not closed". In the non closed case, the end and starting points of a curve may well coincide such as in (2) or (5).

(1) Renner spline, not closed

(2) Renner spline, not closed

(3) Renner spline, closed

(4) Renner spline, not closed,
without rounding

(5) Renner spline, not closed, with rounding

(6) Renner spline, closed, with
rounding of corners

Figure 13.2: Examples for Renner splines

Bibliography for Chapter 13

[AKIM70]; [RENN81]; [RENN82]; [WODI91].

Chapter 14

Numerical Differentiation

14.1 The Task

By using the standard rules of differentiation one can find the derivative of any differentiable function f that can be expressed in terms of elementary functions. Hence an approximate calculation of the derivatives is necessary only in cases where the function f is given empirically as in a table of values.

There are several possibilities to calculate the derivative of a differentiable function f that is known only at discrete nodes numerically:

- differentiate an interpolating polynomial Φ of f,

- differentiate a spline approximation S of f,

- differentiate using Richardson extrapolation (Romberg method), or

- use adaptive numerical differentiation.

The first three methods will be discussed here. Adaptive procedures can be constructed analogously to the adaptive quadrature procedures of section 15.12, see [STEP79], [BJÖR74]. Of the three methods, differentiation of an interpolating polynomial is least recommended. Compared with numerical quadrature, in numerical differentiation one must, however, expect a considerably lower degree of accuracy for all differentiation methods.

14.2 Differentiation Using Interpolating Polynomials

Let f be a sufficiently often differentiable function defined on $[a, b]$ and assume that we know $n + 1$ points $(x_i, y_i = f(x_i))$ for $x_i \in [a, b]$, $i = 0, \ldots, n$. We want to find approximate values for the derivative of f. For this purpose, we can interpolate f with an algebraic polynomial Φ of degree n. This in turn can be differentiated explicitly for any $x \in [a, b]$ instead of f. One can use the interpolating polynomial Φ in any of its representations described in chapter 9. The error in this approximation of the derivative can be found by differentiating the remainder of the interpolation formula. Computational errors due to cancellation take a high toll in this process, even at the nodes x_i themselves where f and the interpolating polynomial Φ agree, as Φ' can differ greatly from f'. Therefore one must at least demand that Φ is a good polynomial approximation for f. Nevertheless, the accuracy of Φ' is in general much worse than that of Φ. This is known as the "roughening" effect of differentiation.

As an illustration for a differentiated interpolation formula, we shall use the Newton interpolation formula $N_+(t)$ for equidistant nodes with $x = x_0 + ht$ and $dx = hdt$, as indicated in section 9.5:

$$\Phi'(x) \equiv N'_+(x) = \frac{1}{h}\tilde{N}'_+(t) = \frac{1}{h}\left\{ \Delta^1_{1/2} + \frac{2t-1}{2!}\Delta^2_1 + \frac{3t^2 - 6t + 2}{3!}\Delta^3_{3/2} + \right.$$

$$\left. + \frac{4t^3 - 18t^2 + 22t - 6}{4!}\Delta^4_2 + \ldots + \frac{1}{n!}\sum_{k=0}^{n-1}\frac{\prod_{i=0}^{n-1}(t-i)}{t-k}\Delta^n_{n/2} \right\}.$$

In a similar fashion one can differentiate any other interpolation formula.

Below we shall give a chart for the approximate calculation of the first and second derivatives at the nodes x_i. We are looking for approximate values Y'_i and Y''_i for the derivatives $y'_i = f'(x_i)$ and $y''_i = f''(x_i)$. Clearly

$$\begin{aligned} y'_i &= Y'_i + &\text{remainder, and} \\ y''_i &= Y''_i + &\text{remainder.} \end{aligned}$$

The chart gives the desired approximate values Y'_i and Y''_i for $i = 0, \ldots, n$, $n = 2, \ldots, 6$ and $n = 2, 3, 4$, respectively, which are derived from the interpolating polynomial. The coefficients of the remainder are smallest near the center of the interpolation interval $[x_0, x_n]$. It is advisable - provided one knows enough interpolation points - to renumber the nodes so that the formulas for the central nodes can be used for finding Y'_i and Y''_i when increasing the index i by 1. But even then computational errors are usually large as

a result of cancellation . For this reason, the derivatives of spline functions described in the following section and the Romberg method of chapter 14.4 are generally preferred over any polynomial interpolation method.

number of the interpolation points	approximate values for Y_i', Y_i''	remainder $(\xi, \xi_1, \xi_2 \in [x_0, x_n])$
3	$Y_0' = \frac{1}{2h}(-3y_0 + 4y_1 - y_2)$	$\frac{h^2}{3} f'''(\xi)$
	$Y_1' = \frac{1}{2h}(-y_0 + y_2)$	$-\frac{h^2}{6} f'''(\xi)$
	$Y_2' = \frac{1}{2h}(y_0 - 4y_1 + 3y_2)$	$\frac{h^2}{3} f'''(\xi)$
	$Y_0'' = \frac{1}{h^2}(y_0 - 2y_1 + y_2)$	$-h f'''(\xi_1) + \frac{h^2}{6} f^{(4)}(\xi_2)$
	$Y_1'' = \frac{1}{h^2}(y_0 - 2y_1 + y_2)$	$-\frac{h^2}{12} f^{(4)}(\xi)$
	$Y_2'' = \frac{1}{h^2}(y_0 - 2y_1 + y_2)$	$h f'''(\xi_1) + \frac{h^2}{6} f^{(4)}(\xi_2)$
4	$Y_0' = \frac{1}{6h}(-11y_0 + 18y_1 - 9y_2 + 2y_3)$	$-\frac{h^3}{4} f^{(4)}(\xi)$
	$Y_1' = \frac{1}{6h}(-2y_0 - 3y_1 + 6y_2 - y_3)$	$\frac{h^3}{12} f^{(4)}(\xi)$
	$Y_2' = \frac{1}{6h}(y_0 - 6y_1 + 3y_2 + 2y_3)$	$-\frac{h^3}{12} f^{(4)}(\xi)$
	$Y_3' = \frac{1}{6h}(-2y_0 + 9y_1 - 18y_2 + 11y_3)$	$\frac{h^3}{4} f^{(4)}(\xi)$

number of the inter-polation points	approximate values for Y_i', Y_i''	remainder $(\xi, \xi_1, \xi_2 \in [x_0, x_n])$
4	$Y_0'' = \frac{1}{6h^2}(12y_0 - 30y_1 + 24y_2 - 6y_3)$	$\frac{11}{12}h^2 f^{(4)}(\xi_1)$ $-\frac{h^3}{10}f^{(5)}(\xi_2),$
	$Y_1'' = \frac{1}{6h^2}(6y_0 - 12y_1 + 6y_2)$	$-\frac{h^2}{12}f^{(4)}(\xi_1)$ $+\frac{h^3}{30}f^{(5)}(\xi_2),$
	$Y_2'' = \frac{1}{6h^2}(6y_1 - 12y_2 + 6y_3)$	$-\frac{h^2}{12}f^{(4)}(\xi_1)$ $-\frac{h^3}{30}f^{(5)}(\xi_2),$
	$Y_3'' = \frac{1}{6h^2}(-6y_0 + 24y_1 - 30y_2 + 12y_3)$	$\frac{11}{12}h^2 f^{(4)}(\xi_1)$ $+\frac{h^3}{10}f^{(5)}(\xi_2).$
5	$Y_0' = \frac{1}{12h}(-25y_0 + 48y_1 - 36y_2 + 16y_3 - 3y_4)$	$\frac{h^4}{5}f^{(5)}(\xi)$
	$Y_1' = \frac{1}{12h}(-3y_0 - 10y_1 + 18y_2 - 6y_3 + y_4)$	$-\frac{h^4}{20}f^{(5)}(\xi)$
	$Y_2' = \frac{1}{12h}(y_0 - 8y_1 + 8y_3 - y_4)$	$\frac{h^4}{30}f^{(5)}(\xi)$
	$Y_3' = \frac{1}{12h}(-y_0 + 6y_1 - 18y_2 + 10y_3 + 3y_4)$	$-\frac{h^4}{20}f^{(5)}(\xi)$
	$Y_4' = \frac{1}{12h}(3y_0 - 16y_1 + 36y_2 - 48y_3 + 25y_4)$	$\frac{h^4}{5}f^{(5)}(\xi)$
	$Y_0'' = \frac{1}{24h^2}(70y_0 - 208y_1 + 228y_2 - 112y_3 + 22y_4)$	$-\frac{5}{6}h^3 f^{(5)}(\xi_1)$ $+\frac{h^4}{15}f^{(6)}(\xi_2),$
	$Y_1'' = \frac{1}{24h^2}(22y_0 - 40y_1 + 12y_2 + 8y_3 - 2y_4)$	$\frac{h^3}{12}f^{(5)}(\xi_1)$ $-\frac{h^4}{60}f^{(6)}(\xi_2),$
	$Y_2'' = \frac{1}{24h^2}(-2y_0 + 32y_1 - 60y_2 + 32y_3 - 2y_4)$	$\frac{h^4}{90}f^{(6)}(\xi)$
	$Y_3'' = \frac{1}{24h^2}(-2y_0 + 8y_1 + 12y_2 - 40y_3 + 22y_4)$	$-\frac{h^3}{12}f^{(5)}(\xi_1)$ $-\frac{h^4}{60}f^{(6)}(\xi_2),$
	$Y_4'' = \frac{1}{24h^2}(22y_0 - 112y_1 + 228y_2 - 208y_3 + 70y_4)$	$\frac{5}{6}h^3 f^{(5)}(\xi_1)$ $+\frac{h^4}{15}f^{(6)}(\xi_2).$

number of the inter-polation points	approximate values for Y_i', Y_i''	remainder $(\xi, \xi_1, \xi_2 \in [x_0, x_n])$
6	$Y_0' = \frac{1}{60h}(-137y_0 + 300y_1 - 300y_2 + 200y_3 - 75y_4 + 12y_5)$	$-\frac{h^5}{6}f^{(6)}(\xi)$
	$Y_1' = \frac{1}{60h}(-12y_0 - 65y_1 + 120y_2 - 60y_3 + 20y_4 - 3y_5)$	$\frac{h^5}{30}f^{(6)}(\xi)$
	$Y_2' = \frac{1}{60h}(3y_0 - 30y_1 - 20y_2 + 60y_3 - 15y_4 + 2y_5)$	$-\frac{h^5}{60}f^{(6)}(\xi)$
	$Y_3' = \frac{1}{60h}(-2y_0 + 15y_1 - 60y_2 + 20y_3 + 30y_4 - 3y_5)$	$\frac{h^5}{60}f^{(6)}(\xi)$
	$Y_4' = \frac{1}{60h}(3y_0 - 20y_1 + 60y_2 - 120y_3 + 65y_4 + 12y_5)$	$-\frac{h^5}{30}f^{(6)}(\xi)$
	$Y_5' = \frac{1}{60h}(-12y_0 + 75y_1 - 200y_2 + 300y_3 - 300y_4 + 137y_5)$	$\frac{h^5}{6}f^{(6)}(\xi)$
7	$Y_0' = \frac{1}{60h}(-147y_0 + 360y_1 - 450y_2 + 400y_3 - 225y_4 + 72y_5 - 10y_6)$	$\frac{h^6}{7}f^{(7)}(\xi)$
	$Y_1' = \frac{1}{60h}(-10y_0 - 77y_1 + 150y_2 - 100y_3 + 50y_4 - 15y_5 + 2y_6)$	$-\frac{h^6}{42}f^{(7)}(\xi)$
	$Y_2' = \frac{1}{60h}(2y_0 - 24y_1 - 35y_2 + 80y_3 - 30y_4 + 8y_5 - y_6)$	$\frac{h^6}{105}f^{(7)}(\xi)$
	$Y_3' = \frac{1}{60h}(-y_0 + 9y_1 - 45y_2 + 45y_4 - 9y_5 + y_6)$	$-\frac{h^6}{140}f^{(7)}(\xi)$
	$Y_4' = \frac{1}{60h}(y_0 - 8y_1 + 30y_2 - 80y_3 + 35y_4 + 24y_5 - 2y_6)$	$\frac{h^6}{105}f^{(7)}(\xi)$
	$Y_5' = \frac{1}{60h}(-2y_0 + 15y_1 - 50y_2 + 100y_3 - 150y_4 + 77y_5 + 10y_6)$	$-\frac{h^6}{42}f^{(7)}(\xi)$
	$Y_6' = \frac{1}{60h}(10y_0 - 72y_1 + 225y_2 - 400y_3 + 450y_4 - 360y_5 + 147y_6)$	$\frac{h^6}{7}f^{(7)}(\xi)$

Bibliography for Section 14.2

[ATKI89], 5.7; [CARN69], 2.11; [CONT80], 4.1; [ENGE87], 8.1; [HÄMM78], 13; [HENR72] vol.2, 12.1,2; [ISAA66], 6.5; [STIE63], 6.1; [STUM82], 3.3-4.

14.3 Differentiation via Interpolating Cubic Splines

The reason for using spline functions S for approximately differentiating a function f is given in theorem 10.8: With its assumptions, S converges to f as $n \to \infty$ and $h_i = x_{i+1} - x_i \to 0$, as does S' towards f' and S'' towards f''. If one formally differentiates the spline function (10.1), one obtains for $x \in [x_i, x_{i+1}]$

$$S'(x) \equiv S_i'(x) = b_i + 2c_i(x - x_i) + 3d_i(x - x_i)^2, \qquad i = 0, \ldots, n-1,$$

$$S''(x) \equiv S_i''(x) = 2c_i + 6d_i(x - x_i), \qquad\qquad i = 0, \ldots, n-1.$$

Thus S_i' and S_i'' approximate f' and f'' for all $x \in [x_i, x_{i+1}]$ and $i = 0, \ldots, n-1$. At the nodes $x = x_i$:

$$S'(x_i) = b_i, \quad S''(x_i) = 2c_i.$$

The accuracy of the approximation of f'' can be increased by using the computed approximate values for $f'(x_i)$. One simply performs another spline interpolation, this time for f', and differentiates the resulting spline ("spline on spline", see [AHLB67], p.43 and 49).

Numerical differentiation using cubic splines generally gives a better agreement of S' and f' than the differentiation of an interpolating polynomial would.

Bibliography for Section 14.3

[AHLB67], 2.3, 2.5; [ENGE87], 8.2; [GREV69], p. 10-15; [RALS67] vol.II, 8.2; [STOE91], 2.4.3; [WERN79], III §5.

14.4 Differentiation by the Romberg Method

Let f be a function in $C^{2n}[a, b]$ and $h = h_0$. We want to find an approximate value for $f'(x)$ at a given point $x = x_0 \in [a, b]$.

By using Richardson extrapolation, the *Romberg method* successively halves the step size h and produces approximate values of high accuracy for $f'(x_0)$ by taking suitable linear weighted sums of approximations to $f'(x_0)$. With

$h_j := h/2^j$ for $j = 0, 1, 2, \ldots$, i.e., $h_{j+1} = h_j/2$, one forms so-called *central difference quotients*

$$(14.1) \qquad D_j^{(0)}(f) := \frac{f(x_0 + h_j) - f(x_0 - h_j)}{2h_j}.$$

With these we can form the sums

$$(14.2) \qquad D_j^{(k)}(f) := D_{j+1}^{(k-1)}(f) + \frac{1}{4^k - 1}\left(D_{j+1}^{(k-1)}(f) - D_j^{(k-1)}(f)\right)$$

for $j = 0, 1, 2, \ldots$ and $k = 1, 2, \ldots$, where k is limited in its range by the assumption that $f \in C^{2n}[a, b]$. Then

$$f'(x_0) = D_j^{(k)}(f) + O(h_j^{2k+2}),$$

so that $D_j^{(k)}(f)$ is an approximate value for $f'(x_0)$ with an error proportional to h_j^{2k+2}. The calculation is carried out rowwise as in the following scheme, where we calculate for each $i = 0, \ldots, m$ successively the elements $D_s^{(t)}$ of row i from the left to the right by means of $D_s^{(t)}$ for $s = i, i-1, \ldots, 1, 0$ and $t = 0, \ldots, i$.

CALCULATION SCHEME 14.1 (*Romberg method for numerical differentiation*).

$D_j^{(0)}$	$D_j^{(1)}$	$D_j^{(2)}$	\cdots	$D_j^{(m-1)}$	$D_j^{(m)}$
$D_0^{(0)}$					
$D_1^{(0)}$	$D_0^{(1)}$				
$D_2^{(0)}$	$D_1^{(1)}$	$D_0^{(2)}$			
\vdots	\vdots	\vdots	\ddots		
$D_{m-1}^{(0)}$	$D_{m-2}^{(1)}$	$D_{m-3}^{(2)}$	\cdots	$D_0^{(m-1)}$	
$D_m^{(0)}$	$D_{m-1}^{(1)}$	$D_{m-2}^{(2)}$	\cdots	$D_1^{(m-1)}$	$D_0^{(m)}$

The scheme has to be continued until for a given $\varepsilon > 0$ the inequality

$$|D_0^{(m)} - D_1^{(m-1)}| < \varepsilon$$

holds, provided $m \leq n$. In this case, $D_0^{(m)}$ is the desired approximate value
for $f'(x_0)$ with

$$f'(x_0) = D_0^{(m)}(f) + O(h^{2m+2}).$$

For values of h that have been chosen too small, the result will be tainted
by computational errors due to cancellation when forming $D_j^{(k)}$. Practically
speaking, as long as the values $D_j^{(k)}$ change monotonically for increasing j,
one can continue the calculations. If they begin to oscillate, the calculation
is to be stopped even if the required accuracy has not been reached. If one
needs a still higher accuracy, the functional values have to be given to a higher
precision. An essential advantage of this procedure is the fact that due to the
continued halving of the step size, there will finally be a value h_j for which
the procedural errors and the rounding errors are equally large. When the
oscillations begin, this optimal value for h_i has been passed, and rounding
errors will cause random results.

Bibliography for Section 14.4

[BJÖR74], 7.22; [CONT80], 4.2; [ENGE87], 8.4; [HÄMM78], 13; [HENR72]
vol.2, 12.3; [WERN79], III §8.

14.5 Decision Hints

Due to the roughening effect of differentiation, one cannot expect the same
nice results in numerical differentiation that one gets when using numerical
quadrature. The calculation of derivatives by means of splines is preferred
over using interpolating polynomials. If the functional values at the nodes
required for the Romberg method can be found or easily evaluated, one ought
to employ this method. Here we recommend to proceed analogously as with
the adaptive quadrature methods (see section 15.12) in order to keep the
number of required function evaluations as low as possible.

Chapter 15

Numerical Integration

In this chapter we describe numerical methods to evaluate definite integrals $\int_a^b f(x)dx$ for functions $f : \mathbb{R} \to \mathbb{R}$ of one variable. This process is often called *numerical quadrature*. Chapter 16 will treat *numerical cubature*, i.e., the evaluation of integrals $\iint_B f(x,y)dxdy$ for functions $f : \mathbb{R}^2 \to \mathbb{R}$ over regions $B \subseteq \mathbb{R}^2$.

15.1 Preliminary Remarks

By the fundamental theorem of calculus, every function $f : \mathbb{R} \to \mathbb{R}$ that is continuous on an interval I_x has an *antiderivative* F on I_x. Thus

$$\frac{dF(x)}{dx} = F'(x) = f(x) \quad \text{for all } x \in I_x.$$

The antiderivative F of f is unique except for an arbitrary additive constant. The *definite integral* of a function f over $[\alpha, \beta]$ is defined for integrable functions f as

$$I(f; \alpha, \beta) := \int_\alpha^\beta f(x)dx \quad \text{for } [\alpha, \beta] \subset I_x.$$

If f is continuous on $[\alpha, \beta] \subset I_x$, then, by the fundamental theorem of calculus,

$$(15.1) \qquad I(f; \alpha, \beta) = \int_\alpha^\beta f(x)dx = F(\beta) - F(\alpha),$$

where F is an antiderivative of f.

In practice most often one cannot evaluate definite integrals $I(f; \alpha, \beta)$ directly, but instead one has to rely on so-called *quadrature formulas* Q. The reasons for this are:

1) Functional expressions for the antiderivative F of a function f are known only for rather few and simple functions f, see e.g. [CRCS91], where over 700 integral formulas are given. While it is possible to differentiate the most complicated function f given by a functional equation $f(x)$ in terms of elementary functions such as $x, \sin x, e^x, \arccos x, \log x$ etc. by using the standard rules of differentiation, i.e., the chain, product and quotient rules, the converse problem of finding a function F with $F' = f$ can be solved only in isolated cases. Several thousand antiderivatives are tabulated in so called *integral tables*, but any seemingly slight deviation of the function to be integrated from one in the table - such as replacing the variable x by e^x or multiplying f by $\sin x$ - will generally lead to an unsolved and possibly unsolvable problem. It is known for example that the functions $f(x) = \frac{\sin x}{x}$ and $g(x) = e^{-x^2}$ do not possess antiderivatives that can be expressed via elementary functions.

2) f is known only empirically at discrete points $x_k \in [\alpha, \beta]$.

3) The antiderivative F can be represented in a closed form; however, to find F or to calculate $F(\alpha)$ and $F(\beta)$ requires too much time or cannot be done accurately due to the complexity of F.

If $I(f; \alpha, \beta)$ cannot be evaluated directly, one has to rely on other methods. Possible alternate methods for integration are weighted sums Q formed from functional values $f(x_k)$ of the integrand at discrete *nodes* $x_k \in [\alpha, \beta]$ that are multiplied or weighted by certain *weights* A_k:

$$Q(f; \alpha, \beta) = \sum_k A_k f(x_k) \approx \int_\alpha^\beta f(x)dx \quad \text{where } x_k \in [\alpha, \beta].$$

Such sums mimic the definition of the *Riemann integral*. Alternatively, a definite integral can be approximated by sums of products involving functional values and those of the derivative of f at certain nodes x_k. Such quadrature formulas Q can supply approximate values for the definite integral $I(f; \alpha, \beta)$. For example

$$Q^R(f; a, b) = (b - a)f(a) \approx \int_a^b f(x)dx$$

is the so-called *rectangle formula*, constructed for the reference interval $[a, b]$. We could have chosen $[-1, 1], [0, h]$, or $[-h, h]$ as reference intervals here. If we want to determine an approximate value for $I(f; \alpha, \beta)$ using this rectangle formula we should partition the interval $[\alpha, \beta]$

$$P : \alpha = t_0 < t_1 < t_2 < \ldots < t_N = \beta$$

into subintervals $[t_k, t_{k+1}]$ of length $h_k := t_{k+1} - t_k$. The set $P = \{t_0, \ldots, t_N\}$ is called a *partition* of the interval of integration. Due to the linearity of integration we have:

$$\int\limits_{\alpha}^{\beta} f(x)dx = \int\limits_{t_0}^{t_1} f(x)dx + \int\limits_{t_1}^{t_2} f(x)dx + \ldots + \int\limits_{t_{N-1}}^{t_N} f(x)dx,$$

and we thus can apply the rectangle formula $Q^R(f; t_k, t_{k+1})$ to each subinterval $[t_k, t_{k+1}]$.

Consequently the *summed rectangle formula*

$$Q_{h_k}^R(f; \alpha, \beta) = \sum_{k=0}^{N-1} (t_{k+1} - t_k)f(t_k) \approx \int\limits_{\alpha}^{\beta} f(x)dx$$

is an approximation for $I(f; \alpha, \beta)$.

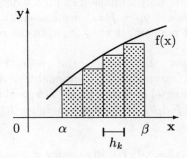

Figure 15.1: Summed rectangle formula

Such a quadrature formula is called a *summed quadrature formula.*

The difference between the true value of an integral and the result given by a quadrature formula for a *subinterval* is called the *local quadrature error*. The difference between the integral value and the result of a summed quadrature formula is called the *global quadrature error*. This error can be reduced to any size by finer partitions of the interval of integration. Thus knowing *one* single quadrature formula would be sufficient in theory to evaluate all definite integrals numerically. But a finer partition will require more functional evaluations $f(t_k)$. Thus one needs to economize, i.e., find optimal weights A_k and possibly nodes x_k so that

$$Q(f; \alpha, \beta) = \sum_{k} A_k f(x_k)$$

approximates $\int_\alpha^\beta f(x)dx$ well for a small number of terms already. We will indicate different quadrature formulas in the following pages. To facilitate the choice between them we will give hints for selecting an appropriate method in section 15.14.

Bibliography for Section 15.1

[ATKI89], 5; [BRON85], III; [ENGE80]; [ENGE87], 9.1; [MAES84]; [McCR72], 5.7; [NIED87], 11.1; [STOE91], 3; [STUM82], p.70.

15.2 Interpolating Quadrature Formulas

For a given reference interval $[a, b]$ we shall construct quadrature formulas Q that evaluate integrals of polynomials of as high a degree as possible exactly. We assume that the values $y_k = f(x_k)$ are known at $n + 1$ distinct but not necessarily equidistant nodes $x_k, k = 0, \ldots, n$, in the reference interval $[a, b]$.

The problem suggests to connect the $n + 1$ interpolation points $(x_k, y_k = f(x_k))$ by an interpolating polynomial Φ of degree at most n and to evaluate the definite integral of Φ over $[a, b] : I(\Phi; a, b)$ is an approximation for the desired integral $I(f; a, b)$. With the remainder $R(x)$ of the interpolation we can express

$$f(x) = \Phi(x) + R(x) \quad \text{for } x \in [a, b].$$

Thus for the integral $I(f; a, b)$:

$$(15.2) \quad \begin{cases} I(f; a, b) & = & \int\limits_a^b f(x)dx = Q(f; a, b) + E(f; a, b) \quad \text{with} \\ Q(f; a, b) & = & I(\Phi; a, b) = \int\limits_a^b \Phi(x)dx \quad \text{and} \\ E(f; a, b) & = & I(R; a, b) = \int\limits_a^b R(x)dx. \end{cases}$$

$Q(f; a, b)$ gives the result of the quadrature formula and $E(f; a, b)$ is the associated *remainder of the quadrature*. The sum of Q and E is called the *integration rule*. The remainder of the quadrature can be written as

$$E(f; a, b) = \int\limits_a^b R(x)dx = \int\limits_a^b (f(x) - \Phi(x))dx.$$

If $f - \Phi$ alternates its sign in $[a, b]$, then the positive and negative errors committed when evaluating $I(f; a, b)$ may neutralize each other so that the resulting integration error may become small even if Φ is *not* a good approximation of f. This suggests that integration can smooth out errors.

An interpolating quadrature formula can be represented in general as

$$(15.3) \quad Q(f; a, b) = \sum_{k=0}^{n} A_k f(x_k) = A_0 f(x_0) + A_1 f(x_1) + \ldots + A_n f(x_n).$$

The demand that (15.3) should integrate all polynomials up to degree $M \geq n$ precisely gives rise to a system of equations by using $f(x) = x^m$ for $m \leq n$ in (15.3):

$$(15.4) \qquad \sum_{k=0}^{n} A_k x_k^m = \frac{1}{m+1}(b^{m+1} - a^{m+1}), \quad m = 0, \ldots, M.$$

This system is linear in the weights A_k and nonlinear in the x_k for $k = 0, \ldots, n$. Depending on whether the x_k or the A_k, or neither the x_k nor the A_k are prescribed, certain types of quadrature formulas will arise with different orders of convergence.

Interpolating quadrature formulas of the type (15.3) can be exact for polynomials up to degree $M = 2n + 1$ at most where n is the number of nodes x_k; for a proof see [NIED87], 11.3.

Local and global error order.

If a quadrature formula Q constructed for the reference interval $[a, b]$ is exact for all polynomials of degree M and if $f \in C^{M+1}[a, b]$, then the local error order of Q is $q = M + 2$.

If $f \in C^{M+1}[\alpha, \beta]$ and if h_{max} is the maximal length of a subinterval of the partition P of $[\alpha, \beta]$, then the error order of the summed quadrature formula Q_{h_k} is $O(h_{max}^{M+1})$, i.e., its global error order is $q = M + 1$, for a proof see [NIED87], 11.2.

If all nodes $x_k, k = 0, \ldots, n$, are prescribed in $[a, b]$, then (15.4) becomes a linear system of $n + 1$ equations for the $n + 1$ weights A_k. This gives rise to the *Newton-Cotes formulas* which integrate polynomials up to the n^{th} degree exactly for any $n + 1$ given nodes when n is odd and up to the $(n+1)^{st}$ degree if n is even.

If $a = -h$, $b = h$ and all weights are chosen as $A_k = \frac{2h}{n+1}$, then (15.4) is a nonlinear system for the nodes x_k. This is the *Chebyshev quadrature formula*

which integrates polynomials exactly up to the $(n + 1)^{st}$ degree, see section 15.6.

If one prescribes neither the nodes x_k nor the weights A_k for the reference interval, but requires that the quadrature formula (15.3) integrate polynomials up to the $(2n + 1)^{st}$ degree exactly, then (15.4) is a nonlinear system for both the nodes x_k and weights A_k. This leads to the *optimal Gauß formulas*, see section 15.7.

All of the above mentioned quadrature methods will be described in detail in the following sections.

Bibliography for Section 15.2

[COLL73] I, 3.2; [CONT80], 5.2; [MEIN79], 6.4; [STIE63], 6.2.

15.3 Newton-Cotes Formulas

The linear system (15.4) gives rise to *quadrature formulas* for equidistant nodes in the reference interval $[a, b]$ in the following way: The end points of the interval $[a, b]$ will always be taken as nodes of the interpolating polynomial used for integration. Quadrature formulas that use the endpoints as nodes are called formulas of *closed type*, while those that do not are called of *open type*. The resulting formulas are called *Newton-Cotes formulas* .
One can find the weights A_k of the Newton-Cotes formulas from (15.4) for $m = 0, \ldots, n$. The linear system (15.4) can be written as:

$$(15.5) \quad \begin{pmatrix} 1 & 1 & 1 & & 1 \\ x_0 & x_1 & x_2 & \cdots & x_n \\ x_0^2 & x_1^2 & x_2^2 & \cdots & x_n^2 \\ \vdots & \vdots & \vdots & & \vdots \\ x_0^n & x_1^n & x_2^n & \cdots & x_n^n \end{pmatrix} \begin{pmatrix} A_0 \\ A_1 \\ A_2 \\ \vdots \\ A_n \end{pmatrix} = \begin{pmatrix} b - a \\ \frac{1}{2}(b^2 - a^2) \\ \frac{1}{3}(b^3 - a^3) \\ \vdots \\ \frac{1}{n+1}(b^{n+1} - a^{n+1}) \end{pmatrix} .$$

The system matrix in (15.5) is called a *Vandermonde matrix*. For distinct numbers x_k its determinant is different from zero, and hence the system (15.4) or (15.5) can be solved uniquely. The weights A_k are uniquely determined by the given nodes x_k.

Thus one can always derive an interpolating quadrature formula for a given reference interval $[a, b]$ and $n+1$ distinct, but not necessarily equidistant nodes $x_k \in [a, b]$. If the functional values $f(x_k)$ are known, then the quadrature formula (15.3) supplies an approximate value for the integral $I(f; a, b)$.

Using the Lagrange remainder of the interpolation for $f \in C^{n+1}[a, b]$, see (9.10), we have the following representation for the error $E(f; a, b)$

$$E(f; a, b) = \frac{1}{(n+1)!} \int_a^b f^{(n+1)}(\xi)\pi(x)dx \quad \text{with}$$

$$\xi = \xi(x, x_0, x_1, \ldots, x_n) \in [a, b] \quad \text{and} \quad \pi(x) = (x - x_0)(x - x_1)\ldots(x - x_n),$$

or

$$(15.6) \quad E(f; a, b) = \frac{1}{(n+1)!} f^{(n+1)}(\xi^*) \int_a^b \pi(x)dx \text{ for some } \xi^* \in [a, b],$$

provided that $\pi(x) \geq 0$ or $\pi(x) \leq 0$ everywhere in $[a, b]$. Since the $(n+1)^{st}$ derivative of any polynomial P of degree n is zero, $E(P; a, b) = 0$ and the associated quadrature formula Q with $n+1$ given nodes in the reference interval integrates polynomials up to the n^{th} degree exactly. If one chooses the nodes x_k suitably, the same quadrature formula can integrate polynomials of even higher degrees precisely, see Simpson's Rule.

In the following, the name of the quadrature formula shall be abbreviated and always appear as an upper index on Q and E, the chosen step size shall be the lower index. We will also stop to indicate the function f that is being integrated in the integration rule $Q(f; a, b)$ or in $E(f; a, b)$.

15.3.1 The Trapezoidal Rule

Trapezoidal rule for [a,b].

To integrate f on the interval $[a, b] = [0, h]$ with end points $x_0 = 0$ and $x_1 = h$ as nodes, we set $n = 1$, $a = 0$, $b = h$. Then the weights in (15.5) are $A_0 = A_1 = \frac{h}{2}$ and thus the quadrature formula (15.3) becomes

$$Q^{TR}(0, h) = A_0 f(x_0) + A_1 f(x_1) = \frac{h}{2}(f(0) + f(h)).$$

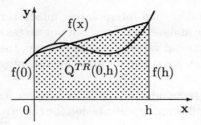

Figure 15.2: Trapezoidal rule

$Q^{TR}(0, h)$ is called the *trapezoidal rule*. For the associated remainder we deduce from (15.6) for $f \in C^2[0, h]$:

$$E^{TR}(0, h) = \frac{1}{2} f''(\xi^*) \int_0^h x(x - h)dx = -\frac{h^3}{12} f''(\xi^*) \text{ for some } \xi^* \in [0, h].$$

Hence, the trapezoidal rule has a *local error order* of $O(h^3)$. It integrates all first degree polynomials exactly. Geometrically, $Q^{TR}(0, h)$ measures the area of the trapezoid defined by the points $(0, 0)$, $(0, f(0))$, $(h, 0)$ and $(h, f(h)) \in \mathbb{R}^2$. The *trapezoidal formula* is:

$$\int_0^h f(x)dx = Q^{TR}(0, h) + E^{TR}(0, h) = \frac{h}{2}(f(0) + f(h)) - \frac{h^3}{12} f''(\xi^*), \quad \xi^* \in [0, h].$$

Summed trapezoidal rule for equidistant nodes

If we want to integrate over a sizeable interval $[\alpha, \beta]$, we can split up the interval $[\alpha, \beta]$ into N subintervals of length $h = \frac{\beta - \alpha}{N}$ each. Here h is the step size. The integral of f over $[\alpha, \beta]$ is approximated by the *summed trapezoidal rule*:

$$\int_\alpha^\beta f(x)dx \quad = \quad Q_h^{TR}(\alpha, \beta) + E_h^{TR}(\alpha, \beta) \quad \text{with}$$

$$Q_h^{TR}(\alpha, \beta) \quad = \quad \frac{h}{2} \left(f(\alpha) + f(\beta) + 2 \sum_{k=1}^{N-1} f(\alpha + kh) \right) \text{ and}$$

$$E_h^{TR}(\alpha, \beta) \quad = \quad -\frac{\beta - \alpha}{12} h^2 f''(\eta) \quad \text{for some } \eta \in [\alpha, \beta], \text{ if } f \in C^2[\alpha, \beta].$$

In this case $Q_h^{TR}(\alpha, \beta)$ is the *summed TR formula* and $E_h^{TR}(\alpha, \beta)$ is the *remainder* of the summed TR formula; its *global error order* is $O(h^2)$.

Figure 15.3: Summed trapezoidal rule

Summed trapezoidal rule for non-equidistant nodes.

We set $h_k = t_{k+1} - t_k$ and $h_{max} := \max_{0 \le k \le N-1} \{h_k\}$ for the partition $\alpha = t_0 < t_1 < ... < t_N = \beta$. Then

$$
\begin{aligned}
\int_\alpha^\beta f(x)dx &= Q_{h_k}^{TR}(\alpha, \beta) + E_{h_k}^{TR}(\alpha, \beta) \quad \text{with} \\
Q_{h_k}^{TR}(\alpha, \beta) &= \tfrac{1}{2} \sum_{k=0}^{N-1} (t_{k+1} - t_k)(f(t_k) + f(t_{k+1})) = \\
&= \frac{t_1 - t_0}{2} f(t_0) + \tfrac{1}{2} \sum_{k=1}^{N-1} (t_{k+1} - t_k) f(t_k) + \frac{t_N - t_{N-1}}{2} f(t_N), \\
E_{h_k}^{TR}(\alpha, \beta) &= O(h_{max}^2).
\end{aligned}
$$

Trapezoidal rule for periodic functions f.

Let $f \in C^{2m}[\alpha, \beta]$ be periodic with period $\beta - \alpha$ and assume $[\alpha, \beta]$ is partitioned into N subintervals of length $h = \frac{\beta - \alpha}{N}$. Then the summed Euler-Maclaurin formula, see chapter 15.5, and the periodicity of f, i.e.,

$$
f^{(2k-1)}(\alpha) = f^{(2k-1)}(\beta)
$$

give the trapezoidal rule as

$$
\int_\alpha^\beta f(x)dx = h \sum_{k=0}^{N-1} f(\alpha + kh) + O(h^{2m}).
$$

15.3.2 Simpson's Rule

Simpson's rule for the interval [0,2h].

If we choose $x_0 = 0$, $x_1 = h$ and $x_2 = 2h$ as nodes, then $n = 2$, $a = 0$, $b = 2h$, and (15.5) gives the weights as $A_0 = A_2 = \frac{h}{3}$, $A_1 = 4h/3$. Thus the quadrature formula (15.3) becomes

$$Q^S(0,2h) = A_0 f(x_0) + A_1 f(x_1) + A_2 f(x_2) = \frac{h}{3}\left(f(0) + 4f(h) + f(2h)\right).$$

$Q^S(0,2h)$ is called *Simpson's rule*. Its remainder is

$$E^S(0,2h) = -\frac{h^5}{90} f^{(4)}(\xi^*) \text{ for some } \xi^* \in [0,2h], \text{ if } f \in C^4[0,2h].$$

Thus Simpson's rule has local error order $O(h^5)$. It integrates polynomials up to the 3^{rd} degree exactly. *Simpson's rule* can be expressed including the error term as

$$\begin{aligned}
\int_0^{2h} f(x)dx &= Q^S(0,2h) + E^S(0,2h) \\
&= \tfrac{h}{3}\left(f(0) + 4f(h) + f(2h)\right) - \tfrac{h^5}{90} f^{(4)}(\xi^*) \text{ for some } \xi^* \in [0,2h].
\end{aligned}$$

Summed Simpson's rule for equidistant nodes.

To integrate f over large interval $[\alpha,\beta]$, one partitions $[\alpha,\beta]$ into $2N$ subintervals of length $h = \frac{\beta-\alpha}{2N}$. Then the *summed Simpson's rule* reads as follows

$$\int_\alpha^\beta f(x)dx = Q_h^S(\alpha,\beta) + E_h^S(\alpha,\beta) \quad \text{with}$$

$$Q_h^S(\alpha,\beta) = \tfrac{h}{3}\left(f(\alpha) + f(\beta) + 4\sum_{k=0}^{N-1} f(\alpha + (2k+1)h) + 2\sum_{k=1}^{N-1} f(\alpha + 2kh)\right),$$

$$E_h^S(\alpha,\beta) = -\frac{\beta-\alpha}{180} h^4 f^{(4)}(\eta) \text{ for some } \eta \in [\alpha,\beta], \text{ if } f \in C^4[\alpha,\beta].$$

Here $Q_h^S(\alpha,\beta)$ is the *summed Simpson's rule* and $E_h^S(\alpha,\beta)$ is its *remainder*. The summed Simpson's rule has a global error order of $O(h^4)$.

One disadvantage of Simpson's rule for equidistant nodes is the requirement that there be an even number of subintervals of length h. However, this disadvantage can be avoided if one combines Simpson's rule with the 3/8 formula in case of an odd number of subintervals, see the remark in section 15.3.3.

Summed Simpson's rule for non-equidistant nodes.

With the partition

$$P : \alpha = t_0 < t_1 < t_2 < \ldots < t_m = \beta$$

of the interval of integration $[\alpha, \beta]$, $h_k := t_{k+1} - t_k$ and $h_{max} = \max\limits_{0 \le k \le m-1}\{h_k\}$, Simpson's rule has the form

$$\int_\alpha^\beta f(x)dx = Q_{h_k}^S(\alpha, \beta) + E_{h_k}^S(\alpha, \beta) \quad \text{with}$$

$$Q_{h_k}^S(\alpha, \beta) = \frac{1}{6}\sum_{k=0}^{m-1}(t_{k+1} - t_k)\left[f(t_k) + 4f\left(\frac{t_k + t_{k+1}}{2}\right) + f(t_{k+1})\right],$$

$$E_{h_k}^S(\alpha, \beta) = O(h_{max}^4).$$

15.3.3 The 3/8 Formula

3/8 Formula for the interval [0,3h].

If we integrate f on $[a, b] = [0, 3h]$ and choose $x_0 = 0$, $x_1 = h$, $x_2 = 2h$ and $x_3 = 3h$ as nodes, then $n = 3$, $a = 0$, $b = 3h$ and (15.5) determines the weights as $A_0 = 3h/8$, $A_1 = 9h/8$, $A_2 = 9h/8$, $A_3 = 3h/8$, hence the name of this method. The quadrature formula (15.3) thus reads

$$\begin{aligned} Q^{3/8}(0, 3h) &= A_0 f(x_0) + A_1 f(x_1) + A_2 f(x_2) + A_3 f(x_3) \\ &= \frac{3h}{8}\left(f(0) + 3f(h) + 3f(2h) + f(3h)\right). \end{aligned}$$

$Q^{3/8}(0, 3h)$ is called the 3/8 *formula*. For the remainder of the 3/8-formula we have

$$E^{3/8}(0, 3h) = -\frac{3}{80}h^5 f^{(4)}(\xi^*) \text{ for some } \xi^* \in [0, 3h], \text{ if } f \in C^4[0, 3h].$$

The local error order is $O(h^5)$, and this quadrature formula integrates polynomials of 3^{rd} degree exactly. The 3/8 *rule* approximates the integral as follows

$$\begin{aligned} \int_0^{3h} f(x)dx &= Q^{3/8}(0, 3h) + E^{3/8}(0, 3h) \\ &= \frac{3h}{8}\left(f(0) + 3f(h) + 3f(2h) + f(3h)\right) - \frac{3}{80}h^5 f^{(4)}(\xi^*), \\ &\qquad \text{for some } \xi^* \in [0, 3h], \text{ if } f \in C^4[0, 3h]. \end{aligned}$$

Summed 3/8 formula with equidistant nodes.

In order to integrate f over a large interval $[\alpha, \beta]$, one partitions $[\alpha, \beta]$ into $3N$ subintervals of length $h = \frac{\beta - \alpha}{3N}$ each, so that the *summed 3/8 formula* arises:

$$\int_\alpha^\beta f(x)dx = Q_h^{3/8}(\alpha, \beta) + E_h^{3/8}(\alpha, \beta) \quad \text{with}$$

$$Q_h^{3/8}(\alpha, \beta) = \frac{3h}{8}\left(f(\alpha) + f(\beta) + 3\sum_{k=1}^N f(\alpha + (3k-2)h) + \right.$$

$$\left. + 3\sum_{k=1}^N f(\alpha + (3k-1)h) + 2\sum_{k=1}^{N-1} f(\alpha + 3kh) \right),$$

$$E_h^{3/8}(\alpha, \beta) = -\frac{\beta - \alpha}{80}h^4 f^{(4)}(\eta) \text{ for some } \eta \in [\alpha, \beta], \text{ if } f \in C^4[\alpha, \beta].$$

In this case, $Q_h^{3/8}(\alpha, \beta)$ is called the *summed 3/8 formula* and $E_h^{3/8}(\alpha, \beta)$ is its *remainder*. The summed 3/8 formula has the global error order $O(h^4)$.

REMARK. If the integral $I(f; \alpha, \beta)$ of f over the interval $[\alpha, \beta]$ is to be calculated with a global error order $O(h^4)$ for a given constant h, and if it is not possible to partition the interval $[\alpha, \beta]$ into $2N$ or $3N$ subintervals of length h, it is recommended to combine Simpson's rule with the 3/8 formula, since both of them have the same error order $O(h^4)$.

Summed 3/8 formula with non-equidistant nodes.

We use the partition

$$P : \alpha = t_0 < t_1 < t_2 < \ldots < t_m = \beta$$

of the interval of integration $[\alpha, \beta]$ and set $h_k := t_{k+1} - t_k$, $h_{max} = \max_{0 \le k \le m-1}\{h_k\}$. Then the summed 3/8 formula is

$$\int_\alpha^\beta f(x)dx = Q_{h_k}^{3/8}(\alpha, \beta) + E_{h_k}^{3/8}(\alpha, \beta) \quad \text{for}$$

$$Q_{h_k}^{3/8}(\alpha, \beta) = \frac{3}{8}\sum_{k=0}^{m-1}(t_{k+1} - t_k)\left[f(t_k) + 3f\left(\frac{2t_k + t_{k+1}}{3}\right) + \right.$$

$$\left. + 3f\left(\frac{t_k + 2t_{k+1}}{3}\right) + f(t_{k+1}) \right] \text{ with}$$

$$E_{h_k}^{3/8}(\alpha, \beta) = O(h_{max}^4).$$

15.3.4 Other Newton-Cotes Formulas

Up to now we have introduced three different Newton-Cotes formulas which were derived by integrating the interpolating polynomial for f with 2, 3 or 4 nodes. We are going to indicate four more such formulas which use 5, 6, 7 and 8 nodes, respectively. We simply give these formulas and their remainders:

4/90 rule (5 nodes for the interval $[0, 4h]$).

$$\int_0^{4h} f(x)dx = \frac{4h}{90}(7f(0) + 32f(h) + 12f(2h) + 32f(3h) + 7f(4h)) +$$

$$-\frac{8h^7}{945}f^{(6)}(\xi^*) \text{ for some } \xi^* \in [0, 4h], \text{ if } f \in C^6[0, 4h].$$

Summed 4/90 rule with equidistant nodes. Set $h = \frac{\beta - \alpha}{4N}$:

$$\int_\alpha^\beta f(x)dx = \frac{4h}{90}\left(7f(\alpha) + 7f(\beta) + 32\sum_{k=1}^N f(\alpha + (4k-3)h) + \right.$$

$$+ 12\sum_{k=1}^N f(\alpha + (4k-2)h) + 32\sum_{k=1}^N f(\alpha + (4k-1)h) +$$

$$\left. + 14\sum_{k=1}^{N-1} f(\alpha + 4kh)\right) - \frac{2(\beta - \alpha)}{945}h^6 f^{(6)}(\eta)$$

for some $\eta \in [\alpha, \beta]$, if $f \in C^6[\alpha, \beta]$.

5/288 rule for the interval [0,5h] (6 nodes).

$$\int_0^{5h} f(x)dx = \frac{5h}{288}(19f(0) + 75f(h) + 50f(2h) + 50f(3h) + 75f(4h) +$$

$$+ 19f(5h)) - \frac{275}{12096}h^7 f^{(6)}(\xi^*)$$

for some $\xi^* \in [0, 5h]$, if $f \in C^6[0, 5h]$.

Summed 5/288 rule for equidistant nodes. Here $h = \frac{(\beta - \alpha)}{5N}$.

$$\int_\alpha^\beta f(x)dx = \frac{5h}{288}\left(19f(\alpha) + 19f(\beta) + 75\sum_{k=1}^N f(\alpha + (5k-4)h) + \right.$$

$$+ 50\sum_{k=1}^N f(\alpha + (5k-3)h) + 50\sum_{k=1}^N f(\alpha + (5k-2)h) +$$

$$\left. + 75\sum_{k=1}^N f(\alpha + (5k-1)h) + 38\sum_{k=1}^{N-1} f(\alpha + 5kh)\right) +$$

$$- \frac{55(\beta - \alpha)}{12096}h^6 f^{(6)}(\eta) \text{ for some } \eta \in [\alpha, \beta], \text{ if } f \in C^6[\alpha, \beta].$$

6/840 rule for the interval [0,6h] (7 nodes).

$$\int_0^{6h} f(x)dx = \frac{6h}{840}(41f(0) + 216f(h) + 27f(2h) + 272f(3h)+$$

$$+ 27f(4h) + 216f(5h) + 41f(6h)) - \frac{9}{1400}h^9 f^{(8)}(\xi^*)$$

for some $\xi^* \in [0, 6h]$, if $f \in C^8[0, 6h]$.

Summed 6/840 rule for equidistant nodes. Set $h = \frac{(\beta-\alpha)}{6N}$.

$$\int_\alpha^\beta f(x)dx = \frac{6h}{840}\left(41f(\alpha) + 41f(\beta) + 216\sum_{k=1}^N f(\alpha + (6k-5)h)+\right.$$

$$+27\sum_{k=1}^N f(\alpha + (6k-4)h) + 272\sum_{k=1}^N f(\alpha + (6k-3)h)+$$

$$+27\sum_{k=1}^N f(\alpha + (6k-2)h) + 216\sum_{k=1}^N f(\alpha + (6k-1)h)+$$

$$\left. + 82\sum_{k=1}^{N-1} f(\alpha + 6kh)\right) - \frac{3(\beta-\alpha)}{2800}h^8 f^{(8)}(\eta)$$

for some $\eta \in [\alpha, \beta]$, if $f \in C^8[\alpha, \beta]$.

7/17280 rule for the interval [0,7h] (8 nodes).

$$\int_0^{7h} f(x)dx = \frac{7h}{17280}(751f(0) + 3577f(h) + 1323f(2h) + 2989f(3h)+$$

$$+2989f(4h) + 1323f(5h) + 3577f(6h) + 751f(7h)) +$$

$$-\frac{8163}{518400}h^9 f^{(8)}(\xi^*) \text{ for some } \xi^* \in [0, 7h], \text{ if } f \in C^8[0, 7h].$$

Summed 7/17280 rule for equidistant nodes. Here $h = \frac{(\beta-\alpha)}{7N}$.

$$\int_\alpha^\beta f(x)dx = \frac{7h}{17280}\left(751f(\alpha) + 751f(\beta) + 3577\sum_{k=1}^N f(\alpha + (7k-6)h)+\right.$$

$$+1323\sum_{k=1}^N f(\alpha + (7k-5)h) + 2989\sum_{k=1}^N f(\alpha + (7k-4)h)+$$

$$+2989\sum_{k=1}^N f(\alpha + (7k-3)h) + 1323\sum_{k=1}^N f(\alpha + (7k-2)h)+$$

$$\left. + 3577\sum_{k=1}^N f(\alpha + (7k-1)h) + 1502\sum_{k=1}^{N-1} f(\alpha + 7kh)\right) +$$

$$-\frac{8163(\beta-\alpha)}{3628800}h^8 f^{(8)}(\eta) \text{ for some } \eta \in [\alpha, \beta], \text{ if } f \in C^8[\alpha, \beta].$$

A derivation of the above Newton-Cotes formulas can be found in [ISAA66], p.323/4 or [KRYL62], 6.1.

15.3.5 The Error Order of Newton-Cotes Formulas

We assume that we are working with $n + 1$ nodes x_k in the interval $[\alpha, \beta]$ and that f is sufficiently often continuously differentiable there. The local error order of the quadrature formula Q used is assumed to be $O(h^q)$.

Then the following statements hold:

1) *For even* $n + 1$: $q = n + 2$, i.e., polynomials up to the n^{th} degree are integrated exactly, the global error order of Q is $O(h^{n+1})$.

 Example 1 (*TR* rule): $n + 1 = 2$ nodes are used here, thus $n = 1$ and first degree polynomials are integrated without error. The local error order is $O(h^3)$, while the global error order is $O(h^2)$.

 Example 2 (3/8 formula): $n + 1 = 4$ nodes, $n = 3$, with local error order $O(h^5)$, and global error order $O(h^4)$. The 3/8 formula integrates all polynomials of degree 3 exactly.

2) *For odd* $n + 1$: $q = n + 3$, polynomials up to the $(n + 1)^{st}$ degree are integrated exactly, the global error order of Q is $O(h^{n+2})$.

 Example (Simpson's rule): $n + 1 = 3$ here, hence $n = 2$ and the method has a local error order is $O(h^5)$, the global error order is $O(h^4)$. For error estimates see section 15.11 .

All the Newton-Cotes formulas mentioned above are formulas of closed type since they use the interval endpoints as nodes. Newton-Cotes formulas of open type that do not use the endpoints as nodes can be found in [CARN69], p.75. If the degree of the integrated interpolating polynomial increases, i.e., for a large number (> 8) of nodes, negative weights will appear so that the convergence of the associated quadrature formulas is no longer assured, see section 15.13. Besides, with higher degrees, the coefficients will vary in size more and more, a fact which can lead to an undesirable accumulation of rounding errors. For this reason, it is better to employ summed formulas of low error order with sufficiently fine partitions or some other procedure instead of formulas of higher order when integrating over large intervals.

Bibliography for Section 15.3

[CARN69], 2.1-2.6; [CONT80], 5.3; [ISAA66], 7.1; [McCA67], 8.2; [NIED87], 11.4; [STOE91], 3.1; [ZURM65], §13.1-5.

15.4 Maclaurin Quadrature Formulas

The nodes of the *Maclaurin formulas* are always chosen as the center of a subinterval of length h. Thus, they are formulas of open type. The weights and remainder can be determined by means of Taylor expansion.

15.4.1 The Tangent Trapezoidal Formula

Tangent trapezoidal formula for the interval [0,h].

Here we consider the integral of f over the interval $[a, b] = [0, h]$ and choose only one node x_0 in $[0, h]$. If we demand x_0 to be chosen so that constants and linear polynomials are integrated exactly, then from (15.4) with $n = 1$ and $M = 1$, we find the solution as $A_0 = h$ and $x_0 = h/2$. The quadrature formula (15.3) becomes

$$Q^{TT}(0, h) = A_0 f(x_0) = h f\left(\frac{h}{2}\right).$$

$Q^{TT}(0, h)$ is called the *tangent trapezoidal formula* since geometrically it measures the area of the trapezoid with one side formed by the tangent for $f(x)$ at $x = h/2$.

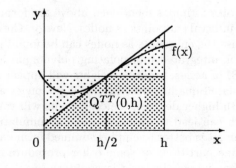

Figure 15.4: The tangent trapezoidal formula

For the associated remainder we obtain

$$E^{TT}(0,h) = \frac{h^3}{24}f''(\xi^*) \text{ for some } \xi^* \in [0,h], \text{ if } f \in C^2[0,h].$$

Its local error order is $O(h^3)$.

The *tangent trapezoidal rule* for the interval $[0,h]$ is

$$\int_0^h f(x)dx = Q^{TT}(0,h) + E^{TT}(0,h) \quad \text{with}$$

$$Q^{TT}(0,h) = hf\left(\frac{h}{2}\right),$$

$$E^{TT}(0,h) = \frac{h^3}{24}f''(\xi^*) \text{ for some } \xi^* \in [0,h], \text{ if } f \in C^2[0,h].$$

Summed tangent trapezoidal formula for equidistant nodes.

If the definite integral of f needs to be found for a large interval $[\alpha, \beta]$, we partition $[\alpha, \beta]$ into N subintervals of length $h = \frac{\beta - \alpha}{N}$. Then the *summed tangent trapezoidal rule* with equidistant nodes becomes

$$\int_\alpha^\beta f(x)dx = Q_h^{TT}(\alpha, \beta) + E_h^{TT}(\alpha, \beta) \quad \text{with}$$

$$Q_h^{TT}(\alpha, \beta) = h \sum_{k=0}^{N-1} f(\alpha + (2k+1)\frac{h}{2}),$$

$$E_h^{TT}(\alpha, \beta) = \frac{h^2}{24}(\beta - \alpha)f''(\eta) \text{ for some } \eta \in [\alpha, \beta], \text{ if } f \in C^2[\alpha, \beta].$$

Here the global error order is $O(h^2)$.

Summed tangent trapezoidal formula for non-equidistant nodes.

For the partition

$$P : \alpha = t_0 < t_1 < t_2 < \ldots < t_N = \beta$$

we set $h_k = t_{k+1} - t_k$ and $h_{\max} := \max_{0 \le k \le N-1}\{h_k\}$ to obtain

$$\int_\alpha^\beta f(x)dx = Q_{h_k}^{TT}(\alpha, \beta) + E_{h_k}^{TT}(\alpha, \beta) \text{ with}$$

$$Q_{h_k}^{TT}(\alpha, \beta) = \sum_{k=0}^{N-1}(t_{k+1} - t_k)f\left(\frac{t_k + t_{k+1}}{2}\right), \text{ and}$$

$$E_{h_k}^{TT}(\alpha, \beta) = O(h_{\max}^2).$$

REMARK. Both trapezoidal formulas TR and TT have the same error order. The remainder coefficient of the TT formula is only half as big as the one for the TR rule. Besides, the TT formula requires one less functional evaluation since the center of the subintervals serve as nodes.

15.4.2 Other Maclaurin Formulas

In the following, we give Maclaurin formulas for 2, 3, 4 and 5 nodes and their associated remainders:

Rule for the interval [0,2h] with 2 nodes:

$$\int_0^{2h} f(x)dx = h\left(f\left(\frac{h}{2}\right) + f\left(\frac{3h}{2}\right)\right) + \frac{h^3}{12}f''(\xi^*), \quad \xi^* \in [0, 2h], \quad f \in C^2[0, 2h].$$

Summed rule with equidistant nodes. Here $h = \frac{(\beta-\alpha)}{2N}$.

$$\int_\alpha^\beta f(x)dx = h\sum_{k=1}^{2N} f\left(\alpha + (2k-1)\frac{h}{2}\right) + \frac{\beta-\alpha}{24}h^2 f''(\eta) \quad \begin{array}{l}\text{for some } \eta \in [\alpha, \beta],\\ \text{if } f \in C^2[\alpha, \beta].\end{array}$$

Rule for the interval [0,3h] with 3 nodes.

$$\int_0^{3h} f(x)dx = \frac{3h}{8}\left(3f\left(\frac{h}{2}\right) + 2f\left(\frac{3h}{2}\right) + 3f\left(\frac{5h}{2}\right)\right) + \frac{21}{640}h^5 f^{(4)}(\xi^*)$$
for some $\xi^* \in [0, 3h]$, if $f \in C^4[0, 3h]$.

Summed rule with equidistant nodes. With $h = \frac{(\beta-\alpha)}{3N}$:

$$\int_\alpha^\beta f(x)dx = \frac{3h}{8}\sum_{k=1}^{N-1}\left(3f\left(\alpha + (6k+1)\frac{h}{2}\right) + 2f\left(\alpha + (6k+3)\frac{h}{2}\right) + \right.$$
$$\left. + 3f\left(\alpha + (6k+5)\frac{h}{2}\right)\right) + \frac{1701}{20480}(\beta-\alpha)h^4 f^{(4)}(\eta),$$
for some $\eta \in [\alpha, \beta]$, if $f \in C^4[\alpha, \beta]$.

Rule for the interval [0,4h] with 4 nodes.

$$\int_0^{4h} f(x)dx = \frac{h}{12}\left(13f\left(\frac{h}{2}\right) + 11f\left(\frac{3h}{2}\right) + 11f\left(\frac{5h}{2}\right) + 13f\left(\frac{7h}{2}\right)\right) +$$

$$+ \frac{103}{1440}h^5 f^{(4)}(\xi^*) \text{ for } \xi^* \in [0,4h], \text{ if } f \in C^4[0,4h].$$

Summed rule with equidistant nodes. Set $h = \frac{(\beta-\alpha)}{4N}$:

$$\int_\alpha^\beta f(x)dx = \frac{h}{12}\sum_{k=0}^{N-1}\left(13f\left(\alpha + (8k+1)\frac{h}{2}\right) + 11f\left(\alpha + (8k+3)\frac{h}{2}\right) +\right.$$

$$+ 11f\left(\alpha + (8k+5)\frac{h}{2}\right) + 13f\left(\alpha + (8k+7)\frac{h}{2}\right)\Big) +$$

$$+ \frac{103}{5760}(\beta-\alpha)h^4 f^{(4)}(\eta) \text{ for } \eta \in [\alpha,\beta], \text{ if } f \in C^4[\alpha,\beta].$$

Rule for the interval [0,5h] with 5 nodes.

$$\int_0^{5h} f(x)dx = \frac{5h}{1152}\left(275f\left(\frac{h}{2}\right) + 100f\left(\frac{3h}{2}\right) + 402f\left(\frac{5h}{2}\right) + 100f\left(\frac{7h}{2}\right) +\right.$$

$$+ 275f\left(\frac{9h}{2}\right)\Big) + \frac{435\,546\,875}{3170\,893\,824}h^7 f^{(6)}(\xi^*)$$

$$\text{for some } \xi^* \in [0,5h], \text{ if } f \in C^6[0,5h].$$

Summed rule with equidistant nodes. With $h = \frac{(\beta-\alpha)}{5N}$:

$$\int_\alpha^\beta f(x)dx = \frac{5h}{1152}\sum_{k=0}^{N-1}\left(275f\left(\alpha + (10k+1)\frac{h}{2}\right) + 100f\left(\alpha + (10k+3)\frac{h}{2}\right) +\right.$$

$$+ 402f\left(\alpha + (10k+5)\frac{h}{2}\right) + 100f\left(\alpha + (10k+7)\frac{h}{2}\right) +$$

$$+ 275f\left(\alpha + (10k+9)\frac{h}{2}\right)\Big) + \frac{87\,109\,375}{3170\,893\,824}(\beta-\alpha)h^6 f^{(6)}(\eta)$$

$$\text{for some } \eta \in [\alpha,\beta], \text{ if } f \in C^6[\alpha,\beta].$$

In this list, we note that the formulas with an odd number of nodes are more advantageous, as is the case with Newton-Cotes formulas . The formula for $n = 6$ is not given since it has the same error order as that for $n = 5$. The formula for $n = 7$ already contains a negative weight, namely A_0, so that the convergence of this rule is not assured anymore, see theorem 15.3 in section 15.13.

Bibliography for Section 15.4

[ENGE80], 6.1; [HÄMM78], 10; [STUM82], 4.1.2.

15.5 Euler-Maclaurin Formulas

Euler-Maclaurin formulas for the interval [0,h].

The Euler-Maclaurin formulas are generated by integrating the Newton interpolation formula $\tilde{N}_+(t)$ for descending differences (see chapter 9.5.2). If $f \in C^{2n}[0, h]$, we consider the integral of f over $[0, h]$ with nodes $x_0 = 0$, and $x_1 = h$. Then an *Euler-Maclaurin formula* (*EM_n formula*) results for each $n \in \mathbb{N}$. If $f \in C^{2n}[0, h]$ then:

$$(15.7) \quad Q^{EM_n}(0,h) = \frac{h}{2}\left(f(0) + f(h)\right) + \sum_{j=1}^{n-1} \frac{B_{2j}}{(2j)!} h^{2j}\left(f^{(2j-1)}(0) - f^{(2j-1)}(h)\right)$$

for the *Bernoulli numbers*

$$B_0 = 1, B_1 = -\tfrac{1}{2}, B_2 = \tfrac{1}{6}, B_4 = -\tfrac{1}{30}, B_6 = \tfrac{1}{42}, \ldots;$$
$$B_{2j+1} = 0 \quad \text{for} \quad j = 1, 2, \ldots.$$

The associated remainder is

$$(15.8) \quad E^{EM_n}(0,h) = -\frac{B_{2n}}{(2n)!} h^{2n+1} f^{(2n)}(\xi^*) \text{ for some } \xi^* \in [0,h],$$

i.e., the local error order is $O(h^{2n+1})$.

For each n there is an *Euler-Maclaurin formula EM_n* defined by (15.7) and (15.8):

$$\int_0^h f(x)dx = Q^{EM_n}(0,h) + E^{EM_n}(0,h).$$

Summed Euler-Maclaurin formulas for equidistant nodes.

If the function is to be integrated over a large interval $[\alpha, \beta]$, one should partition $[\alpha, \beta]$ into N subintervals of length $h = \frac{\beta-\alpha}{N}$ and employ an EM_n formula for each subinterval with the associated remainder. Thus one can obtain the *summed Euler-Maclaurin rule* as:

$$\int\limits_{\alpha}^{\beta} f(x)dx = Q_h^{EM_n}(\alpha,\beta) + E_h^{EM_n}(\alpha,\beta)$$

with the *summed Euler-Maclaurin formula*

$$(15.9) \quad Q_h^{EM_n}(\alpha,\beta) = \frac{h}{2}\left(f(\alpha) + 2\sum_{\nu=1}^{N-1} f(\alpha + \nu h) + f(\beta)\right) +$$

$$+ \sum_{j=1}^{n-1} \frac{B_{2j}}{(2j)!} h^{2j} \left(f^{(2j-1)}(\alpha) - f^{(2j-1)}(\beta)\right)$$

and its remainder:

$$E_h^{EM_n}(\alpha,\beta) = -\frac{\beta - \alpha}{(2n)!} B_{2n} h^{2n} f^{(2n)}(\eta) \text{ for some } \eta \in [\alpha,\beta].$$

Its global error order is $O(h^{2n})$.

REMARK. Using the trapezoidal rule, one can express (15.7) as

$$Q^{EM_n}(0,h) = Q^{TR}(0,h) + \sum_{k=1}^{n-1} \tilde{c}_{2k} h^{2k}.$$

Using the summed trapezoidal formula for (15.9), we obtain

$$Q_h^{EM_n}(\alpha,\beta) = Q_h^{TR}(\alpha,\beta) + \sum_{k=1}^{n-1} c_{2k} h^{2k}.$$

In both expressions, the coefficients \tilde{c}_{2k} and c_{2k} are independent of h. Thus the simple and summed Euler-Maclaurin formula are each a composite formula made up of the simple or summed trapezoidal rule and one correction term. For $n = 1$ the TR rule and the EM_n formula are identical.

Bibliography for Section 15.5

[BRAS77], V.2; [BROS64], §29; [HÄMM78], p.103; [STRO74], 3.11; [STUM82], 4.2.1.

15.6 Chebyshev Quadrature Formulas

When constructing all previously mentioned quadrature formulas of type (15.3), we have assumed that the $n + 1$ nodes $x_k \in [a, b]$ were given. The weights A_k were then obtained as solutions of the linear system (15.4). If the functional values $f(x)$ of the integrand are determined empirically and they all carry similar errors, then the error of the approximate integral will become smallest, if all weights of the quadrature formula are chosen as equal. The *Chebyshev formulas* have the form (15.3) with identical weights.

Chebyshev formulas for the interval [-h,h].

Consider the integral of f over the interval $[-h, h]$ and define the *Chebyshev formula* in the following manner:

$$I(f; -h, h) = \int\limits_{-h}^{h} f(x)dx = Q^{Ch_{n+1}}(-h, h) + E^{Ch_{n+1}}(-h, h),$$

where $n + 1$ is the number of nodes $x_k \in [-h, h]$. $Q^{Ch_{n+1}}(-h, h)$ is called the *Chebyshev formula* (Ch_{n+1} formula) for $n + 1$ nodes and $E^{Ch_{n+1}}(-h, h)$ is the *remainder* of the Ch_{n+1} formula. The weights A_k are given the constant value

$$A_k = \frac{2h}{n+1} \quad \text{for all } k = 0, \ldots, n.$$

If we require that the quadrature formula $Q^{Ch_{n+1}}(-h, h)$ integrates polynomials up to degree $M = n + 1$ exactly, then (15.4) with $m = 1, \ldots, M$ defines $n + 1$ nonlinear equations for the $n + 1$ nodes x_k. Clearly we must assume that the functional values for the nodes x_k can be calculated or found in a table. If one has only a limited table of values for f, then the Chebyshev formulas generally cannot be used.

For $n = 1$ we set $a = -h$, $b = h$ and $m = 1, 2$ in (15.4). Since $A_k = \frac{2h}{(n+1)}$, $A_0 = A_1 = h$. Thus the nodes must be $x_0 = -\frac{h}{\sqrt{3}}$ and $x_1 = \frac{h}{\sqrt{3}}$, and the associated *Chebyshev rule for 2 nodes* is:

$$\int\limits_{-h}^{h} f(x)dx \quad = \quad Q^{Ch_2}(-h, h) + E^{Ch_2}(-h, h) \quad \text{with}$$

$$Q^{Ch_2}(-h, h) \quad = \quad A_0 f(x_0) + A_1 f(x_1) = h \left(f\left(-\frac{h}{\sqrt{3}}\right) + f\left(\frac{h}{\sqrt{3}}\right) \right),$$

$$E^{Ch_2}(-h, h) \quad = \quad O(h^5).$$

It has the local error order $O(h^5)$.

In general, Chebyshev formulas with 2ν or $2\nu + 1$ nodes have an error order of $O(h^{2\nu+3})$.

Table of the Chebyshev nodes for the interval [-h,h]:

n	x_k for $k = 0, \ldots, n$
1	$x_{0,1} = \pm 0.577350\,h$
2	$x_{0,2} = \pm 0.707107\,h, \quad x_1 = 0$
3	$x_{0,3} = \pm 0.794654\,h, \quad x_{1,2} = \pm 0.187592\,h$
4	$x_{0,4} = \pm 0.832498\,h, \quad x_{1,3} = \pm 0.374541\,h, \qquad x_2 = 0$
5	$x_{0,5} = \pm 0.866247\,h, \quad x_{1,4} = \pm 0.422519\,h,$ $x_{2,3} = \pm 0.266635\,h$
6	$x_{0,6} = \pm 0.883862\,h, \quad x_{1,5} = \pm 0.529657\,h,$ $x_{2,4} = \pm 0.323912\,h, \quad x_3 = 0$

The nodes x_k are real numbers for $n = 0, \ldots, 6$ and $n = 8$ only.

Summed Chebyshev formulas for equidistant nodes.

If one has to integrate over a large interval $[\alpha, \beta]$, one can partition $[\alpha, \beta]$ into N subintervals of length $2h$ with $h = \frac{\beta - \alpha}{2N}$ and use the corresponding Ch_{n+1} formula for each subinterval. The nodes x_k of the interval should be transformed as follows:

$$x_k \to \alpha + (2j + 1)h + x_k \quad \text{for } j = 0, \ldots, N - 1 \text{ and } k = 0, \ldots, n.$$

Thus we get for example for $n = 1$ and equidistant nodes the following *summed Chebyshev rule*:

$$\int_{\alpha}^{\beta} f(x)dx = Q_h^{Ch_2}(\alpha,\beta) + E_h^{Ch_2}(\alpha,\beta) \quad \text{with}$$

$$Q_h^{Ch_2}(\alpha,\beta) = h \sum_{j=0}^{N-1} \left(f\left(\alpha + (2j+1)h - \frac{h}{\sqrt{3}}\right) + f\left(\alpha + (2j+1)h + \frac{h}{\sqrt{3}}\right) \right),$$

$$E_h^{Ch_2}(\alpha,\beta) = O(h^4).$$

Here $Q_h^{Ch_2}(\alpha,\beta)$ is the *summed Chebyshev formula* with two nodes and $E_h^{Ch_2}(\alpha,\beta)$ is the *remainder* of the summed Ch_2 formula. Its global error order is $O(h^4)$.

Summed Chebyshev formulas for non-equidistant nodes.

With the partition

$$P : \alpha = t_0 < t_1 < t_2 < \ldots < t_m = \beta$$

and $h_j = t_{j+1} - t_j$, $h_{\max} := \max_{0 \le j \le m-1}\{h_j\}$, the nodes x_k of the reference interval must be transformed as follows:

$$x_k \to t_j + \frac{h_j}{2} + x_k \quad \text{for} \quad j = 0,\ldots,m-1 \text{ and } \quad k = 0,\ldots,n.$$

For $n = 1$ one obtains the summed formula as

$$\int_{\alpha}^{\beta} f(x)dx = Q_{h_j}^{Ch_2}(\alpha,\beta) + E_{h_j}^{Ch_2}(\alpha,\beta) \quad \text{with}$$

$$Q_{h_j}^{Ch_2}(\alpha,\beta) = \sum_{j=0}^{m-1} h_j \left[f\left(t_j + \frac{h_j}{2} - \frac{h_j}{2\sqrt{3}}\right) + f\left(t_j + \frac{h_j}{2} + \frac{h_j}{2\sqrt{3}}\right) \right],$$

$$E_{h_j}^{Ch_2}(\alpha,\beta) = O(h_{\max}^4).$$

For an even number of nodes the Chebyshev formulas have a better error order than the Newton-Cotes formulas. For error estimates see section 15.11.

Bibliography for Section 15.6

[ENGE87], 9.4; [KRYL62], 10.

15.7 Gauß Quadrature Formulas

In order to obtain the *Gauß formulas* of optimal accuracy, we do prescribe neither the nodes x_k nor the weights A_k in (15.3). Hence (15.4) contains $2(n+1) = 2n + 2$ free parameters. The requirement that the quadrature formula (15.3) integrate all polynomials up to degree $M = 2n + 1$ exactly leads to a system of $2n + 2$ equations for the $n + 1$ weights A_k and the $n + 1$ nodes $x_k, k = 0, \ldots, n$, namely:

$$(15.10) \qquad \frac{1}{m+1}(b^{m+1} - a^{m+1}) = \sum_{k=0}^{n} A_k x_k^m, \quad m = 0, \ldots, 2n + 1.$$

(15.10) is a linear system for the weights A_k, and a nonlinear system for the nodes x_k. Hence we must assume here that the functional values $f(x)$ can be calculated or found in a table for the so-called *Gaussian nodes* $x_k \in [a, b]$. If for the function f we have a table of values in which the Gaussian nodes do not appear, we should calculate the integral by means of a Newton-Cotes formula or a Maclaurin formula for equidistant nodes. If the nodes are not equidistant we should calculate the integral by means of a quadrature formula constructed from (15.4).

For the integral of f over the interval $[a, b] = [-1, 1]$ one can show that the $n + 1$ Gaussian nodes x_k are the zeros of the *Legendre polynomials* $P_{n+1}(x)$ in $[-1, 1]$, see section 8.1.2 and e.g. [STRO66], 1.2; [STUM82], p.86/87; [ENGE87], p.277.

Gauß formula for the interval [-h,h].

Consider the integral of f on the interval $[-h, h]$ and set

$$\int_{-h}^{h} f(x)dx = Q^{G_{n+1}}(-h, h) + E^{G_{n+1}}(-h, h) = \sum_{k=0}^{n} A_k f(x_k) + O(h^q).$$

This is called the *Gaussian rule*, $Q^{G_{n+1}}(-h, h)$ is a *Gauß formula* (G_{n+1} formula) and $E^{G_{n+1}}(-h, h)$ is its *remainder* for the $n + 1$ Gaussian nodes.

The interval $[-1, 1]$ should always be transformed into $[-h, h]$. Then we can list the weights A_k and nodes $x_k, k = 0, \ldots, n$, for several Gauß quadrature formulas:

Table of the Gaussian nodes and weights:

n	$x_k, \quad k = 0,\dots,n$	$A_k, \quad k = 0,\dots,n$
0	$x_0 = 0$	$A_0 = 2h$
1	$x_{0,1} = \pm h/\sqrt{3}$	$A_0 = A_1 = h$
2	$x_{0,2} = \pm\sqrt{0.6}\,h$ $x_1 = 0$	$A_0 = A_2 = \frac{5}{9}h$ $A_1 = \frac{8}{9}h$
3	$x_{0,3} = \pm 0.86113631\,h$ $x_{1,2} = \pm 0.33998104\,h$	$A_0 = A_3 = 0.34785485\,h$ $A_1 = A_2 = 0.65214515\,h$
4	$x_{0,4} = \pm 0.90617985\,h$ $x_{1,3} = \pm 0.53846931\,h$ $x_2 = 0$	$A_0 = A_4 = 0.23692689\,h$ $A_1 = A_3 = 0.47862867\,h$ $A_2 = \frac{128}{225}h$
5	$x_{0,5} = \pm 0.93246951\,h$ $x_{1,4} = \pm 0.66120939\,h$ $x_{2,3} = \pm 0.23861919\,h$	$A_0 = A_5 = 0.17132449\,h$ $A_1 = A_4 = 0.36076157\,h$ $A_2 = A_3 = 0.46791393\,h$

Further values are given in [ABRA72], table 25.4.

The remainder has the general form

$$E^{G_{n+1}}(-h,h) = \frac{2^{2n+3}((n+1)!)^4}{(2n+3)((2n+2)!)^3} h^{2n+3} f^{(2n+2)}(\xi^*),$$

for some $\xi^* \in [-h,h]$, if $f \in C^{2n+2}[-h,h]$,

i.e., the *local error order* is $O(h^{2n+3})$.

Next we shall explicitly describe two Gaussian quadrature rules, namely the ones for 2 and 3 nodes $x_k \in [-h,+h]$:

1) n = 1 (2 nodes):

$$\int_{-h}^{h} f(x)dx \quad = \quad Q^{G_2}(-h,h) + E^{G_2}(-h,h) \quad \text{with}$$

$$Q^{G_2}(-h,h) \quad = \quad h\left(f\left(\tfrac{-h}{\sqrt{3}}\right) + f\left(\tfrac{h}{\sqrt{3}}\right)\right),$$

$$E^{G_2}(-h,h) \quad = \quad \tfrac{h^5}{135}f^{(4)}(\xi^*) \text{ for some } \xi^* \in [-h,h], \text{ if } f \in C^4[-h,h].$$

2) n = 2 (3 nodes):

$$\int_{-h}^{h} f(x)dx \quad = \quad Q^{G_3}(-h,h) + E^{G_3}(-h,h) \quad \text{with}$$

$$Q^{G_3}(-h,h) \quad = \quad \tfrac{h}{9}\left(5f(-\sqrt{0.6}h) + 8f(0) + 5f(\sqrt{0.6}h)\right),$$

$$E^{G_3}(-h,h) \quad = \quad \tfrac{h^7}{15750}f^{(6)}(\xi^*) \text{ for some } \xi^* \in [-h,h], \text{ if } f \in C^6[-h,h].$$

with two nodes, the formula has the local error order $O(h^5)$, while for three nodes the local error order is $O(h^7)$. We remark that the Newton-Cotes formulas with a local error order of $O(h^5)$ and $O(h^7)$ require the use of three and five nodes, respectively.

For $n = 4$ and $n = 5$, similar formulas can easily be constructed from the above table. We have

$$E^{G_4} = \tfrac{h^9}{3472875}f^{(8)}(\xi^*), \quad E^{G_5} = \tfrac{h^{11}}{1237732650}f^{(10)}(\xi^*)$$
$$\text{for some } \xi^* \in [-h,h].$$

Summed Gauß quadrature formulas with equidistant nodes.

If the integral of f extends over a large interval $[\alpha, \beta]$, one partitions $[\alpha, \beta]$ into N subintervals of length $2h$ for $h = \tfrac{\beta-\alpha}{2N}$. Next the nodes must be transformed as follows:

$$x_k \to \alpha + (2j+1)h + x_k \quad \text{for } j = 0, \dots, N-1 \text{ and } k = 0, \dots, n.$$

For $n = 1$ and $n = 2$ the following *summed Gauß rules* arise:

$$\int_\alpha^\beta f(x)dx \quad = Q_h^{G_2}(\alpha,\beta) + E_h^{G_2}(\alpha,\beta) \quad \text{with}$$

$$Q_h^{G_2}(\alpha,\beta) \quad = h \sum_{j=0}^{N-1} \left(f(\alpha + (2j+1)h - \frac{h}{\sqrt{3}}) + f(\alpha + (2j+1)h + \frac{h}{\sqrt{3}}) \right),$$

$$E_h^{G_2}(\alpha,\beta) \quad = \frac{\beta-\alpha}{270} h^4 f^{(4)}(\eta) \text{ for some } \eta \in [\alpha,\beta], \text{ if } f \in C^4[\alpha,\beta]. \text{ And}$$

$$\int_\alpha^\beta f(x)dx \quad = Q_h^{G_3}(\alpha,\beta) + E_h^{G_3}(\alpha,\beta) \quad \text{with}$$

$$Q_h^{G_3}(\alpha,\beta) \quad = \frac{h}{9} \sum_{j=0}^{N-1} \left(5f(\alpha + (2j+1)h - \sqrt{\frac{3}{5}}h) + 8f(\alpha + (2j+1)h) + \right.$$

$$\left. + 5f(\alpha + (2j+1)h + \sqrt{\frac{3}{5}}h) \right),$$

$$E_h^{G_3}(\alpha,\beta) \quad = \frac{\beta-\alpha}{31500} h^6 f^{(6)}(\eta) \text{ for some } \eta \in [\alpha,\beta], \text{ if } f \in C^6[\alpha,\beta].$$

Summed Gauß quadrature formulas with non-equidistant nodes.

With the partition

$$P: \alpha = t_0 < t_1 < t_2 < \ldots < t_m = \beta$$

we set $h_j = t_{j+1} - t_j$ and $h_{\max} = \max_{0 \le j \le m-1} \{h_j\}$. Then the nodes x_k of the reference interval must be transformed:

$$x_k \to t_j + \frac{h_j}{2} + x_k \quad \text{for} \quad j = 0, \ldots, m-1, \text{ and } k = 0, \ldots, n.$$

And for $n = 1$ the summed rule is:

$$n = 1: \quad \int_\alpha^\beta f(x)dx = Q_{h_j}^{G_2}(\alpha,\beta) + E_{h_j}^{G_2}(\alpha,\beta) \quad \text{with}$$

$$Q_{h_j}^{G_2}(\alpha,\beta) = \sum_{j=0}^{m-1} h_j \left[f\left(t_j + \frac{h_j}{2} - \frac{h_j}{2\sqrt{3}} \right) + f\left(t_j + \frac{h_j}{2} + \frac{h_j}{2\sqrt{3}} \right) \right],$$

$$E_{h_j}^{G_2}(\alpha,\beta) = O(h_{\max}^4).$$

For n = 2 the summed rule becomes:

$$n = 2: \qquad \int\limits_{\alpha}^{\beta} f(x)dx \quad = Q_{h_j}^{G_3}(\alpha, \beta) + E_{h_j}^{G_3}(\alpha, \beta) \quad \text{with}$$

$$Q_{h_j}^{G_3}(\alpha, \beta) \quad = \sum_{j=0}^{m-1} h_h \left\{ \frac{5h_j}{18} \left[f\left(t_j + \frac{h_j}{2} - \frac{\sqrt{0.6}}{2} h_j \right) + \right.\right.$$

$$\left.\left. + f\left(t_j + \frac{h_j}{2} + \frac{\sqrt{0.6}}{2} h_j \right) \right] + \frac{4}{9} h_j f\left(t_j + \frac{h_j}{2} \right) \right\},$$

$$E_{h_j}^{G_3}(\alpha, \beta) \quad = O(h_{\max}^6).$$

All Gauß quadrature formulas $Q^{G_{n+1}}$ have optimal error order. In comparison with the Newton-Cotes formulas of the same error order, they need only about half of the computational time due to fewer function evaluations. It is difficult to calculate the weights A_k and the Gaussian nodes x_k from the nonlinear system (15.10), so one should rely on our table or the one in [ABRA72], table 25.4. Error estimates are given in section 15.11 .

Bibliography for Section 15.7

[ATKI89], 5.3; [CARN69], 2.10; [COLL73] I, 3.2.3; [CONT80], 4.6; [DAVI82], 2.7; [ENGE80]; [ENGE87], 9.4; [HÄMM78], 12; [ISAA66], 7.3; [NIED87], 11.5; [NOBL65] II, 9.5; [STOE91], 3.5; [STRO66]; [STUM82], 4.3; [WERN79], III §6 V; [ZURM65], §13.6.

15.8 Calculation of Weights and Nodes of Generalized Gaussian Quadrature Formulas

In the following we shall indicate a method to calculate the weights and nodes of *generalized Gauss quadrature formulas* for the reference interval $[a, b]$ which avoids having to solve the nonlinear system (15.10).

A quadrature formula

$$Q(f, g; a, b) = \sum_{i=1}^{n} A_i f(x_i)$$

is to be designed to supply an approximation for the integral $\int\limits_{a}^{b} f(x)g(x)dx$ which integrates all polynomials $f = P$ up to degree $2n - 1$ exactly. Here g is

a weight function with $g(x) > 0$ in (a, b) that is integrable in all subintervals of $[a, b]$.

We can determine a polynomial

$$Q_n(x) = x^n + \sum_{k=0}^{n-1} q_k x^k$$

so that Q_n is orthogonal to all monomials x^j with $0 \le j \le n-1$:

$$\int_a^b Q_n(x)x^j g(x)dx = \int_a^b x^{n+j}g(x)dx + \sum_{k=0}^{n-1} q_k \int_a^b x^{k+j}g(x)dx = 0.$$

This defines a linear system for the coefficients q_k if the integrals $A_j = \int_a^b x^j g(x)dx$ are known for $0 \le j \le 2n-1$. Having solved for the q_k, the zeros x_i, $i = 1, \ldots, n$, of $Q_n(x)$ can subsequently be determined.

An arbitrary polynomial $P(x)$ of degree less than or equal to $2n-1$ can be expressed by long division in the form

$$P(x) = S(x)Q_n(x) + R(x),$$

where $S(x)$ and $R(x)$ are polynomials of degree at most $n-1$. For the zeros x_i of Q_n we have $P(x_i) = R(x_i)$ since $Q_n(x_i) = 0$. Since $R(x)$ has degree at most $n-1$, the Lagrange interpolation is exact, see section 9.2, i.e.,

$$R(x) = \sum_{i=1}^n R(x_i)L_i(x) \quad \text{with} \quad L_i(x) = \prod_{\substack{j=1 \\ j \ne i}}^n \frac{x - x_j}{x_i - x_j}.$$

As degree $S \le n-1$, orthogonality implies that

$$\int_a^b S(x)Q_n(x)g(x)dx = \sum_{j=0}^{n-1} s_j \int_a^b x^j Q_n(x)g(x)dx = 0.$$

Thus:

$$\int_a^b P(x)g(x)dx = \int_a^b R(x)g(x)dx = \sum_{i=1}^n R(x_i)\int_a^b L_i(x)g(x)dx$$

$$=: \sum_{i=1}^n P(x_i)A_i \quad \text{with} \quad A_i = \int_a^b L_i(x)g(x)dx.$$

From

$$Q_n(x) = \prod_{i=1}^{n}(x - x_i)$$

the Lagrange interpolation polynomials (see section 9.2.1) can be determined in the following way:

$$L_i^*(x) \quad := \frac{Q_n(x)}{(x - x_i)}$$

$$L_i(x) \quad = \frac{L_i^*(x)}{L_i^*(x_i)} = \sum_{j=0}^{n-1} L_{i,j}x^j,$$

where the $L_{i,j}$ are the coefficients of the polynomial $L_i(x)$.

Hence, the weights A_i are given as follows:

$$A_i = \int_a^b L_i(x)g(x)dx = \sum_{j=0}^{n-1} L_{i,j}\int_a^b x^j g(x)dx.$$

ALGORITHM 15.1 (*Generalized Gauß formulas*).

Given: An integer n and the exact values of $\int_a^b x^i g(x)dx$ for $i = 0, \ldots, 2n-1$.

Task: Find the Gauß quadrature formula for the interval $[a, b]$:

$$Q(f, g; a, b) = \sum_{i=1}^{n} A_i f(x_i) \quad \text{with} \quad \int_a^b f(x)g(x)dx \approx Q(f, g; a, b),$$

where g is a weight function with $g(x) > 0$ in (a, b).

1) Solve the linear system $\boldsymbol{Aq} - \boldsymbol{b} = \boldsymbol{0}$ with

$$\boldsymbol{A} = \begin{pmatrix} \int g(x)dx & \int xg(x)dx & \cdots & \int x^{n-1}g(x)dx \\ \int xg(x)dx & \int x^2 g(x)dx & & \int x^n g(x)dx \\ \vdots & \vdots & & \vdots \\ \int x^{n-1}g(x)dx & \int x^n g(x)dx & & \int x^{2n-2}g(x)dx \end{pmatrix},$$

$$\boldsymbol{q} = \begin{pmatrix} q_0 \\ q_1 \\ \vdots \\ q_{n-1} \end{pmatrix}, \quad \boldsymbol{b} = \begin{pmatrix} \int x^n g(x)dx \\ \int x^{n+1}g(x)dx \\ \vdots \\ \int x^{2n-1}g(x)dx \end{pmatrix}.$$

2) Calculate the zeros x_i for $i = 1, \ldots, n$ of

$$Q_n(x) = x^n + \sum_{k=0}^{n-1} q_k x^k.$$

(see the methods of chapter 3).

3) (a) Calculate the coefficients $L_{i,j}^*$ of $L_i^*(x)$ using the Horner scheme (Divide Q_n by $x - x_i$),

(b) Calculate $L_i^*(x_i)$, using its coefficients $L_{i,j}^*$ in a Horner scheme,

(c) Calculate $L_{i,j} = L_{i,j}^*/L_i^*(x_i)$.

4) Calculate the weights A_i as

$$A_i = \sum_{j=0}^{n-1} L_{i,j} \int_a^b x^j g(x) dx.$$

REMARK. Due to the fact that the end points of the interval $[a, b]$ are not taken into consideration when calculating the weights A_i, it is possible to employ Gauß quadrature formulas for improper integrals as well.

15.9 Clenshaw-Curtis Quadrature Formulas

If one wants to use quadrature formulas of high order whose nodes can be computed more easily than those of the Gauß quadrature formulas, we recommend to use the *Clenshaw-Curtis formulas* .

Clenshaw-Curtis formulas for the interval [-1,1].

These formulas are interpolation quadrature formulas of the form

$$Q^{CC_n}(f; -1, 1) := \sum_{k=0}^n A_k^{(n)} f(x_k) \text{ for even } n \geq 2$$

for the reference interval $[-1, 1]$ with the Chebyshev nodes

$$x_k = \cos \frac{k\pi}{n}, \qquad k = 0, \ldots, n,$$

which evaluate the integral

$$I(f; -1, 1) = \int_{-1}^{1} f(x)dx$$

for all polynomials of degree $n + 1$ exactly. The weights $A_k^{(n)}$ are all positive, in fact:

$$A_0^{(n)} = A_n^{(n)} = \frac{1}{n^2 - 1}, \text{ and}$$

$$A_k^{(n)} = 2\frac{n^2 - 1 - (-1)^k}{n(n^2 - 1)} - \frac{4}{n} \sum_{j=1}^{n/2-1} \frac{\cos(2jk\pi/n)}{4j^2 - 1} \text{ for } k = 1, \ldots, n - 1.$$

If one wants to integrate over an interval $[a, b]$ instead of $[-1, 1]$, one uses the transformed Clenshaw-Curtis formula

$$Q^{CC_n}(f; a, b) = \frac{b - a}{2} \sum_{k=0}^{n} A_k^{(n)} f\left(\frac{b - a}{2} \cos\left(\frac{k\pi}{n}\right) + \frac{b + a}{2}\right).$$

The local error order is $O(h^{n+3})$ for even n, if $f \in C^{n+2}[a, b]$.

Summed Clenshaw-Curtis formulas.

For a partition P of the interval of integration $[\alpha, \beta]$:

$$P : \alpha = t_0 < t_1 < t_2 < \ldots < t_m = \beta,$$

$$h_j := t_{j+1} - t_j \text{ and } h_{\max} = \max_{0 \leq j \leq m-1} \{h_j\},$$

the nodes x_k of the reference interval $[-1, 1]$ must be transformed as follows:

$$x_k \to \frac{t_{j+1} + t_j}{2} + \frac{h_j}{2} x_k.$$

Then we obtain the *summed Clenshaw-Curtis formula* as:

$$Q_{h_j}^{CC_n}(f; \alpha, \beta) =$$

$$\frac{1}{2}\left(\sum_{j=0}^{m-1}(t_{j+1} - t_j)\right) \cdot \left(\sum_{k=0}^{n} A_k^{(n)} f\left(\frac{t_{j+1} - t_j}{2} \cos\frac{k\pi}{n} + \frac{t_{j+1} + t_j}{2}\right)\right).$$

If $f \in C^{n+2}[\alpha, \beta]$ the global error has the order

$$E_{h_j}^{CC_n}(f; \alpha, \beta) = O(h_{\max}^{n+2}).$$

For error estimates, see section 15.11.

Bibliography for Section 15.9

[BRAS77]; [CLEN60]; [ENGE80], 7.7.4; [NIED87], 11.6 .

15.10 Romberg Integration

Romberg Integration is based upon the approximation of the integral $I(f; \alpha, \beta)$ by the trapezoidal rule. Quadrature formulas of higher error order are produced by successive halving of the step size and by an appropriate linear combination of the resulting approximations for the integral (see [ENGE87], pp.281).

First, one partitions $[\alpha, \beta]$ into N_0 subintervals of length $h_0 = \frac{\beta - \alpha}{N_0}$ and sets

$$N_j = 2^j N_0, \quad h_j = \frac{\beta - \alpha}{2^j N_0} = \frac{h_0}{2^j} \quad \text{for } j = 0, 1, 2, \ldots.$$

This describes the successive halving of the step size.

Next the integral of f over $[\alpha, \beta]$ is expressed as:

$$I(f; \alpha, \beta) = \int\limits_{\alpha}^{\beta} f(x)dx = L_j^{(k)}(f) + O(h_j^{2(k+1)}),$$

where $L_j^{(k)}(f)$ is a quadrature formula with error order $O(h_j^{2(k+1)})$.

The calculations are carried out *row by row* according to the following scheme: Calculate for each row i, $i = 0, \ldots, m$, the elements $L_s^{(t)}$ successively from left to right with $s = i, \ldots, 0$ and $t = 0, \ldots, i$.

CALCULATION SCHEME 15.2 (*Romberg integration*).

$L_j^{(0)} = Q_{h_j}^{TR}(\alpha,\beta)$	$L_j^{(1)}$	$L_j^{(2)}$	\cdots	$L_j^{(m-1)}$	$L_j^{(m)}$
$L_0^{(0)}$					
$L_1^{(0)}$	$L_0^{(1)}$				
$L_2^{(0)}$	$L_1^{(1)}$	$L_0^{(2)}$			
\vdots	\vdots	\vdots	\ddots		
$L_{m-1}^{(0)}$	$L_{m-2}^{(1)}$	$L_{m-3}^{(2)}$	\cdots	$L_0^{(m-1)}$	
$L_m^{(0)}$	$L_{m-1}^{(1)}$	$L_{m-2}^{(2)}$	\cdots	$L_1^{(m-1)}$	$L_0^{(m)}$

Here the $L_j^{(0)}$ are calculated from the formula:

$$L_j^{(0)}(f) := Q_{h_j}^{TR}(\alpha,\beta) = \frac{h_j}{2}\left(f(\alpha) + f(\beta) + 2\sum_{\nu=1}^{N_j-1} f(\alpha + \nu h_j)\right).$$

It is better and faster to use the above formula only for $j = 0$, while for $j = 1, 2, 3, \ldots$ the following formula should be used instead:

$$
\begin{aligned}
L_j^{(0)}(f) &= \tfrac{1}{2}L_{j-1}^{(0)} + h_j\{f(\alpha + h_j) + f(\alpha + 3h_j) + \ldots + f(\beta - h_j)\} \\
&= \tfrac{1}{2}L_{j-1}^{(0)} + h_j\sum_{k=0}^{N_{j-1}-1} f(\alpha + (2k+1)h_j).
\end{aligned}
$$

The $L_j^{(k)}$ are calculated from the formula

$$L_j^{(k)}(f) = L_{j+1}^{(k-1)}(f) + \frac{1}{4^k - 1}\left(L_{j+1}^{(k-1)}(f) - L_j^{(k-1)}(f)\right)$$

for $k \geq 1$ and $j = 0, 1, 2, \ldots$. This scheme is continued until for a given $\varepsilon > 0$ we have $|L_0^{(m)} - L_1^{(m-1)}| < \varepsilon$. Then $L_0^{(m)}(f)$ is used as the best approximate value for $I(f; \alpha, \beta)$. For $m \leq n - 1$ and the Bernoulli numbers B_{ij}, see section 15.5, Romberg integration satisfies:

$$I(f; \alpha, \beta) = \int_{\alpha}^{\beta} f(x)dx = L_0^{(m)}(f) + E^{R_m}(f; \alpha, \beta) \quad \text{with}$$

$$E^{R_m}(f; \alpha, \beta) = (-1)^{m+1}\frac{\beta - \alpha}{2^{m(m+1)}}\frac{B_{2m+2}}{(2m+2)!}h_0^{2m+2}f^{(2m+2)}(\xi), \quad \xi \in [\alpha, \beta].$$

Specifically, if $h_0 = 2^m h_m$, then

$$
\begin{aligned}
E^{R_m}(f;\alpha,\beta) &= (-1)^{m+1}(\beta-\alpha)\frac{B_{2m+2}}{(2m+2)!}2^{m(m+1)}h_m^{2m+2}f^{(2m+2)}(\xi)\\
&= O(h_m^{2m+2}) \text{ for some } \xi \in [\alpha,\beta],
\end{aligned}
$$

i.e., the remainder E^{R_m} has the order $O(h_m^{2m+2})$ for $h_m \to 0$.

For $f \in C^{2n}[\alpha,\beta]$, the columns $L_j^{(k)}$ of the scheme 15.2 converge linearly for each fixed k and $j \to \infty$ towards $I(f;\alpha,\beta)$. If f is infinitely differentiable, the diagonals and subdiagonals $L_j^{(k)}$ of the scheme converge superlinearly towards $I(f;\alpha,\beta)$ for fixed j and $k \to \infty$. It can be proved that the columns, as well as the diagonals and subdiagonals $L_j^{(k)}$ converge towards $I(f;\alpha,\beta)$, even for functions f that are only continuous.

Example: Here is a Romberg scheme with three rows:

h_j	$L_j^{(0)} = Q_{h_j}^{TR}(\alpha,\beta)$	$L_j^{(1)}$	$L_j^{(2)}$
h_0	$L_0^{(0)} = Q_{h_0}^{TR}(\alpha,\beta)$		
$h_1 = \frac{h_0}{2}$	$L_1^{(0)} = Q_{h_1}^{TR}(\alpha,\beta)$	$L_0^{(1)} = L_1^{(0)} + \frac{1}{3}\left(L_1^{(0)} - L_0^{(0)}\right)$	
$h_2 = \frac{h_1}{2}$	$L_2^{(0)} = Q_{h_2}^{TR}(\alpha,\beta)$	$L_1^{(1)} = L_2^{(0)} + \frac{1}{3}\left(L_2^{(0)} - L_1^{(0)}\right)$	$L_0^{(2)} = L_1^{(1)} + \\ + \frac{1}{15}\left(L_1^{(1)} - L_0^{(1)}\right)$

The value of $L_0^{(0)}$ is computed with the trapezoidal formula Q_h^{TR} for the stepsize h_0; $L_i^{(0)}$ can be computed following the simplified rule:

$$
\begin{aligned}
L_1^{(0)} &= \tfrac{1}{2}L_0^{(0)} + h_1 * \left\{ \text{ Sum of additional } f \text{ values at new nodes} \right\},\\
L_2^{(0)} &= \tfrac{1}{2}L_1^{(0)} + h_2 * \left\{ \text{ corresponding sum} \right\}.
\end{aligned}
$$

The rule used for forming $L_0^{(1)}$ and $L_0^{(2)}$ shows that precisely the estimate of the global error of $L_1^{(0)} = Q_{h_1}^{TR}$ is added in column two, and analogously for $L_2^{(0)}$. Thus the global error order of the entries in column two is raised by two powers of h when compared with those of column one.

The same is true for the entries in the third column. As the entries in the first column have errors of order $O(h_j^2)$, the entries in the second column have errors of order $O(h_j^4)$, while those in column three have the error order $O(h_j^6)$. Moreover it is easy to see that the entries in column two of the above scheme correspond to the approximations which would have been computed by using Simpson's rule Q^S for the appropriate step size for the respective row, i.e.,

$L_0^{(1)} = Q_{h_1}^S(f; \alpha, \beta)$ and $L_1^{(1)} = Q_{h_2}^S(f; \alpha, \beta)$.

Bibliography for Section 15.10

[CARN69], 2.7; [CONT80], 4.4; [DAVI82], 6.3; [ENGE80]; [ENGE87], 9.5; [HÄMM78], 11; [HENR72], 13.7; [McCA67], 8.4; [McCR72], 5.11; [MEIN79], 6.3; [STIE63], 6.2.2; [STOE91], 3.2-4; [STRO74], 3.10, 3.11; [STUM82], 4.2.2; [WERN79], III §8.

15.11 Error Estimates and Computational Errors

In general, a precise estimate of the quadrature error cannot be found because the derivatives of the integrand f will not be known, or if known, can be computed and estimated only with considerable effort on $[a, b]$. An exact knowledge of the remainder term is therefore generally only of theoretical use. Knowing the global error order $O(h^q)$ of a quadrature formula is, however, essential: it is enough for *estimating the true error* by using two approximate values for $I(f; \alpha, \beta)$ calculated for different step sizes:

Error estimate for equidistant nodes.

If the integral $I(f; \alpha, \beta)$ has been calculated approximately for the step size h_i by using a quadrature formula with the global error order $O(h_i^q)$, then

$$I(f; \alpha, \beta) = Q_{h_i}(\alpha, \beta) + E_{h_i}(\alpha, \beta) \quad \text{with}$$
$$E_{h_i}(\alpha, \beta) = O(h_i^q).$$

For $i = 1, 2$ and fixed q, we obtain the following error estimate for the global error $E_{h_1}(\alpha, \beta)$ of the approximate value $Q_{h_1}(\alpha, \beta)$ for $I(f; \alpha, \beta)$, which was computed using the step size h_1:

$$(15.11) \qquad E_{h_1}(\alpha, \beta) \approx \frac{Q_{h_1}(\alpha, \beta) - Q_{h_2}(\alpha, \beta)}{\left(\frac{h_2}{h_1}\right)^q - 1} = E_{h_1}^*(\alpha, \beta).$$

From (15.11), we can give an improved approximate value $Q_{h_1}^*(\alpha, \beta)$ for $I(f; \alpha, \beta)$:

$$Q_{h_1}^*(\alpha,\beta) \;=\; Q_{h_1}(\alpha,\beta) + E_{h_1}^*(\alpha,\beta)$$

$$(15.12) \qquad\qquad =\; Q_{h_1}(\alpha,\beta) + \frac{1}{\left(\dfrac{h_2}{h_1}\right)^q - 1}(Q_{h_1}(\alpha,\beta) - Q_{h_2}(\alpha,\beta)).$$

If we choose $h_2 = 2h_1$ and set $h_1 = h$, then (15.11) has the form

$$(15.13) \qquad\qquad E_h(\alpha,\beta) \approx \frac{Q_h(\alpha,\beta) - Q_{2h}(\alpha,\beta)}{2^q - 1},$$

and for $Q_h^*(\alpha,\beta)$ in (15.12) we have the equation

$$(15.14) \qquad Q_h^*(\alpha,\beta) = Q_h(\alpha,\beta) + \frac{1}{2^q - 1}\left(Q_h(\alpha,\beta) - Q_{2h}(\alpha,\beta)\right).$$

Here $Q_h(\alpha,\beta)$ is the approximate value using the step size h, and $Q_{2h}(\alpha,\beta)$ is the approximate value using the doubled step size. $Q_h^*(\alpha,\beta)$ is the approximate value for $I(f;\alpha,\beta)$ obtained in (15.12) as an improvement for $Q_h(\alpha,\beta)$.

For the trapezoidal rule, Simpson's rule and the 3/8 formula, the error estimates corresponding to (15.13) and the improved approximate values Q_h^* corresponding to (15.14) are:

Trapezoidal and tangent trapezoidal rules $(q = 2)$:

$$E_h^{TR} \;\approx\; \tfrac{1}{3}(Q_h^{TR} - Q_{2h}^{TR}) \qquad , \qquad E_h^{TT} \;\approx\; \tfrac{1}{3}(Q_h^{TT} - Q_{2h}^{TT}),$$

$$Q_h^{*TR} \;=\; Q_h^{TR} + \tfrac{1}{3}(Q_h^{TR} - Q_{2h}^{TR}) \quad , \quad Q_h^{*TT} \;=\; Q_h^{TT} + \tfrac{1}{3}(Q_h^{TT} - Q_{2h}^{TT}).$$

Simpson's rule and 3/8 formula $(q = 4)$:

$$E_h^{S} \;\approx\; \tfrac{1}{15}(Q_h^{S} - Q_{2h}^{S}) \qquad , \qquad E_h^{3/8} \;\approx\; \tfrac{1}{15}(Q_h^{3/8} - Q_{2h}^{3/8}),$$

$$Q_h^{*S} \;=\; Q_h^{S} + \tfrac{1}{15}(Q_h^{S} - Q_{2h}^{S}) \quad , \quad Q_h^{*3/8} \;=\; Q_h^{3/8} + \tfrac{1}{15}(Q_h^{3/8} - Q_{2h}^{3/8}).$$

For the Euler-Maclaurin formulas one can show that by using the approximate values Q_h^{*TR} and Q_h^{*S} for the integral instead of Q_h^{TR} and Q_h^{S}, one can gain two powers of h in the error order, namely:

$$I(f;\alpha,\beta) \;=\; Q_h^{*TR}(\alpha,\beta) + O(h^4) \quad \text{and}$$
$$I(f;\alpha,\beta) \;=\; Q_h^{*S}(\alpha,\beta) + O(h^6),$$

see also section 15.10.

Error estimates for non-equidistant nodes.

For all summed quadrature formulas one can find an error estimate by evaluating $I(f; \alpha, \beta)$ once for the partition $P : \alpha = t_0 < t_1 < \dots < t_m = \beta$ and a second time for the partition $P/2$ of $[\alpha, \beta]$ where the centers of each interval $[t_i, t_{i+1}]$ serve as m additional nodes.

For a quadrature formula with global error order q, which integrates polynomials up to degree $q - 1$ exactly, we obtain the global error of $Q_{P/2}$ as:

$$E_{P/2}(f; \alpha, \beta) \approx \frac{Q_{P/2}(f; \alpha, \beta) - Q_P(f; \alpha, \beta)}{2^q - 1} = E^*_{P/2}(f; \alpha, \beta).$$

And thus $Q_{P/2}$ can be replaced by the improved approximation

$$Q^*_{P/2}(f; \alpha, \beta) = Q_{P/2}(f; \alpha, \beta) + E^*_{P/2}(f; \alpha, \beta),$$

which has the global error order $q + 1$.

COMPUTATIONAL ERROR. While the global procedural error decreases for the trapezoidal rule or Simpson's rule quadratically or of fourth order, respectively, as $h \to 0$, the computational error increases in both cases with the order $O(1/h)$ as $h \to 0$. Thus the total error, i.e., the procedural plus the computational error cannot be shrunk arbitrarily. This phenomen holds for other quadrature formulas as well. It is recommended to choose a step size h so that the procedural error and the computational errors have nearly equal magnitudes.

For the TR rule, according to [McCR72], p.173, the following holds for the global computational error r_h:

$$r_h(\alpha, \beta) = \frac{1}{2h}(\beta - \alpha)^2 \varepsilon,$$

where ε is the maximal absolute computational error per step, h is the step size, and the interval of integration is $[\alpha, \beta]$.

Bibliography for Section 15.11

[BRAS77], V.1, V.2; [ENGE80], 7.7.4; [NIED87], 11.6, 11.7; [STET76], 8.2, 8.3 .

15.12 Adaptive Quadrature Methods

Thus far, the mentioned quadrature methods subdivide the integration intervals into subintervals independent of the given function f. The length of the subintervals is mandated by the degree of accuracy desired.

Depending on the graph of the function f that is to be integrated, it may be useful to work with different step sizes while possibly maintaining the same degree of accuracy.

If for instance, one calculates the integral $I(f; \alpha, \beta)$ using adaptive Romberg integration, the successive bisection of the step size of a subinterval can be stopped as soon as the local error indicates sufficient accuracy. Thus one only needs to subdivide those subintervals whose estimated error is above a given accuracy bound. In the worst case, when every subinterval has to be halved, one will get the original Romberg method described in section 15.10. With an adaptive method one tries to keep the number of the subintervals and hence the number of functional evaluations needed as small as possible. Naturally one can also use adaptive Gaussian and other quadrature formulas instead of the trapezoidal rule that forms the basis for Romberg integration. The accompanying program section contains adaptive quadrature methods.

Bibliography for Section 15.12

[KAHA72]; [KAHA83]; [RALS67], 4.11; [STOE91], 3.5.

15.13 Convergence of Quadrature Formulas

THEOREM 15.3. A quadrature formula of the form

(15.15) $$Q^{(n)}(f; \alpha, \beta) = \sum_{k=0}^{n} A_k^{(n)} f(x_k^{(n)}) \text{ with } x_k^{(n)} \in [\alpha, \beta]$$

converges to $I(f; \alpha, \beta)$ as $n \to \infty$ for each continuous function f on $[\alpha, \beta]$, i.e.,

(15.16) $$\lim_{n \to \infty} Q^{(n)}(f; \alpha, \beta) = \lim_{n \to \infty} \sum_{k=0}^{n} A_k^{(n)} f(x_k^{(n)}) = I(f; \alpha, \beta),$$

if and only if

1) (15.16) holds for each polynomial $f \equiv P$ of the form $\sum_{k=0}^{n} c_k x^k$, and

2) there is a constant C, so that $\sum_{k=0}^{n} |A_k^{(n)}| < C$ for every n.

For $f(x) = 1$ in (15.15), condition 1) implies

$$Q^{(n)}(1; \alpha, \beta) = \sum_{k=0}^{n} A_k^{(n)} = \int_{\alpha}^{\beta} dx = \beta - \alpha.$$

Hence if all weights $A_k^{(n)} > 0$, then 2) is satisfied. If, however, negative weights do occur, then $|A_0^{(n)}| + |A_1^{(n)}| + \ldots + |A_n^{(n)}|$ can become arbitrarily large, provided n is sufficiently large.

Bibliography for Section 15.13

[BROS64], §28; [ENGE87], 9.6; [ISAA66], 7.5.2; [WERN79], III §7.

15.14 Hints for Choosing an Appropriate Method

The accuracy of an approximation for a definite integral depends on

- the error order q of the quadrature formula used,

- the fineness of the partition P, and

- the smoothness of the function f to be integrated.

If the integrand is arbitrarily often differentiable, then one can choose among the quadrature formulas according to the error order and adapt the partition used to the graph of the integrand:
In regions where the integrand changes considerably one chooses a finer partition, while larger subintervals suffice where the integrand does not change much, see the examples in [NIED87] 11.2.

For moderate accuracy requirements we recommend Simpson's rule and the 3/8 formula among the mentioned Newton-Cotes formulas. They are preferred over the trapezoidal rule, except for periodic functions, see section 15.3.1, on account of their better error order. The Newton-Cotes formulas of higher order can be recommended only with reservations, because rounding errors produce an unwelcome effect due to the large weights used.

Chebyshev quadrature formulas are nearly immune to the effects of rounding errors because they work with equal weights. Comparing the Chebyshev formulas, the Clenshaw-Curtis formulas and the Gauß formulas, one will realize that the computation of the weights and nodes for the Gauß formulas require the largest amount of computational effort. If one can neglect this disadvantage, the Gauß formulas are the most effective ones amongst the three. For higher order methods, the Clenshaw-Curtis formulas are preferred because their nodes and weights can be very easily computed.

Romberg integration is a very simple and extremely effective method.

Adaptive quadrature methods automatically take the graph of the integrand f into consideration. An adaptive Gauß quadrature formula gives the optimal means to evaluate a definite integral, both with respect to computational effort and accuracy.

Bibliography for Section 15.14

[BRAS7], I.3; [NIED87], 11.13; [STRO74], p. 311 .

Bibliography for Chapter 15

[ATKI89], 5; [BJÖR74], 7.4; [CONT80], 7.; [DAVI82]; [ENGE80]; [HÄMM78]; [HÄMM91], 7.; [HILD78], 8.; [MAES88], 7.; [NIED87], 11.; [QUAD83]; [RALS67], IV.; [SCHM76], 8.; [SHAM73], II. 2; [STOE91], 3.

Chapter 16

Numerical Cubature

16.1 The Problem

In this chapter we try to evaluate definite integrals over bounded regions in the x, y-plane. Let $B \subset \mathbb{R}^2$ be a region of the x, y-plane with a piecewise smooth boundary curve. We assume that the function $f : B \to \mathbb{R}$ is continuous on $B \subset \mathbb{R}^2$. Then

(16.1)
$$I(f; B) = \iint_B f(x, y) \, dx dy$$

is called the *double integral* of f over the region B.

If $f(x, y) \geq 0$ holds for all $(x, y) \in B$, then (16.1) is the *volume* of the cylindrical solid above the region B and below the surface $z = f(x, y)$.

The double integral (16.1) can be calculated by computing two iterated one-dimensional integrals in case B has one of the shapes (2), (3) or (4) in Figure 16.1.

Figure 16.1: Planar regions for surface integrals

If B is a domain of type (2) with

$$B = \{(x,y)|\ a \le x \le b,\ y_1(x) \le y \le y_2(x)\}\quad,$$

then

$$\iint\limits_B f(x,y)dxdy = \int\limits_a^b \left(\int\limits_{y_1(x)}^{y_2(x)} f(x,y)dy \right) dx.$$

If B has the shape (3) with

$$B = \{(x,y)|\ x_1(y) \le x \le x_2(y),\ c \le y \le d\},$$

then

$$\iint\limits_B f(x,y)dxdy = \int\limits_c^d \left(\int\limits_{x_1(y)}^{x_2(y)} f(x,y)dx \right) dy.$$

For a rectangular region $B = R$ (type (4)) with

$$R = \{(x,y)|\ a \le x \le b,\ c \le y \le d\},$$

we have

$$\iint\limits_R f(x,y)dxdy = \int\limits_a^b \left(\int\limits_c^d f(x,y)dy \right) dx.$$

To approximate double integrals, so-called *cubature formulas* $C(f;B)$ are used. This name suggests itself since double integrals often represent volume. One can find an approximate value for $I(f;B)$ if one decomposes B into sub-regions B_j, $j = 1,\ldots,N$, with areas F_j and then sums volumes of each solid over B_j in order to calculate the total volume.

With an arbitrarily chosen point $(x_j,y_j) \in B_j$ for each $j = 1,\ldots,N$ we can thus approximate the double integral as

$$I(f;B) \approx \sum_{j=1}^N F_j f(x_j,y_j).$$

Here the right hand side is a *summed cubature formula*. If the points (x_j,y_j) are chosen as the centers of gravity of the subregions B_j, then one gets a so-called *centroid cubature formula*.

In analogy to the one-dimensional case it is possible to construct cubature formulas for elementary regions B_r of the form

$$(16.2)\quad C(f;B_r) = \sum_{i=0}^m \sum_{k=0}^n a_{ik} f(x_i,y_k) \approx \iint\limits_{B_r} f(x,y)dxdy = I(f;B_r),$$

or

$$(16.3) \quad C(f; B_r) = \sum_{j=1}^{N} A_j f(x_j, y_j) \approx \iint_{B_r} f(x,y)dxdy = I(f; B_r).$$

Thus cubature formulas are weighted sums of $N = (m+1) \cdot (n+1)$ functional values of the integrand f at N discrete nodes in the elementary region B_r.

DEFINITION 16.1 (*Degree of exactness*).
The cubature formula (16.2) or (16.3) approximates the double integral $I(f; B)$ with *degree L*, if it is exact for *all* polynomials P_L with

$$(16.4) \qquad P_L(x,y) = \sum_{0 \le s+t \le L} c_{st} x^s y^t$$

of degree at most L, but is not exact for some polynomial P_{L+1} of degree $L + 1$: i.e.,

$$C(P_L; B) \quad \overset{!}{=} \quad I(P_L; B) \quad \text{for all polynomials } P_i \text{ of degree } \le L, \text{ and}$$
$$C(P_{L+1}; B) \quad \neq \quad I(P_{L+1}; B) \quad \text{for some polynomial } P_{L+1} \text{ of degree } L + 1.$$

If a cubature formula C has the degree of exactness L for a reference domain B_r and if f is sufficiently often differentiable in B_r, then its local error order is $q_l = L + 2$.

If one wants to calculate a double integral over an arbitrary region $B \subseteq \mathbb{R}^2$ one divides B into subregions B_j of the shape of the reference region B_r. Next one applies the cubature formula for B_r to the regions B_j which might only be possible after a nonlinear transformation, and finally one sums over j. The resulting formula is called a *summed cubature formula*; it has the global error order $q_g = L + 1$ if C has the degree of exactness L.

Subsequently, only cubature formulas based on rectangular and triangular reference regions will be treated. As is well known, other regions can be transformed into reference rectangles or reference triangles by nonlinear mappings, see e.g. [ENGE80];[MAES88], 7.4.2; [NIEM91], 7. .

16.2 Interpolating Cubature Formulas

Let B_r be a reference region for which we want to construct a cubature
formula $C(f; B_r)$ by integrating an interpolation polynomial through the
$(m+1) \cdot (n+1)$ nodes

$$(x_i, y_k, f(x_i, y_k)) \text{ for } i = 0, \ldots, m, \quad k = 0, \ldots, n.$$

In this section we choose a rectangle R_r as reference region. Let the associated
interpolation polynomial be Φ which satisfies $\Phi(x_i, y_k) = f(x_i, y_k)$ for all i and
k. For a rectangular grid such an interpolation problem is uniquely solvable.
The Lagrange representation of Φ over R_r is:

$$\Phi(x,y) = \sum_{i=0}^{m} \sum_{k=0}^{n} L_i(x) L_k(y) f(x_i, y_k) \quad \text{with}$$

$$L_i(x) = \prod_{\substack{j=0 \\ j \neq i}}^{m} \frac{x - x_j}{x_i - x_j}, \quad L_k(y) = \prod_{\substack{j=0 \\ j \neq k}}^{n} \frac{y - y_j}{y_k - y_j}.$$

Thus

$$\int_a^b \int_c^d f(x,y) dx dy \approx \int_a^b \int_c^d \Phi(x,y) dx dy$$

$$= \int_a^b \int_c^d \sum_{i=0}^{m} \sum_{k=0}^{n} L_i(x) L_k(y) f(x_i, y_k) dx dy$$

(16.5)
$$= \sum_{i=0}^{m} \sum_{k=0}^{n} \underbrace{\int_a^b L_i(x) dx}_{A_i} \underbrace{\int_c^d L_k(y) dy}_{B_k} f(x_i, y_k)$$

$$= \sum_{i=0}^{m} \sum_{k=0}^{n} A_i B_k f(x_i, y_k)$$

$$=: \sum_{i=0}^{m} \sum_{k=0}^{n} a_{ik} f_{ik} =: C(f; R).$$

The A_i and B_k are called *weights* as in the one-dimensional interpolating
quadrature formulas of chapter 15. The *cubature weights* $a_{ik} := A_i B_k$ can
be written as a matrix product $\boldsymbol{A}\boldsymbol{B}^T$ with a column vector or an $(m+1, 1)$
matrix \boldsymbol{A} and a row vector or a $(1, n+1)$ matrix \boldsymbol{B}^T:

$$A = \begin{pmatrix} A_0 \\ A_1 \\ \vdots \\ A_m \end{pmatrix}, \quad B^T = (B_0, \ B_1, \ \cdots, B_n)$$

formed from the quadrature weights A_i, B_k:

$$(a_{ik}) := \begin{pmatrix} A_0 B_0 & A_0 B_1 & \cdots & A_0 B_n \\ A_1 B_0 & A_1 B_1 & \cdots & A_1 B_n \\ \vdots & & & \\ A_m B_0 & A_m B_1 & \cdots & A_m B_n \end{pmatrix}$$

(16.6)

$$= \begin{pmatrix} A_0 \\ A_1 \\ \vdots \\ A_m \end{pmatrix} (B_0, B_1, \cdots, B_n) = AB^T.$$

The interpolating cubature formula (16.5) thus represents a sum of products of the $(m+1) \cdot (n+1)$ weights a_{ik} and the functional values f_{ik} of f at the $(m+1) \cdot (n+1)$ nodes (x_i, y_k) in the reference rectangle.

The nodes and weights are computed - in complete analogy to the one-dimensional case - with the aim that polynomials P_L of highest possible degree L are integrated exactly. If in (16.5), one prescribes all the nodes of the reference rectangle, then a system with $N = (m + 1) \cdot (n + 1)$ equations arises for the weights of what are usually called the *Newton-Cotes cubature formulas*. If on the other hand one prescribes the weights, the optimal nodes become the solutions of a nonlinear system. The resulting formulas are called the *Chebyshev cubature formulas*.

If one does not prescribe nodes or weights, then the optimal *Gauß cubature formulas* can be found from a nonlinear system.

If the cubature formula (16.5) is to integrate all polynomials P_L up to a maximal degree L exactly, one can use the monomials $x^s y^t$ of degree $s+t \leq L$ to calculate the nodes and weights. And

$$C(x^s y^t; R_r) \overset{!}{=} I(x^s y^t; R_r) \quad \text{for} \quad s + t \leq L$$

implies with (16.5) that

$$\sum_{i=0}^{m} \sum_{k=0}^{n} a_{ik} x_i^s y_k^t \overset{!}{=} \int\limits_a^b \int\limits_c^d x^s y^t \, dy \, dx = \int\limits_a^b x^s \, dx \int\limits_c^d y^t \, dy$$

(16.7)

$$= \frac{1}{(s+1)(t+1)} (b^{s+1} - a^{s+1})(d^{t+1} - c^{t+1})$$

for $s+t = 0, 1, 2, \ldots, L$. Thus for $s+t = 0, 1, 2, \ldots, L$ we have the equations:

$$(16.7a) \qquad \sum_{i=0}^{m}\sum_{k=0}^{n} a_{ik} x_i^s y_k^t = \frac{1}{(s+1)(t+1)}(b^{s+1} - a^{s+1})(d^{t+1} - c^{t+1}).$$

Thus setting $N = (m+1)\cdot(n+1)$:

$$(16.7b) \qquad \sum_{j=1}^{N} A_j x_j^s y_j^t = \frac{1}{(s+1)(t+1)}(b^{s+1} - a^{s+1})(d^{t+1} - c^{t+1}) \ .$$

These give $\frac{1}{2}(L+1)(L+2)$ conditions which must be fulfilled in order to achieve the desired exactness degree L. The monomials $x^s y^t$ associated with each exactness degree L can be found in the following chart:

Exactness degree L	Monomials $x^s y^t$ with $s+t \le L$	$\frac{1}{2}(L+1)(L+2)$, the number of monomials with $s+t \le L$
0	1	1
1	x y	3
2	x^2 xy y^2	6
3	x^3 x^2y xy^2 y^3	10
4	x^4 x^3y x^2y^2 xy^3 y^4	15
5	x^5 x^4y x^3y^2 x^2y^3 xy^4 y^5	21

To achieve an exactness degree of $L = 1$, the cubature formula must integrate all polynomials of degree $s+t \le 1$ exactly. Thus the monomials $1, x, y$ define three conditions for the free parameters. If there should be any free parameters left, they can be determined by demanding exactness for some monomials $x^s y^t$ with $s+t = L+1$. The resulting cubature formula will in general, however, not be exact for all monomials with degree $s+t = L+1$, see section 16.3.

16.3 Newton-Cotes Cubature Formulas for Rectangular Regions

The *Newton-Cotes cubature formulas* for a rectangular region R_r are interpolating formulas of the form

$$C(f; R_r) = \sum_{i=0}^{m}\sum_{k=0}^{n} a_{ik} f(x_i, y_k) \approx \iint_{R_r} f(x,y)\,dx\,dy.$$

Here $N = (m + 1) \cdot (n + 1)$ distinct nodes (x_i, y_k) are given in the reference rectangle R_r with a constant step size h_x in the x-direction and a constant step size h_y in the y-direction. The N weights a_{ik} are determined from the linear system (16.7). The weights can also be found from the corresponding quadrature formulas (16.6).

To begin with, we shall illustrate the trapezoidal rule, Simpson's rule and the 3/8-cubature formulas for the reference rectangles in Figure 16.2:

Figure 16.2: Reference rectangles

Trapezoidal rule for the reference rectangle R_{r1}

The $N = 4$ weights a_{ik}, $i, k = 0, 1$, can be computed from the weights of the corresponding quadrature formula in section 15.3.1 according to (16.6):

$$(a_{ik}) = \boldsymbol{A}\boldsymbol{B}^T = \frac{h_x}{2}\begin{pmatrix} 1 \\ 1 \end{pmatrix}\frac{h_y}{2}(1\,,\,1) = \frac{h_x h_y}{4}\begin{pmatrix} 1 & 1 \\ 1 & 1 \end{pmatrix}.$$

And thus the *trapezoidal cubature formula* is:

$$(16.8)\quad C^T(f; R_{r1}) = \frac{h_x h_y}{4}\left[f(0,0) + f(h_x, 0) + f(0, h_y) + f(h_x, h_y)\right].$$

This cubature formula integrates all two variable polynomials P_1 with $s + t \le 1$ exactly as well as all second degree polynomials with $s = 1$ and $t = 1$. It does not integrate those with $s = 0$, $t = 2$ or $t = 0$, $s = 2$ exactly, hence its exactness degree is $L = 1$.

Simpson's rule for the reference rectangle R_{r2}

The $N = 9$ weights a_{ik}, $i, k = 0, 1, 2$, are given from section 15.3.2 via (16.6) as:

$$(a_{ik}) = \boldsymbol{A}\boldsymbol{B}^T = \frac{h_x}{3}\begin{pmatrix} 1 \\ 4 \\ 1 \end{pmatrix}\frac{h_y}{3}(1,\,4,\,1) = \frac{h_x h_y}{9}\begin{pmatrix} 1 & 4 & 1 \\ 4 & 16 & 4 \\ 1 & 4 & 1 \end{pmatrix}.$$

Hence *Simpson's cubature formula* is:

$$
C^S(f; R_{r2}) = \frac{h_x h_y}{9} \{ f(-h_x, -h_y) + f(-h_x, h_y) + f(h_x, -h_y) +
$$

(16.9)
$$
+ f(h_x, h_y) + 4 [f(0, -h_y) + f(h_x, 0) + f(0, h_y) +
$$

$$
+ f(-h_x, 0)] + 16 f(0, 0) \}.
$$

It is exact for all polynomials P_3 with $s + t \leq L = 3$, i.e., all bicubic polynomials. However it is not exact for any polynomial of degree $L = 4$.

3/8-formula for the reference rectangle R_{r3}

The $N = 16$ weights a_{ik}, $i, k = 0, 1, 2, 3$, are given by section 15.3.3 via (16.6) as:

$$
(a_{ik}) = \boldsymbol{AB}^T = \frac{3h_x}{8} \begin{pmatrix} 1 \\ 3 \\ 3 \\ 1 \end{pmatrix} \cdot \frac{3h_y}{8} (1, \quad 3, \quad 3, \quad 1) = \frac{9 h_x h_y}{64} \begin{pmatrix} 1 & 3 & 3 & 1 \\ 3 & 9 & 9 & 3 \\ 3 & 9 & 9 & 3 \\ 1 & 3 & 3 & 1 \end{pmatrix},
$$

and the 3/8 *cubature formula* becomes:

$$
C^{3/8}(f; R_{r3}) = \frac{9 h_x h_y}{64} \{ f(0, 0) + f(3h_x, 0) + f(0, 3h_y) +
$$

$$
+ \quad f(3h_x, 3h_y) + 3 [f(h_x, 0) + f(2h_x, 0) + f(3h_x, h_y) +
$$

(16.10)
$$
+ \quad f(3h_x, 2h_y) + f(2h_x, 3h_y) + f(h_x, 3h_y) + f(0, 2h_y) +
$$

$$
+ \quad f(0, h_y)] + 9 [f(h_x, h_y) + f(2h_x, h_y) + f(h_x, 2h_y) +
$$

$$
+ \quad f(2h_x, 2h_y)] \}.
$$

The 3/8-formula integrates all polynomials P_3 with $s + t \leq 3$ exactly as well as some (but not all) polynomials with $s + t = 4$, hence it is a method of degree 3.

The *summed Newton-Cotes cubature formulas* come about in the following way: We shall partition the rectangle of integration

$$
R = \{ (x, y) | \ a \leq x \leq b, \ c \leq y \leq d \}
$$

into subrectangles which are used as reference rectangles of the corresponding cubature formula. The sum of the local cubature formulas of all subrectangles gives the summed cubature formula.

Summed trapezoidal rule for a rectangle R

The rectangle R is partitioned into subrectangles R_{pq} as follows: The partition in the x direction is defined as

$$P_x: \quad x_p = a + ph_x, \qquad p = 0, \ldots, P \quad \text{with} \quad h_x = \frac{b-a}{P},$$

the partition in the y direction as

$$P_y: \quad y_q = c + qh_y, \qquad q = 0, \ldots, Q \quad \text{with} \quad h_y = \frac{d-c}{Q}.$$

Thus, we obtain $P \cdot Q$ subrectangles R_{pq} of R:

$$R_{pq} := \{(x,y)| \ x_p \leq x \leq x_{p+1}, \ y_q \leq y \leq y_{q+1}\}$$

for $p = 0, \ldots, P-1, \quad q = 0, \ldots, Q-1$. If we apply the trapezoidal formula (16.8) to each subrectangle R_{pq}, the result is the *summed trapezoidal formula*:

$$
\begin{aligned}
C_{h_x h_y}^T(f; R) \ &= \ \sum_{p=0}^{P-1} \sum_{q=0}^{Q-1} C^T(f; R_{pq}) \\
&= \ \frac{h_x h_y}{4} \{f(a,c) + f(b,c) + f(a,d) + f(b,d)+ \\
&\qquad +2 \sum_{p=1}^{P-1} (f(a+ph_x, c) + f(a+ph_x, d)) + \\
&\qquad +2 \sum_{q=1}^{Q-1} (f(a, c+qh_y) + f(b, c+qh_y)) + \\
&\qquad +4 \sum_{p=1}^{P-1} \sum_{q=1}^{Q-1} f(a+ph_x, c+qh_y)\}.
\end{aligned}
$$

(16.11)

Its global error order is $q_g = 2$. For the trapezoidal formula we have

$$
\iint\limits_R f(x,y)dxdy \ = C_{h_x h_y}^T(f; R) + E_{h_x h_y}^T(f; R), \quad \text{where}
$$

$$
\left|E_{h_x h_y}^T(f; R)\right| \ \leq \frac{(b-a)(d-c)}{12} \left\{h_x^2 \max_{(x,y)\in R} |f_{xx}| + h_y^2 \max_{(x,y)\in R} |f_{yy}|\right\}
$$

$$
= O(h_{\max}^2),
$$

with $h_{\max} = \max\{h_x, h_y\}$.

A proof of this error order is given in [HÄMM91], §6; [MAES88], 7.4.3 .

Summed Simpson's cubature formula

Here we need subrectangles of dimensions $2h_x$ by $2h_y$. Hence we partition R thus:

$$P_x: \quad x_p = a + ph_x, \quad p = 0,\ldots,2P \quad \text{with} \quad h_x = \frac{b-a}{2P},$$

$$P_y: \quad y_q = c + qh_y, \quad q = 0,\ldots,2Q \quad \text{with} \quad h_y = \frac{d-c}{2Q}.$$

Then we have $P \cdot Q$ subrectangles R_{pq}^S of R:

$$R_{pq}^S := \left\{(x,y) \mid x_{2p} \le x \le x_{2(p+1)}, \; y_{2q} \le y \le y_{2(q+1)}\right\}$$

for $p = 0,\ldots,P-1$, $q = 0,\ldots,Q-1$. If we apply Simpson's formula (16.9) to each subrectangle, we obtain the *summed Simpson formula* as:

$$
\begin{aligned}
C_{h_x h_y}^S(f;R) &= \sum_{p=0}^{P-1}\sum_{q=0}^{Q-1} C^S(f;R_{pq}^S) \\
&= \frac{h_x h_y}{9}\Big\{ f(a,c) + f(b,c) + f(a,d) + f(b,d) + \\
&\quad +4\sum_{p=0}^{P-1} \left(f(a+(2p+1)h_x,c) + f(a+(2p+1)h_x,d)\right) + \\
&\quad +4\sum_{q=0}^{Q-1} \left(f(a,c+(2q+1)h_y) + f(b,c+(2q+1)h_y)\right) + \\
&\quad +2\sum_{p=1}^{P-1} \left(f(a+2ph_x,c) + f(a+2ph_x,d)\right) + \\
&\quad +2\sum_{q=1}^{Q-1} \left(f(a,c+2qh_y) + f(b,c+2qh_y)\right) + \\
&\quad +4\sum_{p=1}^{P-1}\sum_{q=1}^{Q-1} f(a+2ph_x,c+2qh_y) + \\
&\quad +16\sum_{p=0}^{P-1}\sum_{q=0}^{Q-1} f(a+(2p+1)h_x,c+(2q+1)h_y) + \\
&\quad +8\sum_{p=1}^{P-1}\sum_{q=0}^{Q-1} f(a+2ph_x,c+(2q+1)h_y) + \\
&\quad +8\sum_{p=0}^{P-1}\sum_{q=1}^{Q-1} f(a+(2p+1)h_x,c+2qh_y)\Big\}.
\end{aligned}
$$

(16.12)

The global error order is $q_g = 4$. This Simpson rule satisfies

$$\iint\limits_{R} f(x,y)dxdy = C^{S}_{h_x h_y}(f;R) + E^{S}_{h_x h_y}(f;R) \qquad \text{with}$$

$$\left| E^{S}_{h_x h_y}(f;R) \right| \leq \frac{(b-a)(d-c)}{180}\left\{ h_x^4 \max|f_{xxxx}| + h_y^4 \max|f_{yyyy}| \right\}$$

$$= O(h_{\max}^4)$$

for $h_{\max} = \max\{h_x, h_y\}$. A proof of this can be found in [MAES88], 7.4.3 .

The same method can be applied analogously to the 3/8-formula.

16.4 Newton-Cotes Cubature Formulas for Triangles

It is possible to use triangles as reference regions as well. Let R_{Δ_1} be a reference triangle whose three vertices shall serve as nodes.

Figure 16.3: Reference triangle

Then the cubature formula becomes:

$$(16.13) \qquad C(f;R_{\Delta_1}) = \frac{h_x h_y}{2} \cdot \frac{1}{3}\left\{ f(0,0) + f(h_x,0) + f(0,h_y) \right\}.$$

It has the exactness degree $L = 1$ and integrates all polynomials P_1 of degree $s + t \leq 1$, i.e., the three monomials $1, x$, and y are integrated exactly. The global error order is $O(h_{\max}^2)$.

Cubature formulas with exactness degree $L = 2$ can be obtained by using the following nodes of the reference triangle:

Figure 16.4: Selection of nodes in the reference triangle

The corresponding formulas are:

(16.14a) $C(f; R_{\Delta_2}) = \dfrac{h_x h_y}{2} \dfrac{1}{3} \left[f\left(\dfrac{h_x}{2}, 0\right) + f\left(\dfrac{h_x}{2}, \dfrac{h_y}{2}\right), f\left(0, \dfrac{h_y}{2}\right) \right]$, and

(16.14b) $C(f; R_{\Delta_3}) = \dfrac{h_x h_y}{2} \dfrac{1}{3} \left[f\left(\dfrac{h_x}{6}, \dfrac{h_y}{6}\right) + f\left(\dfrac{2h_x}{3}, \dfrac{h_y}{6}\right), f\left(\dfrac{h_x}{6}, \dfrac{2h_y}{3}\right) \right]$.

It is noteworthy that these formulas have a higher degree of exactness while using the same number of nodes as formula (16.3). They integrate all polynomials P_2 of degree ≤ 2 exactly and have a global error order of $O(h_{\max}^3)$.
The corresponding summed formulas can be obtained from the following operations: Partition the region B into subtriangles B_j. Then one of the indicated formulas is applied to each subtriangle and the results are summed as before.

Bibliography for Section 16.4

[ENGE80]; [HÄMM91] §6.; [ISAA66]; [MAES88] vol.2, 7.4.

16.5 Romberg Cubature for Rectangular Regions

Richardson extrapolation can be used for double integrals similarly as it has been for one-dimensional ones: Repeated bisecting of intervals in both x- and y-directions and calculating approximations for the double integral via the trapezoidal formula together with extrapolation of the approximations for different step sizes will lead to steadily improved approximations for the double integral. A proof of the convergence of this method is given in [ENGE80] e.g.

The trapezoidal formula for the reference triangle

$$R = \{(x,y)|-1 \le x \le 1, \ -1 \le y \le 1\}$$

for the step size $h_{x_0} = h_{y_0} = 2$ is

$$
\begin{aligned}
C^T(f;R) &= f(-1,-1) + f(1,-1) + f(1,1) + f(-1,1) \\
&=: L_0^{(0)}(f).
\end{aligned}
$$

If one repeatedly halves the step size: $h_j := \frac{h_0}{2^j}$ for $j = 0,1,2,\dots$ in both the x- and the y-direction, one obtains the summed trapezoidal formula for the step size h_j over R as:

$$
\begin{aligned}
L_j^{(0)}(f) &= \tfrac{1}{4^j}\Big\{ f(-1,-1) + f(-1,1) + f(1,-1) + f(1,1) \\
&\quad + 2\sum_{k=1}^{2^j-1}\Big[f(-1,-1+\tfrac{k}{2^{j-1}}) + f(1,-1+\tfrac{k}{2^{j-1}}) \\
&\quad + f(-1+\tfrac{k}{2^{j-1}},-1) + f(-1+\tfrac{k}{2^{j-1}},1)\Big] \\
&\quad + 4\sum_{k=1}^{2^j-1}\sum_{l=1}^{2^j-1} f(-1+\tfrac{k}{2^{j-1}},-1+\tfrac{l}{2^{j-1}})\Big\}.
\end{aligned}
$$

And $L_j^{(0)}(f)$ can be expressed recursively for $j \ge 1$ as:

$$
\begin{aligned}
L_j^{(0)}(f) &= \tfrac{1}{4}L_{j-1}^{(0)} + \tfrac{1}{2^{2j-1}}\Big\{ \sum_{k=0}^{2^{j-1}-1}\Big[f(-1,-1+\tfrac{2k+1}{2^{j-1}}) \\
&\quad + f(1,-1+\tfrac{2k+1}{2^{j-1}}) + f(-1+\tfrac{2k+1}{2^{j-1}},-1) + f(-1+\tfrac{2k+1}{2^{j-1}},1)\Big] \\
&\quad + 2\sum_{k=0}^{2^{j-1}-1}\sum_{l=1}^{2^j-1} f(-1+\tfrac{2k+1}{2^{j-1}},-1+\tfrac{l}{2^{j-1}}) \\
&\quad + 2\sum_{k=0}^{2^{j-1}-2}\sum_{l=0}^{2^{j-1}-1} f(-1+\tfrac{2k+2}{2^{j-1}},-1+\tfrac{2l+1}{2^{j-1}})\Big\}.
\end{aligned}
$$

Analagously to Romberg quadrature, the *Romberg cubature* scheme now has the form:

h_j	$L_j^{(0)}$	$L_j^{(1)}$	$L_j^{(2)}$	\cdots	$L_j^{(m-1)}$	$L_j^{(m)}$
h_0	$L_0^{(0)}$					
$h_1 = \frac{h_0}{2}$	$L_1^{(0)}$	$L_0^{(1)}$				
$h_2 = \frac{h_1}{2}$	$L_2^{(0)}$	$L_1^{(1)}$	$L_0^{(2)}$			
\vdots	\vdots	\vdots	\vdots	\ddots		
$h_{m-1} = \frac{h_{m-2}}{2}$	$L_{m-1}^{(0)}$	$L_{m-2}^{(1)}$	$L_{m-3}^{(2)}$	\cdots	$L_0^{(m-1)}$	
$h_m = \frac{h_{m-1}}{2}$	$L_m^{(0)}$	$L_{m-1}^{(1)}$	$L_{m-2}^{(2)}$	\cdots	$L_1^{(m-1)}$	$L_0^{(m)}$

The extrapolated values $L_j^{(k)}$ can be computed as follows:

$$L_j^{(k)} = L_{j+1}^{k-1} + \frac{1}{4^k - 1}\left(L_{j+1}^{k-1} - L_j^{k-1}\right) \quad \text{for} \quad k \geq 1 \quad \text{and} \quad j = 0,1,2,\ldots.$$

The values $x' \in [-1,1]$, $y' \in [-1,1]$ occurring in the formulas for $L_j^{(k)}$ must be transformed to values (x,y) taken from the actual rectangle using

$$x = \tfrac{h_x}{2}x' + (j_x - 0.5)h_x + a, \qquad j_x = 1,\ldots,j\text{xmax},$$

$$y = \tfrac{h_y}{2}y' + (j_y - 0.5)h_y + c, \qquad j_y = 1,\ldots,j\text{ymax},$$

$$\text{where } h_x = \frac{b-a}{j\text{xmax}}, \qquad h_y = \frac{d-c}{j\text{ymax}}.$$

For $A_x = \tfrac{h_x}{2}$, $A_y = \tfrac{h_y}{2}$, $B_x = (j_x - 0.5)h_x + a$, $B_y = (j_y - 0.5)h_y + c$ the transformation can be expressed as:

$$x = A_x x' + B_x$$
$$y = A_y y' + B_y.$$

Bibliography for Section 16.5

[ENGE80]

16.6 Gauß Cubature Formulas for Rectangles

In complete analogy with the Newton-Cotes formulas, the *Gaussian cubature formulas* can be constructed for rectangular regions by employing the one-dimensional Gaussian quadrature formulas of section 15.7.

The four weights a_{ik}, $i, k = 1, 2$ for the cubature formula are derived from the one-dimensional quadrature formulas $Q^{G_2}(f; -h, h)$ by multiplying their two element weight matrices according to (16.6):

$$(a_{ik}) = \boldsymbol{AB}^T = h_x \begin{pmatrix} 1 \\ 1 \end{pmatrix} h_y(1,1) = h_x h_y \begin{pmatrix} 1 & 1 \\ 1 & 1 \end{pmatrix}.$$

Here the two nodes x_i, y_k for each direction give rise to four planar nodes

$$\text{in x-direction:} \quad x_0 = -\frac{h_x}{\sqrt{3}}, \quad x_1 = \frac{h_x}{\sqrt{3}}$$

$$\text{in y-direction:} \quad y_0 = -\frac{h_y}{\sqrt{3}}, \quad y_1 = \frac{h_y}{\sqrt{3}}$$

for the cubature formula. Thus the Gauß cubature formula for $n = 1, m = 1$, i.e., $N = 4$ is:

$$C^{G_2}(f; R_{r_2}) = h_x h_y \left\{ f\left(-\frac{h_x}{\sqrt{3}}, -\frac{h_y}{\sqrt{3}}\right) + f\left(-\frac{h_x}{\sqrt{3}}, \frac{h_y}{\sqrt{3}}\right) + \right.$$
$$\left. + f\left(\frac{h_x}{\sqrt{3}}, -\frac{h_y}{\sqrt{3}}\right) + f\left(\frac{h_x}{\sqrt{3}}, \frac{h_y}{\sqrt{3}}\right) \right\}.$$

It has the exactness degree $L = 3$ and thus integrates all bicubic polynomials of degree $s + t \leq 3$ exactly. It has the global error order $O(h_{\max}^4)$ with $h_{\max} = max\{h_x, h_y\}$.

For nine weights the corresponding cubature formula derives from the one-dimensional cubature formula $Q^{G_3}(f; -h, h)$ by multiplying

$$(a_{ik}) = \boldsymbol{AB}^T = \frac{h_x}{9} \begin{pmatrix} 5 \\ 8 \\ 5 \end{pmatrix} \frac{h_y}{9}(5, \quad 8, \quad 5) = \frac{h_x h_y}{81} \begin{pmatrix} 25 & 40 & 25 \\ 40 & 64 & 40 \\ 25 & 40 & 25 \end{pmatrix}.$$

The nine nodes of the cubature formula are defined by the three one-dimensional nodes x_i and y_k for $i, k = 0, 1, 2$ as:

$$\text{in x direction:} \quad x_0 = -\sqrt{0.6}h_x, \quad x_1 = 0, \quad x_2 = \sqrt{0.6}h_x$$

$$\text{in y direction:} \quad y_0 = -\sqrt{0.6}h_y, \quad y_1 = 0, \quad y_2 = \sqrt{0.6}h_y.$$

The Gauß cubature formula (16.5) for $n = 2, m = 2$, i.e., $N = 9$ then becomes

$$C^{G_3}(f; R_{r_2}) = \frac{h_x h_y}{81} \{ 25 \left[f(-\sqrt{0.6}h_x, -\sqrt{0.6}h_y) \right.$$
$$+ f(-\sqrt{0.6}h_x, \sqrt{0.6}h_y) + f(\sqrt{0.6}h_x, -\sqrt{0.6}h_y)$$
$$+ \left. f(\sqrt{0.6}h_x, \sqrt{0.6}h_y) \right] + 40 \left[f(0, -\sqrt{0.6}h_y) + f(-\sqrt{0.6}h_x, 0) \right.$$
$$+ \left. f(\sqrt{0.6}h_x, 0) + f(0, \sqrt{0.6}h_y) \right] + 64 f(0, 0) \}.$$

It has the exactness degree $L = 5$ and integrates all polynomials of degree $s + t \leq 5$ exactly. It has the global error order $O(h_{\max}^6)$ with $h_{\max} = \max\{h_x, h_y\}$.

Additional formulas may be derived from their corresponding one-dimensional formulas as given in section 15.7 under "Table of the Gaussian nodes and weights".

Summed formulas can be constructed in the same way as described in section 16.3: In order to compute the volume bounded by a surface above a rectangle $R : \{(x, y) | a \leq x \leq b, c \leq y \leq d\}$, R is partitioned into $P \cdot Q$ subrectangles R_{pq} of sizes $2h_x$ by $2h_y$. A Gauß formula is then applied to each subrectangle. For this the nodes of the reference rectangle must be transformed as follows:

$$x_i \quad \longrightarrow \quad a + (2p + 1)h_x + x_i \quad \text{for } p = 0, \ldots, P-1, \quad i = 0, \ldots, m,$$
$$y_k \quad \longrightarrow \quad c + (2q + 1)h_y + y_k \quad \text{for } q = 0, \ldots, Q-1, \quad k = 0, \ldots, n.$$

For $n = m = 1$ we obtain the summed formula:

$$C^{G_2}(f; R) = h_x h_y \sum_{p=0}^{P-1} \sum_{q=0}^{Q-1} \left[f \left(a + (2p+1)h_x - \frac{h_x}{\sqrt{3}}, c + (2q+1)h_y - \frac{h_y}{\sqrt{3}} \right) \right.$$
$$+ f \left(a + (2p+1)h_x - \frac{h_x}{\sqrt{3}}, c + (2q+1)h_y + \frac{h_y}{\sqrt{3}} \right)$$
$$+ f \left(a + (2p+1)h_x + \frac{h_x}{\sqrt{3}}, c + (2q+1)h_y - \frac{h_y}{\sqrt{3}} \right)$$
$$+ \left. f \left(a + (2p+1)h_x + \frac{h_x}{\sqrt{3}}, c + (2q+1)h_y + \frac{h_y}{\sqrt{3}} \right) \right]$$

with global error order $q_g = 4$.

The 17 nodes and weights (i.e., the 51 parameters x_j, y_j and A_j) of the Gauß formula for $L = 9$ are determined by $\frac{1}{2}(L+1)(L+2) = 55$ nonlinear equations (16.7b) and can be found in [ENGE80], p.257:

I	X(I)	Y(I)	A(I)
1	.96884996636198E+00	.63068011973167E+00	.88879378170200E −01
2	-.96884996636198E+00	-.63068011973167E+00	.88879378170200E −01
3	-.63068011973167E+00	.96884996636198E+00	.88879378170200E −01
4	.63068011973167E+00	-.96884996636198E+00	.88879378170200E −01
5	.75027709997890E+00	.92796164595957E+00	.11209960212960E+00
6	-.75027709997890E+00	-.92796164595957E+00	.11209960212960E+00
7	-.92796164595957E+00	.75027709997890E+00	.11209960212960E+00
8	.92796164595957E+00	-.75027709997890E+00	.11209960212960E+00
9	.52373582021443E+00	.45333982113565E+00	.39828243926207E+00
10	-.52373582021443E+00	-.45333982113565E+00	.39828243926207E+00
11	-.45333982113565E+00	.52373582021443E+00	.39828243926207E+00
12	.45333982113565E+00	-.52373582021443E+00	.39828243926207E+00
13	.76208328192620E −01	.85261572933366E+00	.26905133763978E+00
14	-.76208328192620E −01	-.85261572933366E+00	.26905133763978E+00
15	-.85261572933366E+00	.76208328192620E −01	.26905133763978E+00
16	.85261572933366E+00	-.76208328192620E −01	.26905133763978E+00
17	0.00000000000000E+00	0.00000000000000E+00	.52674897119342E+00

16.7 Gauß Cubature Formulas for Triangles

16.7.1 Right Triangles with Legs Parallel to the Axis

Depending on the desired degree of exactness L, $\frac{1}{2}(L+1)(L+2)$ parameters x_j, y_j, A_j must be determined from the nonlinear system of equations (16.7b) which has as many equations as unknowns for triangular regions as well. For $L = 1$, the following formula has the global error order $O(h_{\max}^2)$:

1. One-point cubature formula for the reference triangle R_Δ, which uses only one node:

$$C^{G_1}(f; R_\Delta) = \frac{h_x h_y}{2} f\left(\frac{h_x}{3}, \frac{h_y}{3}\right).$$

2. Two-point cubature formula for R_Δ, which uses two nodes:

$$C^{G_2}(f; R_\Delta) = \frac{h_x h_y}{4}\left\{f\left(\frac{h_x}{6}, \frac{h_y}{2}\right) + f\left(\frac{h_x}{2}, \frac{h_y}{6}\right)\right\}.$$

For $L = 2$ the three-point formulas (16.14a) (16.14b) from section 16.4 hold as well.

The following seven-point formula with $L = 5$ is taken from [ENGE80]:

$$C^{G_7}(f; R_\triangle) = \frac{h_x h_y}{2400} \left\{ 270 f\left(\frac{h_x}{3}, \frac{h_y}{3}\right) + (155 + \sqrt{15}) \right.$$

$$\left[f\left(\frac{h_x}{21}(6 + \sqrt{15}), \frac{h_y}{21}(6 + \sqrt{15})\right) \right.$$

$$+ f\left(\frac{h_x}{21}(9 - \sqrt{15}), \frac{h_y}{21}(6 + \sqrt{15})\right)$$

$$\left. + f\left(\frac{h_x}{21}(6 + \sqrt{15}), \frac{h_y}{21}(9 - \sqrt{15})\right) \right] +$$

$$+ (155 - \sqrt{15}) \left[f\left(\frac{h_x}{21}(6 - \sqrt{15}), \frac{h_y}{21}(6 - \sqrt{15})\right) + \right.$$

$$+ f\left(\frac{h_x}{21}(9 + 2\sqrt{15}), \frac{h_y}{21}(6 - \sqrt{15})\right)$$

$$\left. \left. + f\left(\frac{h_x}{21}(6 - \sqrt{15}), \frac{h_y}{21}(9 + 2\sqrt{15})\right) \right] \right\}.$$

Its global error order is $O(h_{\max}^6)$.

16.7.2 General Triangles

Thus far we have dealt only with right triangles whose legs are parallel to the two coordinate axis. If one wants to evaluate double integrals over a general triangular region PQR, the following formula arises. We shall illustrate this using the three point formula (16.14b):

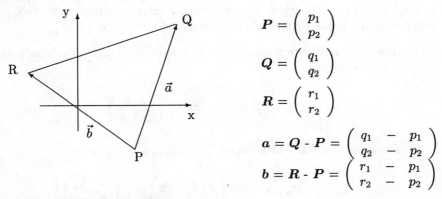

$$P = \begin{pmatrix} p_1 \\ p_2 \end{pmatrix}$$

$$Q = \begin{pmatrix} q_1 \\ q_2 \end{pmatrix}$$

$$R = \begin{pmatrix} r_1 \\ r_2 \end{pmatrix}$$

$$a = Q - P = \begin{pmatrix} q_1 & - & p_1 \\ q_2 & - & p_2 \end{pmatrix}$$

$$b = R - P = \begin{pmatrix} r_1 & - & p_1 \\ r_2 & - & p_2 \end{pmatrix}$$

Figure 16.5: General triangular region

If we express the function f vectorially as $f(x) \in \mathbb{R}^2$ for $x \in \mathbb{R}^2$, we obtain the cubature formula

$$
\begin{aligned}
C^{G_3}(f; R_{PQR}) &= A_{PQR}\tfrac{1}{3}\left[f\left(P + \tfrac{1}{6}a + \tfrac{1}{6}b\right) + \right. \\
&\quad + f\left(P + \tfrac{2}{3}a + \tfrac{1}{6}b\right) + f\left(P + \tfrac{1}{6}a + \tfrac{2}{3}b\right)\Big] \\
&= A_{PQR}\,\tfrac{1}{3}\left[f\left(p_1 + \tfrac{a_1}{6} + \tfrac{b_1}{6}, p_2 + \tfrac{a_2}{6} + \tfrac{b_2}{6}\right) + \right. \\
&\quad + f\left(p_1 + \tfrac{2a_1}{3} + \tfrac{b_1}{6}, p_2 + \tfrac{2a_2}{3} + \tfrac{b_2}{6}\right) + \\
&\quad + f\left(p_1 + \tfrac{a_1}{6} + \tfrac{2b_1}{3}, p_2 + \tfrac{a_2}{6} + \tfrac{2b_2}{6}\right)\Big]
\end{aligned}
$$

where $A_{PQR} := \tfrac{1}{2}\left(p_1(q_2 - r_2) + q_1(r_2 - p_2) + r_1(p_2 - q_2)\right)$.

16.8 Riemann Double Integrals using Bicubic Splines

Here we want to calculate the double integral

$$
I(f; R) = \iint\limits_{R} f(x, y)\,dx\,dy
$$

over a rectangle $R = \{(x, y) \mid a \leq x \leq b,\ c \leq y \leq d\}$. This can be done by means of bicubic splines

$$
S = S(x, y) \equiv f_{ij}(x, y) \quad \text{for} \quad (x, y) \in R_{ij},
$$

as described in algorithms 12.1, 12.2, 12.3 of section 12.1.

For each subrectangle R_{ij} we have the approximation

$$
\begin{aligned}
\int_{x_i}^{x_{i+1}}\int_{y_j}^{y_{j+1}} f(x, y)\,dx\,dy &\approx \int_{x_i}^{x_{i+1}}\int_{y_j}^{y_{j+1}} f_{ij}(x, y)\,dx\,dy \\
&= \int_{x_i}^{x_{i+1}}\int_{y_j}^{y_{j+1}} \sum_{k=0}^{3}\sum_{m=0}^{3} a_{ijkm}(x - x_i)^k (y - y_j)^m\,dx\,dy \\
&= \sum_{k=0}^{3}\sum_{m=0}^{3} a_{ijkm} \int_{x_i}^{x_{i+1}}(x - x_i)^k\,dx \int_{y_j}^{y_{j+1}}(y - y_j)^m\,dy \\
&= \sum_{k=0}^{3}\sum_{m=0}^{3} a_{ijkm} \frac{(x_{i+1} - x_i)^{k+1}}{k+1}\frac{(y_{j+1} - y_j)^{m+1}}{m+1}.
\end{aligned}
$$

The desired integral I can be found approximately as the sum of the subintegrals over all subrectangles, see program fibicu (C) and FIBICU (FORTRAN).

If one wants to calculate the double integral

$$I(f; RT) = \iint\limits_{RT} f(x,y)\,dx\,dy$$

for a subrectangle RT of R with

$$RT = \{(x,y) \mid \alpha \le x \le \beta, \ \gamma \le x \le \delta\},$$

and wants to use the bicubic splines that have been calculated for R, where the end points $\alpha, \beta, \gamma, \delta$ do not lie on the grid used for the calculation of the spline function S, then one can use program fibic2 and FIBIC2.

16.9 Decision Strategy

In this section, we have dealt only with cubature formulas for rectangular regions or for regions composed of triangles. Error estimates for the programs in the annex can be gained analogously as those for the quadrature problem, see section 15.11. The optimal Gaussian formulas are again superior to all other cubature formulas due to the small number of functional evaluations required and their high degree of exactness. In order to apply these, the integrand f must be easily computable everywhere in the region of integration. If functional values are only given at discrete points, the Newton-Cotes formulas should be used.

The Romberg method is also recommended. It uses, however, far more function evaluations than the Gauß formulas for the same degree of exactness. To be most effective its initial step size should not be chosen very small.

Bibliography for Chapter 16

[ENGE80]; [HÄMM91], §6.; [MAES88] vol.2, 7.4; [NIEM91] 7.; [STRO71] .

Chapter 17

Initial Value Problems for Ordinary Differential Equations

17.1 The Problem

For n first order differential equations in one variable involving n functions y_r, $r = 1, \ldots, n$, we shall consider the *initial value problem*

$$(17.1) \qquad \begin{cases} y' & = & f(x, y) \quad \text{for } x \in [x_0, \beta], \text{ where} \\ y(x_0) & = & y_0 \end{cases}$$

with

$$y = \begin{pmatrix} y_1(x) \\ y_2(x) \\ \vdots \\ y_n(x) \end{pmatrix}, \quad f = \begin{pmatrix} f_1(x, y_1, y_2, \ldots, y_n) \\ f_2(x, y_1, y_2, \ldots, y_n) \\ \vdots \\ f_n(x, y_1, y_2, \ldots, y_n) \end{pmatrix}.$$

The interval $I = [x_0, \beta]$ is called the *interval of integration* for the differential equation.

For $n = 1$ the initial value problem is one dimensional:

$$(17.2) \qquad \begin{cases} y'(x) = f(x, y) \quad \text{for } x \in [x_0, \beta], \text{ where} \\ y(x_0) = y_0. \end{cases}$$

Existence and uniqueness conditions.

For the initial value problem (17.1), the following conditions insure a unique solution of the problem:

(1) The functions f_r, $r = 1, \ldots, n$, are continuous in a region D of the $(x, y_1, ..., y_n)$-parameter space.

(2) The functions f_r satisfy a Lipschitz condition for all $(x, y), (x, \tilde{y}) \in D$:

$$\|f(x,y) - f(x,\tilde{y})\| \le L\|y - \tilde{y}\|.$$

Then there exists a region $\tilde{D} \subset D$, where the initial value problem (17.1) has exactly one solution y for the initial condition $y(x_0) = y_0$. Condition (2) is fulfilled if for example the f_r possess bounded partial derivatives with respect to y_k in D, in which case

$$L = \max_{\substack{1 \le r,k \le n \\ (x,y) \in D}} \left| \frac{\partial f_r}{\partial y_k} \right|.$$

Each initial value problem for a single differential equation of n^{th} order in y with n initial conditions

$$\begin{cases} y^{(n)}(x) & = \quad f(x,y,y',\ldots,y^{(n-1)}), \text{ where} \\ y(x_0) & = \quad y_0,\ y'(x_0) = y_0',\ldots,y^{(n-1)}(x_0) = y_0^{(n-1)} \end{cases}$$

can be reduced to a system (17.1) by setting

$$y^{(k)}(x) =: y_{k+1}(x) \quad \text{for } k = 0,\ldots,n-1.$$

The associated initial value conditions become

$$y^{(k)}(x_0) = y_{k+1}(x_0) \quad \text{for} \quad k = 0,\ldots,n-1.$$

Thus all methods of this chapter can be employed equally for the solution of first order initial value problems and n^{th} order initial value problems.

Bibliography for Section 17.1

[ENGE87], 10.1; [FRIE79]; [GEAR71/1], 1.1, 3; [HAIR87] 1; [LUTH87], 1,2; [STET76], 9; [STOE91], 7; [WERN79], §1.

17.2 Principles of the Numerical Methods

The interval $[x_0, \beta]$ of integration for (17.1) shall be partitioned

(17.3) $$x_0 < x_1 < x_2 < \ldots < x_N = \beta$$

with local step sizes $h_i := x_{i+1} - x_i > 0$ for $i = 0,\ldots,N-1$.

At the discrete grid points x_i we have to find approximate values Y_i for the exact solution $y(x_i)$

$$Y(x_i) = Y_i \approx y(x_i) = y_i.$$

Let us first consider the case $n = 1$ and integrate (17.2) from x_i to x_{i+1}:

$$\int_{x_i}^{x_{i+1}} y'(x)dx = \int_{x_i}^{x_{i+1}} f(x, y(x))dx.$$

By the fundamental theorem of calculus

(17.4) $$y(x_{i+1}) = y(x_i) + \int_{x_i}^{x_{i+1}} f(x, y(x))dx \quad \text{for } i = 0, \ldots, N-1.$$

(If $n > 1$, we need to evaluate a vector valued integral, one component at a time).

All numerical procedures for solving (17.1) or (17.2) differ only in the choice of method used to approximate the integral in (17.4). They can be classified into three types:

1) one-step methods,

2) multi-step methods, and

3) extrapolation algorithms.

One-step methods use only *one* preceding value Y_i when calculating the next approximate value Y_{i+1}.

Multi-step methods employ $s + 1$, $s \geq 1$, preceding values $Y_{i-s}, Y_{i-s+1}, \ldots, Y_{i-1}, Y_i$ to calculate Y_{i+1}.

Extrapolation algorithms use the Romberg quadrature method for the numerical solution of initial value problems.

Specialized methods are the so-called *predictor-corrector methods*. These are procedures which first determine an approximate value $Y_{i+1}^{(0)}$ with a one-step or a multi-step method in a *predictor step*. Then the value of $Y_{i+1}^{(0)}$ is improved by a so-called *corrector step*. The corrections are called $Y_{i+1}^{(1)}$, $Y_{i+1}^{(2)}, \ldots$.

The difference

$$\varepsilon_{i+1} := y(x_{i+1}) - Y(x_{i+1}) \quad \text{with} \quad Y(x_{i+1}) \approx y(x_{i+1}), \quad Y(x_i) = y(x_i)$$

is called the *local procedural error* at the grid point x_{i+1}. Here $y(x)$ is assumed to be the exact solution of the problem, and we assume that $Y(x_i)$ has been computed without rounding errors. Thus the local procedural error measures the error associated with integrating (17.4) from x_i to x_{i+1}. $O(h_i^{q_l})$ is the order of the *local procedural error*.

The difference

$$e_{i+1} := y(x_{i+1}) - Y(x_{i+1})$$

is the *global procedural error* at the grid point x_{i+1}. It measures the error at x_{i+1} taking into account all previously made errors. $O(h_{max}^{q_g})$ is the order of the global procedural error, where h_{max} is the maximal step size: $h_{max} = \max_{0 \le i \le N-1} h_i$. A global error order of q_g is achieved by a procedure if the solution y of (17.1) is $(q_g + 1)$-times continuously differentiable.

Bibliography for Section 17.2

[ENGE87], 10.2; [GEAR71/1], 1.2; [HAIR87] 1; [HALL76], 1; [HENR68], 0.3; [LAPI71], 1; [LUTH87], 3; [STET73], 1; [STOE91], 7.2; [STUM82], 11.1, 12.1; [WERN86], chap.1, 2.

17.3 One-Step Methods

17.3.1 The Euler-Cauchy Polygonal Method

One can approximate the integral in (17.4) by the area of a rectangle:

$$\int_{x_i}^{x_{i+1}} f(x)dx = h_i f(x_i) + \frac{h_i^2}{2} f'(\xi_i) \text{ for } \xi_i \in [x_i, x_{i+1}], \ x_{i+1} = x_i + h_i.$$

Thus for $Y_0 = y_0$ we obtain $y_{i+1} = y(x_{i+1})$ as

$$(17.5) \begin{cases} y_{i+1} &= Y_{i+1} + \varepsilon_{i+1}^{EC} \text{ with} \\ Y_{i+1} &= Y_i + h_i f(x_i, Y_i) \text{ if } Y_i = y(x_i), \ i = 0, \ldots, N-1, \\ \varepsilon_{i+1}^{EC} &= \frac{h_i^2}{2} y''(\xi_i) = O(h_i^2) \text{ locally for some } \xi_i \in [x_i, x_{i+1}]. \end{cases}$$

Y_{i+1} is the approximate value for $y(x_{i+1})$, and ε_{i+1}^{EC} is the *local procedural error*. It measures the error of the single Euler-Cauchy step from x_i to x_{i+1} under

the assumption that $Y_i = y(x_i)$ and that y is twice continuously differentiable. The errors of preceding steps are taken into consideration only by the *global procedural error*:

$$e_{i+1}^{EC} := y_{i+1} - Y_{i+1} = O(h_{max}), \ i = 0, \ldots, N-1, \ h_{max} = \max_{0 \le i \le N-1}(x_{i+1} - x_i).$$

For systems (17.1) we can formulate the following algorithm that uses (17.5):

ALGORITHM 17.1 (*Euler-Cauchy method*).

For the solution of (17.1) calculate an approximate value Y_i for $y(x_i)$ at each grid point x_i. With $Y_0 = y(x_0)$ the Y_i are given as

$$Y_{i+1} = Y_i + h_i f(x_i, Y_i), \quad i = 0, \ldots, N-1.$$

The global error order is $O(h_{max})$ for $h_{max} = \max_{0 \le i \le N-1}\{h_i\}$, i.e.,

$$y(x_{i+1}) = Y_{i+1} + O(h_{max}),$$

if y is twice continuously differentiable. For choosing a suitable grid (17.3) and controlling the step sizes, see section 17.3.7.

Figure 17.1: The polygonal method

17.3.2 The Improved Euler-Cauchy Method

The improved Euler-Cauchy method uses the direction field at $(x_i + h_i/2, Y_{i+1/2})$ in order to calculate Y_{i+1}. It thus obtains a much improved fitting of the solution to the direction field of the given differential equation.

ALGORITHM 17.2 (*Improved Euler-Cauchy method*).
For the solution of (17.1) and a suitably chosen grid (17.3), we set
$\boldsymbol{Y}_0 = \boldsymbol{y}(x_0)$ and define

$$
\begin{aligned}
\boldsymbol{Y}_{i+1} &= \boldsymbol{Y}_i + h_i \boldsymbol{f}(x_i + \left(\tfrac{h_i}{2}\right), \boldsymbol{Y}_{i+1/2}) \\
&= \boldsymbol{Y}_i + h_i \boldsymbol{f}(x_i + \left(\tfrac{h_i}{2}\right), \boldsymbol{Y}_i + \left(\tfrac{h_i}{2}\right) \boldsymbol{f}(x_i, \boldsymbol{Y}_i)), \ i = 0, \ldots, N-1,
\end{aligned}
$$

The global error order is $O(h_{\max}^2)$ with $h_{max} := \max\limits_{0 \le i \le N-1} (x_{i+1} - x_i)$, provided that \boldsymbol{y} is three times continuously differentiable. For choosing the proper step size, see section 17.3.7 .

17.3.3 The Predictor-Corrector Method of Heun

Using the trapezoidal rule for evaluating the integral in (17.4) gives an implicit equation for \boldsymbol{Y}_{i+1}, which must be solved iteratively. A first value for $\boldsymbol{Y}_{i+1}^{(0)}$ is determined here by of the Euler-Cauchy method. For a grid (17.3) we have

Heun's method for $n = 1$:

Starting with $Y_0 = y(x_0)$ we calculate for each $i = 0, \ldots, N-1$:

Predictor: $Y_{i+1}^{(0)} = Y_i + h_i f(x_i, Y_i)$,

Corrector: $Y_{i+1}^{(\nu+1)} = Y_i + \frac{h_i}{2}(f(x_i, Y_i) + f(x_{i+1}, Y_{i+1}^{(\nu)})), \ \nu = 0, 1, 2 \ldots$

The local procedural error satisfies:

$$
\varepsilon_{i+1}^{H} = -\frac{h_i^3}{12} y'''(\xi_i) \quad \text{for some } \xi_i \in [x_i, x_{i+1}],
$$

provided the solution $y \in C^3[x_0, \beta]$.

Since we cannot compute Y_{i+1} directly, but only the iterates $Y_{i+1}^{(\nu+1)}$, there is an additional *iteration error*

$$
\delta_{i+1,\nu}^{H} := Y_{i+1} - Y_{i+1}^{(\nu+1)}.
$$

Thus the *proper local procedural error is*

$$
E_{i+1,\nu}^{H} := y_{i+1} - Y_{i+1}^{(\nu+1)} = \varepsilon_{i+1}^{H} + \delta_{i+1,\nu}^{H}.
$$

If $K = h_i L_i < 1$, we can estimate

$$|E_{i+1,\nu}^H| \le \frac{1 - h_i L_i + \left(\frac{h_i L_i}{2}\right)^{\nu+1}}{1 - h_i L_i} \frac{h_i^3}{12}|y'''(\xi_1)| + \frac{\left(\frac{h_i L_i}{2}\right)^{\nu+1}}{1 - h_i L_i} \frac{h_i^2}{2}|y''(\xi_2)|,$$

for some $\xi_1, \xi_2 \in [x_i, x_{i+1}]$, so that we have

$$E_{i+1,0}^H = O(h_i^3),$$

already for $\nu = 0$, provided $y \in C^3[x_0, \beta]$. Here L_i are the local Lipschitz constants for $x \in [x_i, x_{i+1}]$.

The local error order of the corrector step is already attained after one iteration step. Experience shows that, for sufficiently small step sizes h_i, one or two iterations suffice to make $|E_{i+1,\nu}^H|$ essentially equal to $|\varepsilon_{i+1}^H|$. This can be assured if one chooses the local step sizes h_i so that

$$0.05 \le K = h_i L_i \le 0.20.$$

The global procedural error e_{i+1}^H, which takes the errors of previous steps into account, is

$$e_{i+1}^H := y_{i+1} - Y_{i+1} = O(h_{\max}^2) \quad \text{for } h_{\max} = \max_{0 \le i \le N-1} h_i, \ h_i = x_{i+1} - x_i.$$

ALGORITHM 17.3 (*Heun's method*).
For solving (17.1) we must perform the following steps with a suitable grid (17.3) and $Y_0 = y(x_0)$ for each $i = 0, \dots, N-1$:

1. $Y_{i+1}^{(0)} = Y_i + h_i f(x_i, Y_i)$ (predictor).
2. For $\nu = 0, 1$: $Y_{i+1}^{(\nu+1)} = Y_i + \frac{h_i}{2}\left(f(x_i, Y_i) + f(x_{i+1}, Y_{i+1}^{(\nu)})\right)$ (corrector).

The local step size h_i should be chosen so that $K = h_i L_i \le 0.20$. After step 2 we set for each i

$$Y_{i+1} := Y_{i+1}^{(2)}.$$

The global procedural error is

$$e_{i+1}^H = O(h_{\max}^2)$$

with $h_{\max} = \max_{0 \le i \le N-1} h_i$, $h_i = x_{i+1} - x_i$, provided $y \in C^3[x_0, \beta]$.

17.3.4 Explicit Runge-Kutta Methods

17.3.4.1 Construction of Runge-Kutta Methods

The most important category of one-step methods are the Runge-Kutta methods. An explicit Runge-Kutta method of order m has the general form

$$
(17.6) \begin{cases}
\boldsymbol{Y}_{i+1} = \boldsymbol{Y}_i + h_i \sum_{j=1}^{m} A_j \boldsymbol{k}_j(x_i, \boldsymbol{Y}_i, h_i) \text{ with} \\[2mm]
\boldsymbol{k}_1(x, \boldsymbol{Y}, h) := \boldsymbol{f}(x, \boldsymbol{Y}), \\[2mm]
\boldsymbol{k}_j(x, \boldsymbol{Y}, h) := \boldsymbol{f}\left(x + a_j h, \boldsymbol{Y} + h \sum_{s=1}^{j-1} b_{js} \boldsymbol{k}_s(x, \boldsymbol{Y}, h)\right), \; j = 2, \dots, m.
\end{cases}
$$

This approach defines the *explicit* Runge-Kutta formulas.

For explicit m^{th} order Runge-Kutta methods with $m \leq 4$, one can achieve a local error order of $q_\ell = m + 1$ and a global error order of $q_g = m$. For $m > 4$, we have $q_g < m$.

Here is a table of the global error order for the m^{th} order explicit Runge-Kutta methods:

m	1	2	3	4	5	6	7	8	9
q_g	1	2	3	4	4	5	6	6	7

If $m = 1$, the Runge-Kutta method is identical with the Euler-Cauchy method. For $m = 2$ the improved Euler-Cauchy method and Heun's method are obtained. Thus they are Runge-Kutta methods of order two. For $m = 4$ we shall find the so-called *classical Runge-Kutta method:*

17.3.4.2 The Classical Runge-Kutta Method

A Runge-Kutta procedure for $m = 4$ of the form (17.6) is called the *classical Runge-Kutta method.*

ALGORITHM 17.4 (*Classical Runge-Kutta method*).
To solve (17.1) one has to choose a suitable grid (17.3), and starting with $\boldsymbol{Y}_0 = \boldsymbol{y}(x_0)$, one evaluates

$$Y_{i+1} = Y_i + h_i \left\{ \frac{1}{6}k_1 + \frac{1}{3}k_2 + \frac{1}{3}k_3 + \frac{1}{6}k_4 \right\} \text{ with}$$

$$
\begin{aligned}
k_1 &= f(x_i, Y_i), \\
k_2 &= f(x_i + (h_i/2),\ Y_i + h_i(k_1/2)), \\
k_3 &= f(x_i + (h_i/2),\ Y_i + h_i(k_2/2)), \\
k_4 &= f(x_i + h_i,\ Y_i + h_i k_3).
\end{aligned}
$$

At each grid point x_i, Y_i is an approximate value for the exact value $y(x_i)$. The local error order is $O(h_i^5)$, the global error order is $O(h_{max}^4)$ with $h_{max} := \max\limits_{0 \leq i \leq N-1} (x_{i+1} - x_i)$ if $y \in C^5[a,b]$.

For $n = 1$ one can follow the procedure in the following calculation scheme. For $n > 1$ one has to proceed analogously for each component.

CALCULATION SCHEME 17.5 (*Classical Runge-Kutta method for $n = 1$*).

i	x	y	$k_j(x_i, Y_i, h_i),\ j = 1, \ldots, 4$	$k^{(i)}$
0	x_0	y_0	$k_1 = f(x_0, Y_0)$	k_1
	$x_0 + \frac{h_0}{2}$	$y_0 + h_0 \frac{k_1}{2}$	$k_2 = f(x_0 + \frac{h_0}{2}, y_0 + h_0 \frac{k_1}{2})$	$2k_2$
	$x_0 + \frac{h_0}{2}$	$y_0 + h_0 \frac{k_2}{2}$	$k_3 = f(x_0 + \frac{h_0}{2}, y_0 + h_0 \frac{k_2}{2})$	$2k_3$
	$x_0 + h_0$	$y_0 + h_0 k_3$	$k_4 = f(x_0 + h_0, y_0 + h_0 k_3)$	k_4
	$x_1 = x_0 + h_0$	$Y_1 = y_0 + h_0 k^{(0)}$		$k^{(0)} = \frac{1}{6}\sum$
1	x_1	Y_1	$k_1 = f(x_1, Y_1)$	k_1
	$x_1 + \frac{h_1}{2}$	$Y_1 + h_1 \frac{k_1}{2}$	$k_2 = f(x_1 + \frac{h_1}{2}, Y_1 + h_1 \frac{k_1}{2})$	$2k_2$
	$x_1 + \frac{h_1}{2}$	$Y_1 + h_1 \frac{k_2}{2}$	$k_3 = f(x_1 + \frac{h_1}{2}, Y_1 + h_1 \frac{k_2}{2})$	$2k_3$
	$x_1 + h_1$	$Y_1 + h_1 k_3$	$k_4 = f(x_1 + h_1, Y_1 + h_1 k_3)$	k_4
	$x_2 = x_1 + h_1$	$Y_2 = y_1 + h_1 k^{(1)}$		$k^{(1)} = \frac{1}{6}\sum$
2	x_2	Y_2	$k_1 = f(x_2, Y_2)$	k_1
	\vdots	\vdots	\vdots	\vdots

The classical Runge-Kutta method is expensive in computational time. Per Runge-Kutta step, one has to compute four functional values of f since $m = 4$. For higher m more evaluations are necessary. When comparing the calculation time and degree of exactness of the classical Runge-Kutta method or one of higher order with the previously mentioned methods, it is obvious that Runge-Kutta is preferrable to the Euler-Cauchy, the improved Euler-Cauchy method and to Heun's method. A detailed survey can be found in [LUTH87], p.68 .

17.3.4.3 A List of Explicit Runge-Kutta Formulas

In the following we give a coefficient table for explicit Runge-Kutta methods of the form (17.6) for orders $m = 1, \ldots, 8$. For a local error order q_ℓ, the global error order always is $q_g = q_\ell - 1$.

The formulas (17.6) can be written out in detail as:

$$\boldsymbol{Y}_{i+1} = \boldsymbol{Y}_i + h_i(A_1\boldsymbol{k}_1 + A_2\boldsymbol{k}_2 + \ldots + A_m\boldsymbol{k}_m) \text{ with}$$

$$
\begin{aligned}
\boldsymbol{k}_1(x_i, \boldsymbol{Y}_i, h_i) &= \boldsymbol{f}(x_i, \boldsymbol{Y}_i) \\
\boldsymbol{k}_2(x_i, \boldsymbol{Y}_i, h_i) &= \boldsymbol{f}(x_i + a_2 h_i, \boldsymbol{Y}_i + h_i b_{21}\boldsymbol{k}_1) \\
\boldsymbol{k}_3(x_i, \boldsymbol{Y}_i, h_i) &= \boldsymbol{f}(x_i + a_3 h_i, \boldsymbol{Y}_i + h_i(b_{31}\boldsymbol{k}_1 + b_{32}\boldsymbol{k}_2)) \\
&\vdots \\
\boldsymbol{k}_m(x_i, \boldsymbol{Y}_i, h_i) &= \boldsymbol{f}(x_i + a_m h_i, \boldsymbol{Y}_i + h_i(b_{m1}\boldsymbol{k}_1 + b_{m2}\boldsymbol{k}_2 + \ldots + b_{m,m-1}\boldsymbol{k}_{m-1})).
\end{aligned}
$$

The coefficients A_j, a_j and b_{js} are listed for $j = 1, \ldots, m$, $s = 1, \ldots, m-1$ and $m = 1, \ldots, 8$ in the following table.

Further explicit Runge-Kutta formulas can be found in [FEHL60] and in [FEHL66]; see also section 17.3.4.4 for Runge-Kutta embedding formulas.

TABLE 17.6 (*Coefficient table for explicit Runge-Kutta formulas*).

m	j	A_j	a_j	b_{js} for $s = 1, \ldots, m-1$				q_g	name of the method
1	1	1	0					1	Euler-Cauchy
2	1	0	0					2	Improved
	2	1	1/2	1/2					E.-C. method
2	1	1/2	0					2	Heun's
	2	1/2	1	1					method
3	1	1/6	0					3	RK I
	2	2/3	1/2	1/2					
	3	1/6	1	–1	2				3^{rd} order
3	1	1/4	0					3	RK II
	2	0	1/3	1/3					
	3	3/4	2/3	0	2/3				3^{rd} order
4	1	1/8	0					4	3/8–
	2	3/8	1/3	1/3					formula
	3	3/8	2/3	–1/3	1				
	4	1/8	1	1	–1	1			
4	1	1/6	0					4	Classical
	2	1/3	1/2	1/2					RK-method
	3	1/3	1/2	0	1/2				
	4	1/6	1	0	0	1			

s	i	b_i	c_i						order	method
4	1	$\frac{1}{6}$	0						4	RK–Gill
	2	$\frac{2-\sqrt{2}}{6}$	$\frac{1}{2}$	$\frac{1}{2}$						method
	3	$\frac{2+\sqrt{2}}{6}$	$\frac{1}{2}$	$-\frac{1}{2}+\frac{\sqrt{2}}{2}$	$\frac{1-\sqrt{2}}{2}$					
	4	$\frac{1}{6}$	1	0	$-\frac{\sqrt{2}}{2}$	$1+\frac{\sqrt{2}}{2}$				
4	1	1/6	0						4	England
	2	0	1/2	1/2						I
	3	4/6	1/2	1/4	1/4					
	4	1/6	1	0	-1	2				
5	1	$\frac{25}{216}$	0						4	RK-Fehlberg
	2	0	$\frac{1}{4}$	$\frac{1}{4}$						method,
	3	$\frac{1408}{2565}$	$\frac{3}{8}$	$\frac{3}{32}$	$\frac{9}{32}$					4^{th} order
	4	$\frac{2197}{4104}$	$\frac{12}{13}$	$\frac{1932}{2197}$	$-\frac{7200}{2197}$	$\frac{7296}{2197}$				
	5	$-\frac{1}{5}$	1	$\frac{439}{216}$	-8	$\frac{3680}{513}$	$-\frac{845}{4104}$			
6	1	$\frac{16}{135}$	0						5	RK-Fehlberg
	2	0	$\frac{1}{4}$	$\frac{1}{4}$						method,
	3	$\frac{6656}{12825}$	$\frac{3}{8}$	$\frac{3}{32}$	$\frac{9}{32}$					5^{th} order
	4	$\frac{28561}{56430}$	$\frac{12}{13}$	$\frac{1932}{2197}$	$-\frac{7200}{2197}$	$\frac{7296}{2197}$				
	5	$-\frac{9}{50}$	1	$\frac{439}{216}$	-8	$\frac{3680}{519}$	$-\frac{845}{4104}$			
	6	$\frac{2}{55}$	$\frac{1}{2}$	$-\frac{8}{27}$	2	$-\frac{3544}{2565}$	$\frac{1859}{4104}$	$-\frac{11}{40}$		
6	1	$\frac{14}{336}$	0						5	England
	2	0	$\frac{1}{2}$	$\frac{1}{2}$						II
	3	0	$\frac{1}{2}$	$\frac{1}{4}$	$\frac{1}{4}$					
	4	$\frac{35}{336}$	1	0	-1	2				
	5	$\frac{162}{336}$	$\frac{2}{3}$	$\frac{7}{27}$	$\frac{10}{27}$	0	$\frac{1}{27}$			
	6	$\frac{125}{336}$	$\frac{1}{5}$	$\frac{28}{625}$	$-\frac{125}{625}$	$\frac{546}{625}$	$\frac{54}{625}$	$-\frac{378}{625}$		

	i											method
6	1	$\frac{23}{192}$	0								5	Kutta-Nyström method
	2	0	$\frac{1}{3}$	$\frac{1}{3}$								
	3	$\frac{125}{192}$	$\frac{2}{5}$	$\frac{4}{25}$	$\frac{6}{25}$							
	4	0	1	$\frac{1}{4}$	$-\frac{12}{4}$	$\frac{15}{4}$						
	5	$-\frac{81}{192}$	$\frac{2}{3}$	$\frac{6}{81}$	$\frac{90}{81}$	$-\frac{50}{81}$	$\frac{8}{81}$					
	6	$\frac{125}{192}$	$\frac{4}{5}$	$\frac{6}{75}$	$\frac{36}{75}$	$\frac{10}{75}$	$\frac{8}{75}$	0				
6	1	$\frac{31}{384}$	0								5	RK-Fehlberg I (F I)
	2	0	$\frac{1}{6}$	$\frac{1}{6}$								
	3	$\frac{1125}{2816}$	$\frac{4}{15}$	$\frac{4}{75}$	$\frac{16}{75}$							
	4	$\frac{9}{32}$	$\frac{2}{3}$	$\frac{5}{6}$	$-\frac{8}{3}$	$\frac{5}{2}$						
	5	$\frac{125}{768}$	$\frac{4}{5}$	$-\frac{8}{5}$	$\frac{144}{25}$	-4	$\frac{16}{25}$					
	6	$\frac{5}{66}$	1	$\frac{361}{320}$	$-\frac{18}{5}$	$\frac{407}{128}$	$-\frac{11}{80}$	$\frac{55}{128}$				
7	1	$\frac{11}{120}$	0								6	Butcher method
	2	0	$\frac{1}{3}$	$\frac{1}{3}$								
	3	$\frac{27}{40}$	$\frac{2}{3}$	0	$\frac{2}{3}$							
	4	$\frac{27}{40}$	$\frac{1}{3}$	$\frac{1}{12}$	$\frac{1}{3}$	$-\frac{1}{12}$						
	5	$-\frac{4}{15}$	$\frac{1}{2}$	$-\frac{1}{16}$	$\frac{9}{8}$	$-\frac{3}{16}$	$-\frac{3}{8}$					
	6	$-\frac{4}{15}$	$\frac{1}{2}$	0	$\frac{9}{8}$	$-\frac{3}{8}$	$-\frac{3}{4}$	$\frac{1}{2}$				
	7	$\frac{11}{120}$	1	$\frac{9}{44}$	$-\frac{9}{11}$	$\frac{63}{44}$	$\frac{18}{11}$	0	$-\frac{16}{11}$			
8	1	$\frac{7}{1408}$	0								6	RK-Fehlberg II (F II)
	2	0	$\frac{1}{6}$	$\frac{1}{6}$								
	3	$\frac{1125}{2816}$	$\frac{4}{15}$	$\frac{4}{75}$	$\frac{16}{75}$							
	4	$\frac{9}{32}$	$\frac{2}{3}$	$\frac{5}{6}$	$-\frac{8}{3}$	$\frac{5}{2}$						
	5	$\frac{125}{768}$	$\frac{4}{5}$	$-\frac{8}{5}$	$\frac{144}{25}$	-4	$\frac{16}{25}$					
	6	0	1	$\frac{361}{320}$	$-\frac{18}{5}$	$\frac{407}{128}$	$-\frac{11}{80}$	$\frac{55}{128}$				
	7	$\frac{5}{66}$	0	$-\frac{11}{640}$	0	$\frac{11}{256}$	$-\frac{11}{160}$	$\frac{11}{256}$	0			
	8	$\frac{5}{66}$	1	$\frac{93}{640}$	$-\frac{18}{5}$	$\frac{803}{256}$	$-\frac{11}{160}$	$\frac{99}{256}$	0	1		

17.3.4.4 Embedding Formulas

If for two explicit Runge-Kutta formulas of orders m and $\tilde{m} > m$, the values
for k_j coincide for $j = 1, \ldots, m$, one can use both formulas as a pair. The
resulting formula is called an *embedding formula*.

In this sense the improved Euler-Cauchy method ($q_g = 2$) together with the
Runge-Kutta method of order three; the Runge-Kutta Fehlberg methods with
global error orders 4 and 5; the England I and England II formulas with $q_g = 4$
and $q_g = 5$; as well as the Fehlberg I- and Fehlberg II-formulas with $q_g = 5$ or
$q_g = 6$ are embedding formulas. Each of the two formulas of a pair supplies
an approximate value Y and \tilde{Y} for $y(x_i + h)$. These two approximate values
can be used for controlling the step size effectively. In general we have:

$$
\left\{
\begin{array}{ll}
\boldsymbol{Y} & = \boldsymbol{Y}_i + h \sum\limits_{j=1}^{m} A_j \boldsymbol{k}_j \qquad \text{with global error order} \quad q_g \\[2ex]
\tilde{\boldsymbol{Y}} & = \boldsymbol{Y}_i + h \sum\limits_{j=1}^{\tilde{m}} \tilde{A}_j \boldsymbol{k}_j \qquad \text{with global error order} \quad \tilde{q}_g \\[1ex]
\text{for} & \\
\boldsymbol{k}_1 & = \boldsymbol{f}(x_i, \boldsymbol{Y}_i) \\
\boldsymbol{k}_2 & = \boldsymbol{f}(x_i + a_2 h, \boldsymbol{Y}_i + h b_{21} \boldsymbol{k}_1) \\
\vdots & \\
\boldsymbol{k}_j & = \boldsymbol{f}(x_i + a_j h, \boldsymbol{Y}_i + \sum\limits_{s=1}^{j-1} h b_{js} \boldsymbol{k}_s), \quad j = 3, \ldots, \tilde{m}.
\end{array}
\right.
$$

In the following list, we shall give each embedding formula a short code such
as *"rk3(2)"* for the Runge-Kutta embedding formula of 3^{rd} and 2^{nd} order.

RK - embedding formula of 2^{nd} and 3^{rd} order, **rk3(2)**.
($q_g = 2, \tilde{q}_g = 3, m = 2, \tilde{m} = 3$)

List of coefficients:

j	\tilde{A}_j	A_j	a_j	b_{js} for $s = 1, \ldots, \tilde{m} - 1$	
1	$\frac{1}{6}$	0	0		
2	$\frac{2}{3}$	1	$\frac{1}{2}$	$\frac{1}{2}$	
3	$\frac{1}{6}$		1	-1	2

$$(17.7) \quad \begin{cases} \mathbf{Y} & = \mathbf{Y}_i + h\mathbf{k}_2 \\ \tilde{\mathbf{Y}} & = \mathbf{Y}_i + h\left\{\frac{1}{6}\mathbf{k}_1 + \frac{2}{3}\mathbf{k}_2 + \frac{1}{6}\mathbf{k}_3\right\} \\ \text{with} \\ \mathbf{k}_1 & = \mathbf{f}(x_i, \mathbf{Y}_i) \\ \mathbf{k}_2 & = \mathbf{f}(x_i + \frac{1}{2}h, \mathbf{Y}_i + \frac{1}{2}h\mathbf{k}_1) \\ \mathbf{k}_3 & = \mathbf{f}(x_i + h, \mathbf{Y}_i - h\mathbf{k}_1 + 2h\mathbf{k}_2). \end{cases}$$

RK - Fehlberg - embedding formula of 4^{th} *and* 5^{th} *order,* **rkf5(4)**.
($q_g = 4$, $\tilde{q}_g = 5$, $m = 5$, $\tilde{m} = 6$)

Coefficient scheme:

j	\tilde{A}_j	A_j	a_j	b_{js} for $s = 1, \ldots, \tilde{m}-1$				
1	$\frac{16}{135}$	$\frac{25}{216}$	0					
2	0	0	$\frac{1}{4}$	$\frac{1}{4}$				
3	$\frac{6656}{12825}$	$\frac{1408}{2565}$	$\frac{3}{8}$	$\frac{3}{32}$	$\frac{9}{32}$			
4	$\frac{28561}{56430}$	$\frac{2197}{4104}$	$\frac{12}{13}$	$\frac{1932}{2197}$	$-\frac{7200}{2197}$	$\frac{7296}{2197}$		
5	$-\frac{9}{50}$	$-\frac{1}{5}$	1	$\frac{439}{216}$	-8	$\frac{3680}{513}$	$-\frac{845}{4104}$	
6	$\frac{2}{55}$		$\frac{1}{2}$	$-\frac{8}{27}$	2	$-\frac{3544}{2565}$	$\frac{1859}{4104}$	$-\frac{11}{40}$

$$(17.8) \quad \begin{cases} \mathbf{Y} & = \mathbf{Y}_i + h\left\{\frac{25}{216}\mathbf{k}_1 + \frac{1408}{2565}\mathbf{k}_3 + \frac{2197}{4104}\mathbf{k}_4 - \frac{1}{5}\mathbf{k}_5\right\} \\ \tilde{\mathbf{Y}} & = \mathbf{Y}_i + h\left\{\frac{16}{135}\mathbf{k}_1 + \frac{6656}{12825}\mathbf{k}_3 + \frac{28561}{56430}\mathbf{k}_4 - \frac{9}{50}\mathbf{k}_5 + \frac{2}{55}\mathbf{k}_6\right\} \\ \text{with} \\ \mathbf{k}_1 & = \mathbf{f}(x_i, \mathbf{Y}_i) \\ \mathbf{k}_2 & = \mathbf{f}(x_i + \frac{1}{4}h, \ \mathbf{Y}_i + \frac{1}{4}h\mathbf{k}_1) \\ \mathbf{k}_3 & = \mathbf{f}(x_i + \frac{3}{8}h, \mathbf{Y}_i + \frac{3}{32}h\mathbf{k}_1 + \frac{9}{32}h\mathbf{k}_2) \\ \mathbf{k}_4 & = \mathbf{f}(x_i + \frac{12}{13}h, \mathbf{Y}_i + \frac{1932}{2197}h\mathbf{k}_1 - \frac{7200}{2197}h\mathbf{k}_2 + \frac{7296}{2197}h\mathbf{k}_3) \\ \mathbf{k}_5 & = \mathbf{f}(x_i + h, \mathbf{Y}_i + \frac{439}{216}h\mathbf{k}_1 - 8h\mathbf{k}_2 + \frac{3680}{513}h\mathbf{k}_3 - \frac{845}{4104}h\mathbf{k}_4) \\ \mathbf{k}_6 & = \mathbf{f}(x_i + \frac{1}{2}h, \mathbf{Y}_i - \frac{8}{27}h\mathbf{k}_1 + 2h\mathbf{k}_2 - \frac{3544}{2565}h\mathbf{k}_3 + \frac{1859}{4104}h\mathbf{k}_4 - \frac{11}{40}h\mathbf{k}_5). \end{cases}$$

England - formula of 4^{th} and 5^{th} order, **rke5(4)**.
($q_g = 4$, $\tilde{q}_g = 5$, $m = 4$, $\tilde{m} = 6$)

Coefficient scheme:

j	\tilde{A}_j	A_j	a_j	b_{js} for $s = 1, \ldots, \tilde{m} - 1$				
1	$\frac{14}{336}$	$\frac{1}{6}$	0					
2	0	0	$\frac{1}{2}$	$\frac{1}{2}$				
3	0	$\frac{4}{6}$	$\frac{1}{2}$	$\frac{1}{4}$	$\frac{1}{4}$			
4	$\frac{35}{336}$	$\frac{1}{6}$	1	0	-1	2		
5	$\frac{162}{336}$		$\frac{2}{3}$	$\frac{7}{27}$	$\frac{10}{27}$	0	$\frac{1}{27}$	
6	$\frac{125}{336}$		$\frac{1}{5}$	$\frac{28}{625}$	$-\frac{125}{625}$	$\frac{546}{625}$	$\frac{54}{625}$	$-\frac{378}{625}$

(17.9)
$$
\left\{
\begin{aligned}
\boldsymbol{Y} &= \boldsymbol{Y}_i + h\left\{\tfrac{1}{6}\boldsymbol{k}_1 + \tfrac{4}{6}\boldsymbol{k}_3 + \tfrac{1}{6}\boldsymbol{k}_4\right\} \\
\tilde{\boldsymbol{Y}} &= \boldsymbol{Y}_i + h\left\{\tfrac{14}{336}\boldsymbol{k}_1 + \tfrac{35}{336}\boldsymbol{k}_4 + \tfrac{162}{336}\boldsymbol{k}_5 + \tfrac{125}{336}\boldsymbol{k}_6\right\} \\
&\text{with} \\
\boldsymbol{k}_1 &= \boldsymbol{f}(x_i, \boldsymbol{Y}_i) \\
\boldsymbol{k}_2 &= \boldsymbol{f}(x_i + \tfrac{h}{2}, \boldsymbol{Y}_i + \tfrac{h}{2}\boldsymbol{k}_1) \\
\boldsymbol{k}_3 &= \boldsymbol{f}(x_i + \tfrac{h}{2}, \boldsymbol{Y}_i + \tfrac{h}{4}\boldsymbol{k}_1 + \tfrac{h}{4}\boldsymbol{k}_2) \\
\boldsymbol{k}_4 &= \boldsymbol{f}(x_i + h, \boldsymbol{Y}_i - h\boldsymbol{k}_2 + 2h\boldsymbol{k}_3) \\
\boldsymbol{k}_5 &= \boldsymbol{f}(x_i + \tfrac{2}{3}h, \boldsymbol{Y}_i + \tfrac{7}{27}h\boldsymbol{k}_1 + \tfrac{10}{27}h\boldsymbol{k}_2 + \tfrac{1}{27}h\boldsymbol{k}_4) \\
\boldsymbol{k}_6 &= \boldsymbol{f}(x_i + \tfrac{h}{5}, \boldsymbol{Y}_i + \tfrac{28}{625}h\boldsymbol{k}_1 - \tfrac{125}{625}h\boldsymbol{k}_2 + \tfrac{546}{625}h\boldsymbol{k}_3 + \\
&\qquad + \tfrac{54}{625}h\boldsymbol{k}_4 - \tfrac{378}{625}h\boldsymbol{k}_5).
\end{aligned}
\right.
$$

RK – *Fehlberg - embedding formula of* 5^{th} *and* 6^{th} *order,* **rkf6(5)**.
$(q_g = 5,\ \tilde{q}_g = 6,\ m = 6,\ \tilde{m} = 8)$

Coefficient scheme:

j	\tilde{A}_j	A_j	a_j	b_{js} for $s = 1, \ldots, \tilde{m} - 1$						
1	$\frac{7}{1408}$	$\frac{31}{384}$	0							
2	0	0	$\frac{1}{6}$	$\frac{1}{6}$						
3	$\frac{1125}{2816}$	$\frac{1125}{2816}$	$\frac{4}{15}$	$\frac{4}{75}$	$\frac{16}{75}$					
4	$\frac{9}{32}$	$\frac{9}{32}$	$\frac{2}{3}$	$\frac{5}{6}$	$-\frac{8}{3}$	$\frac{5}{2}$				
5	$\frac{125}{768}$	$\frac{125}{768}$	$\frac{4}{5}$	$-\frac{8}{5}$	$\frac{144}{25}$	-4	$\frac{16}{25}$			
6	0	$\frac{5}{66}$	1	$\frac{361}{320}$	$-\frac{18}{5}$	$\frac{407}{128}$	$-\frac{11}{80}$	$\frac{55}{128}$		
7	$\frac{5}{66}$		0	$-\frac{11}{640}$	0	$\frac{11}{256}$	$-\frac{11}{160}$	$\frac{11}{256}$	0	
8	$\frac{5}{66}$		1	$\frac{93}{640}$	$-\frac{18}{5}$	$\frac{803}{256}$	$-\frac{11}{160}$	$\frac{99}{256}$	0	1

$$(17.10)\begin{cases} \boldsymbol{Y} = \boldsymbol{Y}_i + h\left\{ \frac{31}{384}\boldsymbol{k}_1 + \frac{1125}{2816}\boldsymbol{k}_3 + \frac{9}{32}\boldsymbol{k}_4 + \frac{125}{768}\boldsymbol{k}_5 + \frac{5}{66}\boldsymbol{k}_6 \right\} \\[2mm] \tilde{\boldsymbol{Y}} = \boldsymbol{Y}_i + h\left\{ \frac{7}{1408}\boldsymbol{k}_1 + \frac{1125}{2816}\boldsymbol{k}_3 + \frac{9}{32}\boldsymbol{k}_4 + \frac{125}{768}\boldsymbol{k}_5 + \right. \\[2mm] \left. \qquad + \frac{5}{66}\boldsymbol{k}_7 + \frac{5}{66}\boldsymbol{k}_8 \right\} \\[2mm] \text{with} \\[2mm] \boldsymbol{k}_1 = \boldsymbol{f}(x_i, \boldsymbol{Y}_i) \\[2mm] \boldsymbol{k}_2 = \boldsymbol{f}(x_i + \frac{h}{6}, \boldsymbol{Y}_i + \frac{h}{6}\boldsymbol{k}_1) \\[2mm] \boldsymbol{k}_3 = \boldsymbol{f}(x_i + \frac{4}{15}h, \boldsymbol{Y}_i + \frac{4}{75}h\boldsymbol{k}_1 + \frac{16}{75}h\boldsymbol{k}_2) \\[2mm] \boldsymbol{k}_4 = \boldsymbol{f}(x_i + \frac{2}{3}h, \boldsymbol{Y}_i + \frac{5}{6}h\boldsymbol{k}_1 - \frac{8}{3}h\boldsymbol{k}_2 + \frac{5}{2}h\boldsymbol{k}_3) \\[2mm] \boldsymbol{k}_5 = \boldsymbol{f}(x_i + \frac{4}{5}h, \boldsymbol{Y}_i - \frac{8}{5}h\boldsymbol{k}_1 + \frac{144}{25}h\boldsymbol{k}_2 - 4h\boldsymbol{k}_3 + \frac{16}{25}h\boldsymbol{k}_4) \\[2mm] \boldsymbol{k}_6 = \boldsymbol{f}(x_i + h, \boldsymbol{Y}_i + \frac{361}{320}h\boldsymbol{k}_1 - \frac{18}{5}h\boldsymbol{k}_2 + \frac{407}{128}h\boldsymbol{k}_3 + \\[1mm] \qquad - \frac{11}{80}h\boldsymbol{k}_4 + \frac{55}{128}h\boldsymbol{k}_5); \\[2mm] \boldsymbol{k}_7 = \boldsymbol{f}(x_i, \boldsymbol{Y}_i - \frac{11}{640}h\boldsymbol{k}_1 + \frac{11}{256}h\boldsymbol{k}_3 - \frac{11}{160}h\boldsymbol{k}_4 + \frac{11}{256}h\boldsymbol{k}_5) \\[2mm] \boldsymbol{k}_8 = \boldsymbol{f}(x_i + h, \boldsymbol{Y}_i + \frac{93}{640}h\boldsymbol{k}_1 - \frac{18}{5}h\boldsymbol{k}_2 + \frac{803}{256}h\boldsymbol{k}_3 + \\[1mm] \qquad - \frac{11}{160}h\boldsymbol{k}_4 + \frac{99}{256}h\boldsymbol{k}_5 + h\boldsymbol{k}_7). \end{cases}$$

Further embedding formulas can be composed in complete analogy from the following coefficient schemes.

Further embedding formulas:

RK - Fehlberg - embedding formulas of 3^{rd} and 4^{th} order, rkf4(3).
$(q_g = 3, \tilde{q}_g = 4, m = 4, \tilde{m} = 5)$

Coefficient scheme:

j	\tilde{A}_j	A_j	a_j	b_{js} for $s = 1, \ldots, \tilde{m} - 1$			
1	$\frac{229}{1470}$	$\frac{79}{490}$	0				
2	0	0	$\frac{2}{7}$	$\frac{2}{7}$			
3	$\frac{1125}{1813}$	$\frac{2175}{3626}$	$\frac{7}{15}$	$\frac{77}{900}$	$\frac{343}{900}$		
4	$\frac{13718}{81585}$	$\frac{2166}{9065}$	$\frac{35}{38}$	$\frac{805}{1444}$	$-\frac{77175}{54872}$	$\frac{97125}{54872}$	
5	$\frac{1}{18}$		1	$\frac{79}{490}$	0	$\frac{2175}{3626}$	$\frac{2166}{9065}$

Prince-Dormand-embedding formulas of 4^{th} and 5^{th} order, rk5(4)6 m.
$(q_g = 4, \tilde{q}_g = 5, m = 6, \tilde{m} = 6)$

Coefficient scheme:

j	\tilde{A}_j	A_j	a_j	b_{js} for $s = 1, \ldots, \tilde{m} - 1$				
1	$\frac{19}{216}$	$\frac{31}{540}$	0					
2	0	0	$\frac{1}{5}$	$\frac{1}{5}$				
3	$\frac{1000}{2079}$	$\frac{190}{297}$	$\frac{3}{10}$	$\frac{3}{40}$	$\frac{9}{40}$			
4	$-\frac{125}{216}$	$-\frac{145}{108}$	$\frac{3}{5}$	$\frac{3}{10}$	$-\frac{9}{10}$	$\frac{6}{5}$		
5	$\frac{81}{88}$	$\frac{351}{220}$	$\frac{2}{3}$	$\frac{226}{729}$	$-\frac{25}{27}$	$\frac{880}{729}$	$\frac{55}{729}$	
6	$\frac{5}{56}$	$\frac{1}{20}$	1	$-\frac{181}{270}$	$\frac{5}{2}$	$-\frac{266}{297}$	$-\frac{91}{27}$	$\frac{189}{55}$

Prince-Dormand-embedding formulas of 4^{th} and 5^{th} order, **rk5(4)7 m̃.**
$(q_g = 4,\ \tilde{q}_g = 5,\ m = 7,\ \tilde{m} = 7)$

Coefficient scheme:

j	\tilde{A}_j	A_j	a_j	b_{js} for $s=1,\dots,m-1$					
1	$\frac{35}{384}$	$\frac{5179}{57600}$	0						
2	0	0	$\frac{1}{5}$	$\frac{1}{5}$					
3	$\frac{500}{1113}$	$\frac{7571}{16695}$	$\frac{3}{10}$	$\frac{3}{40}$	$\frac{9}{40}$				
4	$\frac{125}{192}$	$\frac{393}{640}$	$\frac{4}{5}$	$\frac{44}{45}$	$-\frac{56}{15}$	$\frac{32}{9}$			
5	$-\frac{2187}{6784}$	$-\frac{92097}{339200}$	$\frac{8}{9}$	$\frac{19372}{6561}$	$-\frac{25360}{2187}$	$\frac{64448}{6561}$	$-\frac{212}{729}$		
6	$\frac{11}{84}$	$\frac{187}{2100}$	1	$\frac{9017}{3168}$	$-\frac{355}{33}$	$\frac{46732}{5247}$	$\frac{49}{176}$	$-\frac{5103}{18656}$	
7	0	$\frac{1}{40}$	1	$\frac{35}{384}$	0	$\frac{500}{1113}$	$\frac{125}{192}$	$-\frac{2187}{6784}$	$\frac{11}{84}$

Prince-Dormand-embedding formulas of 4^{th} and 5^{th} order, **rk5(4)7 m̃.**
$(q_g = 4,\ \tilde{q}_g = 5,\ m = 7,\ \tilde{m} = 7)$

Coefficient scheme:

j	\tilde{A}_j	A_j	a_j	b_{js} for $s=1,\dots,m-1$					
1	$\frac{19}{200}$	$\frac{431}{5000}$	0						
2	0	0	$\frac{2}{9}$	$\frac{2}{9}$					
3	$\frac{3}{5}$	$\frac{333}{500}$	$\frac{1}{3}$	$\frac{1}{12}$	$\frac{1}{4}$				
4	$-\frac{243}{400}$	$-\frac{7857}{10000}$	$\frac{5}{9}$	$\frac{55}{324}$	$-\frac{25}{108}$	$\frac{50}{81}$			
5	$\frac{33}{40}$	$\frac{957}{1000}$	$\frac{2}{3}$	$\frac{83}{330}$	$-\frac{13}{22}$	$\frac{61}{66}$	$\frac{9}{110}$		
6	$\frac{7}{80}$	$\frac{193}{2000}$	1	$-\frac{19}{28}$	$\frac{9}{4}$	$\frac{1}{7}$	$-\frac{27}{7}$	$\frac{22}{7}$	
7	0	$-\frac{1}{50}$	1	$\frac{19}{200}$	0	$\frac{3}{5}$	$-\frac{243}{400}$	$\frac{33}{40}$	$\frac{7}{80}$

> *Prince-Dormand-embedding formula of 5^{th} and 6^{th} order,* **rk6(5)8 \tilde{m}**.
> ($q_g = 5$, $\tilde{q}_g = 6$, $m = 7$, $\tilde{m} = 8$)

Coefficient scheme:

j	\tilde{A}_j	A_j	a_j	b_{js} for $s = 1,\ldots,m-1$						
1	$\frac{61}{864}$	$\frac{821}{10800}$	0							
2	0	0	$\frac{1}{10}$	$\frac{1}{10}$						
3	$\frac{98415}{321776}$	$\frac{19683}{71825}$	$\frac{2}{9}$	$-\frac{2}{81}$	$\frac{20}{81}$					
4	$\frac{16807}{146016}$	$\frac{175273}{912600}$	$\frac{3}{7}$	$\frac{615}{1372}$	$-\frac{270}{343}$	$\frac{1053}{1372}$				
5	$\frac{1375}{7344}$	$\frac{395}{3672}$	$\frac{3}{5}$	$\frac{3243}{5500}$	$-\frac{54}{55}$	$\frac{50949}{71500}$	$\frac{4998}{17875}$			
6	$\frac{1375}{5408}$	$\frac{785}{2704}$	$\frac{4}{5}$	$-\frac{26492}{37125}$	$\frac{72}{55}$	$\frac{2808}{23375}$	$-\frac{24206}{37125}$	$\frac{338}{495}$		
7	$-\frac{37}{1120}$	$\frac{3}{50}$	1	$\frac{5561}{2376}$	$-\frac{35}{11}$	$-\frac{24117}{31603}$	$\frac{899983}{200772}$	$-\frac{5225}{1836}$	$\frac{3925}{4056}$	
8	$\frac{1}{10}$		1	$\frac{465467}{266112}$	$-\frac{2945}{1232}$	$\frac{10513573}{3212352}$	$-\frac{5610201}{14158144}$	$-\frac{424325}{205632}$	$\frac{376225}{454272}$	0

> *Verner - embedding formula of 5^{th} and 6^{th} order,* **rkv6(5)**.
> ($q_g = 5$, $\tilde{q}_g = 6$, $m = 6$, $\tilde{m} = 8$)

Coefficient scheme:

j	\tilde{A}_j	A_j	a_j	b_{js} for $s = 1,\ldots,m-1$						
1	$\frac{57}{640}$	$\frac{3}{80}$	0							
2	0	0	$\frac{1}{18}$	$\frac{1}{18}$						
3	$-\frac{16}{65}$	$\frac{4}{25}$	$\frac{1}{6}$	$-\frac{1}{12}$	$\frac{1}{4}$					
4	$\frac{1377}{2240}$	$\frac{243}{1120}$	$\frac{2}{9}$	$-\frac{2}{81}$	$\frac{4}{27}$	$\frac{8}{81}$				
5	$\frac{212}{320}$	$\frac{77}{160}$	$\frac{2}{3}$	$\frac{40}{33}$	$-\frac{4}{11}$	$-\frac{56}{11}$	$\frac{54}{11}$			
6	0	$\frac{73}{700}$	1	$-\frac{369}{73}$	$\frac{72}{73}$	$\frac{5380}{219}$	$-\frac{12285}{584}$	$\frac{2695}{1752}$		
7	$\frac{891}{8320}$		$\frac{8}{9}$	$-\frac{8716}{891}$	$\frac{656}{297}$	$\frac{39520}{891}$	$-\frac{416}{11}$	$\frac{52}{27}$	0	
8	$\frac{2}{35}$		1	$\frac{3015}{256}$	$-\frac{9}{4}$	$-\frac{4219}{78}$	$\frac{5985}{128}$	$-\frac{539}{384}$	0	$\frac{693}{3328}$

> *Verner - embedding formula of 6^{th} and 7^{th} order.*
> $(q_g = 6, \tilde{q}_g = 7, m = 8, \tilde{m} = 10)$

Coefficient scheme of the embedding formula rkv7(6):

j	\tilde{A}_j	A_j	a_j	b_{js} for $s = 1(1)m-1$								
1	$\frac{2881}{40320}$	$\frac{7}{90}$	0									
2	0	0	$\frac{1}{12}$	$\frac{1}{12}$								
3	0	0	$\frac{1}{6}$	0	$\frac{1}{6}$							
4	$\frac{1216}{2961}$	$\frac{16}{45}$	$\frac{1}{4}$	$\frac{1}{16}$	0	$\frac{3}{16}$						
5	$-\frac{2624}{4095}$	$\frac{16}{45}$	$\frac{3}{4}$	$\frac{21}{16}$	0	$-\frac{81}{16}$	$\frac{9}{2}$					
6	$\frac{24137569}{57482880}$	0	$\frac{16}{17}$	$\frac{1344688}{250563}$	0	$-\frac{1709184}{83521}$	$\frac{1365632}{83521}$	$-\frac{78208}{250563}$				
7	$-\frac{4}{21}$	$\frac{2}{15}$	$\frac{1}{2}$	$-\frac{559}{384}$	0	6	$-\frac{204}{47}$	$\frac{14}{39}$	$-\frac{4913}{78208}$			
8	0	$\frac{7}{90}$	1	$-\frac{625}{224}$	0	12	$-\frac{456}{47}$	$\frac{48}{91}$	$\frac{14739}{136864}$	$\frac{6}{7}$		
9	$\frac{4131}{3920}$		$\frac{2}{3}$	$-\frac{12253}{99144}$	0	$\frac{16}{27}$	$\frac{16}{459}$	$\frac{29072}{161109}$	$-\frac{2023}{75816}$	$\frac{112}{12393}$	0	
10	$-\frac{157}{1260}$		1	$\frac{30517}{2512}$	0	$-\frac{7296}{157}$	$\frac{268728}{7379}$	$\frac{2472}{2041}$	$-\frac{3522621}{10743824}$	$\frac{132}{157}$	0	$-\frac{12393}{4396}$

> *Verner - embedding formula of 7^{th} and 8^{th} order.*
> ($q_g = 7, \tilde{q}_g = 8, m = 11, \tilde{m} = 13$)

Coefficient scheme of the embedding formula rkv8(7):

b_{js} for $s = 1(1)m-1$

j	\tilde{A}_j	A_j	a_j	$s{=}1$	2	3	4	5	6	7	8	9	10	11	12
1	$\frac{31}{720}$	$\frac{13}{288}$	0												
2	0	0	$\frac{1}{4}$	$\frac{1}{4}$											
3	0	0	$\frac{1}{12}$	$\frac{5}{72}$	$\frac{1}{72}$										
4	0	0	$\frac{1}{8}$	$\frac{1}{32}$	0	$\frac{3}{32}$									
5	0	0	$\frac{2}{5}$	$\frac{106}{125}$	0	$-\frac{408}{125}$	$\frac{352}{125}$								
6	$\frac{16}{75}$	$\frac{32}{125}$	$\frac{1}{2}$	$\frac{1}{48}$	0	0	$\frac{8}{33}$	$\frac{125}{528}$							
7	$\frac{16807}{79200}$	$\frac{31213}{144000}$	$\frac{6}{7}$	$-\frac{1263}{2401}$	0	0	$\frac{39936}{26411}$	$-\frac{64125}{26411}$	$\frac{5520}{2401}$						
8	$\frac{16807}{79200}$	$\frac{2401}{12375}$	$\frac{1}{7}$	$\frac{37}{392}$	0	0	0	$\frac{1625}{9408}$	$-\frac{2}{15}$	$\frac{61}{6720}$					
9	$\frac{243}{1760}$	$\frac{1701}{14080}$	$\frac{2}{3}$	$\frac{17176}{25515}$	0	0	$-\frac{47104}{25515}$	$\frac{1325}{504}$	$-\frac{41792}{25515}$	$\frac{20237}{145800}$	$\frac{4312}{6075}$				
10	0	$\frac{2401}{19200}$	$\frac{2}{7}$	$-\frac{23834}{180075}$	0	0	$-\frac{77824}{1980825}$	$-\frac{636635}{633864}$	$\frac{254048}{300125}$	$-\frac{183}{7000}$	$\frac{8}{11}$	$-\frac{324}{3773}$			
11	0	$\frac{19}{450}$	1	$\frac{12733}{7600}$	0	0	$-\frac{20032}{5225}$	$\frac{456485}{80256}$	$-\frac{42599}{7125}$	$\frac{339227}{912000}$	$-\frac{1029}{4180}$	$\frac{1701}{1408}$	$\frac{5145}{2432}$		
12	$\frac{243}{1760}$		$\frac{1}{3}$	$-\frac{27061}{204120}$	0	0	$\frac{40448}{280665}$	$-\frac{1353775}{1197504}$	$\frac{17662}{25515}$	$-\frac{71687}{1166400}$	$\frac{98}{225}$	$\frac{1}{16}$	$\frac{3773}{11664}$	0	
13	$\frac{31}{720}$		1	$\frac{11203}{8680}$	0	0	$-\frac{38144}{11935}$	$\frac{2354425}{458304}$	$-\frac{84046}{16275}$	$\frac{673309}{1636800}$	$\frac{4704}{8525}$	$\frac{9477}{10912}$	$-\frac{1029}{992}$	0	$\frac{729}{341}$

Coefficient scheme of the embedding formula rk8(7)13m:

<div style="border:1px solid">

Prince-Dormand-embedding formulas of 7^{th} and 8^{th} order, **rk8(7)13m.**
$(q_g = 7,\ \tilde{q}_g = 8,\ m = 12,\ \tilde{m} = 13)$

</div>

b_{js} for $s = 1(1)m-1$

j	\hat{A}_j	A_j	a_j	$b_{j,1}$	$b_{j,2}$	$b_{j,3}$	$b_{j,4}$	$b_{j,5}$	$b_{j,6}$	$b_{j,7}$	$b_{j,8}$	$b_{j,9}$	$b_{j,10}$	$b_{j,11}$	$b_{j,12}$
1	$\frac{14005451}{335480064}$	$\frac{13451932}{455176623}$	0												
2	0	0	$\frac{1}{18}$	$\frac{1}{18}$											
3	0	0	$\frac{1}{12}$	$\frac{1}{48}$	$\frac{1}{16}$										
4	0	0	$\frac{1}{8}$	$\frac{1}{32}$	0	$\frac{3}{32}$									
5	0	0	$\frac{5}{16}$	$\frac{5}{16}$	0	$-\frac{75}{64}$	$\frac{75}{64}$								
6	$-\frac{59238493}{1068277825}$	$-\frac{808719846}{976000145}$	$\frac{3}{8}$	$\frac{3}{80}$	0	0	$\frac{3}{16}$	$\frac{3}{20}$							
7	$\frac{181606767}{758867731}$	$\frac{1757004468}{5645159321}$	$\frac{59}{400}$	$\frac{29443841}{614563906}$	0	0	$\frac{77736538}{692538347}$	$-\frac{28693883}{1125000000}$	$\frac{23124283}{1800000000}$						
8	$\frac{561292985}{797845732}$	$\frac{656045339}{265891186}$	$\frac{93}{200}$	$\frac{16016141}{946692911}$	0	0	$\frac{61564180}{158732637}$	$\frac{22789713}{633445777}$	$-\frac{545815736}{2771057229}$	$-\frac{180193667}{1043307555}$					
9	$-\frac{1041891430}{1371343529}$	$-\frac{3867574721}{1518517206}$	$\frac{5490023248}{9719169821}$	$\frac{39632708}{573591083}$	0	0	$-\frac{433636366}{683701615}$	$-\frac{421739975}{2616292301}$	$\frac{100302831}{723423059}$	$\frac{790204164}{839813087}$	$\frac{800635310}{3783071287}$				
10	$\frac{760417239}{1151165299}$	$\frac{465885868}{322736535}$	$\frac{13}{20}$	$\frac{246121993}{1340847787}$	0	0	$-\frac{37695042795}{15268766246}$	$-\frac{309121744}{1061227803}$	$-\frac{12992083}{490766935}$	$\frac{6005943493}{2108947869}$	$\frac{393006217}{1396673457}$	$\frac{123872331}{1001029789}$			
11	$\frac{118820643}{751138087}$	$\frac{53011238}{667516719}$	$\frac{1201146811}{1299901798}$	$-\frac{1028468189}{846180014}$	0	0	$\frac{8478235783}{508512852}$	$\frac{1311729495}{1432422823}$	$-\frac{10304129995}{1701304382}$	$-\frac{48777925059}{3047939560}$	$\frac{15336726248}{1032824649}$	$-\frac{45442868181}{3398467696}$	$\frac{3065993473}{597172653}$		
12	$-\frac{528747749}{2220607170}$	$\frac{2}{45}$	1	$\frac{185892177}{718116043}$	0	0	$-\frac{3185094517}{667107341}$	$-\frac{477755414}{1098053517}$	$-\frac{703635378}{230739211}$	$\frac{5731566787}{1027545527}$	$\frac{5232866602}{850066563}$	$-\frac{4093664535}{808688257}$	$\frac{3962137247}{1805957418}$	$\frac{65686358}{487910083}$	
13	$\frac{1}{4}$	0	1	$\frac{403863854}{491063109}$	0	0	$-\frac{5068492393}{434740067}$	$-\frac{411421997}{543043805}$	$\frac{652783627}{914296604}$	$\frac{11173962825}{925320556}$	$-\frac{13158990841}{6184727034}$	$\frac{3936647629}{1978049680}$	$-\frac{160528059}{685178525}$	$\frac{248638103}{1413531060}$	0

RK-Fehlberg - embedding formulas of 7^{th} and 8^{th} order.
$(q_g = 7, \tilde{q}_g = 8, m = 11, \tilde{m} = 13)$

Coefficient scheme of the embedding formula rkf8(7):

j	\tilde{A}_j	A_j	a_j	b_{js} for $s = 1(1)m-1$											
1	0	$\frac{41}{840}$	0												
2	0	0	$\frac{2}{27}$	$\frac{2}{27}$											
3	0	0	$\frac{1}{9}$	$\frac{1}{36}$	$\frac{1}{12}$										
4	0	0	$\frac{1}{6}$	$\frac{1}{24}$	0	$\frac{1}{8}$									
5	0	0	$\frac{5}{12}$	$\frac{5}{12}$	0	$-\frac{25}{16}$	$\frac{25}{16}$								
6	$\frac{34}{105}$	$\frac{34}{105}$	$\frac{1}{2}$	$\frac{1}{20}$	0	0	$\frac{1}{4}$	$\frac{1}{5}$							
7	$\frac{9}{35}$	$\frac{9}{35}$	$\frac{5}{6}$	$-\frac{25}{108}$	0	0	$\frac{125}{108}$	$-\frac{65}{27}$	$\frac{125}{54}$						
8	$\frac{9}{35}$	$\frac{9}{35}$	$\frac{1}{6}$	$\frac{31}{300}$	0	0	0	$\frac{61}{225}$	$-\frac{2}{9}$	$\frac{13}{900}$					
9	$\frac{9}{280}$	$\frac{9}{280}$	$\frac{2}{3}$	2	0	0	$-\frac{53}{6}$	$\frac{704}{45}$	$-\frac{107}{9}$	$\frac{67}{90}$	3				
10	$\frac{9}{280}$	$\frac{9}{280}$	$\frac{1}{3}$	$-\frac{91}{108}$	0	0	$\frac{23}{108}$	$-\frac{976}{135}$	$\frac{311}{54}$	$-\frac{19}{60}$	$\frac{17}{6}$	$-\frac{1}{12}$			
11	0	$\frac{41}{840}$	1	$\frac{2383}{4100}$	0	0	$-\frac{341}{164}$	$\frac{4496}{1025}$	$-\frac{301}{82}$	$\frac{2133}{4100}$	$\frac{45}{82}$	$\frac{45}{164}$	$\frac{18}{41}$		
12	$\frac{41}{840}$		0	$\frac{3}{205}$	0	0	0	0	$-\frac{6}{41}$	$-\frac{3}{205}$	$-\frac{3}{41}$	$\frac{3}{41}$	$\frac{6}{41}$	0	
13	$\frac{41}{840}$		1	$-\frac{1777}{4100}$	0	0	$-\frac{341}{164}$	$\frac{4496}{1025}$	$-\frac{289}{82}$	$\frac{2193}{4100}$	$\frac{51}{82}$	$\frac{33}{164}$	$\frac{12}{41}$	0	1

Coefficient scheme of the embedding formula rkv9(8):

> **Verner - embedding formulas of 8^{th} and 9^{th} order.**
> $(q_g = 8,\ \bar{g}_g = 9,\ m = 14,\ \tilde{m} = 16)$

Notation for the two-part coefficients: $\dfrac{a,b}{c} \to \dfrac{a+b\sqrt{6}}{c}$

j	\bar{A}_j	A_j	a_j	$b_{j,1}$	$b_{j,2}$	$b_{j,3}$	$b_{j,4}$	$b_{j,5}$	$b_{j,6}$	$b_{j,7}$	$b_{j,8}$	$b_{j,9}$	$b_{j,10}$	$b_{j,11}$	$b_{j,12}$	$b_{j,13}$	$b_{j,14}$	$b_{j,15}$
1	23/525	103/1680	0	0														
2	0	0	1/12	1/12														
3	0	0	1/9	1/27	2/27													
4	0	0	1/6	1/24	0	1/8												
5	0	0	2,2/15	4,94/375	0	−94,−84/125	328,208/375											
6	0	0	6,1/15	9,−1/150	0	0	312,32/1425	69,29/570										
7	0	0	6,−1/15	927,−347/1250	0	0	−16248,7328/9375	−489,179/3750	14268,−5798/9375									
8	171/1400	−27/140	2/3	2/27	0	0	0	16,1/54	16,1/54	−9/256								
9	86/525	76/105	1/2	19/256	0	0	0	118,23/512	118,23/512	−1/16	0							
10	93/280	−201/280	1/3	11/144	0	0	0	266,1/864	266,1/864	0	0	0						
11	−2048/6825	1024/1365	1/4	5034,−271/61440	0	0	0	0	−8/27	7859,−1626/10240	−2232,813/20480	−594,271/960	657,−813/5120					
12	3/18200	3/7280	4/3	5996,−3794/405	0	0	0	0	−4342,−338/9	154922,−40458/135	−4176,3794/405	−340864,242816/405	26304,−15176/45	26624/81				
13	39/175	12/35	5/6	3793,2168/103680	0	0	0	0	4042,2263/13824	−231278,40717/69120	7947,−2168/11520	1048,−542/405	−1383,542/720	2624/1053	3/1664			
14	0	9/280	1	−137/1296	0	0	0	0	5642,−337/864	5642,337/864	−299/48	184/81	−44/9	−5120/1053	−11/468	16/9		
15	9/25	0	1/6	33617,−2168/518400	0	0	0	0	−3846,31/13824	155338,−52807/345600	−12537,2168/57600	92,542/2025	−1797,−542/3600	320/567	−1/1920	4/105	0	
16	233/4200	233/4200	1	−36487,−30352/279600	0	0	0	0	−29666,−4499/7456	2779182,−615973/186400	−94329,91056/93200	−232192,121408/17475	101226,−22764/5825	169984/9087	87/30290	492/1165	0	1260/233

b_{js} for $s = 1(1)m-1$

Each of the formulas of such a compatible embedding pair supplies an approximate value Y and \tilde{Y} for y at $x_i + h$. Since the two formulas of one pair have different error orders and since we have to calculate only one set of k_j values for both formulas, embedding formulas are well suited for step size control and thus for adaptive procedures. Section 17.3.7.2, and algorithms 17.12 and 17.13 indicate procedures with an automatic step size control which relies on the difference $Y - \tilde{Y}$ from the two approximations for y of a compatible pair.

The following expressions for the difference $Y - \tilde{Y}$ can be derived from the embedding formulas (17.7) to (17.10):

For the Runge-Kutta embedding formula of 2^{nd} and 3^{rd} order we have

$$(17.7')\qquad\qquad Y - \tilde{Y} = h\left\{-\tfrac{1}{6}k_1 + \tfrac{1}{3}k_2 - \tfrac{1}{6}k_3\right\} ;$$

for the Runge-Kutta-Fehlberg embedding formula of 4^{th} and 5^{th} order we have

$$(17.8')\qquad Y - \tilde{Y} = h\left\{-\tfrac{1}{360}k_1 + \tfrac{128}{4275}k_3 + \tfrac{2197}{75240}k_4 - \tfrac{1}{50}k_5 - \tfrac{2}{55}k_6\right\};$$

for the England embedding formula of 4^{th} and 5^{th} order we have

$$(17.9')\qquad Y - \tilde{Y} = h\left\{\tfrac{42}{336}k_1 + \tfrac{224}{336}k_3 + \tfrac{21}{336}k_4 - \tfrac{162}{336}k_5 - \tfrac{125}{336}k_6\right\} ;$$

while for the Fehlberg I/II-embedding formula of 5^{th} and 6^{th} order we obtain the especially simple formula

$$(17.10')\qquad\qquad Y - \tilde{Y} = \tfrac{5}{66}h\left\{k_1 + k_6 - k_7 - k_8\right\} .$$

Bibliography for Section 17.3.4

[DORM80]; [DORM81]; [FEHL60]; [FEHL66]; [FEHL69]; [FEHL70]; [FEHL75]; [HAIR87] 2.6; [HULL72]; [LUTH87]; [VERN78].

17.3.5 Implicit Runge-Kutta Methods of Gaussian Type

With an *explicit* Runge-Kutta method for the initial value problem (17.1) and the partition (17.3), one can obtain a local error order of maximally $q_\ell = m+1$ if $m \leq 4$, and of at most $q_\ell = m$, if $m > 4$. Such explicit Runge-Kutta methods use m functional evaluations f_j, $j = 1, \ldots, m$, per step.

With an *implicit* Runge-Kutta method one can obtain a local error order of $q_l = 2m+1$ from m functional evaluations per step if the grid points $x_i + a_j h_i$ are identical with the nodes of the Gaussian quadrature formulas for the interval $[x_i, x_{i+1}]$, see section 15.7. Implicit Runge-Kutta formulas use (17.6) for $s = 1, \ldots, m$ instead of $s = 1, \ldots, j-1$ for the explicit formulas. In the following table, we describe implicit Runge-Kutta formulas of the Gaussian type for $m = 1, 2, 3$:

TABLE 17.7 (*Implicit Runge-Kutta formulas of Gaussian type*).

m		q_ℓ
1	$\begin{aligned} Y_{i+1} &= Y_i + h_i k_1(x_i, Y_i, h_i) \quad \text{with} \\ k_1 &= f(x_i + h_i/2, Y_i + k_1/2) \end{aligned}$	3
2	$\begin{aligned} Y_{i+1} &= Y_i + (h_i/2)(k_1 + k_2) \quad \text{with} \\ k_1 &= f(x_i + (1/2)(1 - 1/\sqrt{3})h_i, \\ & \quad Y_i + (1/4)k_1 + (1/2)(1/2 - 1/\sqrt{3})k_2), \\ k_2 &= f(x_i + (1/2)(1 + 1/\sqrt{3})h_i, \\ & \quad Y_i + (1/2)(1/2 + 1/\sqrt{3})k_1 + (1/4)k_2). \end{aligned}$	5
3	$\begin{aligned} Y_{i+1} &= Y_i + h_i((5/18)k_1 + (4/9)k_2 + (5/18)k_3) \text{ with} \\ k_1 &= f(x_i + (1/2)(1 - \sqrt{3/5})h_i, \ Y_i + (5/36)k_1 + \\ & \quad (2/9 - 1/\sqrt{15})k_2 + (5/36 - 1/(2\sqrt{15}))k_3), \\ k_2 &= f(x_i + h_i/2, \ Y_i + (5/36 + \sqrt{15/24})k_1 + \\ & \quad (2/9)k_2 + (5/36 - \sqrt{15/24})k_3), \\ k_3 &= f(x_i + (1/2)(1 + \sqrt{3/5})h_i, \ Y_i + (5/36 + \\ & \quad 1/(2\sqrt{15}))k_1 + (2/9 + 1/\sqrt{15})k_2 + (5/36)k_3). \end{aligned}$	7

For all $2 \le m \le 20$, [GLAS66] gives tables for the coefficients A_j, a_j, b_{js}. The systems of equations given above for the k_j are nonlinear and must be solved iteratively. Similar systems result for $m > 3$.

The iterative solution of such nonlinear system is demonstrated here for $m = 2$. For this purpose, we use an upper index on k_j as the iteration index. The initial values are

$$k_1^{(0)} = k_2^{(0)} = f(x_i, Y_i).$$

The *iteration rule* is given by:

$$(17.11) \begin{cases} k_1^{(\nu+1)} = f\left(x_i + \frac{1}{2}\left(1 - \frac{1}{\sqrt{3}}\right)h_i, \ Y_i + \frac{1}{4}k_1^{(\nu)} + \frac{1}{2}\left(\frac{1}{2} - \frac{1}{\sqrt{3}}\right)k_2^{(\nu)}\right), \\[2mm] k_2^{(\nu+1)} = f\left(x_i + \frac{1}{2}\left(1 + \frac{1}{\sqrt{3}}\right)h_i, \ Y_i + \frac{1}{2}\left(\frac{1}{2} + \frac{1}{\sqrt{3}}\right)k_1^{(\nu)} + \frac{1}{4}k_2^{(\nu)}\right), \\[2mm] \hspace{9cm} \nu = 1, 2, \ldots \end{cases}$$

The iteration will converge for any initial values $k_1^{(0)}$, $k_2^{(0)}$, see [GRIG77], p. 40, and [SOMM67], p. 31, provided h_i is chosen so that

$$(17.12) \quad \max_{1 \le j \le m} h_i L_i \sum_{s=1}^{m} |b_{js}| < 1, \text{ where } L_i = \max_{x \in [x_i, x_{i+1}]} \left| \frac{\partial f_r}{\partial y_k} \right|, \ 1 \le r, k \le n.$$

To obtain a local error of order $O(h_i^{2m+1})$, $2m - 1$ iteration steps are needed. The step size h_i satisfying (17.12) and the number m of functional evaluations per integration step can be chosen at will. As is shown in [SOMM67], one can minimize the needed computational time $T(\varepsilon, m)$ for a given error threshold ϵ by a proper choice of m. If one knows the optimal m, the step size $h_i = x_{i+1} - x_i = h_i(\varepsilon, m)$ can be found for each integration step.

Corresponding formulas and coefficients for implicit Runge-Kutta methods in which the m arguments $x_i + \alpha_j h_i$ coincide with the nodes of other quadrature formulas such as Newton-Cotes, Maclaurin, etc., can be found in [SOMM67]. Another method of step size control which is based upon two different quadrature formulas is given in [GRIG77], p. 69/70.

Bibliography for Section 17.3.5

[GLAS66]; [GRIG77]; [HAIR87] 2.7; [SOMM67].

17.3.6 Consistence and Convergence of One-Step Methods

When we consider the algorithms of one-step methods for solving the initial value problem (17.1) on a grid (17.3), we note that each procedure can be described by a recursion of the following form:

$$(17.13) \qquad \boldsymbol{Y}_{i+1} = \boldsymbol{Y}_i + h_i \boldsymbol{\Phi}(x_i, \boldsymbol{Y}_i, h_i), \quad i = 0, \ldots, N-1.$$

Each one-step method is uniquely determined by the associated function $\boldsymbol{\Phi}$. For example for the improved Euler-Cauchy method this function is

$$\boldsymbol{\Phi}(x_i, \boldsymbol{Y}_i, h_i) = \boldsymbol{f}(x_i + \frac{h_i}{2}, \boldsymbol{Y}_i + h_i/2 \, \boldsymbol{f}(x_i, \boldsymbol{Y}_i)).$$

And the one for the m^{th} order Runge-Kutta method is:

$$\boldsymbol{\Phi}(x_i, \boldsymbol{Y}_i, h_i) = \sum_{j=1}^{m} A_j \boldsymbol{k}_j(x_i, \boldsymbol{Y}_i, h_i)$$

with the A_j and \boldsymbol{k}_j from the table in section 17.3.4 .

DEFINITION 17.8 (*Local discretization error, truncation error*).
For the defining function $\boldsymbol{\Phi}(x, \boldsymbol{Y}, h)$ of a one-step method using (17.4) and (17.13), we define the *local discretization error* at the grid point x_i as:

$$\tau_i := \frac{1}{h_i}(y(x_{i+1}) - y(x_i)) - \boldsymbol{\Phi}(x_i, y(x_i), h_i).$$

τ_i is also called the *truncation error* at x_i. Here $y(x)$ is the exact solution of the intial value problem $y'(x) = f(x, y)$, $y(x_0) = y_0$.

DEFINITION 17.9 (*Consistency*).
A one-step method is called *consistent,* if the weighted sum of the local discretization errors at all grid points x_i, $i = 0, \ldots, N$,

$$\sum_{i=0}^{N-1} h_i \|\tau_i\|$$

converges to 0 as $h_{\max} \to 0$ with $h_{\max} = \max\limits_{0 \le i \le N-1} h_i$ and $h_i := x_{i+1} - x_i > 0$.

A one-step method is consistent if the maximum local discretization error satisfies

$$\max_{0 \le i \le N-1} \{\|\tau_i\|\} \to 0 \quad \text{as} \quad h_{\max} \to 0.$$

Thus, all the one-step methods of the previous sections are consistent, provided the solution y of (17.1) is sufficiently often continuously differentiable.

DEFINITION 17.10 (*Consistency order*) .
The order $O(h_{max}^q)$, with which the truncation error or local discretization error tends towards zero, is called the *consistency order* of the method.

The consistency order is equal to the global error order. To achieve the consistency order, the exact solution y must be sufficiently often continuously differentiable. To achieve consistency, y must be twice continuously differentiable.

THEOREM 17.11 (*Convergence*).
A consistent one-step method with a consistency order $q > 0$ whose associated function Φ fulfills a Lipschitz condition relative to y is *convergent* of order q. Therefore

$$\lim_{h_{\max} \to 0} \|Y_i - y(x_i)\| = 0.$$

For a proof see [STOE91], 7.2.2; [WERN79] .

17.3.7 Error Estimation and Step Size Control

17.3.7.1 Error Estimation

If $Y_h(x)$ and $Y_{\tilde{h}}(x)$ are approximate values for y at a point $x \in [x_0, \beta]$ computed for step sizes h and \tilde{h} using a method with global error order q_g, then we can estimate the global procedural error as

$$(17.14) \qquad e_h := y(x) - Y_h(x) \approx \frac{Y_h(x) - Y_{\tilde{h}}(x)}{(\tilde{h}/h)^{q_g} - 1} = e_h^*.$$

And

$$Y_h^*(x) = Y_h(x) + e_h^* = \frac{(\tilde{h}/h)^{q_g} Y_h(x) - Y_{\tilde{h}}(x)}{(\tilde{h}/h)^{q_g} - 1}$$

gives an improved approximate value for the exact solution $y(x)$ when compared with $Y_h(x)$. For sufficiently often differentiable solutions y,

$$y(x) = Y_h^*(x) + O(h^{q_g+1}).$$

The global error order is increased by at least one by adding the estimated error to the approximate value, see [STUM82], p.253.

For $\tilde{h} = 2h$ we have

$$e_h(x) \approx \frac{Y_h(x) - Y_{2h}(x)}{2^{q_g} - 1} = e_h^*,$$

$$Y_h^*(x) = \frac{2^{q_g} Y_h(x) - Y_{2h}(x)}{2^{q_g} - 1},$$

where Y_h denotes the approximation for $y(x)$ for the step size h and Y_{2h} the one for the step size $2h$. And the improved approximate value $Y_h^*(x)$ is better by at least one h-power than $Y_h(x)$.

For example, we can obtain the following estimation formulas and improved approximate values for specific one-step procedures with $\tilde{h} = 2h$.

1. Euler-Cauchy method:

$$e_h^{EC}(x) \approx Y_h^{EC}(x) - Y_{2h}^{EC}(x)$$

$$Y_h^*(x) = 2Y_h^{EC}(x) - Y_{2h}^{EC}(x).$$

2. Heun's method and improved Euler-Cauchy method:

$$e_h^H(x) \approx \frac{1}{3}\left(Y_h^H(x) - Y_{2h}^H(x)\right)$$

$$Y_h^{*H}(x) = \frac{1}{3}\left(4Y_h^H(x) - Y_{2h}^H(x)\right).$$

3. Classical Runge-Kutta method:

$$e_h^{RK}(x) \approx \frac{1}{15}\left(Y_h^{RK}(x) - Y_{2h}^{RK}(x)\right)$$

$$Y_h^{*RK}(x) = \frac{1}{15}\left(16Y_h^{RK}(x) - Y_{2h}^{RK}(x)\right).$$

17.3.7.2 Automatic Step Size Control, Adaptive Methods for Initial Value Problems

Generally, it is not appropriate to calculate with a constant step size, i.e., with an equidistant grid. We recommend to adapt the chosen step size locally to the behavior of the solution, for examples see [LUTH87], 4.3. In regions of little change in the solution, one can choose relatively big steps, in regions of large changes of the solution one should proceed with relatively small steps. It is possible to control the step size automatically by using error estimation. For this we shall describe several possibilities and procedures below.

Method A) to control the step sizes.

By using the error estimate (17.14), one can proceed as follows: after every two steps with a step size h from x_{i-1} to x_i and x_{i+1}, one carries out a step with the double step size $2h$ from x_{i-1} using the same method. If the estimated error is far below the given error bound one can increase the step size for the next step. If the estimated error is larger, one integrates over the last half once more with a smaller step size. If the step size has been chosen correctly one can use the error estimate in (17.14) to improve the approximate value, one then continues with the improved value.

Method B) to control the step size.

The following method of automatic step size control is still more effective: One uses two one-step methods with associated functions Φ and $\tilde{\Phi}$, one of which has the global convergence order q_g, and the other at least the order $q_g + 1$. One calculates the approximate values Y and \tilde{Y} at the point $x = x_i + h$ with both methods, starting from an approximate value Y_i for the grid point x_i and a step size h. Depending on the outcome of the error estimation, one can accept the chosen step size h and accept $x_{i+1} := x$ as a new grid point, or one must repeat the integration with a smaller step size. Thus, one works adaptively, as in the following algorithm, see [LUTH87], 4.3.

ALGORITHM 17.12 (*Automatic step size control, adaptive initial value solver*).

Choose two one-step methods with associated functions Φ and $\tilde{\Phi}$ with error orders q_g and $q_g + 1$ at least. Let Y_i be an approximate value for the exact solution y at the grid point x_i. Then proceed with a chosen step size h as follows:

1. Calculate an approximate solution Y with the first method and an approximate solution \tilde{Y} with the second method for $x_i + h$:

$$Y = Y_i + h\Phi(x_i, Y_i, h), \quad \tilde{Y} = Y_i + h\tilde{\Phi}(x_i, Y_i, h).$$

For a given error bound $\varepsilon > 0$ set

$$S := \left(\frac{h\varepsilon}{\|Y - \tilde{Y}\|}\right)^{1/q_g}.$$

2. If $S \geq 1$ then $Y_{i+1} := \tilde{Y}$ is accepted as a new approximation at the grid point $x_{i+1} := x_i + h$. For the next step, carried out as in 1, choose the new step size as

$$h := \min\{2; S\} \cdot h.$$

If $S < 1$, one has to repeat the first step using the step size

$$h := \max\{\tfrac{1}{2}; S\} \cdot h.$$

ALGORITHM 17.13 (*Automatic step size control according to it [HULL72]*).

Choose two one-step methods of orders q_g and $\tilde{q}_g \geq q_g + 1$. Let \tilde{Y}_i be an approximate value for the exact solution y at the grid point x_i.

1. Compute approximate solutions Y and \tilde{Y} at $x_i + h$ by using the two one-step methods.

2. Compute

$$S := 0.9h\left(\frac{\varepsilon}{\|Y - \tilde{Y}\|}\right)^{1/(q_g + 1)},$$

where $\varepsilon = \|\tilde{Y}\|$ RELERR + ABSERR.

3. If $\|Y - \tilde{Y}\| < \varepsilon$, then $Y_{i+1} := Y$ is accepted as a new approximation for the grid point $x_{i+1} := x_i + h$. The next step uses the step size

$$h := \min\{S; 4h\}.$$

If $\|Y - \tilde{Y}\| > \varepsilon$, the first step has to be repeated with the new step size

$$h := \max\{S; \tfrac{1}{4}h\}.$$

REMARK concerning the error estimates when using embedding formulas:

Section 17.3.4.4 described embedding formulas. They are especially suited for adaptive methods using step size control as done in algorithm 17.12 or 17.13, because the approximate value \tilde{Y} can be found with very little computational effort once the approximate value Y has been computed: All the k_j-values necessary for computing Y can be used for finding \tilde{Y}.

As examples, in section 17.3.4.4 we have given the Runge-Kutta formulas of 2^{nd} and 3^{rd} order, the Runge-Kutta-Fehlberg formulas of 4^{th} and 5^{th} order, the England formulas of 4^{th} and 5^{th} order and the Fehlberg formulas of 5^{th} and 6^{th} order. The differences $Y - \tilde{Y}$ that are needed for algorithms 17.12 or 17.13 are given in the formulas $(17.7')$ up to $(17.10')$.

REMARK: In the program section of this book we include a program IVP which works adaptively using automatic step size control, giving the user the choice between the Runge-Kutta embedding formulas of second and third order, or the England formulas of fourth and fifth order. If for an initial value problem (17.1) one wants to calculate approximate values for the solution $y(x)$ at the points $x_k = x_0 + kh$ for $k = 1, \ldots, k_{end}$, it is useful to call the program IVP in a loop in such a way that the solution is calculated at the point x_{k+1}, i.e., in each loop the initial values x_k and $Y(x_k)$ are used to compute $Y(x_{k+1})$.

When integrating (17.1) from x_k to x_{k+1}, one can use a step size control according to algorithm 17.12. The mixed error test (1.6) is to be used and the error bound ε should be set as

$$\varepsilon = \text{ABSERR} + \text{RELERR} \, \|\tilde{Y}\|,$$

so that the S of algorithm 17.12 has the form

$$S = \left(h_k \frac{\text{ABSERR} + \text{RELERR} \, \|\tilde{Y}\|}{\|Y - \tilde{Y}\|} \right)^{1/q}.$$

REMARK concerning the choice of a suitable embedding formula:

Naturally, it is possible to use any other pair of embedding formulas of section 17.3.4.4 adaptively. Section 17.8 will deal with those as well as give test results on algorithms 17.12 and 17.13 for automatic step size control and decision hints.

Bibliography for Section 17.3

[BJÖR74], 8.1-8.3; [COLL66], II, §2; [CONT80], 6; [ENGE87], 10.3, 11.1; [GEAR71/1], 2; [GRIG77], vol.1; [HAIR87] 2; [HENR68], part I; [LAPI71],

2,3; [LUTH87], 4.1-4.3; [NOBL65], II, 10.2-10.5; [RALS79] vol.1, 9; [RICE77],
p.257-276; [SCHW89], 9.1; [STET73], 3; [STUM82], 11; [WERN79], IV, §6,7;
[ZURM65], §25, 27 .

17.4 Multi-Step Methods

17.4.1 The Principle of Multi-Step Methods

Multi-step methods use $s + 1$ preceding values $\boldsymbol{Y}_{i-s}, \boldsymbol{Y}_{i-s+1}, \ldots, \boldsymbol{Y}_{i-1}, \boldsymbol{Y}_i$
and calculate an approximate value \boldsymbol{Y}_{i+1} for $\boldsymbol{y}(x_{i+1})$ on a given grid.

One considers the initial value problem

$$(17.15) \quad \begin{cases} \boldsymbol{y}'(x) = \boldsymbol{f}(x, \boldsymbol{y}) = \boldsymbol{f}(x, \boldsymbol{y}(x)) = \boldsymbol{f}(x, y_1, y_2, \ldots, y_n) \text{ for } x \in [x_{-s}, \beta] \\ \boldsymbol{y}(x_{-s}) = \boldsymbol{y}_{-s}. \end{cases}$$

On the interval $[x_{-s}, \beta]$ of integration for the differential equation we define
a partition

$$x_{-s} < x_{-s+1} < \ldots < x_{N-s} = \beta$$

with local step sizes $h_i := x_{i+1} - x_i > 0$ for $i = -s, \ldots, N-s$, where $N > s$.

Initially, we assume that the values of \boldsymbol{y} and those of $\boldsymbol{f}(x, \boldsymbol{y})$ are known at
the points $x_{-s}, x_{-s+1}, \ldots, x_{-1}, x_0$. The points $(x_i, \boldsymbol{f}(x_i, \boldsymbol{y}_i))$ for $i = -s, \ldots, 0$
form the *starting values* for computing the approximate values $\boldsymbol{Y}_i = \boldsymbol{Y}(x_i)$
for $\boldsymbol{y}_i = \boldsymbol{y}(x_i), i = 1, \ldots, N - s$, at the remaining $N - s$ grid points
$x_1, x_2, \ldots, x_{N-s}$. The starting values of \boldsymbol{y} are either given (exactly or approx-
imately) or they must be calculated approximately by means of a one-step
method (e.g. by the classical Runge-Kutta method). In the following we de-
note the starting values by $(x_i, \boldsymbol{f}(x_i, \boldsymbol{y}_i)) = (x_i, \boldsymbol{f}_i)$ for $i = -s, \ldots, 0$.

We will proceed from the integral equation (17.4) associated with the differ-
ential equation (17.15) in $[x_i, x_{i+1}]$. With one class of multi-step methods,
the function \boldsymbol{f} in (17.4) is replaced by the interpolating polynomial $\boldsymbol{\Phi}_s$ of
degree s through the $s + 1$ interpolation points $(x_j, \boldsymbol{f}_j), j = i-s, \ldots, i$, and
$\boldsymbol{\Phi}_s$ is integrated over the interval $[x_i, x_{i+1}]$. Thus one can obtain an approx-
imate value \boldsymbol{Y}_{i+1} for \boldsymbol{y}_{i+1}. If $i = 0$, these interpolation points are identical
with the starting points, for $i > 0$ some starting points and points (x_j, \boldsymbol{f}_j) for
$j = 1, \ldots, i$ jointly determine the approximate value \boldsymbol{Y}_{i+1} using the computed
approximations $\boldsymbol{Y}_1, \boldsymbol{Y}_2, \ldots, \boldsymbol{Y}_i$. Since the right hand side of (17.4) contains

only the values Y_{i-s} to Y_i, this yields an *explicit formula* for calculating the approximate value Y_{i+1}, and the integration step is an extrapolation step.

Analogously we can obtain an *implicit formula* if we use the node x_{i+1} in the interpolation polynomial for f together with the nodes $x_{i-s}, x_{i-s+1}, \ldots, x_i$. Then the right hand side of (17.4) also contains Y_{i+1}, besides $Y_{i-s}, Y_{i-s+1}, \ldots, Y_i$. A formula of this type is the corrector formula of Heun's method.

If one uses an explicit and an implicit formula as a pair, the explicit formula is called a *predictor*, and the implicit one a *corrector*, while the procedure is called a *predictor-corrector method*.

References: For multi-step methods we recommend particularly the book of Shampine and Gordon, see [SHAM75]. It emphasizes the Adams methods and gives very efficient algorithms and FORTRAN programs.

17.4.2 The Adams-Bashforth Method

The *Adams-Bashforth method* results from (17.4) by replacing $f(x, y(x))$ by its interpolation polynomial $\Phi_s(x)$ and the associated remainder $R_{s+1}(x)$ at the $s + 1$ interpolation points (x_j, f_j), $j = (i-s), \ldots, i$. Integrating from x_i to x_{i+1} yields

$$y_{i+1} = Y_{i+1} + \varepsilon_{i+1}^{AB} \quad \text{with} \quad Y_{i+1} = Y_i + \int\limits_{x_i}^{x_{i+1}} \Phi_s(x)dx \quad \text{with}$$

$$\varepsilon_{i+1}^{AB} := y_{i+1} - Y_{i+1} = \int\limits_{x_i}^{x_{i+1}} R_{s+1}(x)dx.$$

ε_{i+1}^{AB} is the local procedural error which results from integrating over $[x_i, x_{i+1}]$, assuming that Y_i is exact.

Thus for every s and given points (x_j, f_j), $j = i-s, \ldots, i$, one obtains an Adams-Bashforth formula that computes Y_{i+1} by integrating from x_i to x_{i+1} with an associated local procedural error of $\varepsilon_{i+1}^{AB} = O(h^{q_t})$.

Next we give the Adams-Bashforth formulas for $s = 3, 4, 5, 6$ and equidistant grid points: With $h_i = h = \text{const}$ we have

$$s = 3 \ (q_\ell = 5): \quad \boldsymbol{Y}_{i+1} \ = \ \boldsymbol{Y}_i + \tfrac{h}{24}(55\boldsymbol{f}_i - 59\boldsymbol{f}_{i-1} + 37\boldsymbol{f}_{i-2} - 9\boldsymbol{f}_{i-3}),$$
$$i = 0, \ldots, n - 4,$$
$$\varepsilon_{i+1}^{AB} \ = \ \tfrac{251}{720}h^5 \boldsymbol{y}^{(5)}(\eta_i) = O(h^5), \ \eta_i \in [x_i, x_{i+1}];$$

$$s = 4 \ (q_\ell = 6): \quad \boldsymbol{Y}_{i+1} \ = \ \boldsymbol{Y}_i + \tfrac{h}{720}(1901\boldsymbol{f}_i - 2774\boldsymbol{f}_{i-1} + 2616\boldsymbol{f}_{i-2} +$$
$$-1274\boldsymbol{f}_{i-3} + 251\boldsymbol{f}_{i-4}),$$
$$\varepsilon_{i+1}^{AB} \ = \ \tfrac{95}{288}h^6 \boldsymbol{y}^{(6)}(\eta_i) = O(h^6), \ \eta_i \in [x_i, x_{i+1}],$$
$$i = 0, \ldots, n - 5;$$

$$s = 5 \ (q_\ell = 7): \quad \boldsymbol{Y}_{i+1} \ = \ \boldsymbol{Y}_i + \tfrac{h}{1440}(4277\boldsymbol{f}_i - 7923\boldsymbol{f}_{i-1} + 9982\boldsymbol{f}_{i-2} +$$
$$-7298\boldsymbol{f}_{i-3} + 2877\boldsymbol{f}_{i-4} - 475\boldsymbol{f}_{i-5}),$$
$$\varepsilon_{i+1}^{AB} \ = \ \tfrac{19087}{60480}h^7 \boldsymbol{y}^{(7)}(\eta_i) = O(h^7), \ \eta_i \in [x_i, x_{i+1}],$$
$$i = 0, \ldots, n - 6;$$

$$s = 6 \ (q_\ell = 8): \quad \boldsymbol{Y}_{i+1} \ = \ \boldsymbol{Y}_i + \tfrac{h}{60480}(198721\boldsymbol{f}_i - 447288\boldsymbol{f}_{i-1} +$$
$$+705549\boldsymbol{f}_{i-2} - 688256\boldsymbol{f}_{i-3} + 407139\boldsymbol{f}_{i-4} +$$
$$-134472\boldsymbol{f}_{i-5} + 19087\boldsymbol{f}_{i-6}),$$
$$\varepsilon_{i+1}^{AB} \ = \ \tfrac{5257}{17280}h^8 \boldsymbol{y}^{(8)}(\eta_i) = O(h^8), \ \eta_i \in [x_i, x_{i+1}],$$
$$i = 0, \ldots, n - 7.$$

The global error order is $O(h^{q_g})$ with $q_g = q_\ell - 1$.

For the Adams-Bashforth formulas, one always needs $s+1$ given points (x_j, \boldsymbol{f}_j), which have to be determined by another method. This method should have the same local error order. This could be achieved by a suitable Runge-Kutta method and would be a good reason to employ the Runge-Kutta method for the entire interval $[x_{-s}, \beta]$, instead of combining the Adams-Bashforth formula with the Runge-Kutta formula.

Since for an Adams-Bashforth step from x_i to x_{i+1}, one has to calculate only one new functional value \boldsymbol{f}_i in contrast to m new functional values for a Runge-Kutta step of order m, the Adams-Bashforth formula works much faster than the Runge-Kutta formula. This would be a good reason to combine the Runge-Kutta method with the Adams-Bashforth formula.

However, the Adams-Bashforth formula should not be employed by itself, but rather as a predictor together with an implicit formula as a corrector. The reason for this is that in the Adams-Bashforth formula the interpolation interval is $[x_{i-s}, x_i]$ for $\boldsymbol{\Phi}_s$, while $[x_i, x_{i+1}]$ is the integration interval of $\boldsymbol{\Phi}_s$, so that the integration is evaluated in an extrapolation step. As is well known, the remainder \boldsymbol{R}_{s+1} of an interpolation has large values for points outside of the interpolation interval, see section 9.6. We therefore have to be aware

that the local procedural error ε_{i+1}^{AB} will increase strongly for large h and will become bigger than the local procedural error of the Runge-Kutta method of the same error order. For error estimates, consult section 17.4.5.

One can construct other multi-step formulas by again replacing $f(x, y(x))$ in (17.4) with an interpolating polynomial Φ_s for $s+1$ interpolation points $(x_j, f_j), j = i - s, \ldots, i$, and by integrating over $[x_{i-r}, x_{i+1}]$ for integers $0 \leq r \leq s$. The case $r = 0$ gives the above Adams-Bashforth formulas. For further such methods see [COLL66], p.86-88; [HENR68], p.199-201, 241; [SHAM75]; [STUM82], p.273-276; [WERN79], p.290-294.

17.4.3 The Predictor-Corrector Method of Adams-Moulton

This method combines an Adams-Bashforth extrapolation formula with an implicit corrector formula of at least the same error order. We recommend to choose a corrector formula with an error order one higher than that of the predictor formula. In this case we obtain a predictor-corrector method. One can obtain a corrector of higher error order if one replaces $f(x, y(x))$ in (17.4) by its interpolation polynomial for the $s + 2$ interpolation points $(x_j, f_j), j = i - s, \ldots, i+1$, and then proceeds in an analogous way to section 17.4.2.

If $s = 3$, we obtain for an integration step from x_i to x_{i+1} and an equidistant partition:

$$y_{i+1} = Y_{i+1} + \varepsilon_{i+1}^{AM_3} \quad \text{with}$$

$$Y_{i+1} = Y_i + \frac{h}{720}(251f_{i+1} + 646f_i - 264f_{i-1} + 106f_{i-2} - 19f_{i-3}),$$

$$\varepsilon_{i+1}^{AM_3} = -\frac{3}{160}h^6 y^{(6)}(\eta_i) = O(h^6) \text{ for some } \eta_i \in [x_i, x_{i+1}].$$

Since $f_{i+1} = f(x_{i+1}, Y_{i+1})$, the formula for Y_{i+1} is implicit so that Y_{i+1} must be determined iteratively. This iteration shall be labelled with an upper index ν. Then the *Adams-Moulton formula* for $s = 3$ becomes:

$$(17.16) \quad Y_{i+1}^{(\nu+1)} = Y_i + \frac{h}{720}(251f(x_{i+1}, Y_{i+1}^{(\nu)}) + 646f_i - 264f_{i-1} + \\ +106f_{i-2} - 19f_{i-3}).$$

It is employed as a corrector formula together with the Adams-Bashforth formula for $s = 3$ as a predictor. The corrector formula will converge if

$$\frac{251}{720}hL = \kappa < 1, \quad \text{where} \quad L = \max_{1 \leq k, r \leq n} \left| \frac{\partial f_r}{\partial y_k} \right|.$$

If the step size h is sufficiently small, one or two iterations in (17.16) will suffice.

ALGORITHM 17.14 (*Predictor-corrector method of Adams-Moulton for $s = 3$*).

Given: The differential equation $y'(x) = f(x, y), x \in [x_{-3}, \beta = x_{N-3}]$, with the initial condition $y(x_{-3}) = y_{-3}$, the step size $h > 0$, the nodes $x_i = x_0 + ih$, $i = -3, \ldots, N - 3$, and the starting values (x_i, f_i), $i = -3, \ldots, 0$.

Task: Compute approximations Y_i for $y(x_i), i = 1, \ldots, N - 3$ by performing the following steps for each integration step from x_i to x_{i+1}:

1^{st} step: Calculate $Y_{i+1}^{(0)}$ from the Adams-Bashforth formula (predictor-formula with $q_\ell = 5$)

$$Y_{i+1}^{(0)} = Y_i + \frac{h}{24}(55 f_i - 59 f_{i-1} + 37 f_{i-2} - 9 f_{i-3}).$$

2^{nd} step: Calculate $f(x_{i+1}, Y_{i+1}^{(0)})$.

3^{rd} step: Calculate $Y_{i+1}^{(\nu+1)}$ for $\nu = 0$ and $\nu = 1$ from the Adams-Moulton formula (17.16) (corrector formula with $q_\ell = 6$).

In general two iteration steps will suffice if h is chosen so that $K = hL \leq 0.20$. Then one can accept

$$Y_{i+1}^{(\nu+1)} = Y_{i+1} \approx y_{i+1}.$$

If in the course of computations, it is necessary to decrease the step size before reaching the node x_j, it is generally recommended to halve h. In this case, one has to recalculate the initial values needed for the ensuing calculations for $i = j - 2, j - \frac{3}{2}, j - 1$ and $j - \frac{1}{2}$.

CALCULATION SCHEME 17.15 (*Adams-Moulton method for $s = 3$ and $n = 1$*).

	i	x_i	$Y_i = Y(x_i)$	$f_i = f(x_i, Y_i)$
Starting Values	-3	x_{-3}	$Y_{-3} = y_{-3}$	f_{-3}
	-2	x_{-2}	Y_{-2}	f_{-2}
	-1	x_{-1}	Y_{-1}	f_{-1}
	0	x_0	Y_0	f_0
Extrapolation with Adams-Bashforth	1	x_1	$Y_1^{(0)}$	$f(x_1, Y_1^{(0)})$
Interpolation with Adams-Moulton	1	x_1	$Y_1^{(1)}$	$f(x_1, Y_1^{(1)})$
	1	x_1	$Y_1^{(2)} =: Y_1$	$f(x_1, Y_1)$
Extrapolation with A.–B.	2	x_2	$Y_2^{(0)}$	$f(x_2, Y_2^{(0)})$
Interpolation with A.–M.	2	x_2	$Y_2^{(1)}$	$f(x_2, Y_2^{(1)})$
	2	x_2	$Y_2^{(2)} =: Y_2$	

Further Adams-Moulton methods.

In the following we shall indicate further Adams-Moulton methods in which the error order of the predictor is one less than that of the corrector. We shall use the abbreviation $f_{i+1}^{(\nu)} := f(x_{i+1}, Y_{i+1}^{(\nu)})$.

$$s = 4: \quad Y_{i+1}^{(0)} = Y_i + \tfrac{h}{720}(1901 f_i - 2774 f_{i-1} + 2616 f_{i-2} + $$
$$-1274 f_{i-3} + 251 f_{i-4}),$$

$$Y_{i+1}^{(\nu+1)} = Y_i + \tfrac{h}{1440}(475 f_{i+1}^{(\nu)} + 1427 f_i - 798 f_{i-1} + 482 f_{i-2} + $$
$$-173 f_{i-3} + 27 f_{i-4}),$$

$$\varepsilon_{i+1}^{AM_4} = -\tfrac{863}{60480} h^7 y^{(7)}(\eta_i) = O(h^7), \quad \eta_i \in [x_i, x_{i+1}],$$
$$i = 0, \ldots, n - 4.$$

$$s = 5: \quad \boldsymbol{Y}_{i+1}^{(0)} \quad = \quad \boldsymbol{Y}_i + \tfrac{h}{1440}(4277f_i - 7923f_{i-1} + 9982f_{i-2} +$$

$$-7298f_{i-3} + 2877f_{i-4} - 475f_{i-5}),$$

$$\boldsymbol{Y}_{i+1}^{(\nu+1)} \quad = \quad \boldsymbol{Y}_i + \tfrac{h}{60480}(19087f_{i+1}^{(\nu)} + 65112f_i - 46461f_{i-1} +$$

$$+37504f_{i-2} - 20211f_{i-3} + 6312f_{i-4} - 863f_{i-5}),$$

$$\varepsilon_{i+1}^{AM_5} \quad = \quad -\tfrac{275}{24192}h^8 y^{(8)}(\eta_i) = O(h^8), \ \eta_i \in [x_i, x_{i+1}],$$

$$i = 0, \ldots, n-5.$$

$$s = 6: \quad \boldsymbol{Y}_{i+1}^{(0)} \quad = \quad \boldsymbol{Y}_i + \tfrac{h}{60480}(198721f_i - 447288f_{i-1} + 705549f_{i-2} +$$

$$-688256f_{i-3} + 407139f_{i-4} - 134472f_{i-5} +$$

$$+19087f_{i-6}),$$

$$\boldsymbol{Y}_{i+1}^{(\nu+1)} \quad = \quad \boldsymbol{Y}_i + \tfrac{h}{120960}(36799f_{i+1}^{(\nu)} + 139849f_i - 121797f_{i-1} +$$

$$+123133f_{i-2} - 88547f_{i-3} + 41499f_{i-4} +$$

$$-11351f_{i-5} + 1375f_{i-6}),$$

$$\varepsilon_{i+1}^{AM_6} \quad = \quad -\tfrac{33953}{3628800}h^9 y^{(9)}(\eta_i) = O(h^9), \ \eta_i \in [x_i, x_{i+1}],$$

$$i = 0, \ldots, n-6.$$

Since the error order of the corrector is always one larger than that of the predictor, one or two iteration steps are sufficient in most cases. In general, a predictor-corrector method whose predictor part has the error order r_1 and whose corrector has the error order r_2, has the following local procedural error \mathbf{E}_{i+1}^{PK} after $\nu + 1$ iteration steps:

$$\boldsymbol{E}_{i+1}^{PK} := y_{i+1} - \boldsymbol{Y}_{i+1}^{(\nu+1)} = O(h^{\min(r_2, r_1 + \nu + 1)}).$$

For $r_1 = r_2 - 1$ we thus attain the error order of the corrector after one iteration step. For an arbitrary $r_1 < r_2$, we can achieve the error order $O(h^{r_2})$ after $\nu = r_2 - r_1 - 1$ iteration steps. Since, however, the error of the predictor exceeds that of the corrector for $s \geq 3$ by a factor greater than 10, one or more iterations can be required in practice to reduce the total error to the error of the corrector. If one is, however, satisfied with obtaining an overall error order equal to the one for the corrector, then if $r_1 = r_2 - 1$ only one iteration is required. If $r_1 = r_2$, one must be satisfied with one iteration, see also [HENR68], p.196; [STUM82], p.271; [WERN79], p.299. If still more iterations are needed, it is better to decrease the step size than to continue iterating.

In the following we give an Adams-Moulton method whose predictor formula is the Adams-Bashforth formula for $s = 3$ and whose corrector is the Adams-Moulton formula for $s = 2$. They have the same local error order $O(h^5)$:

Predictor: $\quad\quad\quad Y_{i+1}^{(0)} = Y_i + \frac{h}{24}(55f_i - 59f_{i-1} + 37f_{i-2} - 9f_{i-3})$,
(Adams – Bashforth
for $s = 3$)

Corrector: $\quad\quad\quad Y_{i+1}^{(\nu+1)} = Y_i + \frac{h}{24}(9f_{i+1}^{(\nu)} + 19f_i - 5f_{i-1} + f_{i-2})$.
(Adams – Moulton
for $s = 2$)

This procedure requires only one iteration step for each grid point and thus it saves computing time. This predictor-corrector pair, has an especially simple error estimate, see also section 17.4.5, formulas (17.18), (17.19), so that without much calculation time and without requiring additional calculations with different step sizes, each value Y_i can be improved immediately.

Instead of the Adams-Moulton formulas as corrector, one can also use formulas with an especially advantageous error propagation. For this purpose, we define the corrector with $q_\ell = m + 3$ in the form

$$(17.17) \quad\quad Y_{i+1} = \sum_{k=0}^{m} a_{i-k} Y_{i-k} + h \sum_{k=-1}^{m} b_{i-k} f(x_{i-k}, Y_{i-k}).$$

If e_{i+1}^F stands for the global procedural error of a formula (17.17), and e_{i+1}^{AM} for the same error for the Adams-Moulton formula of the same error order, then $\|e_{i+1}^F\|/\|e_{i+1}^{AM}\|$ is a measure for the quality of the corrector (17.17) with respect to error propagation. According to [FEHL61],

$$Y_{i+1}^{(\nu+1)} = \frac{243}{1000}Y_i + \frac{1}{8}Y_{i-2} + \frac{79}{125}Y_{i-5} + \frac{h}{400}\Big(120f(x_{i+1}, Y_{i+1}^{(\nu)}) +$$
$$+567f(x_i, Y_i) + 600f(x_{i-2}, Y_{i-2}) + 405f(x_{i-4}, Y_{i-4}) +$$
$$+72f(x_{i-5}, Y_{i-5})\Big)$$

is a corrector with $q_\ell = 7$ for which $\|e_{i+1}^F\|/\|e_{i+1}^{AM}\|$ amounts to only about 8% of the global procedural error of the Adam-Moulton formula of the same error order. To be used as a predictor, one would need an extrapolation formula with $q_\ell = 6$. Hence one can use the Adams-Bashforth formula for $s = 4$. Due to the very small error terms in (17.17), we recommend to iterate more than twice.

17.4.4 The Adams-Störmer Method

In this section we describe a multi-step method that treats an initial value problem of the form

$$y''(x) = g(x, y, y') \quad \text{with}$$
$$y(x_0) = y_0, \quad y'(x_0) = y_0'$$

directly without reducing it to an initial value problem (17.1).

ALGORITHM 17.16 (*Adams-Störmer method*).

Assume an initial value problem $y''(x) = g(x, y, y')$ is given with $y(x_{-3}) = y_{-3}$, $y'(x_{-3}) = y_{-3}'$. If $x_i = x_0 + ih$, $i = -3, \ldots, N-3$, are the nodes of the integration interval $[x_{-3}, x_{N-3} = \beta]$, one carries out the following steps in order to calculate the approximate value Y_{i+1} for y_{i+1} for each $i = 1, \ldots, N-2$, after having calculated the starting values from the triples (x_i, Y_i, Y_i'), $i = -3, \ldots, 0$, possibly by using a Runge-Kutta method:

1^{st} step: Calculate the values $Y_{i+1}^{(0)}$, $Y_{i+1}'^{(0)}$ from the predictor formula of local error order $O(h^5)$:

$$Y_{i+1}^{(0)} = Y_i + hY_i' + \frac{h^2}{360}(323g_i - 264g_{i-1} + 159g_{i-2} - 38g_{i-3}),$$

$$Y_{i+1}'^{(0)} = Y_i' + \frac{h}{24}(55g_i - 59g_{i-1} + 37g_{i-2} - 9g_{i-3}),$$

with $g_i := g(x_i, Y_i, Y_i')$.

2^{nd} step: Calculate $g(x_{i+1}, Y_{i+1}^{(0)}, Y_{i+1}'^{(0)})$.

3^{rd} step: Calculate $Y_{i+1}^{(\nu+1)}$ and $Y_{i+1}'^{(\nu+1)}$ for $\nu = 0$ and $\nu = 1$ according to the corrector formulas ($q_\ell = 6$):

$$Y_{i+1}^{(\nu+1)} = Y_i + hY_i' + \frac{h^2}{1440}\Big(135g(x_{i+1}, Y_{i+1}^{(\nu)}, Y_{i+1}'^{(\nu)}) + 752g_i - 246g_{i-1} +$$
$$+ 96g_{i-2} - 17g_{i-3}\Big),$$

$$Y_{i+1}'^{(\nu+1)} = Y_i' + \frac{h}{720}\Big(251g(x_{i+1}, Y_{i+1}^{(\nu)}, Y_{i+1}'^{(\nu)}) + 646g_i - 264g_{i-1} +$$
$$+ 106g_{i-2} - 19g_{i-3}\Big).$$

NOTE. There is no easy rule to decide whether it is more advantageous to treat an initial value problem of a second or higher order differential equation by:

(1) using the Adams-Störmer method (direct procedure) directly or another direct method for differential equations of higher order, or

(2) reducing the problem to an initial value problem for a system of first order differential equations (indirect procedure).

According to [RUTI60], to proceed with (1) for problems with many integration steps can lead to a detrimental accumulation of rounding errors, see also [ENGE87], 11.4. Thus the approach (2) is generally preferred.

For high order differential equations, it has been shown that the direct methods that correspond to the classical Runge-Kutta method and the Adams-Moulton method achieve a smaller global error only if the derivative $y^{(n-1)}$ does not occur in f. For a problem of the form $y^{(n)} = f(x, y, y', \ldots, y^{(n-1)})$, the indirect methods generally have the smaller global error.

17.4.5 Error Estimates for Multi-Step Methods

The error estimates given in section 17.3.7 can also be used for multi-step methods. For instance, the estimates and improved approximations from (17.14) are for:

1) the Adams-Bashforth method for $s = 3$:

$$e_h^{AB}(x) \approx \frac{1}{15}\left(Y_h(x) - Y_{2h}(x)\right),$$

$$Y_h^*(x) = \frac{1}{15}\left(16Y_h(x) - Y_{2h}(x)\right);$$

2) the Adams-Moulton method for $s = 3$:

$$e_h^{AM}(x) \approx \frac{1}{31}\left(Y_h(x) - Y_{2h}(x)\right),$$

$$Y_h^*(x) = \frac{1}{31}\left(32Y_h(x) - Y_{2h}(x)\right).$$

If the Adams-Bashforth formula for $s = 3$ (with a local error order of $O(h^5)$) is used as a predictor, and the Adams-Moulton formula for $s = 2$ (with local error order of $O(h^5)$) is used as a corrector, then the following estimate for the global procedural error holds, see [CONT80] p.237:

$$q_\ell = 5: \quad e_h^{AM}(x) := y(x) - Y_h^{(1)}(x) \approx -\tfrac{1}{14}\left(Y_h^{(1)}(x) - Y_h^{(0)}(x)\right).$$

$$Y^*(x) = Y_h^{(1)}(x) - \tfrac{1}{14}\left(Y_h^{(1)}(x) - Y_h^{(0)}(x)\right)$$

$$= \tfrac{1}{14}\left(13 Y_h^{(1)}(x) + Y_h^{(0)}(x)\right).$$

This estimate is very simple to use because it does not require any calculations for a doubled step size. It also can be used to determine whether the chosen step size is sufficient for the desired precision.

Analogously, one can combine an Adams-Bashforth formula with an Adams-Moulton formula of equal error order for $q_\ell = 6, 7, 8$ to form a predictor-corrector pair. Then only one iteration step is required, and the following estimates for the global procedural error hold:

$$q_\ell = 6: \quad e_h^{AM}(x) \quad := y(x) - Y_h^{(1)}(x) \approx -\tfrac{1}{18}\left(Y_h^{(1)}(x) - Y_h^{(0)}(x)\right),$$

$$q_\ell = 7: \quad e_h^{AM}(x) \quad := y(x) - Y_h^{(1)}(x) \approx -\tfrac{1}{22}\left(Y_h^{(1)}(x) - Y_h^{(0)}(x)\right),$$

$$q_\ell = 8: \quad e_h^{AM}(x) \quad := y(x) - Y_h^{(1)}(x) \approx -\tfrac{1}{26}\left(Y_h^{(1)}(x) - Y_h^{(0)}(x)\right).$$

REMARK: An effective automatic step size control, like the one in section 17.3.7 for one-step methods, is also possible for multi-step methods by using much more involved procedures. Further information can be found in [SHAM75] with its quite elaborate FORTRAN programs.

17.4.6 Computational Error of One-Step and Multi-Step Methods

While the global procedural error of one-step and multi-step methods decreases with $h \to 0$ of order q_g, the global computational error increases with decreasing step size. The total error, i.e., the sum of procedural error and computational error, can thus not be reduced arbitrarily. Hence the step size h should be chosen so that procedural and computational errors have about the same order of magnitude.

If $r_h(x)$ denotes the global computational error at x, the following crude estimate ($n = 1$) is valid for one-step methods:

$$|r_h(x)| \leq \begin{cases} \dfrac{\varepsilon}{h_{\max}}(x - x_0) & \text{if} \quad C = 0, \\[2mm] \dfrac{\varepsilon}{h_{\max}}\left(e^{C(x-x_0)} - 1\right) & \text{otherwise.} \end{cases}$$

Here ε is the maximum of the absolute computational error of each calculation step and $C = L$, the Lipschitz constant, in case of the Euler-Cauchy method and $C \approx L$ for the classical Runge-Kutta method.

For multi-step and predictor-corrector methods we have

$$|r_h(x)| \le \frac{\varepsilon}{h_{\max}} \cdot \frac{x_i - x_0}{1 - C_2 h_{\max} L} \cdot e^{\frac{C_1 (x - x_0)}{1 - C_2 h_{\max} L}},$$

where C_1 and C_2 depend on the coefficients of the formulas used, see [HENR68], 5.3, 5.4. The global computational error is thus of order $O(1/h_{max})$ for both one-step and multi-step methods.

Bibliography for Section 17.4

[CARN69], 6.8-6.12; [COLL73], II, 4.2; [CONT80], 6.6-6.8, 6.11; [ENGE87], 10.4; [GEAR71/1], 7-10; [HAIR87] 3; [HENR64], 14.6-14.7; [HENR68], 5,6; [LAPI71], 4.7; [LUTH87], 4.4; [McCA67], 9.2; [RALS67] vol.1, 8.; [SCHW89], 9.2; [STET73], 4; [STOE91], 7.2.6 - 13; [STUM82], 12; [WERN79], IV, §§8-10.

17.5 Bulirsch-Stoer-Gragg Extrapolation

We consider the initial value problem (17.1). W.l.o.g. we can assume $n = 1$:

$$y'(x) = f(x, y), \qquad y(x_0) = y_0.$$

The extrapolation method of Bulirsch-Stoer-Gragg can easily be expanded to systems.

We want to find an approximation $Y(\bar{x})$ for the exact solution $y(\bar{x})$ of the initial value problem at \bar{x} with

$$\bar{x} := x_0 + Nh \quad \text{for} \quad h := \frac{\bar{x} - x_0}{N} \quad \text{and} \quad N > 0.$$

Gragg's function $S(\bar{x}; h)$ supplies an approximate value for $y(\bar{x})$ with a global error order of $O(h^2)$. It is calculated according to [STOE91], 7.2.14 as follows:

With

$$(17.18) \quad \begin{cases} z_0 & := y_0 \\ z_1 & := z_0 + hf(x_0, y_0), \quad x_1 \quad := x_0 + h \\ z_{i+1} & := z_{i-1} + 2hf(x_i, z_i), x_{i+1} := x_i + h \text{ for } i = 1, \ldots, N-1, \end{cases}$$

S is defined as

$$(17.19) \qquad S(\bar{x}; h) := \frac{1}{2}[z_N + z_{N-1} + hf(x_N, z_N)].$$

In the extrapolation method of Bulirsch and Stoer, one chooses a sequence of positive integers

$$(17.20) \qquad \{n_0, n_1, n_2, \ldots\}$$

with $0 < n_0 < n_1 < \ldots$ and computes Gragg's function $S(\bar{x}; h_j)$ for each

$$(17.21) \qquad h_j := \frac{\bar{x} - x_0}{n_j}.$$

The numbers n_j must all be even or all be odd. The values $S(\bar{x}; h_j)$ are computed as in (17.18) and (17.19). One can obtain the value

$$S(\bar{x}; h_j) := \frac{1}{2}[z_{n_j} + z_{n_j-1} + h_j f(x_{n_j}, z_{n_j})].$$

for each $j = 0, 1, \ldots$ with

$$\begin{array}{ll} z_0 & := y_0 \\ z_1 & := z_0 + h_j f(x_0, y_0), \qquad x_1 \qquad := x_0 + h_j \\ z_{i+1} & := z_{i-1} + 2h_j f(x_i, z_i), \qquad x_{i+1} := x_i + h_j, \quad i = 1, \ldots, n_j - 1. \end{array}$$

Then $y(\bar{x}) = S(\bar{x}; h_j) + O(h_j^2)$.

Since S has an asymptotic expansion in powers of h_j^2, one can construct approximations with a higher error order by using Richardson extrapolation, just as in Romberg integration. For this one forms a "Romberg scheme" as follows, see section 14.10:

The numbers of its first column are given as

$$L_j^{(0)} := S(\bar{x}; h_j) \quad \text{for} \quad j = 0, 1, \ldots.$$

Then one calculates the values for the columns with $k = 1, 2, \ldots$ using the formula

$$(17.22) \qquad L_j^{(k)} = \frac{\left(\dfrac{h_j}{h_{j+k}}\right)^{2k} L_{j+1}^{(k-1)} - L_j^{(k-1)}}{\left(\dfrac{h_j}{h_{j+k}}\right)^2 - 1} \quad \text{for} \quad j = 0, 1, \ldots$$

The columns of this Romberg scheme converge towards $y(\bar{x})$ for all functions y that are sufficiently often differentiable:

$$\lim_{j\to\infty} L_j^{(k)} = y(\bar{x}) \quad \text{for} \quad k \quad \text{fixed};$$

The convergence of the k^{th} column has the order $q_g = 2k+2$ for $k = 0,1,2,\ldots$.

NOTE: The number of columns used should be limited so that no oscillations occur. Oscillations can be triggered by the beginning influences of rounding errors as well as the possibility that f is not sufficiently smooth, i.e., not differentiable sufficiently often.

Romberg sequence.

If we choose $n_j = 2^j N$ for an even integer N, the sequence (17.20) becomes

$$N \cdot \{1,2,4,8,16,32,\ldots\}.$$

This is called the *Romberg sequence.* For these n_j, (17.21) becomes

(17.23)
$$h_j := \frac{\bar{x} - x_0}{2^j N} = \frac{h}{2^j}.$$

The $L_j^{(k)}, k = 1,2,\ldots$, are computed from (17.22) and (17.23) as

$$L_j^{(k)} = \frac{2^{2k} L_{j+1}^{(k-1)} - L_j^{(k-1)}}{2^{2k} - 1} \quad \text{for} \quad j = 0,1,\ldots.$$

Bulirsch sequence.

If we choose the Bulirsch sequence $\{2,4,6,8,12,16,\ldots\}$ in (17.20), we obtain for $j > 0$ and $h_0 := h$ with

$$h_j = \begin{cases} \dfrac{h_0}{2^{(j+1)/2}} & \text{for odd } j \\[2mm] \dfrac{h_0}{3 \cdot 2^{(j-2)/2}} & \text{for even } j \end{cases}$$

the h_j values as $\{h_0, h_1, \ldots\} = h \cdot \{1, \frac{1}{2}, \frac{1}{3}, \frac{1}{4}, \frac{1}{6}, \frac{1}{8}, \frac{1}{12}, \frac{1}{16}, \ldots\}$.

With these h_j, the computations in (17.22) are significantly simplified.

Step size control.

In extrapolation methods one should also utilize step size control. This can be done by means of one of the methods in section 17.3.7, see specifically the algorithms 17.12 and 17.13.

The program DESEXT in the program section uses the above extrapolation method with the Bulirsch sequence. Its step size control is realized with the method of [HALL76], p.113.

NOTE. Instead of the above method based upon polynomial extrapolation or the Richardson principle, see e.g. [BJÖR74], 7.22; [STUM82], p. 253; [WERN79] III, §7, one can also use algorithms based on rational extrapolation. Test examples for such procedures have given even better numerical results, see [BULI66], [GRAG65].

Bibliography for Section 17.5

[GEAR71/1], 6; [GRIG77], 5,2; [HAIR87] 2.9; [HALL76], 6; [LAPI71], 5; [STET73], 6.3; [STOE91], 7.2.14.

17.6 Stability

17.6.1 Preliminary Remarks

Integrating the initial value problem (17.1) numerically supplies approximate solutions $Y_i = Y(x_i)$ at the grid points $x_0 < x_1 < \ldots < x_n = \beta$ for the unknown exact solution $y_i = y(x_i)$. For all previously mentioned methods one can prove that the approximate values Y_i converge towards the exact values $y(x_i)$ as $h_{max} \to 0$ under the assumption that the associated function Φ satisfies a Lipschitz condition and that the calculations have no rounding errors.

However, it is necessary to examine a method in the presence of rounding errors as well. The only useful algorithms are so-called *stable algorithms*. According to section 1.4, an algorithm is *stable* if the error committed in one calculation step does not increase in the following steps. It is called *unstable* if, even for arbitrarily many steps, the approximations Y_i differ unboundedly from the exact values y_i, so that the computed solution is in essence useless.

The cause of an instability can lie within the differential equation itself or it can originate in the numerical procedure. In the first case, the instability is a result of the physical process described by the differential equation. In the second case, the instability can be avoided by chosing a more appropriate numerical method. All our investigations in this section will again only be made for $n = 1$, i.e., for an initial value problem $y'(x) = f(x, y)$ with $y(x_0) = y_0$.

17.6.2 Stability of Differential Equations

Let y be the solution of the initial value problem (17.1) with $n = 1$:

$$y'(x) = f(x, y), \quad y(x_0) = y_0,$$

and u be a solution close to y that satisfies the same differential equation as y. However assume that the initial condition for u is slightly altered by rounding errors and procedural errors. Such errors can originate for example if the initial value y_0 of the problem has been calculated numerically.

We can express u as:

$$u(x) := y(x) + \varepsilon \eta(x).$$

Here η is a so-called *error function* and ε is a parameter with $0 < \varepsilon \ll 1$.

Then η satisfies the so-called *differential variational equation:*

$$\eta' = f_y \eta.$$

Under the assumption $f_y = c = $ const, it has the solution

$$\eta(x) = \eta_0 e^{c(x - x_0)}, \qquad \eta(x_0) = \eta_0.$$

If $f_y = c < 0$, then $\eta(x)$ decreases for increasing x. In this case, the differential equation is called *stable*, otherwise it is *unstable*. For a stable differential equation, the distance of the solutions for different initial values diminishes for growing x, so that an error made at one point, such as a rounding error in the initial condition, will fade away.

17.6.3 Stability of the Numerical Method

For $n = 1$ one-step and multi-step methods have the general form

$$(17.24) \qquad \sum_{k=0}^{M} a_{M-k} Y_{i+1-k} = h \sum_{k=0}^{M} b_{M-k} f_{i+1-k}$$

for an equidistant grid . They are explicit, if $b_M = 0$, and implicit for $b_M \neq 0$.
Their coefficients must satisfy

$$\sum_{k=0}^{M} a_k = 0, \quad \sum_{k=0}^{M} b_k = 1.$$

For example for the Euler-Cauchy method and $M = 1$, we have seen:

$$a_0 = -1, \quad a_1 = 1, \quad b_0 = 1, \quad b_1 = 0,$$

while for Heun's method with $M = 2$:

$$a_0 = 0, \quad a_1 = -1, \quad a_2 = 1, \quad b_0 = 0, \quad b_1 = 1/2, \quad b_2 = 1/2.$$

An equation of the form (17.24) is called a *difference equation* of *order M*.

For multi-step methods we have $M > 1$, since several preceding values
$Y_i, Y_{i-1}, \ldots,$ are used to calculate Y_{i+1}. The Runge-Kutta methods are not
multi-step procedures although $M = 3$, for the values Y_i, Y_{i-1} associated with
b_1, b_2 are determined by $f(x_{i-2}, Y_{i-2})$. The reason for this is that in algorithm
17.4 the value Y_i corresponds to Y_{i-2}, the value $Y_i + \frac{h_i k_1}{2}$ to Y_{i-1}, and the
value $Y_i + h_i k_2/2$ to Y_i.

The differential equation $y'(x) = f(x, y)$ can thus be replaced by a differ-
ence equation of the form (17.24), which gives rise to a *difference variational
equation* with $U_i := Y_i + \varepsilon H_i$:

$$(17.25) \quad \sum_{k=0}^{M} a_{M-k} H_{i+1-k} = h \sum_{k=0}^{M} b_{M-k} H_{i+1-k} f_y(x_{i+1-k}, U_{i+1-k}).$$

Here U_i is another approximate solution close to the approximate solution
Y_i, H_i is the error solution, and ε is a parameter with $0 < \varepsilon \ll 1$. To solve
(17.25) we set

$$H(x_j) := H_j = \lambda^j \quad \text{for an integer } j.$$

(Strictly speaking, the following considerations hold only for linear differential
equations with constant coefficients. For a more general theory see [DAHL74]).

Under the assumption $f_y = c = \text{const}$, we can insert the above expression for H_j into (17.25):

$$(17.26) \qquad \sum_{k=0}^{M} a_{M-k}\lambda^{i+1-k} - hc\sum_{k=0}^{M} b_{M-k}\lambda^{i+1-k} = 0.$$

Clearly $\lambda \neq 0$, for if $\lambda = 0$ then $H = 0$, i.e., the approximate values Y_i would not be subject to any perturbations caused by rounding errors or truncation errors at preceding nodes. Multiplying (17.26) by λ^{M-1-i}, we obtain a polynomial equation for λ:

$$P(\lambda) := \sum_{k=0}^{M} a_{M-k}\lambda^{M-k} - hc\sum_{k=0}^{M} b_{M-k}\lambda^{M-k} = 0.$$

With

$$\varrho(\lambda) \quad := \quad \sum_{k=0}^{M} a_{M-k}\lambda^{M-k} = \sum_{k=0}^{M} a_k\lambda^k, \quad \text{and}$$

$$\sigma(\lambda) \quad := \quad \sum_{k=0}^{M} a_{M-k}\lambda^{M-k} = \sum_{k=0}^{M} b_k\lambda^k$$

this can be written in the form

$$(17.27) \qquad\qquad P(\lambda) = \varrho(\lambda) - hc\sigma(\lambda) = 0.$$

$P(\lambda) := \varrho(\lambda) - hc\sigma(\lambda)$ is called the *characteristic polynomial* of the multi-step method.

The M zeros of (17.27) are λ_ν, $\nu = 1, \ldots, M$. For each value of ν there is an *error solution* $(H_\nu)_j = (\lambda_\nu)^j$ of (17.25). The error solutions $(H_\nu)_j$ will not increase for increasing j precisely when $|\lambda_\nu| \leq 1$ for all ν.

If $|\lambda_\nu| < 1$ for all ν, we have *strong stability*. If there is an index ν with $|\lambda_\nu| = 1$, then λ_ν may only be a simple zero of (17.27) for asymptotic stability.

Since the equation (17.25) is linear in H_j, every linear combination of the functions $(H_\nu)_j$ is also a solution. If all λ_ν are different, then the M solutions $(H_\nu)_j = (\lambda_\nu)^j$, $\nu = 1, \ldots, M$, are linearly independent, and the general solution of (17.25) has the form

$$H_j = \sum_{v=1}^{M} c_\nu (\lambda_\nu)^j$$

with arbitrary real coefficients c_ν. The class of solutions of (17.25) thus has cardinality ∞^M, where M denotes the order of the difference equation.

One consequence of this is: If $M > 1$ (except for the Runge-Kutta methods), the order of the difference equation is higher than that of the differential equation. Thus the difference equation will have a larger solution space than that of the underlying differential equation. And the numerical method will produce "parasitic solutions". Only one of the solutions of the difference equation will converge for $h \to 0$ towards the solution of the initial value problem (17.1), provided that the multi-step procedure on which it is based does converge. The higher the order M of the difference equation is, the better its local error order $O(h^q)$. A difference equation (17.24) of M^{th} order can give rise to a numerically stable integration method with the local error order $O(h^{M+3})$ for even M, and with $O(h^{M+2})$ for odd M at most. Procedures with a higher error order must be numerically unstable ([WERN79],§9). In the following we shall only deal with local behavior, and locally we shall set $f_y = c = $ const. We can distinguish the following types of stability:

- Asymptotic stability,

- Absolute stability ,

- A-stability, and

- Stiff stability .

DEFINITION 17.17 (*Asymptotic stability, stability for $h \to 0$*).
An algorithm for the numerical integration of an initial value problem (17.1) is called *asymptotically stable* for $h \to 0$, if the polynomial $\varrho(\lambda)$ fulfills the *root condition*, i.e., if it has only roots λ_ν with $|\lambda_\nu| \leq 1$ and if $|\lambda_\nu| = 1$, then λ_ν is a simple root of ϱ.

Consistent one-step methods are always asymptotically stable, if the associated function Φ is Lipschitz bounded. Since $\varrho(1) = \Sigma a_k = 0$, each method has at least one root $\lambda_\nu = 1$, thus one can never speak of a strongly asymptotically stable method here. During computations, however, it cannot be guaranteed that one works with a sufficiently small h for which the method satisfies the above indicated criterion. For this reason, one needs stability statements also for $h \neq 0$ in order to form a valid assessment of the stability behavior of the method used for the chosen step size h:

DEFINITION 17.18 (*Absolute stability, stability for $h \neq 0$*).
An algorithm for the numerical integration of (17.1) is *strongly absolutely stable* for a fixed $h \neq 0$, if all roots λ_ν of (17.27) satisfy $|\lambda_\nu| < 1$. It is called *weakly absolutely stable*, if the root condition of Definition 17.17 is fulfilled, otherwise it is *unstable*.
A curve in \mathbb{R}^2 which bounds the region of absolute stability of a method is called its *stability boundary*, see also [ENGE87], p. 458.

With absolutely stable procedures we are guaranteed that stable solutions of the given differential equation are approximated by stable solutions of the corresponding difference equation. The following phenomenon is, however, possible:

$$\lim_{x \to \infty} y(x) = 0, \quad \text{but} \quad \lim_{\substack{x \to \infty \\ h \neq 0}} Y(x_0 + ih) = d > 0.$$

If $\lim_{x \to \infty} y(x) = 0$ must imply $\lim_{\substack{i \to \infty \\ h \neq 0}} Y(x_0 + ih) = 0$ as well, then one has to require stability for arbitrary values hc with $Re(hc) < 0$ and $|hc| \to \infty$. Such a behavior is called *A-stability*.

DEFINITION 17.19 (*A-stability, stability for arbitrary hc with $|hc| \to \infty$*).
A procedure for the numerical integration of (17.1) is called *A-stable*, if for arbitrary hc with $Re(hc) < 0$, the root condition of Definition 17.17 is fulfilled for $P(\lambda) = \varrho(\lambda) + hc\sigma(\lambda) = 0$.

For A-stable methods the stability region must therefore contain the left half-plane of \mathbb{C}, i.e., the entire negative real axis of the hc-plane must be part of the stability region. The following statements hold about A-stability of one-step and multi-step methods:

(1) Explicit one-step methods are not A-stable.

(2) There exist A-stable implicit one-step method such as the implicit Runge-Kutta formulas of Gaussian type or the Rosenbrock methods, see [HALL76], p. 148, but not all implicit Runge-Kutta methods are A-stable.

(3) An explicit multi-step procedure cannot be A-stable.

(4) The global error order (or consistency order) of an A-stable multi-step method can be at most two.

(5) Among the A-stable multi-step methods with consistency order two, Heun's method has the smallest error.

The extrapolation method of section 17.5 is not A-stable. This summary shows that, except for the methods under (2), no method is known with a global error order $q_g > 2$ that fulfills the necessary condition of A-stability for all values c with $Re(hc) < 0$.

The term "stiff stability" will be defined in section 17.7.

Bibliography for Section 17.6

[ATKI78], 6.8; [HAIR87]; [HALL76], 2; [JELT76]; [JELT78]; [LINI77]; [LUTH87], 12; [RUTI52]; [STOE91], 7.2; [WERN79], §9-11 .

17.7 Stiff Systems of Differential Equations

17.7.1 The Problem

There is one class of initial value problems (17.1) for whose numerical treatment only very specific methods are useful. This is the class of *stiff differential equations* $y'(x) = f(x, y)$.

DEFINITION 17.20. A system of differential equations $y' = f(x, y)$ is called *stiff*, if the component functions $y_i(x)$ of the solution y of (17.1) have a very different growth behavior: For increasing x, there are strongly decreasing, weakly decreasing, as well as increasing solutions y_i and the increasing contributions grow much less quickly than the decreasing ones attenuate.

Requirements for a method to solve stiff systems of differential equations.

A method that is to be useful for stiff systems must meet the requirement that components of an approximate solution that have decreased below a certain threshold cannot have any more influence on the solution when the integration process is continued. This means that the stiff components must be integrated with a method in which for arbitrary $h > 0$ and all complex c with $Re(c) < 0$ we have

$$\lim_{\substack{i \to \infty \\ h \neq 0}} Y(x_0 + ih) = 0.$$

A-stability is (at least in a limited way) necessary for a numerical method to integrate stiff differential equations, since the numerical solution must reproduce correctly the required diminishing of the stiff components. In this respect the implicit Runge-Kutta methods of Gaussian type in section 17.3.5 are well suited for stiff systems.

17.7.2 Criteria for the Stiffness of a System

(A) For the system (17.1) of differential equations we assume at first that $f(x, y) = Ay$ for a constant (n, n) matrix $A = (a_{ik})$, $i, k = 1, \ldots, n$. Thus we assume that a system

$$(17.28) \qquad\qquad y'(x) = Ay(x)$$

of linear differential equations with constant coefficients is given.

If A is diagonalizable, then the problem can be transformed to diagonal form. This is always possible if A has n distinct eigenvalues. The system (17.28) then separates into n scalar differential equations $y_i' = \lambda_i y_i$ for $i = 1, \ldots, n$. In general A can be reduced by similarity to its Jordan normal form. If the Jordan normal form of A is nondiagonal, then two eigenvalues of A would have to be equal, which is very improbable. So the assumption that A is diagonalizable is not as restrictive as it might seem, see chapter 7.2. Hence let us assume that the system (17.28) can be separated into n scalar differential equations. And in the remainder of this section we shall investigate scalar model problems for which we want to find integration methods with an appropriate stability behavior.

A system (17.28) is called *stiff* if for the eigenvalues λ_i of A:

$$(17.29) \qquad\qquad -\frac{\min_{i, x \in I} Re(\lambda_i(x, y))}{|\max_{i, x \in I} Re(\lambda_i(x, y))|} \gg 1.$$

Criterion (17.29) states that a system (17.28) can only be stiff if one of the eigenvalues of A lies in the left half-plane of \mathbb{C} and moreover the real part of the left-most eigenvalue of A is significantly larger than the maximal real part of all the eigenvalues of A.

(B) To a given system of the form (17.1) we can locally associate a system of the form (17.28) for every $x \in I$. The matrix A can be taken as the Jacobi

matrix of (17.1):

$$(17.30) \quad \begin{cases} A \ := \left(\dfrac{\partial f_i(x, y)}{\partial y_k} \right) = A(x, y), \quad i, k = 1, \dots, n, \\ f_i \ = f_i(x, y) = f_i(x, y_1, y_2, \dots, y_n). \end{cases}$$

If λ_i, $i = 1, \dots, n$, are the eigenvalues of A in (17.30) at (x, y), then (17.29) is a criterion whether the system (17.1) is stiff in a neighborhood of (x, y). We note that the matrix A in (17.30) can vary strongly in the interval of integration.

A detailed analysis of stiffness is given in [HAIR91].

17.7.3 Gear's Method for Integrating Stiff Systems

The stability region of an A-stable method includes the negative hc-half-plane. Demanding A-stability restricts the global error order of methods suitable for integrating stiff systems. For this reason, several modified stability notions have been introduced which are related to A-stability, but allow to increase the global error order of a method. See [GEAR71/2], [GEAR71/1], [GRIG77] vol. 2).

Gear's method is based on the characteristic polynomial

$$(17.31) \qquad\qquad P(\lambda) = \varrho(\lambda) - hc\sigma(\lambda).$$

At first $\sigma(x)$ is chosen in such a way that $P(\lambda) \to 0$ as $|hc| \to \infty$: Division of (17.31) by hc and taking the limit $|hc| \to \infty$ leads to $\sigma(\lambda) = 0$. The simplest choice is

$$(17.32) \qquad\qquad \sigma(\lambda) = \lambda^M.$$

For this reason, $P(\lambda)$ has the best possible stability property for $hc = \infty$: If $|hc| = \infty$, P has an M-fold root at $\lambda = 0$. The condition of strong absolute stability (definition 17.18) is fulfilled at $|hc| = \infty$. In order to determine the behavior at finite points of the hc-plane, one has to
1) find the polynomial $P(\lambda)$ and thus $\varrho(\lambda)$, and
2) to determine its stability region.

1) For (17.32) one can calculate the associated $\varrho(x)$ from the consistency conditions: A linear multi-step method has the consistency order q, if (see [GRIG77] vol. 2, p. 334)

$$(17.33) \quad \begin{cases} (i) & \sum_{k=0}^{M} \left(a_k k^j - j b_k k^{j-1}\right) = 0, \ j = 1,\dots,q, \text{ and} \\ (ii) & \sum_{k=0}^{M} a_k = 0. \end{cases}$$

(17.33) is a system of $q+1$ linear equations for the coefficients a_k, $k = 0,\dots,q$, of the multi-step procedure.

2) The stability region of the multi-step method obtained by solving the linear system in 1) is determined for $\lambda := re^{i\varphi}$ with $|\lambda| = r \le 1$ from

$$hc = \frac{\varrho(\lambda)}{\sigma(\lambda)} = \frac{\varrho(re^{i\varphi})}{\sigma(re^{i\varphi})} = u + iv, \ \varphi \in [0, 2\pi].$$

The stability boundary is obtained by setting $r = 1$.

Corrector formulas of Gear's method for $q = 1,\dots,6$.

The iteration rules for the various methods are given below for $\sigma(\lambda) = \lambda^M$, $M = 1,\dots,6$, with $M = q$ as consistency order:

M	Corrector formulas of Gear's method for $q = 1,\dots,6$	
1	$Y_{i+1}^{(\nu+1)} = Y_i + h f_{i+1}^{(\nu)} = Y_i + h f(x_{i+1}, Y_{i+1}^{(\nu)})$,	$\nu = 0,1,2,\dots$
2	$Y_{i+1}^{(\nu+1)} = \frac{1}{3}\left(4Y_i - Y_{i-1} + 2h f_{i+1}^{(\nu)}\right)$,	$\nu = 0,1,2,\dots$
3	$Y_{i+1}^{(\nu+1)} = \frac{1}{11}\left(18Y_i - 9Y_{i-1} + 2Y_{i-2} + 6h f_{i+1}^{(\nu)}\right)$,	$\nu = 0,1,2,\dots$
4	$Y_{i+1}^{(\nu+1)} = \frac{1}{25}\left(48Y_i - 36Y_{i-1} + 16Y_{i-2} + -3Y_{i-3} + 12h f_{i+1}^{(\nu)}\right)$,	$\nu = 0,1,2,\dots$
5	$Y_{i+1}^{(\nu+1)} = \frac{1}{137}\left(300Y_i - 300Y_{i-1} + 200Y_{i-2} + -75Y_{i-3} + 12Y_{i-4} + 60h f_{i+1}^{(\nu)}\right)$,	$\nu = 0,1,2,\dots$
6	$Y_{i+1}^{(\nu+1)} = \frac{1}{147}\left(360Y_i - 450Y_{i-1} + 600Y_{i-2} + -225Y_{i-3} + 72Y_{i-4} - 10Y_{i-5} + 60h f_{i+1}^{(\nu)}\right)$,	$\nu = 0,1,2,\dots$

For $q = 1, \ldots, 6$ only one functional evaluation of \boldsymbol{f} is necessary per iteration step. The stability regions for $q = 1, \ldots, 6$ are given in the following picture (see [GEAR71/1], p. 212). Every boundary curve of a stability region passes through the point $|hc| = \infty$ of the hc-plane, see figures 17.2 and 17.3. For $M = 2$ the method is weakly A-stable, as is Heun's method. The formulas are stiffly stable for $M = 3, \ldots, 6$.

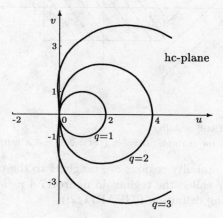

Figure 17.2: Stability regions for Gear's method for $q = 1, 2, 3$

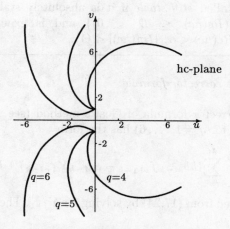

Figure 17.3: Stability regions for Gear's method for $q = 4, 5, 6$

One can represent the *stability regions* for the methods of Gear in a simplified way as follows:

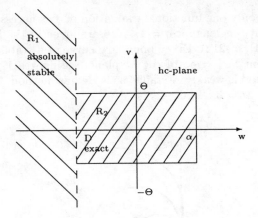

Figure 17.4: Regions of stiff stability
(for $q = 6$, for example, $D = -6.1$, $\Theta = 0.5$, and $\alpha \approx 0$)

The borders of the stability regions are tangents to the curves in the figures 17.2 and 17.3. Gear called the regions in figure 17.4 *regions of stiff stability*. He gave the following definition ([GEAR71/2]):

DEFINITION 17.21 (*Stiff stability*).
A procedure is called *stiff-stable* if it is absolutely stable in the region $R_1 = \{hc \mid (Re(hc) \le D < 0)\}$ and is exact in the region $R_2 = \{hc \mid D \le Re(hc) < \alpha, |Im(hc)| < \Theta\}$.

Convergence of the corrector formula.

In general, the corrector formula of Gear's method (see the earlier table of corrector formulas for $q = 1, \ldots, 6$) has the form

$$\boldsymbol{Y}_{i+1}^{(\nu+1)} = \sum_{k=1}^{M}(hb_{M-k}\boldsymbol{f}_{i+1-k} - a_{M-k}\boldsymbol{Y}_{i+1-k}) + hb_M\boldsymbol{f}_{i+1}^{(\nu)},$$

which can be derived from (17.24) by solving for \boldsymbol{Y}_{i+1}. The corrector iteration converges if

$$\left\| hb_M \left(\frac{\partial f_i}{\partial y_k} \right) \right\| < 1,$$

where $\left(\frac{\partial f_i}{\partial y_k} \right)$ is the Jacobi matrix and $\| \cdot \|$ a matrix norm. For stiff single differential equations, as well as for separated differential equations, the convergence condition becomes

$$h \left| b_M \frac{\partial f}{\partial y} \right| < 1.$$

The convergence condition for the iteration might demand a smaller h than is required by the stiffly-stable method used. In this case, one should use Newton's method instead of the general iteration procedure for solving the corrector equation. To save computational time it may suffice to work with the simplified Newton's method - provided its convergence is not jeopardized, see section 6.2.

The predictor.

As initial values (predictors) for the corrector iteration, the values $Y_{i+1}^{(0)} = Y_i$ were suggested by Gear. The corrector iteration converges if h is chosen according to the convergence conditions just mentioned. In general, three iterations are suggested. Predictors such as the Adams-Bashforth formulas for the Adams-Moulton formulas can be constructed for the procedures of M^{th} order with $M = 3, \ldots, 6$: Starting with the equations (17.33), one constructs an explicit procedure ($b_M = 0$) by setting $\sigma(\lambda) = \lambda^{M-1}$.

Remarks for Gear's method.

The whole procedure was implemented by Gear, see SUBROUTINE DIFSUB in [GEAR71/3]. The program works with automatic step size control and automatically selects the global error order for a given error bound. Besides DIFSUB there exist newer versions Gear 3 or Gear Rev. 3 and a program package developed by Byrne and Hindmarsh in 1975 from DIFSUB. It uses the MS-methods as given by Nordsiek, see [GRIG77], where both q_g and h can be controlled. At the same time, this package contains a numerical method with which one can avoid calculating the starting values. One begins the calculation with a one-step method of the class of chosen methods ([GRIG77], p. 90 ff.). When modifying step sizes one must be careful so that stability is not jeopardized (i.e., one should not routinely halve the step size). For details consult [GEAR74]. Further procedures for systems of stiff differential equations are given in [GRIG77], p. 236.

In general, Gear's method does not give useful results for stiff problems where the eigenvalues λ_i of the Jacobi matrix (f_y) are close to the imaginary axis of the hc-plane. For then some eigenvalues are outside of or close to the boundary of the region of absolute stability. In this case, Rosenbrock's or the modified Rosenbrock's methods have proven useful ([KAPS81]. Here, instead of the notion of stiff stability, other modifications of A-stability have to be used according to the structure of the procedure ([GRIG77, p.236 ss.; [JELT76];

[JELT78]). A-stable methods are the implicit Runge-Kutta formulas with Gaussian nodes (see section 17.3.5), as well as certain types of Rosenbrock methods. The amount of computational time, however, is considerably higher here than with Gear's method.

REMARK. Effective algorithms for solving stiff systems and their FORTRAN programs based on the work of Gear, Krogh and Hull are contained in the program package DEPAC which is available in coded form, see [SHAM75]. Another effective procedure for stiff systems is the Enright method, see [ENGE87], p. 491.

Bibliography for Section 17.7

[AHLB67]; [GEAR71/1]; [GEAR71/2]; [GEAR71/3]; [GEAR74]; [GRIG77] vol.2, 3.3; [HAIR91]; [HALL76], Part 2; [HULL72]; [HULL75]; [LAPI71], 6; [LIND77]; [LINI77]; [LUTH87], 13; [SHAM75]; [WERN86], chap.5.

17.8 Suggestions for Choosing among the Methods

None of the methods of this chapter will perform better than any other for every problem. One must, therefore, learn about the advantages and disadvantages of each method in order to decide for each particular problem which method to use. One cannot expect that the theory alone will be able to furnish strict criteria for choosing the optimal method for a desired accuracy. One can only formulate the following general guidelines on comparing the class of one-step methods and extrapolation methods on the one side with the multi-step methods on the other.

As one typical representative of one-step methods we shall choose the *classical Runge-Kutta method*:

Advantages: 1) Self-starting, fixed local error order $O(h^5)$, simple to handle.
2) Automatic step size control easily possible.

Disadvantage: Each integration step requires the calculation of four functional values.

Extrapolation method:

Advantages: 1) Self-starting, no fixed error order.
2) Step size control possible.

Disadvantage: Considerable computational time per step.

Multi-step methods:

Advantages: In general, each integration step requires only two or three functional evaluations per iteration step. Formulas of arbitrarily high order can easily be constructed.

Disadvantages:1) Not self-starting. Calculation of starting values required.
2) Step size control possible ([GRIG77], vol.2, p.92, 98.), but very time consuming since the starting values must be recalculated if the step size is changed (see [GEAR80]).

These statements are independently true for all types of initial value problems. To help us decide on the merits of the methods we must moreover distinguish between non-stiff and stiff initial value problems:

Suggestions for non-stiff systems of differential equations.

The literature contains extensive numerical comparisons of the known numerical methods for integrating initial value problems for systems of differential equations of first order. A general test program DETEST was developed to investigate and compare all of the methods for solving initial value problems of ordinary differential equations in [HALL 73] by Hall, Enright, Hull and Sedgivida . A short description of this program and test results can be found in [HULL72] and [ENRI76].

There the test problems were divided into five classes:

1) Single first order differential equations.
2) Small systems of first order.
3) Medium sized systems of first order.
4) Systems of first order for calculating orbits.
5) Large order systems.

In each instance an error bound ε is given for the local procedural error so that rounding errors have no effect if computations are carried out in double precision. The step size is adjusted if the given bound ε demands it. The amount of

computational effort serves as a measure of the suitability of a method under
the given conditions (i.e., type of problem, size of ε). The computational effort
is subdivided into:

(a) Time for the calculation of functional values of f, which depends
 on their number per step and the complexity of the function f.

(b) Time for the other operations, apart from (c) (overhead).

(c) Time for necessary changes in the step size.

The following values for ε were used:

$$(i)\ \varepsilon = 10^{-3}, \quad (ii)\ \varepsilon = 10^{-6}, \quad (iii)\ \varepsilon = 10^{-9}.$$

Test results from [ENRI76], p. 626-635:

If the time for (a) is not significant such as it would be for a simple function
f, the Bulirsch-Stoer-Gragg extrapolation method is best, see section 17.5. If,
however, the needed number of f-values is large and their calculation is time
consuming, then the Adams procedures are more advantageous, although the
amount of calculation time in (b) is larger here. Implicit Adams type meth-
ods of variable order are particularly advantageous, see [KROG68], where the
order of the method is automatically determined at the beginning of calcu-
lations from the given ε. h is chosen as large as possible while maintaining
stability. Depending on whether the accuracy bound ε is exceeded or not, the
error order of the method used locally is increased or decreased. Runge-Kutta
methods are only of advantage if the amount of computational time in (a) is
low and a modest accuracy is required, such as $\varepsilon = 10^{-3}$. Time tests for single
differential equations are also included in [ENRI76].

Suggestions for Runge-Kutta embedding formulas.

Adaptive initial value solvers are the methods of choice here in every situa-
tion. Since adaptive methods require the calculation of two approximations
Y and \tilde{Y} for the solution y in each step if step size control is to be used,
we must try and minimize the computational effort to find \tilde{Y} in order to be
competitive with other methods. When using Runge-Kutta embedding for-
mulas this computational effort is minimal, since Y and \tilde{Y} are formed from
the same k values and the computation of \tilde{Y} can be performed quickly, see
section 17.3.4.4.

Our own extensive tests of embedding formulas have lead to the following
results:

(1) The method **rk5(4)6m** (Prince-Dormand embedding formula of 4^{th} and
 5^{th} order) can be especially recommended among the formulas of 4^{th} and
 5^{th} order. **rk5(4)7m** and **rke5(4)** are less suitable. **rke5(4)** requires a
 rather large amount of computational time.

(2) Among the formulas of 6^{th} and 7^{th} order, **rkv6(5)** is to be preferred for its reliability and exactness. The formulas **rk6(5)8m** and **rkv7(6)** can be highly recommended due to their short computational times.

(3) Among the formulas of 8^{th} and 9^{th} order, **rk8(7)13m** is preferred due to its high reliability, its exactness and its short computational time; **rkv8(7)** is a close second. However, **rkv9(8)** requires a large amount of computations and has no other advantages over the above methods of 8^{th} order.

(4) The formulas **rkf6(5)** and **rkf8(7)** did not give comparably good results in the tests.

(5) For stiff problems none of the explicit Runge-Kutta methods gave acceptable results.

Suggestions for stiff systems.

Before deciding upon a method one should try to verify whether the given system is stiff. In general, one can decide this question with the criteria of section 17.7.2. For small systems, it is sometimes possible to recognize stiffness directly from the given differential equations. In general a system will reveal its stiffness if when using an explicit method the step size has to be chosen very small due to an already irrelevant component of the solution. A verification of the stiffness criteria in section 17.7.2 can, however, require considerable effort. But if one does not test for stiffness or falsely diagnoses a system as stiff and then uses one of the implicit methods for stiff systems of section 17.7, one might get very inexact results for a non-stiff system despite the long calculation time of the procedures there. For this reason, attempts are being made to develop numerical tests that find out easily whether a given problem is stiff or not.

Such a test has been proposed by L.F. Shampine. The given initial value problem is integrated first by an explicit Runge-Kutta method, and then by the Euler-Cauchy method with a prescribed level of exactness for the local procedural error. The Runge-Kutta method is normally chosen as the one given by Fehlberg with $m = 5$, and $q_\ell = 5$ with coefficients in table 17.6. Formula (17.8) provides a good approximation for the local procedural error at x_{i+1}.

If such a test indicates that the problem is stiff, then suggestions for the choice of the method can be drawn from the examination in [HULL75]. This paper describes systematic numerical tests in which five different methods that are suitable for solving stiff problems are applied to 25 specific known stiff problems. To test procedures for stiff systems, a test program STIFF DETEST was developed to test all known useful methods. The program can be found in a technical report of Bedet, Enright, Hull of the Department of Computer

Science of the University of Toronto. The results are published in [HULL75].
A measure of the usefulness of a method for a certain problem class is taken as
the computational time required for obtaining a given accuracy bound. This
time is composed of computer overhead, the time for functional evaluations
and the computing of Jacobians as well as for matrix inversions. Here are
some general suggestions: The stiff-stable methods of Enright and Gear have
proved well suited for all problems in which the eigenvalues of every Jacobi
matrix do not lie close to the imaginary axis of the hc-plane. The Enright
procedure is stiffly-stable. The Enright method generally gives more accurate
results, but it takes more time than Gear's method.

For problems in electrical engineering, where eigenvalues often appear close to
the imaginary axis, Gear's method is nearly useless, while Enright's method
has only limited use. A stiffly-stable procedure is obviously of no use here
either, since $Im(hc) \gg 1$ and $Re(hc)$ is small. For this reason, one should
uses an A-stable method. Implicit Runge-Kutta methods with Gaussian nodes
have this property. It is, however, easier to deal with Rosenbrock methods for
the integration of stiff systems. Suitable Rosenbrock methods are given in
[KAPS81]. Kaps indicated modified Rosenbrock methods up to order 6 in
1977; they were implemented by Kaps and Wanner in [KAPS81].

A short description of the tests in [HULL75] can be found in [GRIG77], vol.
2. FORTRAN-Codes for Gear's method can be found in [SHAM75], see also
our remark at the end of section 17.4.1 .

Bibliography for Section 17.8

[ENRI76]; [GEAR71/1], 12; [HAIR87]; [HAIR91]; [HULL72]; [HULL75];
[KAPS81]; [KROG66]; [KROG68]; [WERN86], 4.8, 5.5.

Chapter 18

Boundary Value Problems for Ordinary Differential Equations

18.1 Statement of the Problem

In this chapter we want to solve a differential equation of the form

(18.1) $$y''(x) = g(x, y, y')$$

with boundary conditions

(18.2) $$\begin{cases} \alpha_1 y(a) + \alpha_2 y'(a) = A, \\ \beta_1 y(b) + \beta_2 y'(b) = B, \end{cases}$$

where $|\alpha_1| + |\beta_1| > 0$ (two-point boundary problem) and $\alpha_i \neq 0$ and $\beta_j \neq 0$ for at least one index. The equations (18.2) are called *linear boundary conditions*, the differential equation (18.1) together with the boundary conditions (18.2) form a *boundary value problem* of second order. The boundary conditions (18.2) can be replaced by nonlinear conditions $r_1(y(a), y'(a)) = 0$ and $r_2(y(b), y'(b)) = 0$. Such conditions will not be treated explicitly here. The subroutine BVP in the program section can handle such conditions as well.

A boundary value problem of n^{th} order is given by a differential equation of the form

(18.3) $$y^{(n)}(x) = f(x, y, y', \ldots, y^{(n-1)})$$

and n (linear or nonlinear) boundary conditions involving the quantities $y, y', \ldots, y^{(n-1)}$ at two points at least. In practice, boundary value problems of second and fourth order occur most often.

For a boundary value problem (18.3) with a differential equation of first order

$$y'(x) = f(x, y), \quad x \in [a, b], \quad r(y(a), y(b)) = 0$$

with linear or nonlinear boundary conditions, the subroutine BVP determines
an approximation YANF for the initial value $y(a)$ of a solution y of the
boundary value problem via the shooting method (see [LUTH87], 15.1). The
desired solution y can then be determined approximately by means of an
initial value problem solver of chapter 17 applied to

$$y'(x) = f(x, y), \quad x \in [a, b], \quad y(a) = \text{YANF}.$$

The nonlinear system which arises in the shooting method is solved iteratively
by Newton's method. In this process, one repeatedly solves initial value prob-
lems associated with the given boundary value problem. To do so, we shall use
the subroutine IVP which works with step size control. This subroutine, how-
ever, is not useful for solving stiff differential equations. Furthermore, even for
seemingly "harmless" boundary value problems, the associated initial value
problems can react very sensitively to perturbations in the initial values, so
that Newton's method may converge only if the given initial value agrees very
accurately with the true initial value $y(a)$.

18.2 Reduction of Boundary Value Problems to Initial Value Problems

18.2.1 Boundary Value Problems for Nonlinear Differential Equations of Second Order

The existence and uniqueness of solutions of a boundary value problem (18.1),
(18.2) is governed by:
For the boundary value problem (18.1) with (18.2), assume that the right
hand side function g has continuous partial first derivatives for $a \le x \le b$ and
that $y^2 + y'^2$ is finite in $[a, b]$. Assume that for $x \in [a, b]$ and some constants
$0 < L, M < \infty$ the following inequalities hold:

$$0 < \frac{\partial g}{\partial y} \le L, \quad \left| \frac{\partial g}{\partial y'} \right| \le M,$$

and that the products in the boundary conditions satisfy $\alpha_1 \alpha_2 \le 0$, $\beta_1 \beta_2 \ge 0$.
Then the boundary value problem (18.1), (18.2) has a unique solution for all
$x \in [a, b]$ (see [KELL68], p. 9 and p. 50).

A solution of the boundary value problem can be constructed as follows: Start
from the initial value problem given by (18.1) and the initial condition

(18.4)
$$\begin{cases} \alpha_1 y(a) \;+\; \alpha_2 y'(a) \;=\; A, \\ \gamma_1 y(a) \;+\; \gamma_2 y'(a) \;=\; s. \end{cases}$$

In (18.4), the constants γ_1, γ_2 are chosen so that $\alpha_2 \gamma_1 - \alpha_1 \gamma_2 = 1$.

Then the parameter s must satisfy the equation

(18.5)
$$f(s) \equiv \beta_1\, y(b,s) + \beta_2\, y'(b,s) - B = 0.$$

Without loss of generality we can assume that $\alpha_1 \geq 0$, $\alpha_2 \leq 0$ and $\beta_i \geq 0$ for $i = 1, 2$ and that $\alpha_1 + \beta_1 > 0$. The solution of the initial value problem (18.1), (18.4) clearly depends on s. We shall denote the solution by $y = y(x, s)$. From (18.4) we can calculate $y(a)$ and $y'(a)$ as

(18.6)
$$y(a) = \alpha_2 s - \gamma_2 A, \quad y'(a) = \gamma_1 A - \alpha_1 s.$$

Note that the initial value problem (18.1), (18.6) is equivalent to that given by (18.1) and (18.4). For each value of s, it can be solved by one of the methods of chapter 17. Hence we can get approximate values $Y(x_i, s)$ and $Y'(x_i, s)$ for $y(x_i, s)$ and $y'(x_i, s)$ for each s with $x_i = a + i\frac{b-a}{N}$, $x_i \in [a, b]$, $i = 1, \ldots, N$. Thus $Y(b, s) \approx y(b, s)$ and $Y'(b, s) \approx y'(b, s)$.

The parameter s can be determined from (18.5) iteratively. For this, we need the values of $y(b, s)$ and $y'(b, s)$ which can only be computed numerically, so that the equation $f(s) = 0$ in (18.5) does not really fall under those treated in chapter 2.

Starting with an arbitrary initial value $s^{(0)}$ and solving the initial value problem (18.1), (18.6), the computed values $Y(b, s^{(0)}), Y'(b, s^{(0)})$ are used to calculate an improved approximation $s^{(1)}$ for s according to the following rule:

(18.7)
$$\begin{cases} s^{(\nu+1)} = s^{(\nu)} - mF(s^{(\nu)}) \quad \text{for } \nu = 0, 1, 2, \ldots \quad \text{with} \\[4pt] F(s) = \beta_1\, Y(b, s) + \beta_2\, Y'(b, s) - B \\[2pt] \qquad \text{and} \quad m = \frac{2}{\Gamma + \gamma}, \text{ where} \\[4pt] \gamma = \beta_1 \left(\alpha_1 \dfrac{1 - e^{-M(b-a)}}{M} - \alpha_2 \right) + \alpha_1 \beta_2 e^{-M(b-a)}, \\[6pt] \Gamma = \dfrac{e^{(M/2)(b-a)}}{2\sigma} \Big\{ (\alpha_1 - \alpha_2(\sigma - M/2))(\beta_1 + \\[4pt] \qquad + \beta_2(\sigma + M/2))e^{\sigma(b-a)} + \\[4pt] \qquad - (\alpha_1 + \alpha_2(\sigma + M/2))(\beta_1 - \beta_2(\sigma - M/2))e^{-\sigma(b-a)} \Big\}, \\[6pt] \sigma = \tfrac{1}{2}\sqrt{4L + M^2} \quad \text{with} \quad \dfrac{\partial g}{\partial y} \leq L, \quad L > 0, \quad M > 0. \end{cases}$$

With $s^{(1)}$ one solves the problem (18.1), (18.6) again and determines $s^{(2)}$ via (18.7) etc.. Thus for each new value $s^{(\nu)}$, the initial value problem (18.1), (18.6) must be solved anew. The procedure is stopped if two successive approximations $s^{(\nu)}, s^{(\nu+1)}$ do not differ by more than a given ϵ. Under the quoted conditions for existence and uniqueness of the solution of the boundary value problem there is exactly one solution $\bar{s} = \lim_{\nu \to \infty} s^{(\nu)}$ of the equation $F(s) = 0$ in (18.7), independent of the initially chosen $s^{(0)}$.

When using a method of global error order $O(h^q)$ for solving (18.1), (18.6), we obtain the following rules of convergence (see [KELL68], §2.2) with $\lambda := (\Gamma - \gamma)/(\Gamma + \gamma) < 1$ for any $i = 0, \dots, N$:

$$\left| \bar{s} - s^{(\nu)} \right| \quad\quad \leq \quad \lambda^\nu \left| F(s^{(0)}) \right| / \gamma, \quad \nu = 1, 2, 3, \dots,$$

$$\left| Y(x_i, s^{(\nu)}) - y(x_i) \right| \quad \leq \quad O(h^q) + O(\lambda^\nu),$$

$$\left| Y'(x_i, s^{(\nu)}) - y'(x_i) \right| \quad \leq \quad O(h^q) + O(\lambda^\nu).$$

REMARK: Using Newton's method for solving (18.5) iteratively with the iteration rule $s^{(\nu+1)} = s^{(\nu)} - F(s^{(\nu)})/F'(s^{(\nu)})$ presupposes knowledge of $F'(s^{(\nu)})$. An approximate formula for $F'(s^{(\nu)})$ is given in [KELL68], p. 53. One can also employ the damped Newton method and approximate the derivative by forward difference quotients.

SPECIAL CASE: If $\alpha_2 = \beta_2 = 0$, we can set $\alpha_1 = \beta_1 = 1$ w.l.o.g. Then the boundary value problem (18.1), (18.2) reduces to the initial value problem with the differential equation (18.1) and the initial condition $y(a) = A$, $y'(a) = s$. Choose an initial value $s^{(0)}$. The numerical solution of this initial value problem gives the result $Y(b, s^{(0)})$. For another initial value $s^{(1)}$ we will obtain the value $Y(b, s^{(1)})$. Then the regula falsi gives an improved value for $s^{(2)}$ as

$$s^{(2)} = s^{(0)} + (s^{(1)} - s^{(0)}) \frac{B - Y(b, s^{(0)})}{Y(b, s^{(1)}) - Y(b, s^{(0)})}.$$

One should try to choose $s^{(0)}, s^{(1)}$, so that $Y(b, s^{(0)}) - B$ and $Y(b, s^{(1)}) - B$ have opposite signs. The procedure ought to be continued until $\left| Y(b, s^{(\nu)}) - B \right| < \varepsilon$ for a given $\varepsilon > 0$.

18.2.2 Boundary Value Problems for Systems of Differential Equations of First Order

Boundary value problems for differential equations (18.3) with linear boundary conditions can be reduced via the substitutions $y_{k+1} := y^{(k)}$ for $k = 0, \dots, n-1$ to first order boundary value problems of the form

$$(18.8) \quad \begin{cases} \boldsymbol{y}'(x) = \boldsymbol{f}(x, \boldsymbol{y}), \quad x \in [a, b], \\[2mm] \boldsymbol{A}\boldsymbol{y}(a) + \boldsymbol{B}\boldsymbol{y}(b) - \boldsymbol{g} = \boldsymbol{0} \\[2mm] \text{with} \quad \boldsymbol{y} = (y_1, y_2, \ldots, y_n)^T, \quad \boldsymbol{g} = (a_1, a_2, \ldots, a_n)^T. \end{cases}$$

Here $\boldsymbol{A} = (a_{jk})$, and $\boldsymbol{B} = (b_{jk})$ are (n, n) matrices with constant coefficients and \boldsymbol{g} is a constant vector (see section 18.1).

Regarding existence and uniqueness of a solution of (18.8), the same conditions hold as do for the boundary value problems (18.1), (18.2), see [KELL68], p. 16. For large systems these conditions can be checked only with considerable effort, and besides, they are not always fulfilled in important real life problems. Hence one should use the method described next, or the multiple shooting method of section 18.2.3 instead.

To find a solution of the boundary value problem (18.8), one starts with the initial value problem

$$(18.9) \qquad \boldsymbol{y}'(x) = \boldsymbol{f}(x, \boldsymbol{y}), \qquad \boldsymbol{y}(a) = \boldsymbol{s} = \begin{pmatrix} s_1 \\ \vdots \\ s_n \end{pmatrix}.$$

Denote its solution by $\boldsymbol{y} = \boldsymbol{y}(x, \boldsymbol{s})$. The vector \boldsymbol{s} must be determined in such a way that

$$(18.10) \qquad \boldsymbol{f}(\boldsymbol{s}) = \boldsymbol{A}\boldsymbol{s} + \boldsymbol{B}\boldsymbol{y}(b, \boldsymbol{s}) - \boldsymbol{g} = \boldsymbol{0}.$$

In general (18.10) is a nonlinear system of n equations for the n components s_k of \boldsymbol{s}. The initial value problem (18.9) can be solved for each vector \boldsymbol{s} by one of the methods given in chapter 17 so that one can obtain approximations $\boldsymbol{Y}(x_i, \boldsymbol{s})$ for the solution vectors $\boldsymbol{y}(x_i, \boldsymbol{s})$ of the initial value problem at $x_i = a + i(b - a)/N$, $i = 1, \ldots, N$, depending on \boldsymbol{s}. Thus one can find $\boldsymbol{Y}(b, \boldsymbol{s}) \approx \boldsymbol{y}(b, \boldsymbol{s})$. The system (18.10) must be solved iteratively. Starting with an initial vector $\boldsymbol{s}^{(0)}$, one solves the problem (18.9) for $\boldsymbol{s}^{(0)}$. The solution $\boldsymbol{Y}(b, \boldsymbol{s}^{(0)})$ is used to calculate an improved approximation $\boldsymbol{s}^{(1)}$ for \boldsymbol{s} according to the rule

$$(18.11) \quad \begin{cases} \boldsymbol{s}^{(\nu+1)} = \boldsymbol{s}^{(\nu)} - (\boldsymbol{A} + \boldsymbol{B})^{-1}(\boldsymbol{A}\boldsymbol{s}^{(\nu)} + \boldsymbol{B}\boldsymbol{Y}(b, \boldsymbol{s}^{(\nu)}) - \boldsymbol{g}), \\[2mm] \qquad\qquad\qquad\qquad\qquad\qquad\qquad \nu = 0, 1, 2, \ldots. \end{cases}$$

This can only be done if $\det(\boldsymbol{A} + \boldsymbol{B}) \neq 0$. This will be the case e.g. if the row or column sum criterion (5.5) or (5.6) is satisfied for the matrix $\boldsymbol{A} + \boldsymbol{B}$. With $\boldsymbol{s}^{(1)}$ one solves the initial value problem (18.9) again and determines $\boldsymbol{s}^{(2)}$ from (18.11), etc. The procedure is stopped if the inequality

$\max_{1 \le j \le n} \left| s_j^{(\nu+1)} - s_j^{(\nu)} \right| < \varepsilon$ is fulfilled. For error estimates, see [KELL68], p.56.

NOTE. The program BVP in the program section solves initial value problems (18.8) with linear and nonlinear boundary conditions.

18.2.3 The Multiple Shooting Method

The methods of sections 18.2.1 and 18.2.2 are only useful for obtaining an approximate solution of boundary value problems if the equations (18.5) and (18.10) are well-conditioned, see section 4.15.1. The notions developed there can also be applied to nonlinear systems. Otherwise - or if not all the sufficient conditions for existence and uniqueness are satisfied - the approximate solution of the initial value problems (18.1), (18.6) or (18.9) can grow strongly during the integration over $[a, b]$ or even show singularities for certain $x \in [a, b]$ and certain values of the iteration sequence $s^{(\nu)}$. In this case, one can either use *difference methods*, see section 18.3, or the so-called *multiple shooting method*, see [KELL68], p.61 or the detailed description in [STOE91], 7.3.4 - 7.3.7. The multiple shooting method uses a partition of the interval of integration $[a, b]$ into subintervals $[\bar{x}_k, \bar{x}_{k+1}]$:

$$(18.12) \qquad a = \bar{x}_0 < \bar{x}_1 < \ldots < \bar{x}_m < \bar{x}_{m+1} = b.$$

Each \bar{x}_k for $k = 0, \ldots, m+1$ is one of the nodes $x_i, i = 0, \ldots, N$. Here $m + 1 \le N$. The shorter the length of the subintervals $[\bar{x}_k, \bar{x}_{k+1}]$, the less will be the effects of possible ill-conditioning of the equations (18.5) or (18.10).

ALGORITHM 18.1 (*Multiple shooting method*).

Given a boundary value problem (18.8). If $x_i = a + ih$, $h = (b - a)/N$, $i = 0, \ldots, N$, are the nodes in the interval of integration $[a, b]$ for the boundary value problem, one has to follow the following steps in order to calculate approximate values $\boldsymbol{Y}(x_i)$ of the solution $\boldsymbol{y}(x_i)$:

I. Determining the nodes $\bar{x}_k, k = 0, \ldots, m+1$.

1^{st} step. At the nodes x_i, starting with x_1, determine approximations $\tilde{\boldsymbol{Y}}_1(x_i)$ for the solutions of the initial value problem

$$\boldsymbol{y}'(x) = \boldsymbol{f}(x, \boldsymbol{y}), \quad \boldsymbol{y}(a) = \tilde{\boldsymbol{s}}$$

for an arbitrary starting vector \tilde{s}. For each i test whether for a given R with $2 \leq R \leq 10^3$ the inequality

$$(18.13) \qquad \left\| \tilde{Y}_1(x_i) \right\| > R \left\| \tilde{Y}_1(a) \right\|$$

holds, where $\| \ldots \|$ denotes any one of the vector norms of chapter 5.2. The calculation should be stopped as soon as (18.13) is fulfilled for $i = j_1$.

REMARK. It is better to determine approximations $\bar{Y}(x_i)$ for $y(x_i)$ with the method described in section 18.2.2, and to set $\tilde{s} = \bar{Y}(a)$ instead of working with the arbitrary starting vector \tilde{s}.

2^{nd} **step.** Set $\bar{x}_1 := x_{j_1}$, and determine approximations $\tilde{Y}_2(x_i)$ for the solution of the initial value problem

$$y'(x) = f(x,y), \quad y(\bar{x}_1) = \tilde{Y}_1(\bar{x}_1)$$

at the nodes x_i for indices $i > j_1$. For each i check whether the following inequality is fulfilled for R as chosen in the first step:

$$\left\| \tilde{Y}_2(x_i) \right\| > R \left\| \tilde{Y}_2(\bar{x}_1) \right\|.$$

If this happens for the first time for $i = j_2$, the calculation should be stopped.

3^{rd} **step.** Set $\bar{x}_2 := x_{j_2}$, and determine approximations $\tilde{Y}_3(x_i)$ for the solutions of the initial value problem

$$y'(x) = f(x,y), \quad y(\bar{x}_2) = \tilde{Y}_2(\bar{x}_2).$$

at the nodes x_i with $i > j_2$. Continue analogously until for an index j_k we have $x_{j_k} =: \bar{x}_k = x_{N-1}$. Set $m = j_k$, so that all points of the partition (18.12) are now fixed. If the function $f(x,y)$ or its derivative are only piecewise continuous, the points of discontinuity must be added to the nodes and the nodes rearranged to satisfy (18.12).

II. **Calculation of** $Y(x_i) \approx y(x_i)$.

4^{th} **step.** In each of the subintervals $[\bar{x}_k, \bar{x}_{k+1}]$ we introduce a new variable t by setting

$$(18.14) \qquad t = \frac{x - \bar{x}_k}{h_k} \quad \text{with} \quad \bar{h}_k = \bar{x}_{k+1} - \bar{x}_k, \quad k = 0, \ldots, m.$$

For $x \in [\bar{x}_k, \bar{x}_{k+1}]$ we set

$$(18.15) \quad \begin{cases} y(x) = y(\bar{x}_k + t\bar{h}_k) =: y_k(t), \\ f_k(t, y_k(t)) := \bar{h}_k f(\bar{x}_k + t\bar{h}_k, y_k(t)), \end{cases}$$

so that

(18.16) $\dfrac{d\boldsymbol{y}_k(t)}{dt} = \boldsymbol{f}_k(t, \boldsymbol{y}_k(t))$ for $0 \le t \le 1$, $k = 0, \dots, m$

with the connecting conditions

(18.17) $\boldsymbol{y}_{k+1}(0) - \boldsymbol{y}_k(1) = \boldsymbol{0}$ for $k = 0, \dots, m-1$,

and the boundary condition

(18.18) $A\boldsymbol{y}_0(0) + B\boldsymbol{y}_m(1) - \boldsymbol{g} = \boldsymbol{0}$.

5th step. The $m+1$ systems of first order differential equations (18.16) can be rearranged into one system of $(m+1) \cdot n$ differential equations of first order:

(18.19) $\begin{cases} \dfrac{d}{dt}\hat{\boldsymbol{y}}(t) = \boldsymbol{F}(t, \hat{\boldsymbol{y}}(t)) & \text{for } 0 \le t \le 1 \quad \text{with} \\[2mm] \hat{\boldsymbol{y}}(t) = \begin{pmatrix} \boldsymbol{y}_0(t) \\ \boldsymbol{y}_1(t) \\ \vdots \\ \boldsymbol{y}_m(t) \end{pmatrix}, \quad \boldsymbol{F}(t, \hat{\boldsymbol{y}}) = \begin{pmatrix} \boldsymbol{f}_0(t, \boldsymbol{y}_0) \\ \boldsymbol{f}_1(t, \boldsymbol{y}_1) \\ \vdots \\ \boldsymbol{f}_m(t, \boldsymbol{y}_m) \end{pmatrix}. \end{cases}$

The conditions (18.17) and (18.18) can be combined to form one boundary condition

(18.20) $\hat{A}\hat{\boldsymbol{y}}(0) + \hat{B}\hat{\boldsymbol{y}}(1) = \hat{\boldsymbol{g}}$ with

$$\hat{A} = \begin{pmatrix} A & 0 & 0 & \dots & 0 \\ 0 & I & 0 & \dots & 0 \\ 0 & 0 & I & \dots & 0 \\ \cdot & \cdot & \cdot & \ddots & \cdot \\ 0 & 0 & 0 & \dots & I \end{pmatrix},$$

$$\hat{B} = \begin{pmatrix} 0 & 0 & 0 & \dots & 0 & B \\ -I & 0 & 0 & \dots & 0 & 0 \\ 0 & -I & 0 & \dots & 0 & 0 \\ \cdot & \cdot & \cdot & \dots & \cdot & \cdot \\ 0 & 0 & 0 & \dots & -I & 0 \end{pmatrix}, \quad \hat{\boldsymbol{g}} = \begin{pmatrix} g \\ 0 \\ 0 \\ \vdots \\ 0 \end{pmatrix}.$$

Here $\boldsymbol{0}$ denotes the (n, n) zero matrix and I the (n, n) identity matrix. The vector $\hat{\boldsymbol{g}}$ has $(m+1) \cdot n$ components, \hat{A}, \hat{B} are $(m+1, m+1)$ block matrices, where each block is itself an (n, n) matrix.

The boundary value problem (18.19), (18.20) for $\hat{\boldsymbol{y}}(t)$, $t \in [0, 1]$, is equivalent to the boundary value problem (18.8) for $\boldsymbol{y}(x)$, $x \in [a, b]$. Clearly \boldsymbol{y} and $\hat{\boldsymbol{y}}$ are related via (18.15) and (18.19).

6^{th} **step**. The boundary value problem (18.19), (18.20) can be solved by reduction to the initial value problem

$$\frac{d\hat{y}(t)}{dt} = F(t, \hat{y}), \quad \hat{y}(0) = \hat{s} = \begin{pmatrix} s_0 \\ s_1 \\ \vdots \\ s_m \end{pmatrix},$$

which can be solved iteratively with the method described in section 18.2.2.

REMARK. A modification of the multiple shooting method that avoids the transformation (18.14) is given in [STOE91], 7.3.5. A bibliography for the multiple shooting method can be found there as well as references to ALGOL and FORTRAN codes.

Bibliography for Section 18.2

[ATKI89], 6.11; [CARN69], 6.14; [COLL66], III §4.3; [CONT80], 7.3; [GOOS79]; [HALL76], 15, 17; [ISAA66], 8.7.1; [KELL68], 2; [KIOU78]; [LUTH87], 15.1; [STOE91], 7.3; [WERN86], 6; [ZURM65], §29.

18.3 Difference Methods

18.3.1 The Ordinary Difference Method

This method is primarily used for linear boundary value problems, i.e., where instead of differential equations (18.1) or (18.3) we are given a linear differential equation of the form

$$L(y) \equiv y^{(n)} + p_{n-1}(x)y^{(n-1)} + \ldots + p_1(x)y' + p_0(x)y = q(x).$$

For example a *linear second order boundary value problem* has the form

$$(18.21) \quad \begin{cases} L(y) & \equiv & y'' + p_1(x)y' + p_0(x)y = q(x) \text{ with} \\ \alpha_1 y(a) & + & \alpha_2 y'(a) = A, \\ \beta_1 y(b) & + & \beta_2 y'(b) = B. \end{cases}$$

If $q(x) = 0$, the differential equation is *homogeneous*, for $A = B = 0$ the boundary conditions are homogeneous. If the differential equation or the boundary conditions are homogeneous, we deal with a *semihomogeneous* boundary value problem. If we have $q(x) \equiv 0$ and $A = B = 0$ at the same time, we have a *completely homogeneous* problem. The solvability of inhomogeneous and homogeneous boundary value problems, as known from the theory of linear differential equations, is based on knowing a fundamental system of solutions for the homogeneous differential equation and a particular solution for the inhomogeneous differential equation. For the numerical solution, solvability hinges on the solvability of a linear system associated with the given boundary value problem via *discretization*. The numerical solution can be obtained from

ALGORITHM 18.2. (*Ordinary difference method for second order linear boundary value problems*).

We want to find approximate values Y_i for the solution $y(x_i)$ of the boundary value problem (18.21) at the nodes $x_i = x_0 + ih$, $h = (b - a)/N$, $i = 0, \ldots, N$, in the interval of integration $[a, b]$.

1^{st} step. Replace the derivatives at x_i : $y_i' := y'(x_i)$, $y_i'' := y''(x_i)$ by the difference quotients of error order $O(h^2)$ formed with the approximate values $Y_j \approx y(x_j)$, $j = i - 1, i, i + 1$:

$$y_i' = \frac{1}{2h}(-Y_{i-1} + Y_{i+1}) + O(h^2), \quad i \neq 0, \quad i \neq N,$$

$$y_i'' = \frac{1}{h^2}(Y_{i-1} - 2Y_i + Y_{i+1}) + O(h^2).$$

This is called a *discretization*.

2^{nd} step. Form a linear system associated with the boundary value problem (18.21). For this we form the discretized differential equation at the points x_i, $i = 1, \ldots, N-1$, multiply by h^2 and rearrange according to Y_i. With $p_{ki} := p_k(x_i)$ for $k = 0, 1$, $q_i := q(x_i)$, there are $N-1$ linear equations for the $N+1$ unknown values Y_i, $i = 0, \ldots, N$:

$$(18.22) \qquad (1 - \tfrac{h}{2}p_{1i})Y_{i-1} + (-2 + h^2 p_{0i})Y_i + (1 + \tfrac{h}{2}p_{1i})Y_{i+1} = h^2 q_i$$
$$\text{for } i = 1, \ldots, N - 1.$$

An equation of the form (18.22) is called a *difference equation*.
The discretization of the boundary condition becomes

$$\alpha_1 Y_0 + \alpha_2 \frac{Y_1 - Y_{-1}}{2h} = A, \quad \beta_1 Y_n + \beta_2 \frac{Y_{N+1} - Y_{N-1}}{2h} = B.$$

To eliminate the values Y_{-1}, Y_{N+1}, one can form the discretized differential equation at the points x_0, x_N in so-called *additional boundary conditions*,

see [KREI72]:

$$(1 - \tfrac{h}{2}p_{10})Y_{-1} + (-2 + h^2 p_{00})Y_0 + (1 + \tfrac{h}{2}p_{10})Y_1 = h^2 q_0,$$

$$(1 - \tfrac{h}{2}p_{1N})Y_{N-1} + (-2 + h^2 p_{0N})Y_N + (1 + \tfrac{h}{2}p_{1N})Y_{N+1} = h^2 q_N.$$

The first of these equations is solved for Y_{-1}, the second for Y_{N+1}. These values are then inserted into the discretized boundary conditions. With the abbreviations

$$\tilde{\alpha}_1 = \alpha_1 h - \tilde{\alpha}_2 \frac{2 - h^2 p_{00}}{2}, \quad \tilde{\alpha}_2 = \frac{2\alpha_2}{2 - hp_{10}}, \quad \tilde{A} = A + \alpha_2 \frac{hq_0}{2 - hp_{10}},$$

$$\tilde{\beta}_1 = \beta_1 h + \tilde{\beta}_2 \frac{2 - h^2 p_{0N}}{2}, \quad \tilde{\beta}_2 = \frac{2\beta_2}{2 + hp_{1N}}, \quad \tilde{B} = B - \beta_2 \frac{hq_N}{2 + hp_{1N}},$$

we thus obtain the equations

$$\tilde{\alpha}_1 Y_0 + \tilde{\alpha}_2 Y_1 = \tilde{A}h, \quad -\tilde{\beta}_2 Y_{N-1} + \tilde{\beta}_1 Y_N = \tilde{B}h.$$

Together with (18.22), we now have a system of $N+1$ linear equations for the $N+1$ approximate values Y_i of the form

(18.23) $$Ay = g$$

with

$$y = (Y_0, Y_1, \ldots, Y_n)^T \text{ and } g = (\tilde{A}h, h^2 q_1, \ldots, h^2 q_{N-1}, \tilde{B}h)^T, \text{ where}$$

$$A = \begin{pmatrix} \tilde{\alpha}_1 & \tilde{\alpha}_2 & 0 & 0 & \cdots & 0 & 0 \\ 1 - \tfrac{h}{2}p_{11} & -2 + h^2 p_{01} & 1 + \tfrac{h}{2}p_{11} & 0 & \cdots & 0 & 0 \\ \vdots & & & & & & \vdots \\ 0 & \cdots & & 1 - \tfrac{h}{2}p_{1N-1} & -2 + h^2 p_{0N-1} & 1 + \tfrac{h}{2}p_{1N-1} \\ 0 & \cdots & & & 0 & -\tilde{\beta}_2 & \tilde{\beta}_1 \end{pmatrix}$$

3^{rd} **step.** The matrix A of the linear system (18.23) is tridiagonal. The system can be solved with one of the algorithms given in section 4.10. The solvability conditions for (18.23) are equivalent to those of the boundary value problem (18.21). If the boundary value problem is inhomogeneous ($q(x) \not\equiv 0$, or at least $A \neq 0$ or $B \neq 0$), the system has a uniquely determined solution, provided $\det A \neq 0$. If $q(x) \equiv 0$ then all $q_i = 0$. If in this case we also have $A = B = 0$, then the boundary value problem is homogeneous. Only if $\det A = 0$, will the system have non-trivial solutions. The solutions Y_i of the linear system converge for $h \to 0$ towards the exact values $y(x_i)$ of the solution of the boundary value problem (18.21).

If the step size h satisfies the conditions $h|p_1(x)| < 2$ and $p_0(x) < 0$ for all $x \in [a, b]$, then the system matrix \boldsymbol{A} is tridiagonal, diagonally dominant and thus $\det \boldsymbol{A} \neq 0$, see section 4.10, or [ISAA66], p. 445.

The global total error (procedural plus rounding error) can be estimated for $y \in C^4[a, b]$ as

$$|Y_i - y(x_i)| \leq Ch^2 + \frac{D}{h^2}, \quad i = 1, \ldots, N - 1.$$

However, the constant C cannot be determined in general. D is proportional to the maximal local rounding error ϱ. In order to limit the total error to $O(h^2)$, ϱ must be of order $O(h^4)$ (see [COLL66], p. 145, 177 or [ISAA66], 8.7.2).

Error estimation. If $Y_h(x_i)$ is the approximate value for $y(x_i)$, calculated with the step size h and $Y_{2h}(x_i)$ is the approximate value calculated with step size $2h$ for y at x_i, then the global procedural error $e_{i,h}$ of the approximate value $Y_h(x_i)$ is:

(18.24) $$e_{i,h} := y(x_i) - Y_h(x_i) \approx \frac{1}{3}(Y_h(x_i) - Y_{2h}(x_i)).$$

An improved approximate value for $y(x_i)$ is

$$Y_h^*(x_i) = Y_h(x_i) + e_{i,h} = y(x_i) + O(h^4).$$

PARTICULAR CASE: If $\alpha_2 = \beta_2 = 0$ in the boundary conditions, then Y_0 becomes $\frac{A}{\alpha_1}$ and $Y_N = \frac{B}{\beta_1}$. In this case, (18.23) reduces to a tridiagonal system of $N-1$ equations for the $N-1$ values Y_i, $i = 1, \ldots, N-1$.

A *fourth order linear boundary value problem* is given by a fourth order linear differential equation and four boundary conditions that are pairwise independent:

$$y^{(4)} + p_3(x)y''' + p_2(x)y'' + p_1(x)y' + p_0(x)y = q(x); \text{ and}$$

$$\alpha_{k0}y(a) + \alpha_{k1}y'(a) + \alpha_{k2}y''(a) + \alpha_{k3}y'''(a) = A_k,$$

$$\beta_{k0}y(b) + \beta_{k1}y'(b) + \beta_{k2}y''(b) + \beta_{k3}y'''(b) = B_k \text{ for } k = 1, 2.$$

In order to form a linear system associated with such a boundary value problem, one needs approximate values for y_i' and y_i'' at x_i and also for $y_i''' := y'''(x_i)$ and $y_i^{(4)} := y^{(4)}(x_i)$.

Depending on whether the node x_i is an interior point or an end point of $[a, b]$, i.e., whether $i \neq 0, i \neq N$, or $i = 0$ or $i = N$, we must use different formulas:

$$y_i''' = \frac{1}{2h^3}(-Y_{i-2} + 2Y_{i-1} - 2Y_{i+1} + Y_{i+2}) + O(h^2), \quad i \neq 0, \quad i \neq N,$$

$$y_0''' = \frac{1}{2h^3}(-3Y_{-1} + 10Y_0 - 12Y_1 + 6Y_2 - Y_3) + O(h^2),$$

$$y_N''' = \frac{1}{2h^3}(Y_{N-3} - 6Y_{N-2} + 12Y_{N-1} - 10Y_N + 3Y_{N+1}) + O(h^2),$$

$$y_i^{(4)} = \frac{1}{h^4}(Y_{i-2} - 4Y_{i-1} + 6Y_i - 4Y_{i+1} + Y_{i+2}) + O(h^2), \quad i \neq 0, \quad i \neq N,$$

$$y_0^{(4)} = \frac{1}{h^4}(Y_{-1} - 4Y_0 + 6Y_1 - 4Y_2 + Y_3) + O(h),$$

$$y_N^{(4)} = \frac{1}{h^4}(Y_{N-3} - 4Y_{N-2} + 6Y_{N-1} - 4Y_N + Y_{N+1}) + O(h).$$

The expressions for $y^{(4)}$ differently in the remainder terms. By using the approximate values for the derivatives in the discretized differential equation at the points x_i, $i = 3, \ldots, N-3$, one can obtain a system of $N-5$ linear equations for $N+1$ unknowns Y_i, $i = 0, \ldots, N$. Four more equations are obtained by forming the discretized differential equation for $i = 1, 2, N-1, N-2$. For $i = 1$ and $i = N-1$ we use the values given above for y_0''', y_N''' and $y_0^{(4)}, y_N^{(4)}$. These four equations, called "additional boundary conditions", see [KREI72], contain the unknown values Y_{-1}, Y_{N+1}. By using the approximations for the derivatives in the first and second or third and fourth boundary condition above, one can obtain two linear equations each in $Y_0, Y_{-1}, Y_1, Y_2, Y_3$ and in $Y_N, Y_{N+1}, Y_{N-1}, Y_{N-2}, Y_{N-3}$ that can be solved for Y_0, Y_{-1} and Y_N, Y_{N+1}, respectively:

$$Y_0 = A_0 Y_1 + B_0 Y_2 + C_0 Y_3 + D_0,$$

$$Y_{-1} = A_{-1} Y_1 + B_{-1} Y_2 + C_{-1} Y_3 + D_{-1},$$

$$Y_N = A_n Y_{N-1} + B_N Y_{N-2} + C_N Y_{N-3} + D_N,$$

$$Y_{N+1} = A_{N+1} Y_{N-1} + B_{N+1} Y_{N-2} + C_{N+1} Y_{N-3} + D_{N+1}.$$

The constants A_j, B_j, C_j, D_j, $j = 0, -1, N, N+1$, depend on the coefficients of the given boundary conditions and on h. If one uses the values for Y_0, Y_{-1} and Y_N, Y_{N+1} in the $N-1$ discretized equations of the differential equations, the following system of $N-1$ linear equations for the $N-1$ unknown values $Y_i \approx y(x_i)$ arises:

$$\tilde{d}_1 Y_1 + \tilde{e}_1 Y_2 + \tilde{f}_1 Y_3 = \tilde{q}_1,$$

$$\tilde{c}_2 Y_1 + \tilde{d}_2 Y_2 + \tilde{e}_2 Y_3 + f_2 Y_4 = \tilde{q}_2,$$

$$b_i Y_{i-2} + c_i Y_{i-1} + d_i Y_i + e_i Y_{i+1} + f_i Y_{i+2} = 2h^4 q_i$$
$$\text{for } i = 3, \ldots, N-3,$$

$$b_{N-2} Y_{N-4} + \tilde{c}_{N-2} Y_{N-3} + \tilde{d}_{N-2} Y_{n-2} + \tilde{e}_{N-2} Y_{N-1} = \tilde{q}_{N-2},$$

$$\tilde{b}_{N-1} Y_{N-3} + \tilde{c}_{N-1} Y_{N-2} + \tilde{d}_{N-1} Y_{N-1} = \tilde{q}_{N-1},$$

where

$$
\left.
\begin{aligned}
b_i &= 2 - hp_{3i}, \quad f_i = 2 + hp_{3i}, \\
c_i &= -8 + 2hp_{3i} + 2h^2 p_{2i} - h^3 p_{1i}, \\
d_i &= 12 - 4h^2 p_{2i} + 2h^4 p_{0i}, \\
e_i &= -8 - 2hp_{3i} + 2h^2 p_{2i} + h^3 p_{1i},
\end{aligned}
\right\} \quad \text{for } i = 1, \dots, N-1;
$$

$$
\begin{aligned}
\tilde{d}_1 &= d_1 + b_1 A_{-1} + c_1 A_0, \\
\tilde{e}_1 &= e_1 + b_1 B_{-1} + c_1 B_0, \\
\tilde{f}_1 &= f_1 + b_1 C_{-1} + c_1 C_0, \\
\tilde{c}_2 &= c_2 + b_2 A_0, \quad \tilde{d}_2 = d_2 + b_2 B_0, \\
\tilde{e}_2 &= e_2 + b_2 C_0, \\
\tilde{c}_{N-2} &= c_{N-2} + f_{N-2} C_N, \\
\tilde{d}_{N-2} &= d_{N-2} + f_{N-2} B_N, \\
\tilde{e}_{N-2} &= e_{N-2} + f_{N-2} A_N, \\
\tilde{b}_{N-1} &= b_{N-1} + f_{N-1} C_{N+1} + e_{N-1} C_N, \\
\tilde{c}_{N-1} &= c_{N-1} + f_{N-1} B_{N+1} + e_{N-1} B_n, \\
\tilde{d}_{N-1} &= d_{n-1} + f_{N-1} A_{N+1} + e_{N-1} A_N;
\end{aligned}
$$

$$
\begin{aligned}
p_{ji} &= p_j(x_i), \quad j = 1, 2, 3; \quad q_i = q(x_i), \quad i = 1, \dots, N-1; \\
\tilde{q}_1 &= 2h^4 q_1 - b_1 D_{-1} - c_1 D_0, \\
\tilde{q}_2 &= 2h^4 q_2 - b_2 D_0, \\
\tilde{q}_{N-1} &= 2h^4 q_{N-1} - f_{N-1} D_{N+1} - e_{N-1} D_n, \\
\tilde{q}_{N-2} &= 2h^4 q_{N-2} - f_{N-2} D_N.
\end{aligned}
$$

The matrix A of the linear system is the following five-diagonal matrix:

$$
A = \begin{pmatrix}
\tilde{d}_1 & \tilde{e}_1 & \tilde{f}_1 & 0 & 0 & 0 & 0 & \cdot & \cdot & \cdot & \cdot & \cdot & \cdot & 0 \\
\tilde{c}_2 & \tilde{d}_2 & \tilde{e}_2 & f_2 & 0 & 0 & 0 & \cdot & \cdot & \cdot & \cdot & \cdot & \cdot & 0 \\
b_3 & c_3 & d_3 & e_3 & f_3 & 0 & 0 & \cdot & \cdot & \cdot & \cdot & \cdot & \cdot & 0 \\
0 & b_4 & c_4 & d_4 & e_4 & f_4 & 0 & \cdot & \cdot & \cdot & \cdot & \cdot & \cdot & 0 \\
 & & & \cdot & \cdot & \cdot & & & & & & & & \\
 & & & & \cdot & \cdot & \cdot & & & & & & & \\
 & & & & & \cdot & \cdot & \cdot & & & & & & \\
0 & \cdot & \cdot & \cdot & \cdot & \cdot & 0 & b_{N-4} & c_{N-4} & d_{N-4} & e_{N-4} & f_{N-4} & & 0 \\
0 & \cdot & \cdot & \cdot & \cdot & \cdot & 0 & 0 & b_{N-3} & c_{N-3} & d_{N-3} & e_{N-3} & f_{N-3} \\
0 & \cdot & \cdot & \cdot & \cdot & \cdot & 0 & 0 & 0 & b_{N-2} & \tilde{c}_{N-2} & \tilde{d}_{N-2} & \tilde{e}_{N-2} \\
0 & \cdot & \cdot & \cdot & \cdot & \cdot & 0 & 0 & 0 & 0 & \tilde{b}_{N-1} & \tilde{c}_{N-1} & \tilde{d}_{N-1}
\end{pmatrix}
$$

One can solve the system

$$
Ay = g = \begin{pmatrix} a_1 \\ \vdots \\ a_{N-1} \end{pmatrix} \quad \text{with } a_i = 2h^4 q_i \quad \text{for} \quad i = 3, \dots, N-3, \text{ and}
$$

$$
a_1 = \tilde{q}_1, \quad a_2 = \tilde{q}_2, \quad a_{N-2} = \tilde{q}_{N-2}, \quad a_{N-1} = \tilde{q}_{N-1}
$$

with the methods of section 4.10 or 4.12.

With

$$
\begin{aligned}
\alpha_1 &= \tilde{d}_1, \alpha_2 = \tilde{d}_2 - \delta_2 \beta_1, \alpha_{N-2} = \tilde{d}_{N-2} - b_{N-2}\gamma_{N-4} - \delta_{N-2}\beta_{N-3}, \\
\alpha_{N-1} &= \tilde{d}_{N-1} - \tilde{b}_{N-1}\gamma_{N-3} - \delta_{N-1}\beta_{N-2}, \\
\beta_1 &= \tilde{e}_1/\alpha_1, \quad \beta_2 = (\tilde{e}_2 - \delta_2\gamma_1)/\alpha_2, \\
\beta_{N-2} &= (\tilde{e}_{N-2} - \delta_{N-2}\gamma_{N-3})/\alpha_{N-2}, \quad \beta_{N-1} = 0, \\
\gamma_1 &= \tilde{f}_1/\alpha_1, \quad \gamma_2 = f_2/\alpha_2, \quad \gamma_{N-2} = \gamma_{N-1} = 0, \\
\delta_2 &= \tilde{c}_2, \delta_{N-2} = \tilde{c}_{N-2} - b_{N-2}\beta_{N-4}, \quad \delta_{N-1} = \tilde{c}_{N-1} - \tilde{b}_{N-1}\beta_{N-3}, \\
\alpha_i &= d_i - b_i\gamma_{i-2} - \delta_i\beta_{i-1}, \quad \beta_i = (e_i - \delta_i\gamma_{i-1})/\alpha_i, \quad i = 3, \dots, N-3, \\
\gamma_i &= f_i/\alpha_i, \quad \delta_i = c_i - b_i\beta_{i-2}, \quad i = 3, 4, \dots, N-3, \\
g_0 &= 0, \quad g_1 = a_1/\alpha_1, \\
g_{N-2} &= (a_{N-2} - b_{N-2}g_{N-4} - \delta_{N-2}g_{N-3})/\alpha_{N-2}, \\
g_{N-1} &= (a_{N-1} - \tilde{b}_{N-1}g_{N-3} - \delta_{N-1}g_{N-2})/\alpha_{N-1}, \\
g_i &= (a_i - b_ig_{i-2} - \delta_ig_{i-1})/\alpha_i, \quad i = 2, \dots, N-3,
\end{aligned}
$$

one can write the solution as:

$$
\begin{aligned}
Y_{N-1} &= g_{N-1}, \\
Y_{N-2} &= g_{N-2} - \beta_{N-2} Y_{N-1}, \quad \text{and} \\
Y_i &= g_i - \beta_i Y_{i+1} - \gamma_i Y_{i+2}, \quad \text{for } i = N-3, N-4, \ldots, 2, 1.
\end{aligned}
$$

The error order is $O(h^2)$, or $y(x_i) = Y_i + O(h^2)$. The error estimate is given by (18.24).

The procedural error of the approximate values Y_i obtained via this ordinary difference method depends on the step size h. Decreasing h will, however, lead to a larger sized linear system for the Y_i and hence to an increase of rounding errors affecting the solution.

18.3.2 Higher Order Difference Methods

If, instead of the first order approximations for the derivatives $y_i^{(k)}$ in section 18.3.1, one uses higher order approximations for the derivatives, one can obtain more accurate approximate values Y_i without having to reduce h, i.e., without increasing the size of the system matrix for the linear system of equations. When using *higher order difference methods*, each equation in the linear system associated with the boundary value problem will, however, contain more unknown values Y_i than was the case with the ordinary difference method.

Finite expressions for the derivatives of error order $O(h^4)$ and $O(h^6)$ are given in the following table. For the k^{th} derivative we have:

$$
y_i^{(k)} := y^{(k)}(x_i) = \quad \text{"finite expression"} \quad + O(h^q) \quad \text{with } q = 4 \quad \text{or} \quad 6
$$

and $|Y_i - y(x_i)| = O(h^q)$.

$y_i^{(k)}$	Finite expression for the derivative	error order
y_i'	$\dfrac{1}{12h}(Y_{i-2} - 8Y_{i-1} + 8Y_{i+1} - Y_{i+2})$	$O(h^4)$
y_i''	$\dfrac{1}{12h^2}(-Y_{i-2} + 16Y_{i-1} - 30Y_i + 16Y_{i+1} - Y_{i+2})$	$O(h^4)$
y_i'''	$\dfrac{1}{8h^3}(Y_{i-3} - 8Y_{i-2} + 13Y_{i-1} - 13Y_{i+1} + 8Y_{i+2} - Y_{i+3})$	$O(h^4)$
$y_i^{(4)}$	$\dfrac{1}{6h^4}(-Y_{i-3} + 12Y_{i-2} - 39Y_{i-1} + 56Y_i - 39Y_{i+1} + \\ +12Y_{i+2} - Y_{i+3})$	$O(h^4)$
y_i'	$\dfrac{1}{60h}(-Y_{i-3} + 9Y_{i-2} - 45Y_{i-1} + 45Y_{i+1} - 9Y_{i+2} + Y_{i+3})$	$O(h^6)$
y_i''	$\dfrac{1}{180h^2}(2Y_{i-3} - 27Y_{i-2} + 270Y_{i-1} - 490Y_i + 270Y_{i+1} + \\ -27Y_{i+2} + 2Y_{i+3})$	$O(h^6)$

If one uses finite expressions of error order $O(h^4)$ for discretized boundary value problems of second order, one will obtain a linear system with a five-diagonal system matrix. When applying finite expressions of error order $O(h^4)$ to a linear boundary value problem of fourth order, one gets a linear system with a seven-diagonal system matrix. To obtain the system matrix \boldsymbol{A}, the differential equation - discretized with the $O(h^4)$ or $O(h^6)$ finite expressions - is formed for x_i and $i = 3, \ldots, N-3$ for second order boundary value problems and an $O(h^4)$ discretization. One uses the x_i for $i = 4, \ldots, N-4$ for fourth order boundary problems with $O(h^4)$ or for second order boundary with $O(h^6)$ discretizations. When discretizing the boundary conditions, so called "fictitious values" Y_i with $i < 0$ or $i > N$ will appear which lie outside of the integration interval $[a, b]$. "Additional boundary conditions" can serve to eliminate them. They are obtained by using the discretized differential equation with an $O(h^q)$ discretization for $q < 4$ or $q < 6$ for the points x_i with

$$
\begin{aligned}
i &= 0, 1, 2, N-2, N-1, N \quad \text{or} \\
i &= 0, 1, 2, 3, N-3, N-2, N-1, N.
\end{aligned}
$$

Here q has to be chosen so that, after eliminating the fictitious values Y_i, precisely $N+1$ linear equations remain for the $N+1$ approximate values Y_i, $i = 0, \ldots, N$, see section 18.3.1. The error order of the resulting global method needs to be examined in more detail. For general statements on this

see [ESSE77] or [KREI72] for example. The error order will depend on the differentiability of the solution as well as on the discretization order chosen. For example the $O(h^4)$ discretization of a second order boundary value problem with $\alpha_2 = \beta_2 = 0$ has the global error order $O(h^4)$, see [ESSE77], [KREI72].

If one applies a difference method to a nonlinear boundary value problem such as (18.1), (18.2), then one usually will obtain a system of nonlinear equations for the approximate values Y_i.

For linear differential equations of various special types one can obtain an error order of $O(h^4)$ or $O(h^6)$ already with a smaller number of function values Y_i occurring in the individual equations of the linear system: One simply expresses linear sums of derivatives at neighboring nodes x_i as linear sums of functional values (*Hermite method* or multinode method, see [COLL66], III,§2.4).

18.3.3 Iterative Solution of Linear Systems for Special Boundary Value Problems

If, when solving boundary value problems with difference methods of high order, large linear systems with a large band width ($m \geq 3$) occur, we recommend to solve them iteratively with a relaxation method, see section 5.5. This can happen with ordinary boundary value problems for large intervals of integration. But most often this occurs for boundary value problems for partial differential equations which are not treated in this book, see [STOE91], 8 for example and the references there, as well as the references listed at the end of our bibliography. In some cases, it is not necessary to check the convergence conditions for the coefficients of the linear system, because one can often deduce from the given boundary value problem whether an iterative solution is possible or not. One such example is the inhomogeneous boundary value problem for a linear second order differential equation in self-adjoint form (see e.g. [COLL66], p. 208; [COLL68], §7.3; [ZURM65], p. 476):

$$-(fy')' + gy = r, \quad y(a) = A, \quad y(b) = B.$$

If $f(x) > 0$ and $g(x) \geq 0$ holds for all $x \in [a, b]$, then the ordinary difference method can be used with both the Jacobi and Gauß-Seidel iterations, see sections 5.3 and 5.4 ([COLL66], p. 173 ff or [COLL68] §23.2). Moreover, for the associated discretization matrix A: det $A \neq 0$.

18.3.4 Linear Eigenvalue Problems

The homogeneous linear boundary value problem

$$y'' + p_1(x)y' + (p_0(x) - \lambda)y = 0,$$
$$y(a) = 0, \qquad y(b) = 0$$

is to be solved numerically with a parameter λ by using the ordinary difference method. This results in a homogeneous linear system for Y_i, $i = 1, \ldots, N-1$ with $Y_0 = Y_N = 0$:

$$(1 - \tfrac{h}{2}p_{1i})Y_{i-1} + (-2 + h^2 p_{0i})Y_i + (1 + \tfrac{h}{2}p_{1i})Y_{i+1} = h^2 \lambda Y_i,$$
(18.25)
$$i = 1, \ldots, N-1.$$

The system matrix \boldsymbol{A} of (18.25) consists of the rows $2, \ldots, N-1$ of the matrix \boldsymbol{A} of algorithm 18.2. Thus we can rewrite (18.25) in the form

$$\boldsymbol{Ay} = \lambda \boldsymbol{y} \quad \text{with} \quad \boldsymbol{y} = \begin{pmatrix} Y_1 \\ \vdots \\ Y_{N-1} \end{pmatrix}.$$

And consequently we have to solve a matrix eigenvalue problem. \boldsymbol{A} is a tridiagonal matrix, which is symmetric only for $p_1(x) \equiv 0$, $x \in [a, b]$. In this case we can use the QD method of section 7.6 if h is sufficiently small, for then the matrix \boldsymbol{A} will be positive definite as well. Otherwise, we must use one of the other eigenvalue algorithms of chapter 7 in order to determine its eigenvalues λ and their eigenvectors \boldsymbol{y}.

Bibliography for Section 18.3

[ATKI89], 6.11; [COLL66], III, §§1-3; [CONT80], 7.1-2; [ESSE80]; [GOOS79]; [HENR68], part III; [ISAA66], 8.7.2-3; [KELL68], 3, 5.3; [KREI72]; [LUTH87], 15.2; [STIE63], 6.52; [STOE91], 7.4, 8; [TÖRN79] vol.2, 15; [WERN86], 6; [ZURM65], §30.

A Appendix

Standard FORTRAN 77 Subroutines

Jürgen Dietel, Guido Dubois, Gisela Engeln-Müllges, Thomas Eul, Michael Grönheim, Eberhard Heyne, Ina Hinze, Volker Krüger, Klaus Niederdrenk, Günter Palm, Elmar Pohl, Richard Reuter, Hermann-Josef Rheinbach, Hartmut Turowski and others.

A.1 Preface of the Appendix

Almost all of the algorithms described in the text part of this book have been
realized in the programming language

- **Standard FORTRAN 77**
 (by J. Dietel, G. Dubois, G. Engeln–Müllges, Th. Eul, M. Grönheim,
 E. Heyne, I. Hinze, V. Krüger, K. Niederdrenk, G. Palm, E. Pohl, R.
 Reuter, H.–J. Rheinbach, H. Turowski and others).

The source codes for the associated programs as well as those for several
further algorithms that are not described in the text portion are contained
(with references) on the enclosed CD with test programs and test examples
and data for all routines. Also enclosed on the CD are a compiler and some
utilities.

Additionaly we have translated the FORTRAN 77 programs into column inde-
pendent notation, so that they can be readily inserted into Fortran 90 subrou-
tines. These quasi Fortran 90 codes including test programs and test examples
are also included on the CD.

This appendix is intended to give a printed account of the actual contents of
the CD–ROM and of the enclosed source codes.

Section A.2 contains information about campus and site licenses for the Nu-
merical Library in FORTRAN 77 and ANSI C. Section A.3 contains the table
of contents for the CD.

Section A.4 gives information on the FORTRAN subroutines using the chapter
numbering system of the text part with a leading F and the identical headings.
We include all declarations and parameters and a short desciption for each of
the functions.

The majority of the algorithms in this book were originally written as FOR-
TRAN 77 codes, which were later translated into other programming lan-
guages. These translations were developed with special care for the individual
properties and qualities of the different computer languages. All our decla-
rations are clearly annotated to help the users adapt our programs for their
individual needs. Finally we have used the ANSI standard for FORTRAN in

order to achieve maximal compatability of our codes. In coding the algorithms
we have tried to follow the development of the algorithms as detailed in the
text portion of the book for the ensuing readability of the codes, and thus we
have often deliberatedly avoided more slick programming techniques.

For assistance with testing in FORTRAN, site licensing and organizing the
distribution of our software, we are indebted to Guido Dubois. The quasi–
Fortran 90 codes were developed by Andreas Sohn, while Jürgen Dietel pre-
pared the codes and software for the enclosed CD.

Despite all our joint efforts, we can not exclude the possibility of errors. There-
fore we invite and welcome all our users to get in contact with us if such should
be detected.

June 1996 Gisela Engeln–Müllges
 Frank Uhlig

A.2 Information on Campus and Site Licenses, as well as on Other Software Packages

With the purchase of the book **Numerical Algorithms**, the buyer has obtained a single user license of the Numerical Library in FORTRAN, encoded on the enclosed CD.

Upon request we can furnish a price list for campus and site licenses. The purchase of such a license entails free access of all faculty and students or members of the respective organization to all source codes of the Numerical Library. The campus and site licenses are priced depending on the number of potential users. Using Numerical Library algorithms based on a site license insures all users have access to the most up to date codes, as we maintain our programs continually and will supply the newest versions of codes to subscribers of site or campus licences in between new editions of the book 'Numerical Algorithms'.

At the moment over 50 campus licenses (for the German version of the codes, of course) have been issued to various German universities and Fachhochschulen. All members of these academic communities have access to the Numerical Library at their respective schools without any further fees, even for home use.

In addition to the software encoded on the CD, we can supply floppy disks with the source codes of our programs in ANSI C with English annotations, and for Turbo Pascal and QuickBasic with German annotations only.

Moreover we have floppy disks available with programs for:

> 2D graphics
> 3D graphics
> Symbolic math interpreter
> Curve tracing
> Numerical expert system

These last items, however, are only available with annotations in German at the moment.[1]

[1] If there is sufficient interest, we shall translate the German code annotations into English as well; we refer interested users to our WWW page for announcements on progress with the Numerical Library.

To illustrate, we give one example of a practical application for our numerical expert system using curve tracing:

Contour maps are useful in geography to descibe the lay of the land graphically. Such contour maps can likewise be used to map scalar quantities, such as temperature or the concentration of a chemical over a multidimensional region. Here each contour describes the curve with constant level, be it height above the ocean level, temperature, or concentration.

Starting with a set of measured quantities z_1, z_2, \ldots, z_n of a scalar at arbitrary measurement locations $(x_1, y_1), (x_2, y_2), \ldots, (x_n, y_n) \in \mathbf{R}^2$, we can trace level curves for z following the following steps with the expert system:

1. The numeric expert system suggests a method to construct a smooth surface or surface spline for the n data points (x_i, y_i, z_i) for $i = 1, 2, \ldots, n$. Depending on the method used, the resulting surface will contain the measured points precisely or will approximate them smoothly.

2. Using the curve tracing program, we can now plot the level curves with constant z values on the surface, giving a 3–dimensional picture of the situation.

Using the curve tracer, one can also determine the curves of intersection for highly complex surfaces in diverse representations. One necessary attribute for any curve tracer is to use a fine enough discretization to allow it to follow all branches of a level curve well. This then allows a curve tracer to find all points of intersection of two curves for example.

Other possible applications are: create the boundary curves to synthesize certain variable transmissions, compute the curve of intersection of two surfaces in \mathbf{R}^3, determine the shadow boundary of parametrized surfaces, find contour lines for surfaces given by a table of values, etc.

The numerical expert system helps with splines for interpolation and approximation of curves and surfaces and with several other areas, such as: linear and nonlinear equations, interpolation, approximation, numerical quadrature and cubature, systems of DEs, fast Fourier transform and eigenvalue problems.

It also contains tools to transfer and analyze data, to easily link the programs needed, as well as for printing planar and three dimensional graphics.

Information

on campus and site licenses for the Numerical Library in the programming languages FORTRAN 77/90, Turbo Pascal, ANSI C, QuickBASIC, as well as on further software packages can be obtained by FAX or e–mail:

Prof. Dr. Gisela Engeln–Müllges
Kesselstraße 88
D–52076 Aachen–Lichtenbusch

FAX: +49 2408 7812
e–mail: engeln–muellges@fh–aachen.de

or

Frank Uhlig
706 Cary Dr.
Auburn, AL 36830
USA

FAX: (334) 844 6555
e–mail: uhligfd@mail.auburn.edu

Further information together with current updates are to be found on the WWW at:

http://www.fh–aachen.de/home/engeln–muellges/num–alg.html
or
http://www.auburn.edu/num–alg.html

A.3 Contents of the Enclosed CD

The following items are contained on the CD:

- Standard FORTRAN 77 subroutines with test programs and test examples
- Quasi Fortran 90 subroutines with test programs and test examples
- an HTML version of the source codes combined with a table of contents and a list of key words
- GNU FORTRAN (newest version) for OS/2 with GNU C as a basis for the FORTRAN compiler
- DMAKE 3.80 (source codes and executable files for MS–DOS and OS/2)

Before actually using the CD, we recommend to read the `readme.txt` file in the root file of the CD. Under MS–DOS the essential parts of the directory tree on the CD might look like this:

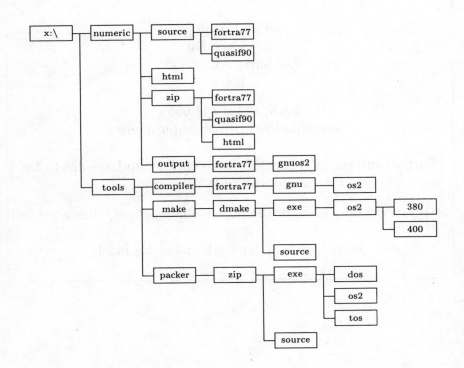

Contents of the Appendix

A.4 FORTRAN 77 Subroutines

All of our FORTRAN 77 subroutines are using the ANSI standard. They have been developed and tested on IBM PCs with math co–processors under MS–DOS and with the Microsoft FORTRAN 77 compiler 5.0. Further testing was done under OS/2 and LINUX, and under UNIX on workstations such as HP9000/700 and IBM RS/6000.

The FORTRAN 77 subroutines on the enclosed CD all use double precision arithmetic for real numbers. This was done to counteract unnecessary bad results due to rounding errors on some computers, which use only 24 bit mantissas for single precision numbers, and correspondingly carry only about 7 digits.

The machine constant is computed in all programs that rely on it by using the following lines of code:

```
      FMACHN = 1.
   10 FMACHN = 0.5 * FMACHN
      IF (MASCHD (1. + FMACHN) .EQ. 1) GOTO 10
      FMACHN = 2. * FMACHN
```

This code uses the function:

```
      INTEGER FUNCTION MASCHD (X)
      DOUBLE PRECISION X
      MASCHD = 0
      IF (1.D0 .LT. X) MASCHD = 1
      RETURN
      END
```

We found this rather awkward way to compute the machine constant necessary, in order to avoid the compiler dependent optimization of math co–processors on most PCs.

The following standard FORTRAN 77 subroutines are contained on the CD:

F 2 Nonlinear Equations in One Variable

F 2.5.1 The Newton Method for Simple Roots

```
        SUBROUTINE NEWPSZ(FCT,FDER,X0,MAXIT,ABSERR,RELERR,X,F,
      +                   NUMIT,IERR)
C
C********************************************************************
C                                                                  *
C   This SUBROUTINE determines an approximate value X for a zero   *
C   of the function FCT with known derivative FDER by applying     *
C   Newton's method for simple zeros.                              *
C   If FCT is an algebraic polynomial the                          *
C   SUBROUTINE NEWPOL should be applied instead.                   *
C   For multiple zeros the modified Newton method                  *
C   (SUBROUTINE NEWMOD) is recommended.                            *
C********************************************************************
```

F The Newton Method for Polynomials

```
        SUBROUTINE NEWPOL(A,N,X0,MAXIT,ABSERR,RELERR,X,PN,NUMIT,
      +                   IERR,WORK)
C
C********************************************************************
C                                                                  *
C   This SUBROUTINE determines an approximation X for a zero of    *
C   the algebraic polynomial                                       *
C       PN(X)=A(0)+A(1)*X+A(2)*X**2+...+A(N)*X**N                   *
C   with real coefficients A(I) using Newton's method for simple   *
C   roots.                                                         *
C********************************************************************

        SUBROUTINE NEWPOH(A,N,X,PN,PNDER,WORK)
C
C********************************************************************
C                                                                  *
C   This SUBROUTINE uses the Horner scheme in order to calculate   *
C   the functional value PN(X) and the 1st derivative PNDER(X) of  *
C   an Nth degree algebraic polynomial with real coefficients      *
C       PN(X)=A(0)+A(1)*X+A(2)*X**2+...+A(N)*X**N                   *
C   at X.                                                          *
C********************************************************************
```

F 2.5.3 Newton's Method for Multiple Zeros; a Modified Newton's Method

```
        SUBROUTINE NEWMOD(FCT,FDER,F2DER,X0,MAXIT,LUN,ABSERR,
      +                   RELERR,X,F,J,NUMIT,IERR)
C
C********************************************************************
C                                                                  *
C   This SUBROUTINE determines an approximation X for a J-fold     *
C   zero of the function FCT by applying the modified Newton       *
C   method for the starting value X0.                              *
C********************************************************************
```

F 2.8.2 Pegasus Method

```
      SUBROUTINE PEGASU ( FCT,A,B,ABSERR,RELERR,MAXIT,XSI,X1,X2,
     +                    NUMIT,IERR)
C
C*******************************************************************
C                                                                 *
C We assume that the given continuous function FCT(X) satisfies:*
C        FCT(A) * FCT(B) < 0.0  on the interval [A,B].           *
C Thus there is at least one zero of odd order in [A,B].         *
C The SUBROUTINE PEGASU determines one of these zeros by         *
C applying the Pegasus-method.                                   *
C The method always converges provided the following condition   *
C is met:                                                        *
C        FCT(A) * FCT(B) < 0.0                                   *
C*******************************************************************
```

F 2.8.4 The King and the Anderson-Björck-King Methods, the Illinois Method

```
      SUBROUTINE ZERORF(FCT,ABSERR,RELERR,FAB,MAXIT,IMETH,IEXTRA,
     +                  DELX,X1,X2,X3,F1,F2,F3,NUMIT,IHELP1,
     +                  IHELP2,INCL,IERR)
C
C*******************************************************************
C                                                                 *
C The SUBROUTINE ZERORF is the governing program for several     *
C subroutines that determine zeros of a continuous real-valued   *
C function FCT. The methods used are primarily intended for      *
C simple zeros and zeros of odd order.                          *
C This SUBROUTINE adapts its approach depending on whether      *
C the starting values enclose a zero of the function FCT or not.*
C If a zero is enclosed, an iteration sequence is constructed    *
C using one of several iteration methods: Pegasus,             *
C Anderson/Bjoerck, or King. These methods ensure that inclusion*
C is preserved. All methods used work without derivatives and    *
C each possesses a convergence order of P > 1.6.               *
C If no two starting estimates are known that enclose a         *
C functional zero, then two identical starting values X1, X2     *
C may be specified.                                             *
C*******************************************************************

      SUBROUTINE PEG(IHELP1,F1,F2,F3)
C
C*******************************************************************
C                                                                 *
C This SUBROUTINE uses the Pegasus-method to calculate a new     *
C functional value F1 for the governing program ZERORF.         *
C*******************************************************************

      SUBROUTINE PEGK(IHELP1,IHELP2,F1,F2,F3)
C
C*******************************************************************
C                                                                 *
C This SUBROUTINE uses an improved version of the Pegasus        *
C method by King for calculating a new functional value F1      *
C for use in the governing program ZERORF.                     *
C*******************************************************************

      SUBROUTINE ANDBJ(IHELP1,F1,F2,F3)
C
C*******************************************************************
C                                                                 *
C This SUBROUTINE uses the Anderson / Bjoerck method to calcu-  *
C late a new functional value F1 for the governing program      *
```

```
C ZERORF.                                                         *
C*********************************************************************

      SUBROUTINE ANDBJK(IHELP1,IHELP2,F1,F2,F3)
C
C*********************************************************************
C                                                                 *
C  This SUBROUTINE uses a combination of the Anderson/Bjoerck     *
C  and King methods suggested by King to calculate a new func-    *
C  tional value F1 for use in the governing program ZERORF.       *
C*********************************************************************

      SUBROUTINE EXCHG(X,Y,FX,FY)
C
C*********************************************************************
C                                                                 *
C  This SUBROUTINE exchanges X and Y, and FX and FY.              *
C                                                                 *
C*********************************************************************

      SUBROUTINE SUB1(F1,F2,F3)

      SUBROUTINE SUB2(F1,F2,F3)
```

F 2.8.5 Zeroin Method

```
      SUBROUTINE ZEROIN(FCT,ABSERR,RELERR,MAXFCT,ITAPE,A,B,FB,
     +                  NUMFCT,IERR)
C
C*********************************************************************
C                                                                 *
C  The SUBROUTINE ZEROIN finds a zero of odd order of a           *
C  continuous function FCT in the interval [A,B] provided that    *
C  FCT(A) and FCT(B) have different signs, i.e.,                  *
C  FCT(A)*FCT(B) < 0.                                             *
C  The Zeroin method combines the bisection and the secant method*
C  with inverse quadratic interpolation.                         *
C*********************************************************************

      SUBROUTINE BI(XM,ITAPE,D,E)
C
C*********************************************************************
C                                                                 *
C  The SUBROUTINE BI is a bisection algorithm for the SUBROUTINE  *
C  ZEROIN.                                                        *
C                                                                 *
C*********************************************************************
```

F 3 Roots of Polynomials

F 3.3.2 Muller's Method

```
      SUBROUTINE MULLRP(NPOL,POLYNM,ITERMX, NZ,ZERO,P,PHELPD)
C
C*********************************************************************
C                                                                 *
C  This SUBROUTINE computes all zeros of a polynomial by using    *
C  the method of Muller.                                          *
```

```
C*****************************************************************

      SUBROUTINE MULLER(N,P,NFND,XR,XI,PHELPD)

      SUBROUTINE HORNC(A,NA,XR,XI,FR,FI,B)

      SUBROUTINE HORNCE(XEPS,XEFR,XEFI,XEB1,XEBO,A,NA,XR,XI,FR,FI,B)

      SUBROUTINE POLDIV(A,NA,B,NB,C,NC,R,NR)

      DOUBLE PRECISION FUNCTION YEPS()

      SUBROUTINE COMPAR(M,A,B)
```

F 3.3.3 Bauhuber's Method

```
      SUBROUTINE BAUPOL(COEFRE,COEFIM,N,COMPL,ROOTRE,ROOTIM,NUMIT)
C
C*****************************************************************
C                                                               *
C  Without knowing any approximations of the roots, this        *
C  SUBROUTINE finds N approximate values Z(I), I=1, ..., N for   *
C  the N zeros of a polynomial PN of degree N with real or       *
C  complex coefficients.                                         *
C*****************************************************************

      SUBROUTINE BAUZRO(XO,YO,N,GAMMA,XNEW,YNEW,NUMIT)
C
C*****************************************************************
C                                                               *
C  This SUBROUTINE calculates a zero of a polynomial PN with     *
C  complex coefficients.                                         *
C*****************************************************************

      SUBROUTINE COMHOR(X,Y,N,GAMMA,U,V,UD,VD,UDD,VDD,BDP,BDPD)
C
C*****************************************************************
C                                                               *
C  This SUBROUTINE calculates the complex functional value       *
C  PN(X+I*Y)=(U+I*V), the complex valued derivatives             *
C  PN'(X+I*Y)=(UD+I*VD) and PN''(X+I*Y)=(UDD+I*VDD) of a polymial*
C  PN(Z) of degree N (N>0) with complex coefficients by using     *
C  the Horner-scheme.                                            *
C  Additionally bounds BDP and BDPD are computed for the         *
C  rounding errors in computing DABS(PN(Z)) and DABS(PN'(Z)).     *
C  The complex coefficients of PN are stored in a 2-dimensional   *
C  array A(1:2*(N+1)), arranged in descending order of the       *
C  powers (they will remain unchanged by this subroutine).        *
C  The complex coefficients of the polynomial Q(Z) of degree N-1  *
C  are stored in the array B(1:2*(N+1)), arranged in descending   *
C  order. Here Q(Z) is defined by PN(Z)=Q(Z)*(Z-ZO)+PN(ZO).       *
C  The array C(1:2*N) is used as an auxiliary array.              *
C*****************************************************************

      DOUBLE PRECISION FUNCTION ABSCOM(X,Y)
C
C*****************************************************************
C                                                               *
C  This FUNCTION-subroutine calculates the absolute value of a   *
C  complex number (X+I*Y).                                        *
C*****************************************************************

      SUBROUTINE MCONST(FMACHP,INFINY,SMALNO,BASE)
C
C*****************************************************************
C                                                               *
```

```
C   This subroutine sets up some constants that are machine      *
C   dependent.                                                   *
C*******************************************************************

        DOUBLE PRECISION FUNCTION SCALFC(NN,PT,FMACHP,INFINY,SMALNO,
       +                            BASE)
C
C*******************************************************************
C                                                                 *
C   This FUNCTION-subroutine calculates a scaling factor which    *
C   is used to scale the polynomial coefficients.                 *
C*******************************************************************

        SUBROUTINE CDIV(A,B,C,D,X,Y)
C
C*******************************************************************
C                                                                 *
C   This SUBROUTINE performs a complex division                   *
C        (X+I*Y) := (A+I*B)/(C+I*D).                              *
C*******************************************************************
```

F 3.3.5 The Laguerre Method

```
        SUBROUTINE LAGUER (A,N,ABSERR,RELERR,MAXIT,XI,NITER,INUM,
       +                   WORK,IERR)
C
C*******************************************************************
C                                                                 *
C   For a real polynomial PN of degree N, this program computes   *
C   all real roots using Laguerre's method, provided PN has only  *
C   real roots.                                                   *
C*******************************************************************

        SUBROUTINE HORN1 (A,N,XO,NACT)
C
C*******************************************************************
C                                                                 *
C   If P has degree NACT, this subroutine finds the remainder     *
C   polynomial PAB of degree NACT-1 that results from dividing off*
C   one known root of P. It is neeeded in the SUBROUTINE LAGUER.  *
C*******************************************************************

        SUBROUTINE HORN2 (A,N,X,NACT,P,P1,P2,WORK)
C
C*******************************************************************
C                                                                 *
C   This subroutine calculates the function value and that of the *
C   first and second derivative of a given polynom P of degree    *
C   NACT at X. It is needed in SUBROUTINE LAGUER.                 *
C*******************************************************************

        SUBROUTINE QUADRA (A,B,C,EPS,X1,X2,IERR)
C
C*******************************************************************
C                                                                 *
C   This subroutine solves a real quadratic equation              *
C        A*X**2 + B*X + C = 0  with A not equal to 0,            *
C   that has real roots. It serves the subroutine LAGUER.         *
C*******************************************************************
```

F 4 Direct Methods for Solving Systems of Linear Equations

F 4.5.1 Gauß Algorithm with Column Pivot Search

```
      SUBROUTINE GAUSS(N,A,LDA,Y,X,MARK,D,IPIVOT)
C
C**********************************************************************
C                                                                    *
C Solving a linear system of equations  A * X = Y  by applying       *
C the Gauss-elimination method with scaling and column pivot         *
C search.                                                            *
C**********************************************************************

      SUBROUTINE GAUSSP(N,A,LDA,IPIVOT,MARK,D)
C
C**********************************************************************
C                                                                    *
C Factoring the matrix A into the product of two matrices L and      *
C R so that  P * A = L * R, where P = permutation matrix,            *
C L = unit lower triangular matrix and R = upper triangular          *
C matrix by applying the Gauss-elimination method with               *
C scaling and column pivot search.                                   *
C**********************************************************************

      SUBROUTINE GAUSSS(N,A,LDA,IPIVOT,Y,X)
C
C**********************************************************************
C                                                                    *
C Calculating the solution X of a linear system of equations         *
C A * X = Y, where A has been factored via Gauss-elimination         *
C in SUBROUTINE GAUSSP.                                              *
C**********************************************************************
```

F 4.5.4 Gauß Algorithm for Systems with Several Right Hand Sides

```
      SUBROUTINE GAUSRS(N,A,LDA,M,RS,XL,MARK,D,IPIVOT)
C
C**********************************************************************
C                                                                    *
C Solving a linear systems of equations  A * XL = RS  for M          *
C right hand sides using the Gauss-elimination method with           *
C scaling and column pivot search .                                  *
C If the system has the form                                         *
C      A * XL = I   , where I = identity matrix and A, XL, I         *
C are all (NxN) matrices, then the solution XL is the matrix         *
C inverse of A.                                                      *
C**********************************************************************
```

F 4.7.1 The Cholesky Decomposition

```
      SUBROUTINE CHOKY (N,A,LDA,RS,X,Z,MARK)
C
C**********************************************************************
C                                                                    *
C      Solving a linear system of equations                          *
C                A * X = RS                                          *
C      for a symmetric, positive definite matrix A the               *
C      Cholesky-method.                                              *
C**********************************************************************

      SUBROUTINE CHOKYP (N,A,LDA,Z,MARK)
C
```

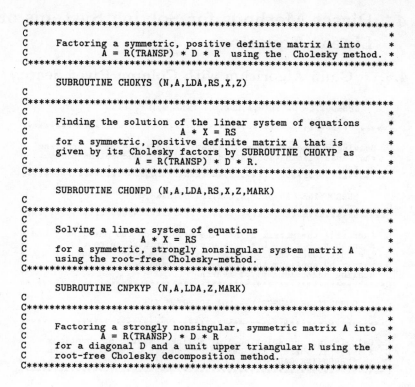

```
C**********************************************************************
C                                                                    *
C       Factoring a symmetric, positive definite matrix A into       *
C               A = R(TRANSP) * D * R  using the  Cholesky method.   *
C**********************************************************************

        SUBROUTINE CHOKYS (N,A,LDA,RS,X,Z)
C
C**********************************************************************
C                                                                    *
C       Finding the solution of the linear system of equations       *
C                       A * X = RS                                    *
C       for a symmetric, positive definite matrix A that is          *
C       given by its Cholesky factors by SUBROUTINE CHOKYP as        *
C                       A = R(TRANSP) * D * R.                        *
C**********************************************************************

        SUBROUTINE CHONPD (N,A,LDA,RS,X,Z,MARK)
C
C**********************************************************************
C                                                                    *
C       Solving a linear system of equations                         *
C                       A * X = RS                                    *
C       for a symmetric, strongly nonsingular system matrix A        *
C       using the root-free Cholesky-method.                         *
C**********************************************************************

        SUBROUTINE CNPKYP (N,A,LDA,Z,MARK)
C
C**********************************************************************
C                                                                    *
C       Factoring a strongly nonsingular, symmetric matrix A into    *
C               A = R(TRANSP) * D * R                                 *
C       for a diagonal D and a unit upper triangular R using the     *
C       root-free Cholesky decomposition method.                     *
C**********************************************************************
```

F 4.7.2 The Conjugate Gradient Method

```
        SUBROUTINE CG (A, N, IA, Y, X, IERR, D, G, AMULD)
C
C**********************************************************************
C                                                                    *
C This SUBROUTINE solves a linear system of equations AX = Y         *
C using the conjugate gradient method.                               *
C                                                                    *
C ASSUMPTION:                                                         *
C ===========                                                         *
C           A must be a symmetric and positive definite N by N       *
C           matrix                                                    *
C**********************************************************************
```

F 4.9 The Matrix Inverse via Exchange Steps

```
        SUBROUTINE PIVOT (A,LDA,N,B,S1,S2,IERR,MX,MY,VAL)
C
C**********************************************************************
C                                                                    *
C This subroutine calculates the inverse of a real square NxN        *
C matrix  A  by applying exchange steps, also called the method      *
C of pivotization.                                                   *
C**********************************************************************
```

F 4.10.1 Systems with Tridiagonal Matrices

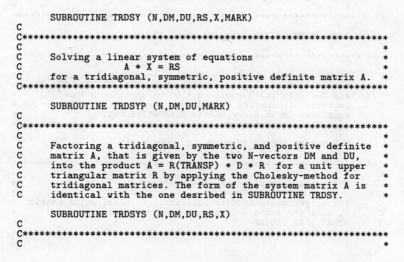

```
      SUBROUTINE TRDIG (N,DL,DM,DU,RS,X,MARK)
C
C*********************************************************************
C                                                                   *
C     This routine solves a linear system of equations              *
C                   A * X = RS                                       *
C     for a tridiagonal, strongly nonsingular matrix A.             *
C*********************************************************************

      SUBROUTINE TRDIGP (N,DL,DM,DU,MARK)
C
C*********************************************************************
C                                                                   *
C     Factor the tridiagonal, strongly nonsingular matrix A,        *
C     that is given by the vectors DL, DM and DU of co-diagonal     *
C     and diagonal entries into the product  L * R  for a bi-       *
C     diagonal lower triangular matrix  L  and a unit bidiagonal    *
C     upper triangular matrix R. The form of the system matrix A    *
C     is identical to that given in SUBROUTINE TRDIG.               *
C*********************************************************************

      SUBROUTINE TRDIGS (N,DL,DM,DU,RS,X)
C
C*********************************************************************
C                                                                   *
C     Solving a linear system of equations                          *
C                   A * X = RS                                       *
C     for a tridiagonal, strongly nonsingular matrix A, once the    *
C     tringular factorization has been calculated by SUBROUTINE     *
C     TRDIGP. The factors are used as known input matrices and      *
C     are stored in the three N-vectors DL, DM and DU.              *
C*********************************************************************
```

F 4.10.2 Systems with Tridiagonal Symmetric Strongly Nonsingular Matrices

```
      SUBROUTINE TRDSY (N,DM,DU,RS,X,MARK)
C
C*********************************************************************
C                                                                   *
C     Solving a linear system of equations                          *
C                   A * X = RS                                       *
C     for a tridiagonal, symmetric, positive definite matrix A.     *
C*********************************************************************

      SUBROUTINE TRDSYP (N,DM,DU,MARK)
C
C*********************************************************************
C                                                                   *
C     Factoring a tridiagonal, symmetric, and positive definite     *
C     matrix A, that is given by the two N-vectors DM and DU,       *
C     into the product A = R(TRANSP) * D * R  for a unit upper      *
C     triangular matrix R by applying the Cholesky-method for       *
C     tridiagonal matrices. The form of the system matrix A is      *
C     identical with the one desribed in SUBROUTINE TRDSY.          *
C*********************************************************************

      SUBROUTINE TRDSYS (N,DM,DU,RS,X)
C
C*********************************************************************
C                                                                   *
```

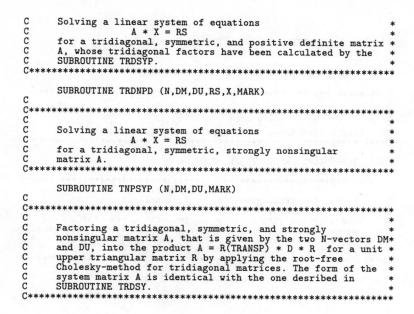

```
C     Solving a linear system of equations                      *
C             A * X = RS                                        *
C     for a tridiagonal, symmetric, and positive definite matrix *
C     A, whose tridiagonal factors have been calculated by the   *
C     SUBROUTINE TRDSYP.                                         *
C***************************************************************

      SUBROUTINE TRDNPD (N,DM,DU,RS,X,MARK)
C
C***************************************************************
C                                                              *
C     Solving a linear system of equations                     *
C             A * X = RS                                        *
C     for a tridiagonal, symmetric, strongly nonsingular        *
C     matrix A.                                                 *
C***************************************************************

      SUBROUTINE TNPSYP (N,DM,DU,MARK)
C
C***************************************************************
C                                                              *
C     Factoring a tridiagonal, symmetric, and strongly          *
C     nonsingular matrix A, that is given by the two N-vectors DM*
C     and DU, into the product A = R(TRANSP) * D * R  for a unit *
C     upper triangular matrix R by applying the root-free        *
C     Cholesky-method for tridiagonal matrices. The form of the  *
C     system matrix A is identical with the one desribed in      *
C     SUBROUTINE TRDSY.                                          *
C***************************************************************
```

F 4.11.1 Systems with a Symmetric Cyclically Tridiagonal Matrix

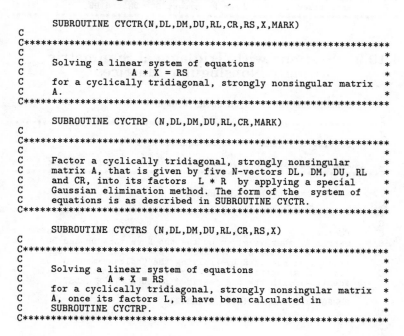

```
      SUBROUTINE CYCTR(N,DL,DM,DU,RL,CR,RS,X,MARK)
C
C***************************************************************
C                                                              *
C     Solving a linear system of equations                     *
C             A * X = RS                                        *
C     for a cyclically tridiagonal, strongly nonsingular matrix  *
C     A.                                                        *
C***************************************************************

      SUBROUTINE CYCTRP (N,DL,DM,DU,RL,CR,MARK)
C
C***************************************************************
C                                                              *
C     Factor a cyclically tridiagonal, strongly nonsingular     *
C     matrix A, that is given by five N-vectors DL, DM, DU, RL   *
C     and CR, into its factors  L * R  by applying a special     *
C     Gaussian elimination method. The form of the  system of    *
C     equations is as described in SUBROUTINE CYCTR.             *
C***************************************************************

      SUBROUTINE CYCTRS (N,DL,DM,DU,RL,CR,RS,X)
C
C***************************************************************
C                                                              *
C     Solving a linear system of equations                     *
C             A * X = RS                                        *
C     for a cyclically tridiagonal, strongly nonsingular matrix  *
C     A, once its factors L, R have been calculated in           *
C     SUBROUTINE CYCTRP.                                         *
C***************************************************************
```

F 4.11.2 Systems with Symmetric Cyclically Tridiagonal Strongly Nonsingular Matrices

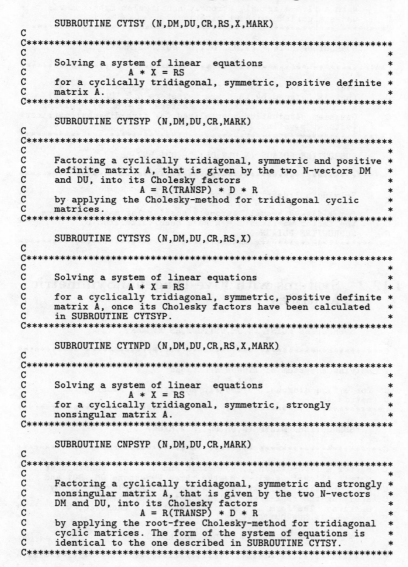

```
        SUBROUTINE CYTSY (N,DM,DU,CR,RS,X,MARK)
C
C*********************************************************************
C                                                                   *
C       Solving a system of linear  equations                       *
C                      A * X = RS                                    *
C       for a cyclically tridiagonal, symmetric, positive definite  *
C       matrix A.                                                    *
C*********************************************************************

        SUBROUTINE CYTSYP (N,DM,DU,CR,MARK)
C
C*********************************************************************
C                                                                   *
C       Factoring a cyclically tridiagonal, symmetric and positive  *
C       definite matrix A, that is given by the two N-vectors DM     *
C       and DU, into its Cholesky factors                           *
C                      A = R(TRANSP) * D * R                         *
C       by applying the Cholesky-method for tridiagonal cyclic      *
C       matrices.                                                   *
C*********************************************************************

        SUBROUTINE CYTSYS (N,DM,DU,CR,RS,X)
C
C*********************************************************************
C                                                                   *
C       Solving a system of linear equations                        *
C                      A * X = RS                                    *
C       for a cyclically tridiagonal, symmetric, positive definite  *
C       matrix A, once its Cholesky factors have been calculated    *
C       in SUBROUTINE CYTSYP.                                       *
C*********************************************************************

        SUBROUTINE CYTNPD (N,DM,DU,CR,RS,X,MARK)
C
C*********************************************************************
C                                                                   *
C       Solving a system of linear  equations                       *
C                      A * X = RS                                    *
C       for a cyclically tridiagonal, symmetric, strongly           *
C       nonsingular matrix A.                                       *
C*********************************************************************

        SUBROUTINE CNPSYP (N,DM,DU,CR,MARK)
C
C*********************************************************************
C                                                                   *
C       Factoring a cyclically tridiagonal, symmetric and strongly  *
C       nonsingular matrix A, that is given by the two N-vectors    *
C       DM and DU, into its Cholesky factors                        *
C                      A = R(TRANSP) * D * R                         *
C       by applying the root-free Cholesky-method for tridiagonal   *
C       cyclic matrices. The form of the system of equations is     *
C       identical to the one described in SUBROUTINE CYTSY.         *
C*********************************************************************
```

F 4.12.1 Systems with Five-Diagonal Matrices

```
        SUBROUTINE FDIAG (N,DL2,DL1,DM,DU1,DU2,RS,X,MARK)
C
```

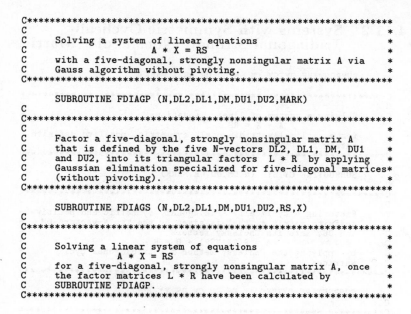

```
C*********************************************************************
C                                                                   *
C      Solving a system of linear equations                         *
C                  A * X = RS                                        *
C      with a five-diagonal, strongly nonsingular matrix A via      *
C      Gauss algorithm without pivoting.                            *
C*********************************************************************

       SUBROUTINE FDIAGP (N,DL2,DL1,DM,DU1,DU2,MARK)
C
C*********************************************************************
C                                                                   *
C      Factor a five-diagonal, strongly nonsingular matrix A        *
C      that is defined by the five N-vectors DL2, DL1, DM, DU1      *
C      and DU2, into its triangular factors  L * R  by applying     *
C      Gaussian elimination specialized for five-diagonal matrices* 
C      (without pivoting).                                          *
C*********************************************************************

       SUBROUTINE FDIAGS (N,DL2,DL1,DM,DU1,DU2,RS,X)
C
C*********************************************************************
C                                                                   *
C      Solving a linear system of equations                         *
C                  A * X = RS                                        *
C      for a five-diagonal, strongly nonsingular matrix A, once     *
C      the factor matrices L * R have been calculated by            *
C      SUBROUTINE FDIAGP.                                           *
C*********************************************************************
```

F 4.12.2 Systems with Five-Diagonal Symmetric Matrices

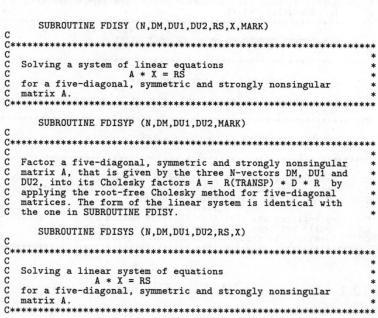

```
       SUBROUTINE FDISY (N,DM,DU1,DU2,RS,X,MARK)
C
C*********************************************************************
C                                                                   *
C  Solving a system of linear equations                             *
C                  A * X = RS                                        *
C  for a five-diagonal, symmetric and strongly nonsingular          *
C  matrix A.                                                        *
C*********************************************************************

       SUBROUTINE FDISYP (N,DM,DU1,DU2,MARK)
C
C*********************************************************************
C                                                                   *
C  Factor a five-diagonal, symmetric and strongly nonsingular       *
C  matrix A, that is given by the three N-vectors DM, DU1 and       *
C  DU2, into its Cholesky factors A =  R(TRANSP) * D * R  by        *
C  applying the root-free Cholesky method for five-diagonal         *
C  matrices. The form of the linear system is identical with        *
C  the one in SUBROUTINE FDISY.                                     *

       SUBROUTINE FDISYS (N,DM,DU1,DU2,RS,X)
C
C*********************************************************************
C                                                                   *
C  Solving a linear system of equations                             *
C                  A * X = RS                                        *
C  for a five-diagonal, symmetric and strongly nonsingular          *
C  matrix A.                                                        *
C*********************************************************************
```

F 4.13 Linear Systems with Band Matrices

F Linear Systems with Band Matrices using Pivots

```
      SUBROUTINE BAND(AP,LDAP,MB,N,ML,MU,B,IERR,IP)
C
C**********************************************************************
C                                                                     *
C     BAND solves a linear system of equations for a banded           *
C     system matrix AP and a right hand side B. This program          *
C     should not be used if several systems with the same system      *
C     matrix AP are to be solved. In this case it is better to        *
C     call BANDP below once and BANDS separately for each right       *
C     hand side.                                                       *
C**********************************************************************

      SUBROUTINE BANDP(AP,LDAP,MB,N,ML,MU,IP,ISIG,IERR)
C
C**********************************************************************
C                                                                     *
C     BANDP transforms a banded matrix AP to upper or lower           *
C     triangular form by applying a Gauss algorithm with              *
C     partial pivot search.                                           *
C**********************************************************************

      SUBROUTINE BANDS(AP,LDAP,MB,N,ML,MU,B,IP)
C
C**********************************************************************
C                                                                     *
C     After calling BANDP, the subroutine BANDS solves the linear*
C     system of equations consisting of a banded, condensed          *
C     coefficient matrix AP and a right hand side B.                  *
C**********************************************************************
```

F The Cholesky Decomposition for Band Matrices

```
      SUBROUTINE CHOBND (N, M, AP, RS, X, IFLAG, Z)
C
C**********************************************************************
C                                                                     *
C CHOBND solves a linear system of equations                          *
C                A * X = RS                                            *
C for a symmetric, positive definite banded matrix A in               *
C condensed form using the Cholesky method.                           *
C**********************************************************************

      SUBROUTINE CHOBDL (N, M, AP, RS, X, Z)
C
C**********************************************************************
C                                                                     *
C CHOBDL solves a linear system of equations                          *
C                A * X = RS                                            *
C for a symmetric, positive definite banded matrix A in               *
C condensed form using the Cholesky method.                           *
C Here the matrix A has been overwritten in AP by the subroutine*
C CHOBDZ with its Cholesky factors  A = R(TRANS)*D*R   in             *
C condensed form.                                                     *
C**********************************************************************

      SUBROUTINE CHOBDZ (N, M, AP, IFLAG, Z)
C
```

```
C*********************************************************************
C                                                                   *
C  CHOBDZ factors a symmetric, positive definite banded matrix AP*
C  given in condensed form into  R(TRANS)*D*R  using the           *
C  Cholesky method. Here D is a diagonal matrix and R is a unit     *
C  diagonal upper tringular banded matrix with as many             *
C  codiagonals as the original A. The output is again stored in     *
C  condensed form with D in the first column and the codiagonal     *
C  bands of R in subsequent columns.                                *
C*********************************************************************
```

F 4.14 Solving Linear Systems via Householder Transformations

```
       SUBROUTINE SHOUSE (A, LDA, M, N, D, X, IERR)

C*********************************************************************
C                                                                   *
C  SUBROUTINE SHOUSE   solves a linear minimization problem by       *
C  applying Householder transformations, i.e., the euclidean        *
C  norm of A(ORG)*X-B is minimized. Here A is an (M+1)x(N+2)         *
C  matrix A(0:M,0:N+1) with M >= N that contains A(ORG) of          *
C  dimensions  (M+1)x(N+1) in the columns 0, ..., N; B(0:M) is a     *
C  vector of length M+1 stored in the (N+1)st column of A, and X    *
C  is the solution vector of length N+1.                            *
C*********************************************************************
```

F 4.15.1 Error and the Condition Number

```
       SUBROUTINE HACOND(N,AO,A,LDA,MARK,HCOND)
C
C*********************************************************************
C                                                                   *
C  SUBROUTINE HACOND calculates the Hadamard condition number       *
C  of the matrix AO with AO = A(ORG).   The determinant of AO is    *
C  calculated via the product of the diagonal elements of the       *
C  upper triangular factor R from SUBROUTINE GAUSSP, where          *
C  P * A(ORG) = L * R, for  P a permutation matrix.                 *
C*********************************************************************
```

F 4.15.2 Condition Estimates

F Condition Estimate according to Cline

```
       SUBROUTINE CLINE (L,R,N,IA,COND,X,Y,Z,ZSUM,NA)
C
C*********************************************************************
C                                                                   *
C  This subroutine computes an estimate for the condition number*
C  of a matrix A whose  LR decoposition is known, where L is a     *
C  unit lower triangular and R is a nonsingular upper triangular*
C  matrix.                                                          *
C*********************************************************************

       SUBROUTINE BACK (L,B,N,IL,X)
C
C*********************************************************************
```

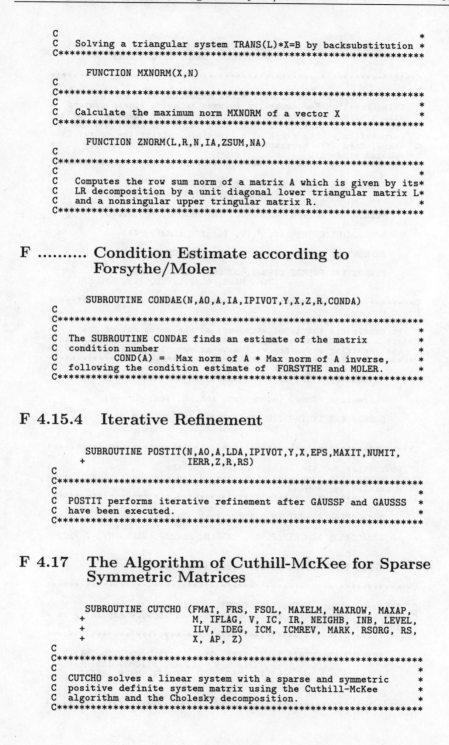

```
C                                                                    *
C    Solving a triangular system TRANS(L)*X=B by backsubstitution  *
C********************************************************************

      FUNCTION MXNORM(X,N)
C
C********************************************************************
C                                                                    *
C    Calculate the maximum norm MXNORM of a vector X                 *
C********************************************************************

      FUNCTION ZNORM(L,R,N,IA,ZSUM,NA)
C
C********************************************************************
C                                                                    *
C    Computes the row sum norm of a matrix A which is given by its*
C    LR decomposition by a unit diagonal lower triangular matrix L*
C    and a nonsingular upper tringular matrix R.                   *
C********************************************************************
```

F Condition Estimate according to Forsythe/Moler

```
      SUBROUTINE CONDAE(N,AO,A,IA,IPIVOT,Y,X,Z,R,CONDA)
C
C********************************************************************
C                                                                    *
C    The SUBROUTINE CONDAE finds an estimate of the matrix          *
C    condition number                                                *
C          COND(A) =  Max norm of A * Max norm of A inverse,         *
C    following the condition estimate of  FORSYTHE and MOLER.        *
C********************************************************************
```

F 4.15.4 Iterative Refinement

```
      SUBROUTINE POSTIT(N,AO,A,LDA,IPIVOT,Y,X,EPS,MAXIT,NUMIT,
     +                  IERR,Z,R,RS)
C
C********************************************************************
C                                                                    *
C    POSTIT performs iterative refinement after GAUSSP and GAUSSS  *
C    have been executed.                                             *
C********************************************************************
```

F 4.17 The Algorithm of Cuthill-McKee for Sparse Symmetric Matrices

```
      SUBROUTINE CUTCHO (FMAT, FRS, FSOL, MAXELM, MAXROW, MAXAP,
     +                   M, IFLAG, V, IC, IR, NEIGHB, INB, LEVEL,
     +                   ILV, IDEG, ICM, ICMREV, MARK, RSORG, RS,
     +                   X, AP, Z)
C
C********************************************************************
C                                                                    *
C    CUTCHO solves a linear system with a sparse and symmetric     *
C    positive definite system matrix using the Cuthill-McKee       *
C    algorithm and the Cholesky decomposition.                      *
C********************************************************************
```

```
      SUBROUTINE CUTHIL (NNODES, NEIGHB, INB, IDEG,
     +                   ICM, ICMREV, MARK, LEVEL, ILV)
C
C*****************************************************************
C                                                               *
C  CUTHIL computes the Cuthill-McKee numbering of a graph.       *
C  This Cuthill-McKee numbering is used to solve linear syatems  *
C  with sparse and symmetric system matrices and saves storage  *
C  space and computation al time. When using the Cuthill-McKee   *
C  permutation on the graph of a symmetric matrix, the matrix is *
C  transformed into a symmetric band matrix with a generally     *
C  reduced bandwidth.                                            *
C*****************************************************************

      SUBROUTINE CUTH1K (IROOT, ISTART, NNODES,
     +                   NEIGHB, INB, IDEG, MARK, ICM)

      SUBROUTINE BLDGPH (NROW, IC, IR, NEIGHB, INB, IDEG)

      SUBROUTINE CUTPAK (N, M, V, IR, IC, ICMREV, AP)

      SUBROUTINE CUTPK2 (N, M, V, IR, IC, ICMREV, AP)

      SUBROUTINE FNDROO (IROOT, NNODES, NEIGHB, INB,
     +                   IDEG, MARK, NLV, LEVEL, ILV, LNODES)
C
C*****************************************************************
C                                                               *
C  Subroutine for the Cuthill-McKee algorithm.                   *
C  It constructs the level structure of the component of the     *
C  graph of IROOT and attempts to choose a starting node so that *
C  the resulting structure will have as many levels as possible. *
C*****************************************************************

      INTEGER FUNCTION IBDWID (NNODES, NEIGHB, INB, NOLD, NNEW)

      subroutine ckbdwd (n, neighb, inb, m, icm, icmrev)

      SUBROUTINE LVSTRU (IROOT, NNODES, NEIGHB, INB,
     +                   MARK, NLV, LEVEL, ILV, LNODES)
C
C*****************************************************************
C                                                               *
C  Subroutine for the Cuthill-McKee algorithm.                   *
C  It constructs the level structure of the component of a graph *
C  generated by IROOT.                                           *
C*****************************************************************

      SUBROUTINE PERMUT (N, IPERM, XOLD, XNEW)

      SUBROUTINE RDMTRX (FLNAM, MAXELM, MAXROW, NROW, NV, V, IC,
     +                   IR)

      SUBROUTINE SRTDEG (NODE, IDEG, IBEG, IEND)
C
C*****************************************************************
C                                                               *
C  Subroutine for  SUBROUTINE CUTHIL.                            *
C  Sorts a partial list of nodes in the vector NODE according to *
C  increasing degrees.                                           *
C*****************************************************************

      SUBROUTINE CUTGAU (FMAT, FRS, FSOL, MAXELM, MAXROW, MAXAP,
     +                   M, IFLAG, V, IC, IR, NEIGHB, INB, LEVEL,
     +                   ILV, IDEG, ICM, ICMREV, MARK, RSORG, RS,
     +                   AP, IP)
C
C*****************************************************************
C                                                               *
```

```
C  CUTGAU solves a linear system with a sparse and symmetric     *
C  system matrix using the Cuthill-McKee algorithm and Gauß      *
C  elimination with column pivot search.                         *
C****************************************************************
```

F 5 Iterative Methods for Linear Systems

F 5.4 The Gauß-Seidel Iteration

```
      SUBROUTINE ADSOR(A,N,IA,B,X,KADAPT,EPS,KMAX,IMETH,ISWITC,
     *                 OMEGA,WORK,RES,ITNUMB,IERR)
C
C****************************************************************
C                                                              *
C  This program solves an inhomogeneous linear system AX = B of *
C  equations with a nonsingular system matrix A. The method of  *
C  Jacobi is used jointly with relaxation, where the relaxation *
C  parameter OMEGA is adjusted during the iteration (adaptive   *
C  SOR method).                                                 *
C****************************************************************
      SUBROUTINE GAUSEI(A,N,IA,B,OMEGA,X)
C
C****************************************************************
C                                                              *
C  This subroutine performs one iteration with the Gauß-Seidel  *
C  method for a given relaxation parameter.                     *
C****************************************************************
      SUBROUTINE MNORM(X,N,XNORM)
C
C****************************************************************
C                                                              *
C  This subroutine calculates the maximum norm XNORM of an      *
C  N-vector X.                                                  *
C                                                              *
C-----------------------------------------------------------*
      SUBROUTINE CONV(ISWITC,A,N,IA,IERR)
C
C****************************************************************
C                                                              *
C  This subroutine helps check convergence.                     *
C****************************************************************
      SUBROUTINE RESID(A,N,IA,B,X,RES)
C
C****************************************************************
C                                                              *
C  This subroutine computes the residuum  RES = B - AX, where   *
C  both A and B are given in modified form.                     *
C                                                              *
C-----------------------------------------------------------*
```

F 6 Systems of Nonlinear Equations

F 6.2.1.2 Damped Newton Method for Systems

```
      SUBROUTINE SMNEWD (FX,N,MAXIT,IERR,KMAX,LUN,IUPD,
     1                   EPS,RNORM2,F,X,DF,LDDF,IWORK,WORK)
```

```
C
C*******************************************************************
C                                                                 *
C      SMNEWD finds a solution of a nonlinear system of equations *
C                                                                 *
C                   F1(X(1),...,X(N))=0                           *
C                   F2(X(1),...,X(N))=0                           *
C                   - - - - - - - - -                             *
C                   FN(X(1),...,X(N))=0                           *
C                                                                 *
C      by the damped Newton method, if this method converges for  *
C      the starting vector. Here the Jacobi matrix is replaced    *
C      by the forward difference quotients.                       *
C      Hence the user does not need to supply the partial         *
C      derivatives. The parameter IUPD can be used to predetermine*
C      after how many iterations the Jakobi matrix is to be re-   *
C      computed and LR decomposed.                                *
C*******************************************************************

       SUBROUTINE FDIFQU (FX,N,X,F,FXJPH,DF,LDDF)
C
C*******************************************************************
C                                                                 *
C Approximates the  Jacobi matrix for the function                *
C                                                                 *
C                   F1 (X(1),...,X(N))                            *
C                   F2 (X(1),...,X(N))                            *
C                   - - - - - - - - - -                           *
C                   FN (X(1),...,X(N))                            *
C                                                                 *
C by forward difference quotients.                                *
C*******************************************************************

       SUBROUTINE SMNEWT (FX,DFX,N,MAXIT,IERR,KMAX,LUN,EPS,
      1               RNORM2,F,X,DF,LDDF,IWORK,WORK)
C
C*******************************************************************
C                                                                 *
C      SMNEWT finds a solution of the nonlinear system of         *
C      equations                                                  *
C                   F1(X(1),...,X(N))=0                           *
C                   F2(X(1),...,X(N))=0                           *
C                   - - - - - - - - -                             *
C                   FN(X(1),...,X(N))=0                           *
C                                                                 *
C      via the damped Newton method, if it converges for the      *
C      starting vector.                                           *
C*******************************************************************

       DOUBLE PRECISION FUNCTION FENORM (N,X)
C
C*******************************************************************
C                                                                 *
C Determining the euclidean norm of a vector X.                   *
C*******************************************************************

       SUBROUTINE NLESYS (FX,M,N,JACMA,DFX,DF,LDDF,MAXIT,EPS,
      +               KMAX,LUN,X,F,RNORM2,IERR,D,S,WORK)
C
C*******************************************************************
C                                                                 *
C  SUBROUTINE NLESYS solves a nonlinear system of equations       *
C  consisting of M+1 equations in N+1 unknowns by applying the    *
C  damped Newton method provided that it converges for the        *
C  starting vector.                                               *
```

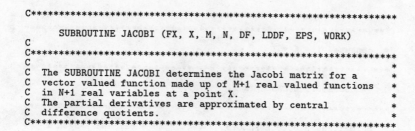

```
C*******************************************************************
      SUBROUTINE JACOBI (FX, X, M, N, DF, LDDF, EPS, WORK)
C
C*******************************************************************
C                                                                  *
C The SUBROUTINE JACOBI determines the Jacobi matrix for a         *
C vector valued function made up of M+1 real valued functions      *
C in N+1 real variables at a point X.                              *
C The partial derivatives are approximated by central              *
C difference quotients.                                            *
C*******************************************************************
```

F 6.2.4 Brown's Method for Nonlinear Systems

```
      SUBROUTINE BROWN (N,FCT,XO,EPS,EPSM,IOUT,MAXIT,IHF,LDIHF,
     1                  HF,LDHF,X1,NUMIT,IERR)
C
C*******************************************************************
C                                                                  *
C SUBROUTINE BROWN determines the zero of a nonlinear system of    *
C equations consisting of N equations in N unknowns by applying    *
C the Brown-method.                                                *
C*******************************************************************

      SUBROUTINE ITER4 (N,FCT,EPSM,IHF,LDIHF,HF,LDHF,X1,SING)
C
C*******************************************************************
C                                                                  *
C SUBROUTINE ITER4 computes an approximation by Brown's            *
C algorithm.                                                       *
C*******************************************************************

      SUBROUTINE SUBST (K,N,IHF,LDIHF,HF,LDHF,X1)
C
C*******************************************************************
C                                                                  *
C SUBROUTINE SUBST solves a linear system of equations.            *
C*******************************************************************
```

F 7 Eigenvalues and Eigenvectors of Matrices

F 7.3.1 Vector Iteration for the Dominant Eigenvalue and the Associated Eigenvector of a Matrix

```
      SUBROUTINE EIGVAL(A,N,LDA,MO,EPSI,X,Y,Z,EV,IERR)
C
C*******************************************************************
C*                                                                 *
C* Determining the eigenvalue of largest magnitude of an n by n    *
C* matrix A with the corresponding eigenvector by vector           *
C* iteration                                                       *
C*******************************************************************

      SUBROUTINE MAVE(A,X,Y,N,LDA)
```

```
      C
      C********************************************************************
      C*                                                                  *
      C*  compute  Y = A * X                                              *
      C*                                                                  *
      C********************************************************************

            SUBROUTINE DBNORM(X,N)
      C
      C********************************************************************
      C*                                                                  *
      C*  Normalizes a DOUBLE PRECISION vector to euclidean length 1      *
      C*                                                                  *
      C********************************************************************

            DOUBLE PRECISION FUNCTION QUOT(X,Y,N)

            SUBROUTINE QSCAL(X,Y,R,N)
      C
      C********************************************************************
      C*                                                                  *
      C*  computes the dot product of two DOUBLE PRECISION vectors        *
      C*                                                                  *
      C********************************************************************
```

F 7.8 Eigenvalues and Eigenvectors of a Matrix via the QR Algorithm

```
            INTEGER FUNCTION EIGEN (VEC,ORTHO,EVNORM,BASIS,LD,N,MAT,
           *                        SCAL,D,EIVEC,VALR,VALI,CNT,LOW,HIGH)
      C
      C********************************************************************
      C                                                                   *
      C   PROGRAM OBJECTIVE:                                              *
      C   ==================                                              *
      C   This FUNCTION-subroutine of type INTEGER determines all         *
      C   eigenvalues and eigenvectors of a real square matrix MAT.       *
      C********************************************************************

            INTEGER FUNCTION BALAN (LD,N,MAT,SCAL,LOW,HIGH,BASIS)
      C
      C********************************************************************
      C                                                                   *
      C   The procedure BALAN balances a given real matrix with           *
      C   respect to the column sum norm.                                 *
      C********************************************************************

            INTEGER FUNCTION BALBAK (LD,N,LOW,HIGH,SCAL,EIVEC)
      C
      C********************************************************************
      C                                                                   *
      C   BALBAK transforms all right eigenvectors of a balanced          *
      C   matrix back into the eigenvectors of the original matrix.       *
      C   The balanced matrix was generated by calling the procedure      *
      C   BALAN.                                                          *
      C********************************************************************

            INTEGER FUNCTION ELMHES (LD,N,LOW,HIGH,MAT,PERM)
      C
      C********************************************************************
      C   Reduction of a general square eigenvalue matrix a(1:N,1:N) to    *
      C   Hessenberg form by similarity with non-orthogonal Gaussian      *
      C   elimination matrices.                                           *
      C********************************************************************

            INTEGER FUNCTION ELMTRA (LD,N,LOW,HIGH,MAT,PERM,H)
```

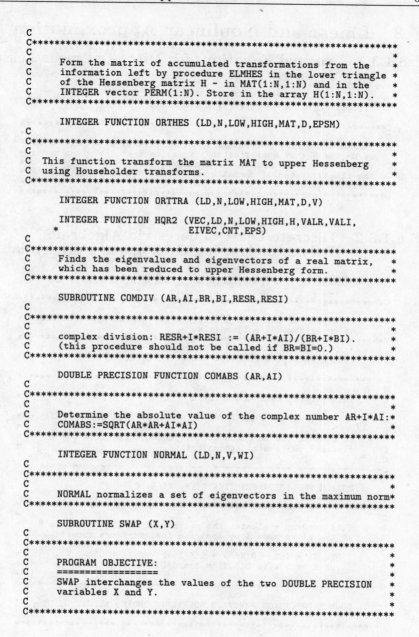

```
C
C********************************************************************
C                                                                  *
C       Form the matrix of accumulated transformations from the    *
C       information left by procedure ELMHES in the lower triangle *
C       of the Hessenberg matrix H - in MAT(1:N,1:N) and in the    *
C       INTEGER vector PERM(1:N). Store in the array H(1:N,1:N).    *
C********************************************************************

        INTEGER FUNCTION ORTHES (LD,N,LOW,HIGH,MAT,D,EPSM)
C
C********************************************************************
C                                                                  *
C  This function transform the matrix MAT to upper Hessenberg      *
C  using Householder transforms.                                   *
C********************************************************************

        INTEGER FUNCTION ORTTRA (LD,N,LOW,HIGH,MAT,D,V)

        INTEGER FUNCTION HQR2 (VEC,LD,N,LOW,HIGH,H,VALR,VALI,
       *                       EIVEC,CNT,EPS)
C
C********************************************************************
C                                                                  *
C       Finds the eigenvalues and eigenvectors of a real matrix,   *
C       which has been reduced to upper Hessenberg form.           *
C********************************************************************

        SUBROUTINE COMDIV (AR,AI,BR,BI,RESR,RESI)
C
C********************************************************************
C                                                                  *
C       complex division: RESR+I*RESI := (AR+I*AI)/(BR+I*BI).      *
C       (this procedure should not be called if BR=BI=0.)          *
C********************************************************************

        DOUBLE PRECISION FUNCTION COMABS (AR,AI)
C
C********************************************************************
C                                                                  *
C       Determine the absolute value of the complex number AR+I*AI:*
C       COMABS:=SQRT(AR*AR+AI*AI)                                   *
C********************************************************************

        INTEGER FUNCTION NORMAL (LD,N,V,WI)
C
C********************************************************************
C                                                                  *
C       NORMAL normalizes a set of eigenvectors in the maximum norm*
C********************************************************************

        SUBROUTINE SWAP (X,Y)
C
C********************************************************************
C                                                                  *
C       PROGRAM OBJECTIVE:                                         *
C       ==================                                         *
C       SWAP interchanges the values of the two DOUBLE PRECISION   *
C       variables X and Y.                                         *
C                                                                  *
C********************************************************************
```

F 8 Linear and Nonlinear Approximation

F 8.1.3.1 Normal Equations for Discrete Linear Least Squares

```
        SUBROUTINE GADESM(M,X,F,W,LDA,N,C,A,B,Y,Z,IERR)
C
C*********************************************************************
C                                                                    *
C  SUBROUTINE GADESM determines the coefficients of a polynomial     *
C  of degree N that approximates a function f at the given nodes      *
C  in the discrete Gaussian least squares sense.                      *
C  The linear system of normal equations is solved using the          *
C  Cholesky method.                                                    *
C*********************************************************************
```

F 8.1.3.2 Discrete Least Squares via Algebraic Polynomials using Orthogonal Polynomials

```
        SUBROUTINE POLFIT (N,X,F,W,NDEG, ALPHA,B,C,SUMERS,IERR)
C
C*********************************************************************
C                                                                    *
C     This program performs a discrete polynomial approximation      *
C     using orthogonal polynomials.                                   *
C*********************************************************************

        DOUBLE PRECISION FUNCTION SCALP (K, N, X, F, W, B, C,CHOICE)
C
C*********************************************************************
C                                                                    *
C     This program uses the formula                                  *
C                                                                    *
C      ( G1 , G2 ) = SUMME( J=0 to N ) W(J)*G1(X(J))*G2(X(J))        *
C                                                                    *
C     to compute the weighted scalar products required for the       *
C     discrete polynomial approximation.                             *
C                                                                    *
C-------------------------------------------------------------------*

        DOUBLE PRECISION FUNCTION Q (K, X, B, C)
C
C*********************************************************************
C                                                                    *
C     This program evaluates the polynomial Q(K,X) of degree K       *
C     at the given node X by applying the two-stage recursion        *
C                                                                    *
C     Q(K,X) = ( X - B(K) ) * Q(K-1,X) - C(K) * Q(K-2,X),            *
C     for K >= 2 with Q(0,X) = 1 and Q(1,X) = X - B(1).              *
C                                                                    *
C-------------------------------------------------------------------*

        DOUBLE PRECISION FUNCTION POEVAL (X, NDEG, ALPHA, B, C)
C
C*********************************************************************
C                                                                    *
C     In a Horner like way, this program evaluates the               *
C     approximating polynomial                                       *
C                                                                    *
C     POLY(X) = SUMME( K=0 to NDEG ) ALPHA(K) * Q(K,X)               *
C                                                                    *
C     of degree NDEG, that was determined by SUBROUTINE POLFIT,      *
C     at the given node X.                                           *
C*********************************************************************
```

F 8.1.3.4 Solving Linear Least Squares Problems using Householder Transformations

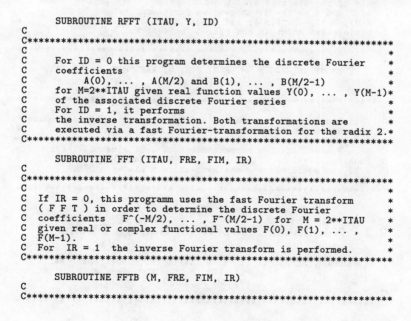

```
      SUBROUTINE SLFIT (X, Y, W, IWFL, FCT, PSI, LDA, M, N, INUM,
     +                  A, D, C, FV, SQERR, IERR)
C
C********************************************************************
C                                                                  *
C  SUBROUTINE SLFIT solves the linear discrete least squares       *
C  problem for M+1 nodes (X(I), Y(I)) by computing a linear        *
C  approximation function                                          *
C         F(X) = C(0) * F1(X) + ... + C(N) * FN(X).                *
C  The model functions F1(X), ..., FN(X) must be provided by the   *
C  user in form of a SUBROUTINE.                                   *
C********************************************************************

      SUBROUTINE SLPRE (X, W, IWFL, FCT, LDA, M, N, F, A)
C
C********************************************************************
C                                                                  *
C  The SUBROUTINE SLPRE forms the matrix A made up of function     *
C  values as needed for a discrete linear least square problem.    *
C********************************************************************

      DOUBLE PRECISION FUNCTION SENORM (X, N)
C
C********************************************************************
C                                                                  *
C  The FUNCTION-subroutine SENORM computes the square of the       *
C  euclidean norm of a real vector X of length N+1                 *
C********************************************************************
```

F 8.1.5.3 Complex Discrete Fourier Transformation

```
      SUBROUTINE RFFT (ITAU, Y, ID)
C
C********************************************************************
C                                                                  *
C  For ID = 0 this program determines the discrete Fourier         *
C  coefficients                                                    *
C         A(0), ... , A(M/2) and B(1), ... , B(M/2-1)              *
C  for M=2**ITAU given real function values Y(0), ... , Y(M-1)*
C  of the associated discrete Fourier series                       *
C  For ID = 1, it performs                                         *
C  the inverse transformation. Both transformations are            *
C  executed via a fast Fourier-transformation for the radix 2.*
C********************************************************************

      SUBROUTINE FFT (ITAU, FRE, FIM, IR)
C
C********************************************************************
C                                                                  *
C  If IR = 0, this programm uses the fast Fourier transform        *
C  ( F F T ) in order to determine the discrete Fourier            *
C  coefficients   F~(-M/2), ... , F~(M/2-1)  for  M = 2**ITAU      *
C  given real or complex functional values F(0), F(1), ... ,       *
C  F(M-1).                                                          *
C  For  IR = 1  the inverse Fourier transform is performed.        *
C********************************************************************

      SUBROUTINE FFTB (M, FRE, FIM, IR)
C
C********************************************************************
```

```
C                                                                    *
C   For IR = 0 this programm uses the fast Fourier transform        *
C   ( F F T ) in order to determine the discrete Fourier           *
C   coefficients F^(-M/2), ... , F^(M/2-1) for an arbitrary        *
C   number M of given real or complex functional values F(0),      *
C   F(1), ... , F(M-1).                                            *
C********************************************************************

        SUBROUTINE FDICHT (M, FRE, FIM, P, TETA)
C
C********************************************************************
C                                                                    *
C   This program  computes the values of the trigonometric         *
C   interpolating polynomial (i.e. the discrete partial Fourier    *
C   sum) using the Fast Fourier Transform (FFT) for a given set    *
C   of function values F(0), F(1), ... , F(M-1) of a P-periodic    *
C   function F for equidistant nodes t(0), t(1), ... , t(M-1),     *
C   t(j) = j*P/M at a set of shifted nodes t(j) + TETA  for        *
C   j = 0, 1, ... , M-1.  ("Increase of number of nodes")          *
C********************************************************************

        SUBROUTINE FOURN (M, FRE, FIM, A, DELTAX)
C
C********************************************************************
C                                                                    *
C   This program uses the fast Fourier transform (FFT) to compute  *
C   approximate values of the Fourier transform                    *
C   Here the nonperiodic function  F is known by its functional    *
C   values F(0), F(1), ... , F(M-1) at equidistant nodes with      *
C   uniform distance DELTAX. These nodes lie in the support of F.  *
C********************************************************************

        SUBROUTINE FFAKO (M, FRE, FIM, HRE, HIM)
C
C********************************************************************
C                                                                    *
C   This program uses the fast Fourier transform (FFT) to compute  *
C   the discrete values of the convolution                         *
C   and of the discrete cyclic correlation                         *
C   of F and H at the given complex functional values              *
C   F(0), F(1), ... F(M-1) and  H(0), H(1), ... , H(M-1)   at      *
C   equidistant nodes in the period interval for                   *
C   j = 0, 1, ... , M-1 .                                          *
C********************************************************************

        SUBROUTINE FFAKON (M, FRE, FIM, N, HRE, HIM, L, DELTAX,
       +                  IND, IERR)
C
C********************************************************************
C                                                                    *
C   This program uses the fast Fourier transform (FFT) to compute  *
C   approximations for the convolution                             *
C   or approximations for the correlation                          *
C   Here F and H are two nonperiodic functions given by their      *
C   funktional values  F(0), F(1), ... F(M) and  H(0), H(1), ... , *
C   H(N)  at equidistant nodes with uniform distance DELTAX.       *
C   These nodes lie in the support of F and G.                     *
C********************************************************************
```

F 8.2.2 Nonlinear Root-Mean-Square Fitting

```
        SUBROUTINE SNLFIT (X, Y, W, IWFL, PHI, JNDVT, DVT, MAXIT,
       +                  PSI, LDA, M, N, INUM, EPS, A, D, S, C,
       +                  FV, SQERR, IERR)
C
C********************************************************************
```

```
C                                                                      *
C    The SUBROUTINE SNLFIT computes a non-linear discrete fitting      *
C    for M+1 given pairs of values (X(I), Y(I)), I=0, ..., M,          *
C    possibly using weights, by finding a non-linear approximating     *
C    function defined by N+1 parameters C(K), K=0, ..., M.             *
C**********************************************************************

         SUBROUTINE SNLPRE (X, W, PHI, DVT, JNDVT, C, LDA, M, N,
        +                   EPS, F, A)
C
C**********************************************************************
C                                                                      *
C    The SUBROUTINE SNLPRE computes the Jacobi matrix (as it is        *
C    needed in SUBROUTINE SNLFIT), either using a subroutine,          *
C    denoted by DVT in the program that is provided by the user        *
C    and which computes the partial derivatives, or by central        *
C    difference quotients.                                             *
C**********************************************************************
```

F 9 Polynomial and Rational Interpolation

F 9.5.1 Newton Formula for Arbitrary Nodes

```
         SUBROUTINE NEWTIP(N,X,Y,B,IERR)
C
C**********************************************************************
C                                                                      *
C    NEWTIP determines the coefficients of the interpolation           *
C    polynomial in Newton's form.                                      *
C**********************************************************************

         DOUBLE PRECISION FUNCTION NIPFCT(XO,X,B,N)
C
C**********************************************************************
C                                                                      *
C    NIPFCT determines the functional value of the Newton              *
C    interpolation polynomial with the coefficients B(I) for           *
C    I=0,1, ...,N at XO by a generalized Horner scheme.                *
C**********************************************************************
```

F 9.7 Rational Interpolation

```
         SUBROUTINE RALGIP(X,A,N,L,L1,IERR,NLEFT,Z)
C
C**********************************************************************
C                                                                      *
C    Rational Lagrange interpolation. The interpolation                *
C    function defined by the nodes X1, X2, X3, ... and the             *
C    functional values is the ratio of two polynomials. The           *
C    degree of the nominator polynomial can be chosen by the           *
C    user.                                                             *
C**********************************************************************

         DOUBLE PRECISION FUNCTION RLIFCT(XW,X,A,L1,N)
C
C**********************************************************************
C                                                                      *
C    Evaluate the rational interpolation function                      *
```

```
C     (see SUBROUTINE RALGIP)                                           *
C***********************************************************************

      SUBROUTINE RLIORD (X,A,N,XJ,AJ)
C
C***********************************************************************
C                                                                      *
C     Determines the functional value of least magnitude               *
C***********************************************************************
```

F 9.8.2 Shepard Interpolation

```
      SUBROUTINE SHPGLO(X,Y,FX,FY,F,W,R,N,DMUE,PHI)

C***********************************************************************
C   This subroutine computes one functional value at (X,Y) for given*
C   nodes using the global Shepard method.                           *
C***********************************************************************

      SUBROUTINE SHPLOK(X,Y,FX,FY,F,PPHI,W,R,N,DMUE,RR,PHI)

C***********************************************************************
C   This subroutine computes one functional value at (X,Y) for given*
C   nodes using the local Shepard method.                           *
C***********************************************************************

      SUBROUTINE SHPFLG(X,Y,FX,FY,F,W,R,EPS,N,DMUE,RR,PHI,IERR)

C***********************************************************************
C   This subroutine computes one functional value at (X,Y) for given*
C   nodes using the local Shepard method and Franke-Little weights. *
C***********************************************************************
```

F 10 Interpolating Polynomial Splines for Constructing Smooth Curves

F 10.1.2 Non-Parametric Cubic Splines

```
      SUBROUTINE ISPLNP (N,XN,FN,IB,ALPHA,BETA,B,C,D,DUMMY,IERR)
C
C***********************************************************************
C                                                                      *
C   ISPLNP computes the coefficients B(I), C(I), D(I) for I=0,1,...,    *
C   N-1 of a nonparametric cubic interpolating spline for various end   *
C   point conditions.                                                  *
C***********************************************************************

      SUBROUTINE ISPL1D (N,XN,FN,ALPHA,BETA,MREP,B,C,D,
     +                   H,DU,DM,RS,IERR)
C
C***********************************************************************
C                                                                      *
C   ISPL1D computes the coefficients B(I), C(I), D(I) for I=0,1,..,     *
C   N-1 of a cubic interpolating spline with prescribed first end      *
C   point derivatives.                                                 *
C***********************************************************************
```

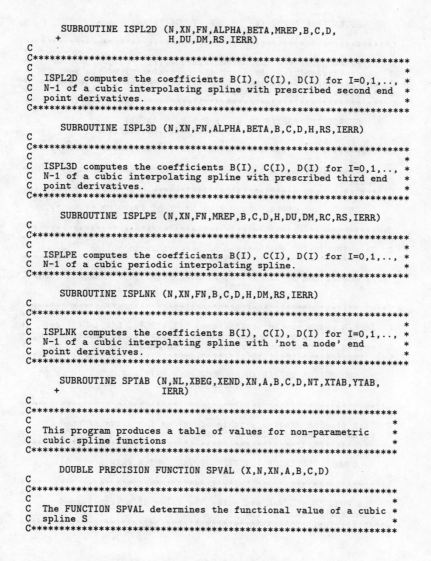

```
      SUBROUTINE ISPL2D (N,XN,FN,ALPHA,BETA,MREP,B,C,D,
     +                   H,DU,DM,RS,IERR)
C
C*********************************************************************
C                                                                   *
C  ISPL2D computes the coefficients B(I), C(I), D(I) for I=0,1,..,  *
C  N-1 of a cubic interpolating spline with prescribed second end   *
C  point derivatives.                                               *
C*********************************************************************

      SUBROUTINE ISPL3D (N,XN,FN,ALPHA,BETA,B,C,D,H,RS,IERR)
C
C*********************************************************************
C                                                                   *
C  ISPL3D computes the coefficients B(I), C(I), D(I) for I=0,1,..,  *
C  N-1 of a cubic interpolating spline with prescribed third end    *
C  point derivatives.                                               *
C*********************************************************************

      SUBROUTINE ISPLPE (N,XN,FN,MREP,B,C,D,H,DU,DM,RC,RS,IERR)
C
C*********************************************************************
C                                                                   *
C  ISPLPE computes the coefficients B(I), C(I), D(I) for I=0,1,..,  *
C  N-1 of a cubic periodic interpolating spline.                    *
C*********************************************************************

      SUBROUTINE ISPLNK (N,XN,FN,B,C,D,H,DM,RS,IERR)
C
C*********************************************************************
C                                                                   *
C  ISPLNK computes the coefficients B(I), C(I), D(I) for I=0,1,..,  *
C  N-1 of a cubic interpolating spline with 'not a node' end        *
C  point derivatives.                                               *
C*********************************************************************

      SUBROUTINE SPTAB (N,NL,XBEG,XEND,XN,A,B,C,D,NT,XTAB,YTAB,
     +                  IERR)
C
C*********************************************************************
C                                                                   *
C  This program produces a table of values for non-parametric       *
C  cubic spline functions                                           *
C*********************************************************************

      DOUBLE PRECISION FUNCTION SPVAL (X,N,XN,A,B,C,D)
C
C*********************************************************************
C                                                                   *
C  The FUNCTION SPVAL determines the functional value of a cubic *
C  spline S                                                         *
C*********************************************************************
```

F 10.1.3 Parametric Cubic Splines

```
      SUBROUTINE ISPLPA (N,XN,FN,T,MT,IB,ALPHA,BETA,BX,CX,DX,
     +                   BY,CY,DY,DUMMY,IERR)
C
C*********************************************************************
C                                                                   *
C  ISPLPA computes the coefficients BX(I), CX(I), DX(I), BY(I),     *
C  CY(I), DY(I) for I=0,1,...,N-1 of a parametric cubic inter-      *
C  polating spline for various end point conditions.               *
C*********************************************************************
```

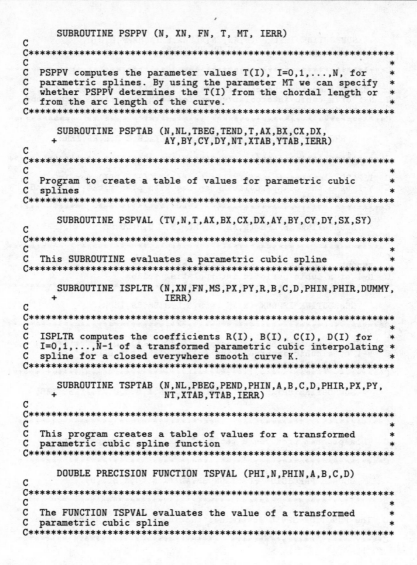

```
      SUBROUTINE PSPPV (N, XN, FN, T, MT, IERR)
C
C*********************************************************************
C                                                                   *
C   PSPPV computes the parameter values T(I), I=0,1,...,N, for       *
C   parametric splines. By using the parameter MT we can specify     *
C   whether PSPPV determines the T(I) from the chordal length or     *
C   from the arc length of the curve.                                *
C*********************************************************************

      SUBROUTINE PSPTAB (N,NL,TBEG,TEND,T,AX,BX,CX,DX,
     +                   AY,BY,CY,DY,NT,XTAB,YTAB,IERR)
C
C*********************************************************************
C                                                                   *
C   Program to create a table of values for parametric cubic         *
C   splines                                                          *
C*********************************************************************

      SUBROUTINE PSPVAL (TV,N,T,AX,BX,CX,DX,AY,BY,CY,DY,SX,SY)
C
C*********************************************************************
C                                                                   *
C   This SUBROUTINE evaluates a parametric cubic spline              *
C*********************************************************************

      SUBROUTINE ISPLTR (N,XN,FN,MS,PX,PY,R,B,C,D,PHIN,PHIR,DUMMY,
     +                   IERR)
C
C*********************************************************************
C                                                                   *
C   ISPLTR computes the coefficients R(I), B(I), C(I), D(I) for      *
C   I=0,1,...,N-1 of a transformed parametric cubic interpolating    *
C   spline for a closed everywhere smooth curve K.                   *
C*********************************************************************

      SUBROUTINE TSPTAB (N,NL,PBEG,PEND,PHIN,A,B,C,D,PHIR,PX,PY,
     +                   NT,XTAB,YTAB,IERR)
C
C*********************************************************************
C                                                                   *
C   This program creates a table of values for a transformed         *
C   parametric cubic spline function                                 *
C*********************************************************************

      DOUBLE PRECISION FUNCTION TSPVAL (PHI,N,PHIN,A,B,C,D)
C
C*********************************************************************
C                                                                   *
C   The FUNCTION TSPVAL evaluates the value of a transformed         *
C   parametric cubic spline                                          *
C*********************************************************************
```

F 10.2.2 Non-Parametric Hermite Splines

```
      SUBROUTINE HERMIT (N,MARG,X,Y,Y1,BC1,BCN,IREP,A,B,C,D,E,F,
     1                   IERR,H,SUPER,AINFER,PRAEC,DEXT,AR1,AR2,AR3)
C
C*********************************************************************
C                                                                   *
C   HERMIT computes the coefficients of a hermite polynomial         *
C   spline of fifth degree.                                          *
C*********************************************************************

      DOUBLE PRECISION FUNCTION FDEXT (A1,A2,A3,B1,B2,B3,REC1,REC2)
C
```

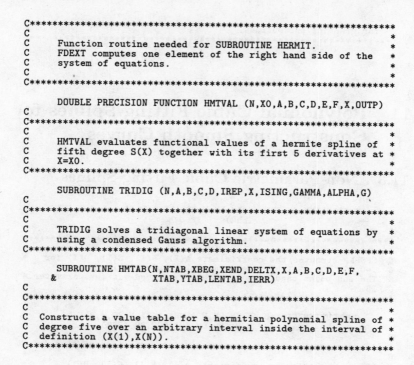

```
C*********************************************************************
C                                                                    *
C      Function routine needed for SUBROUTINE HERMIT.                *
C      FDEXT computes one element of the right hand side of the      *
C      system of equations.                                          *
C                                                                    *
C*********************************************************************

       DOUBLE PRECISION FUNCTION HMTVAL (N,XO,A,B,C,D,E,F,X,OUTP)
C
C*********************************************************************
C                                                                    *
C      HMTVAL evaluates functional values of a hermite spline of     *
C      fifth degree S(X) together with its first 5 derivatives at    *
C      X=XO.                                                          *
C*********************************************************************

       SUBROUTINE TRIDIG (N,A,B,C,D,IREP,X,ISING,GAMMA,ALPHA,G)
C
C*********************************************************************
C                                                                    *
C      TRIDIG solves a tridiagonal linear system of equations by     *
C      using a condensed Gauss algorithm.                            *
C*********************************************************************

       SUBROUTINE HMTAB(N,NTAB,XBEG,XEND,DELTX,X,A,B,C,D,E,F,
      &                 XTAB,YTAB,LENTAB,IERR)
C
C*********************************************************************
C                                                                    *
C  Constructs a value table for a hermitian polynomial spline of     *
C  degree five over an arbitrary interval inside the interval of     *
C  definition (X(1),X(N)).                                           *
C*********************************************************************
```

F 10.2.3 Parametric Hermite Splines

```
       SUBROUTINE PARMIT (N,MARG,X,Y,XDIREC,YDIREC,IDIREC,BCND1,
      1                   BCNDN,AX,BX,CX,DX,EX,FX,AY,BY,CY,DY,EY,
      2                   FY,T,IERR,XT,YT,SUP,DXT,AINF,PRC,AR1,
      3                   AR2,AR3,H )
C
C*********************************************************************
C                                                                    *
C      PARMIT computes the coefficients of a parametric hermite      *
C      spline. The end point conditions can be specified via MARG.*
C*********************************************************************

       SUBROUTINE PMTVAL (N,TO,T,AX,BX,CX,DX,EX,FX,
      1                   AY,BY,CY,DY,EY,FY,SX,SY,OUTP)
C
C*********************************************************************
C                                                                    *
C      PMTVAL computes functional values of a parametric hermite     *
C      polynomial spline of fifth degree ( SX(T),SY(T) ) and its     *
C      derivatives at T=TO.                                          *
C*********************************************************************

       SUBROUTINE PMTAB(N,NTAB,TBEG,TEND,DELT,T,AX,BX,CX,DX,EX,FX,
      &                 AY,BY,CY,DY,EY,FY,XTAB,YTAB,LENTAB,IERR)
C
C*********************************************************************
C                                                                    *
C  Tabulates a parametric Hermite spline. The nodes of the           *
C  spline are retained, as long as they lie in the table interval*
C  [TBEG, TEND]. This allows the program to create input data        *
```

```
C  for graphics subroutines.                                        *
C*******************************************************************
```

F 11 Polynomial Cubic Fitting Splines for Constructing Smooth Curves

F 11.3 Non-Parametric Cubic Fitting Splines

```
      SUBROUTINE CFSPNP (N,XN,FN,W,IB,ALPHA,BETA,A,B,C,D,AUXF,
     +                   IERR)
C
C*******************************************************************
C                                                                  *
C  CFSPNP computes the coefficients A(I), B(I), C(I), D(I) for      *
C  I=0, 1, ..., N-1 of a nonparametric cubic fitting spline.        *
C  The end point condition is to be prescribed via the parameter    *
C  IB.                                                              *
C*******************************************************************
      SUBROUTINE CFSP1D (N,XN,FN,W,ALPHA,BETA,MREP,A,B,C,D,
     +                   H,H1,H2,DM,DU1,DU2,RS,IERR)
C
C*******************************************************************
C                                                                  *
C  CFSP1D computes the coefficients A(I), B(I), C(I), D(I),         *
C  I=0, 1. ..., N-1, of a cubic fitting spline with prescribed      *
C  first end point derivative.                                      *
C*******************************************************************
      SUBROUTINE CFSP2D (N,XN,FN,W,ALPHA,BETA,MREP,A,B,C,D,
     +                   H,H1,H2,DM,DU1,DU2,RS,IERR)
C
C*******************************************************************
C                                                                  *
C  CFSP2D computes the coefficients A(I), B(I), C(I), D(I),         *
C  I=0, 1. ..., N-1, of a cubic fitting spline with prescribed      *
C  second end point derivative.                                     *
C*******************************************************************
      SUBROUTINE CFSP3D (N,XN,FN,W,ALPHA,BETA,A,B,C,D,DL1,DL2,
     +                   DU1,DU2,RS,H1,H,IERR)
C
C*******************************************************************
C                                                                  *
C  CFSP3D computes the coefficients A(I), B(I), C(I), D(I),         *
C  I=0, 1. ..., N-1, of a cubic fitting spline with prescribed      *
C*******************************************************************
      SUBROUTINE CFSPPE (N,XN,FN,W,MREP,A,B,C,D,H,H1,H2,H3,RS,
     +                   WORK,IERR)
C
C*******************************************************************
C                                                                  *
C  'CFSPPE' computes the coefficients A(I), B(I), C(I), D(I) for    *
C  I = 0, 1, ..., N-1 of a periodic cubic least squares spline.     *
C*******************************************************************
      SUBROUTINE NCYFSY (N,DM,DU1,DU2,RS,X,DML,DL1L,DL2L,RL2L,
     +                   RL1L,DU1U,DU2U,RC2U,RC1U,IERR)
C
```

```
C***********************************************************************
C                                                                     *
C  NCYFSY computes the solution X of a linear system of equations*
C  A*X=RS  for a symmetric, almost cyclic five-diagonal matrix A.*
C***********************************************************************

      SUBROUTINE NCYFSP (N,DM,DU1,DU2,DML,DL1L,DL2L,RL2L,RL1L,
     +                   DU1U,DU2U,RC2U,RC1U,IERR)
C
C***********************************************************************
C                                                                     *
C  Factorization of a symmetric, almost cyclic five-diagonal          *
C  matrix A into a product of a lower triangular matrix C and an      *
C  upper triangular matrix B.                                         *
C***********************************************************************

      SUBROUTINE NCYFSS (N,RS,X,DML,DL1L,DL2L,RL2L,RL1L,DU1U,
     +                   DU2U,RC2U,RC1U)
C
C***********************************************************************
C                                                                     *
C  Find the solution of a linear system of equations A*X=RS for       *
C  a symmetric, cyclic five-diagonal matrix A, once the factor        *
C  matrizes C and B have been determind in SUBROUTINE NCYFSP.         *
C***********************************************************************
```

F 11.4 Parametric Cubic Fitting Splines

```
      SUBROUTINE CFSPPA (N,XN,FN,WX,WY,T,MT,IB,ALPHA,BETA,MW,
     +                   AX,BX,CX,DX,AY,BY,CY,DY,AUXF,IERR)
C
C***********************************************************************
C                                                                     *
C  CFSPPA computes the coefficients AX(I), BX(I), CX(I), DX(I),       *
C  AY(I), BY(I), CY(I), DY(I) for I=0, 1, ..., N-1 of a              *
C  parametric cubic fitting spline for various end point con-         *
C  ditions. The end point conditions are prescribed via the           *
C  parameter IB.                                                      *
C***********************************************************************

      SUBROUTINE CFSPTR (N,XN,FN,W,MS,PX,PY,A,B,C,D,PHIN,R,PHIR,
     +                   AUXF,IERR)
C
C***********************************************************************
C                                                                     *
C  CFSPTR computes the coefficients A(I), B(I), C(I), D(I) for        *
C  I=0, 1, ..., N-1 of a transformed parametric cubic fitting         *
C  spline for a closed smooth curve K.                                *
C***********************************************************************
```

F 12 Two-Dimensional Splines, Surface Splines, Bézier Splines, B-Splines

F 12.1 Interpolating Two-Dimensional Cubic Splines for Constructing Smooth Surfaces

```
      SUBROUTINE BICSP1(N,M,A,X,Y,F,IERR)
C
```

```
C*********************************************************************
C                                                                   *
C  Determination of bicubic splines.                                *
C*********************************************************************

      SUBROUTINE BIC1S1(N,M,A,X,H,SA,SB,SC,SD,SX,SGAMM,
     +                  SALPH,SG,IERR)

      SUBROUTINE BIC1S2(N,M,A,Y,H,SA,SB,SC,SD,SX,SGAMM,
     +                  SALPH,SG,IERR)

      SUBROUTINE BIC1S3(N,M,A,X,H,SA,SB,SC,SD,SX,SGAMM,
     +                  SALPH,SG,IERR)

      SUBROUTINE BIC1S4(N,M,A,Y,H,SA,SB,SC,SD,SX,SGAMM,
     +                  SALPH,SG,IERR)

      SUBROUTINE BIC1S5(I,G,N,X)

      SUBROUTINE BIC1S6(J,G,M,Y)

      SUBROUTINE BIC1S7(N,M,A,I,J,W)

      SUBROUTINE BIC1S8(N,M,A,X,Y,I,J,AA)

      SUBROUTINE BIC1S9(N,M,A,I,J,AA)

      SUBROUTINE BICSP2(N,M,A,X,Y,F,IERR)

      SUBROUTINE BIC2S1(N,M,A,X,H,IERR)

      SUBROUTINE BIC2S2(N,M,A,Y,H,IERR)

      SUBROUTINE BIC2S3(N,M,A,X,H,IERR)

      SUBROUTINE BICSP3(N,M,A,X,Y,FN,F,IERR)
C
C*********************************************************************
C                                                                   *
C  Determining bicubic splines for given functional values and      *
C  surface normals at all points.                                   *
C*********************************************************************

      SUBROUTINE BIC3S1(N,M,A,FN,IERR)

      SUBROUTINE BSPVAL(N,M,A,X,Y,XX,YY,VALUE,IERR)
C
C*********************************************************************
C                                                                   *
C  Used for computing the functional value of a bicubic spline       *
C  at the point (XX,YY).                                             *
C*********************************************************************

      SUBROUTINE XYINTV(N,M,X,Y,I,J,XI,ETA,XX,YY,IERR)
C
C*********************************************************************
C                                                                   *
C  Determines the interval,in which the point (XX,YY) lies.          *
C*********************************************************************
```

F 12.2 Two-Dimensional Interpolating Surface Splines

```
      SUBROUTINE PROB2 (NX,X,Y,F,M,MARK,C,A,IWORK,WK)
C
```

```
C*********************************************************************
C                                                                   *
C PROB2 determines 2-dimensional surface splines for a given set    *
C of triples (X(I),Y(I), F(X(I),Y(I)), I=1, ..., NX. The points     *
C (X(I),Y(I)) must be distinct. For each pair (X,Y) there must      *
C be precisely one value F=F(X,Y), i.e., F has to be a function.    *
C The nodes do not have to lie on a rectangular grid and may be     *
C ordered in any way. It is advisable to transform the nodes        *
C (X(I),Y(I)) to the unit circle. A program for this is included.*
C*********************************************************************

      SUBROUTINE ALPHA2 (NX,X,Y,M,A,IDX,IDY)
C
C*********************************************************************
C                                                                   *
C Determine the polynomial components P of the syatem matrix.       *
C*********************************************************************

      SUBROUTINE NEXT2 (I,IDX,IDY,IXY,K)
C
C*********************************************************************
C                                                                   *
C Auxiliary routine for efficiently determining all 2-dimensional*
C monomials up to degree M; see SUBROUTINE ALPHA2.                  *
C*********************************************************************

      SUBROUTINE GAMMA2 (NX,X,Y,M,A)
C
C*********************************************************************
C                                                                   *
C Initializing the kernel function components G of the matrix.      *
C*********************************************************************

      DOUBLE PRECISION FUNCTION E2 (X,Y,M)
C
C*********************************************************************
C                                                                   *
C Evaluation of E at (X,Y). (See formula (12.19))                   *
C*********************************************************************

      SUBROUTINE APPRX2 (X0,Y0,NX,M,X,Y,C,AP)
C
C*********************************************************************
C                                                                   *
C Evaluation function for the interpolation.                        *
C*********************************************************************

      SUBROUTINE PR2TRA (NX,X,Y,F,M,MARK,C,A,IWORK,WK,
     F                   XQUER,YQUER,R)
C
C*********************************************************************
C                                                                   *
C PR2TRA determines a 2-dimensional surface spline for any set      *
C of triples (X(I),Y(I),F(X(I),Y(I))), I=1, ..., NX. The pairs      *
C (X(I),Y(I)) must be distinct, i.e., for each value (X,Y) there*
C must be exactly one value F=F(X,Y), or F must represent a         *
C functional relation.                                              *
C The nodes (X(I),Y(I)) do not have to lie on a rectangular         *
C grid. In fact they can be given in any order. The nodes           *
C (X(I),Y(I)) are transformed onto the unit circle.                 *
C*********************************************************************

      SUBROUTINE APPRT2 (X0,Y0,NX,M,X,Y,C,AP,XQUER,YQUER,R)
C
C*********************************************************************
C                                                                   *
C Evaluation function used for interpolation of the surface         *
C spline.                                                           *
C*********************************************************************
```

```fortran
      SUBROUTINE TRCIRC (X,Y,NX,XQUER,YQUER,R)
C
C*********************************************************************
C                                                                   *
C Transformation of the nodes (X(I),Y(I)) to the unit circle.       *
C*********************************************************************

      SUBROUTINE PROB3(NX, X, Y, Z, F,M, MARK, C, A, IWORK, WK)
C
C*********************************************************************
C                                                                   *
C PROB3 computes three-dimensional surface splines for arbitrary    *
C given points (X(I),Y(I),Z(I),F(X(I),Y(I)), I=1, ..., NX.          *
C The nodes (X(I),Y(I),Z(I)) must be distinct and F must be a       *
C function, i.e., for each (X,Y,Z) in the node set there must       *
C correspond a unique  F=F(X,Y,Z). The nodes need not be ordered.*
C*********************************************************************

      SUBROUTINE ALPHA3(NX,X,Y,Z,M,A,IDX,IDY,IDZ)
C
C*********************************************************************
C                                                                   *
C Computes the polynomial part P of the system matrix.              *
C*********************************************************************

      SUBROUTINE NEXT3(I,IDX,IDY,IDZ,ID,K)
C
C*********************************************************************
C                                                                   *
C SUBROUTINE that efficiently determines all three-dimensional      *
C monomials up to degree M; refer to SUBROUTINE ALPHA3              *
C*********************************************************************

      SUBROUTINE GAMMA3(NX,X,Y,Z,M,A)
C
C*********************************************************************
C                                                                   *
C Initialize the G part of the system matrix.                       *
C*********************************************************************

      DOUBLE PRECISION FUNCTION APPRX3(XO,YO,ZO,NX,M,X,Y,Z,C)
C
C*********************************************************************
C                                                                   *
C Evaluation FUNCTION for the interpolation                         *
C*********************************************************************

      DOUBLE PRECISION FUNCTION E3(X,Y,Z,M)
C
C*********************************************************************
C                                                                   *
C We evaluate the function F at (X,Y,Z).                            *
C*********************************************************************

      SUBROUTINE PR3TRA(NX, X, Y, Z, F,M, MARK, C, A, IWORK, WK,
     F                               XMEAN, YMEAN, ZMEAN, R)
C
C*********************************************************************
C                                                                   *
C PR3TRA computes three-dimensional surface splines for arbitrary*
C given points (X(I),Y(I),Z(I),F(X(I),Y(I)), I=1, ..., NX.          *
C The nodes (X(I),Y(I),Z(I)) must be distinct and F must be a       *
C function, i.e., for each (X,Y,Z) in the node set there must       *
C correspond a unique  F=F(X,Y,Z). The nodes need not be ordered.*
C*********************************************************************

      DOUBLE PRECISION FUNCTION APPTR3(XO,YO,ZO,NX,M,X,Y,Z,C,
     F                               XMEAN,YMEAN,ZMEAN,R)
```

```
C
C**********************************************************************
C                                                                    *
C Evaluation function for the interpolation                          *
C**********************************************************************

      SUBROUTINE TRBALL (X,Y,Z,NX,XMEAN,YMEAN,ZMEAN,R)
C
C**********************************************************************
C                                                                    *
C Transform all nodes (X(I),Y(I),Z(I)) to the unit ball.             *
C**********************************************************************

      SUBROUTINE SESSPM (AZ,N,KPVT,B)
C
C**********************************************************************
C                                                                    *
C      SESSPM solves the symmetric system of equations A*X = B       *
C      by using the factorization of A produced by SUBROUTINE        *
C      CEPSPM, ZSPMOK or ZSPMMK.                                      *
C**********************************************************************

      SUBROUTINE CEPSPM(AP,N,B,KPVT,RCOND,Z,WK)
C
C**********************************************************************
C                                                                    *
C      Condition estimate for a symmetric matrix AP which is given*
C      in factored and condensed form from SUBROUTINE ZSPMMK.        *
C**********************************************************************

      SUBROUTINE ZSPMMK (AP,N,B,KPVT,WK,IERR)
C
C**********************************************************************
C                                                                    *
C      Factorization of a symmetric matrix, given in condensed       *
C      format, by elimination that relies on symmetric pivoting.     *
C**********************************************************************

      SUBROUTINE VECADD (N,SA,SX,SY)
C
C**********************************************************************
C                                                                    *
C      Multiplies the vector SX by the constant SA and then adds     *
C      the vector SX to the vector SY.                               *
C                                                                    *
C--------------------------------------------------------------------*

      SUBROUTINE VECXCH (N,SX,SY)
C
C**********************************************************************
C                                                                    *
C      Swaps the vectors SX and SY.                                  *
C                                                                    *
C--------------------------------------------------------------------*

      INTEGER FUNCTION INDMAX (N,SX)
C
C**********************************************************************
C                                                                    *
C      Determines the index of the element of the vector SX that     *
C      has the largest entry in magnitude.                           *
C                                                                    *
C--------------------------------------------------------------------*

      SUBROUTINE PCOSOL (IPOINT,AP,N,Z,KPVT,S,EK,YNORM)
C
C**********************************************************************
C                                                                    *
C      Subroutine of CEPSPM.                                         *
C      This subroutine helps to solve a system of equations          *
C                  U*D*W  = E   or   U*D*V = Y.                       *
```

```
C       It is called twice by ZSPM.. .                              *
C*****************************************************************

        SUBROUTINE VECMWC (N,SA,SX)
C
C*****************************************************************
C                                                                *
C       Multiplies the vector SX by the constant SA.             *
C                                                                *
C----------------------------------------------------------------*

        SUBROUTINE PCOLTG(AP,N,Z,KPVT,S)
C
C*****************************************************************
C                                                                *
C       Subroutine of CEPSPM.                                    *
C       This subroutine helps to solve the system the equations  *
C       TRANS(U) * Y = W   or  TRANS(U) * Z = V.                 *
C*****************************************************************

        DOUBLE PRECISION FUNCTION SCAPRO (N,SX,SY)
C
C*****************************************************************
C                                                                *
C       Determines the scalar product of two vectors SX and SY.  *
C                                                                *
C----------------------------------------------------------------*

        DOUBLE PRECISION FUNCTION ABSSUM (N,SX)
C
C*****************************************************************
C                                                                *
C       Forms the sum of the absolute values of the entries in SX. *
C                                                                *
C----------------------------------------------------------------*

        SUBROUTINE ZSPMOK (AP,N,B,KPVT,INFO)
C
C*****************************************************************
C                                                                *
C       Decomposition of a symmetric matrix A given in condensed *
C       form using symmetric pivoting for elimination.          *
C*****************************************************************
```

F 12.3.1 Bézier Spline Curves

```
        SUBROUTINE CUBBEZ (B,D,M)
C
C*****************************************************************
C                                                                *
C       This subroutine determines BEZIER points of a curve by   *
C       the cubic BEZIER-method.                                 *
C*****************************************************************
```

F 12.3.3 Bézier Spline Surfaces

```
        SUBROUTINE BEZIER (B,D,WORK,ITYP,M,N,EPS)
C
C*****************************************************************
C                                                                *
C       This subroutine performs                                 *
C       for ITYP = 0   the bicubic BEZIER-method. Here we determine*
C                       interpolation points from the input data for*
```

```
C                            a spline surface that is constructed using  *
C                            the bicubic BEZIER-method;                  *
C             for ITYP = 1   the modified bicubic BEZIER-method. In it   *
C                            the input interpolation points are regarded *
C                            as weight points at first, for which pseudo-*
C                            interpolation points are determined. These  *
C                            are shifted until they coincide with the    *
C                            true interpolation points, except for an    *
C                            accuracy EPS.                                *
C**********************************************************************

      SUBROUTINE INTPOL (DIFF,J,I,B,M,N)

      SUBROUTINE BEZPNT (B,D,M,N)
C
C**********************************************************************
C                                                                     *
C      This Subroutine computes the BEZIER points, that are           *
C      needed for determining the surface when using the bicubic      *
C      BEZIER-method.                                                 *
C**********************************************************************

      SUBROUTINE BEZBRD (B,D,M,N)
C**********************************************************************
C                                                                     *
C      Auxiliary program for SUBROUTINE BEZIER.                       *
C      The boundary BEZIER points are determined from the prede-      *
C      termined boundary weight points using 1-dimensional BEZIER     *
C      splines.                                                       *
C**********************************************************************

      SUBROUTINE CALCP (B,M,N,VP,WP,POINT)
C
C**********************************************************************
C                                                                     *
C      This subroutine computes a point on the spline surface         *
C      at the intersection (VV,WW) of two parameter lines that        *
C      are defined by WP and VP.                                      *
C**********************************************************************

      SUBROUTINE CALCVP (B,M,N,VP,ICALC,POINTS)
C
C**********************************************************************
C                                                                     *
C      This subroutine determines points, ICALC in number, on the     *
C      parameter line defined by VP.                                  *
C**********************************************************************

      SUBROUTINE CALCWP (B,M,N,WP,ICALC,POINTS)
C
C**********************************************************************
C                                                                     *
C      This subroutine determines points, ICALC in number, on the     *
C      parameter line defined by WP.                                  *
C**********************************************************************
```

F 12.3.3 Modified (Interpolating) Cubic Bézier Splines

```
      SUBROUTINE MOCUBE (D,B,M,EPS)
C
C**********************************************************************
C                                                                     *
C      SUBROUTINE MOCUBE determines the coefficients of a             *
C      modified BEZIER spline from the weight points so that the      *
C      weight points will be located on the computed curve, except*
C      for an accuracy of EPS.                                        *
```

```
C*******************************************************************
```

F 12.4.1 B-Spline Curves

```
      SUBROUTINE CURVP (NP,N,M,K,KV,DP,X,XP,D,E,IERR,IX)
C
C*******************************************************************
C                                                                 *
C   The subroutine CURVP computes at most NP+1 points on an open  *
C   uniform B-Spline curve of order K.                            *
C*******************************************************************

      SUBROUTINE KNOTVO (N,K,KV)
C
C*******************************************************************
C                                                                 *
C                                                                 *
C    This program computes the node vector KV of an open uniform  *
C    B spline.                                                     *
C*******************************************************************

      SUBROUTINE DEBOOR (N,M,DP,K,KV,T,IR,D,E,X,IERR)
C
C*******************************************************************
C                                                                 *
C   This program computes the point corresponding to the parameter*
C   T of a uniform b spline whose nodes are known by using the    *
C   algorithm of DE BOOR.                                         *
C*******************************************************************

      SUBROUTINE CURVPG (NP,N,M,K,KV,DP,X,XP,D,E,IERR)

C********************************************************************
C                                                                  *
C   The subroutine CURVPG computes NP points of a closed uniform   *
C   b spline curve of order K.                                     *
C********************************************************************

      SUBROUTINE KNOTVG (N,K,KV)

C********************************************************************
C                                                                  *
C   This program computes the node vector KV of a closed uniform B *
C   spline.                                                        *
C********************************************************************
```

F 12.4.2 B-Spline Surfaces

```
      SUBROUTINE BSPLFL (KK,NU,NV,M,N,KU,KV,KVU,KVV,DP,XP,UPOL,D1,DO
     f ,E1,EO,HILF,X,IERR)

C********************************************************************
C                                                                  *
C   This subroutine generates a mesh for a b spline surface from   *
C   given u and v curves.                                          *
C********************************************************************
```

F 13 Akima and Renner Subsplines

F 13.1 Akima Subsplines

```
        SUBROUTINE AKIMA1(N,XN,FN,NK,BETA,B,C,D,ISWTCH,HELP,IERR)
C
C*********************************************************************
C                                                                   *
C   The program AKIMA1 computes the coefficients  B(I), C(I) and     *
C   D(I) for I=0, ... , N-1 of an interpolating cubic AKIMA          *
C   spline, which can be either periodic or nonperiodic.            *
C*********************************************************************

        SUBROUTINE AKIMA2(N,XN,FN,NK,BETA,B,C,D,ISWTCH,HELP,IERR)
C
C*********************************************************************
C                                                                   *
C   The program AKIMA2 computes the coefficients  B(I), C(I) and     *
C   D(I) for I=0, ... , N-1 of an interpolating cubic AKIMA          *
C   spline, which can be either periodic or nonperiodic.            *
C   The computation of slopes in AKIMA2 follows the suggestion of    *
C   R. Wodicka, see the literature quote below.                     *
C*********************************************************************
```

F 13.2 Renner Subsplines

```
        SUBROUTINE RENN2D (N,XN,YN,NK,BETA,B,C,D,T,HELP,IMARK,IERR)
C
C*********************************************************************
C                                                                   *
C   The programm RENN2D computes the coefficient vectors B, C and    *
C   D, as well as the lengths T(I) of the parameter intervals of     *
C   a closed or open 2-dimensional parametric Renner spline.        *
C*********************************************************************

        SUBROUTINE RENN3D (N,XN,YN,ZN,NK,BETA,B,C,D,T,HELP,IMARK,IERR)
C
C*********************************************************************
C                                                                   *
C   The programm RENN3D computes the coefficient vectors B, C and    *
C   D, as well as the lengths T(I) of the parameter intervals of     *
C   a closed or open parametric 3-dimensional Renner spline.        *
C*********************************************************************
```

F 14 Numerical Differentiation

F 14.3 Differentiation by Using Interpolating Polynomials

```
        SUBROUTINE SPLFVD (X,N,XN,A,B,C,D,S,S1,S2,S3)
C
C*********************************************************************
C                                                                   *
C   Subroutine SPLFVD determines the functional value as well as     *
C   the 1st, 2nd and 3rd derivatives of a cubic spline function S    *
C*********************************************************************

        SUBROUTINE TSPANA (PHI,N,PHIN,A,B,C,D,PHIR,PX,PY,S,S1,S2,S3,
```

```
      +                  XK,YK,C1,CCR,IERR1,IERR2)
C
C**********************************************************************
C                                                                    *
C  Evaluation program for transformed parametric cubic spline        *
C  functions                                                         *
C**********************************************************************

      SUBROUTINE PSPANA (TV,N,T,AX,BX,CX,DX,AY,BY,CY,DY,
      +                  SX,SY,XDVT,YDVT,C1,C2,IERR)
C
C**********************************************************************
C                                                                    *
C  A program to evaluate parametric cubic splines                    *
C**********************************************************************
```

F 14.4 Differentiation by the Romberg Method

```
      SUBROUTINE DIFROM (FCT,XO,EPS,N,H,RES,ERREST,NEND,
      +                  HEND,IERR,D)
C
C**********************************************************************
C                                                                    *
C  The subroutine DIFROM approximately computes the first            *
C  derivative of a given function FCT at XO by using the             *
C  ROMBERG-method.                                                   *
C**********************************************************************
```

F 15 Numerical Integration

F 15.3 Newton-Cotes Formulas

```
      SUBROUTINE  QUANEC  (A,B,N,NRP,FCT,VAL,IEG,ERREST,IERR)
C
C**********************************************************************
C                                                                    *
C  This subroutine integrates a given function FCT(X) over the       *
C  interval [A,B] using the summed NEWTON-COTES formulas.            *
C**********************************************************************
```

F 15.7 Gauß Quadrature Formulas

```
      SUBROUTINE ORTOGP(N,AINT,X,W,IERR,WK,WKD,IWK)
C
C**********************************************************************
C                                                                    *
C  This subroutine determines the nodes and weights of the           *
C  generalized GAUSSIAN quadrature formula.                          *
C**********************************************************************

      SUBROUTINE ORTPOL(N,AINT,X,W,IERR,
      F                  Q,RS,DFG,IPFG,GL,WKR,ETA,Z,WKD)
```

F 15.9 Clenshaw-Curtis Quadrature Formulas

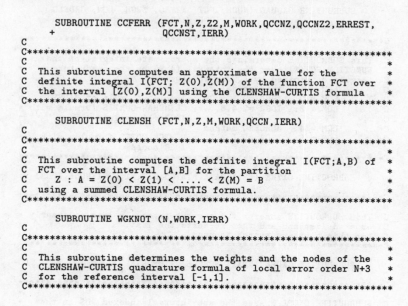

```
      SUBROUTINE CCFERR (FCT,N,Z,Z2,M,WORK,QCCNZ,QCCNZ2,ERREST,
     +                   QCCNST,IERR)
C
C*********************************************************************
C                                                                   *
C This subroutine computes an approximate value for the             *
C definite integral I(FCT; Z(0),Z(M)) of the function FCT over      *
C the interval [Z(0),Z(M)] using the CLENSHAW-CURTIS formula        *
C*********************************************************************

      SUBROUTINE CLENSH (FCT,N,Z,M,WORK,QCCN,IERR)
C
C*********************************************************************
C                                                                   *
C This subroutine computes the definite integral I(FCT;A,B) of      *
C FCT over the interval [A,B] for the partition                     *
C    Z : A = Z(0) < Z(1) < .... < Z(M) = B                          *
C using a summed CLENSHAW-CURTIS formula.                           *
C*********************************************************************

      SUBROUTINE WGKNOT (N,WORK,IERR)
C
C*********************************************************************
C                                                                   *
C This subroutine determines the weights and the nodes of the       *
C CLENSHAW-CURTIS quadrature formula of local error order N+3       *
C for the reference interval [-1,1].                                *
C*********************************************************************
```

F 15.10 Romberg Integration

```
      SUBROUTINE  QUAROM  (A, B, EPS, N, H, FCT, EL,
     !                     RESULT, ERREST, IERR)
C
C*********************************************************************
C                                                                   *
C     This subroutine determines an approximation for the           *
C     integral of the FUNCTION FCT(X) over the interval [A,B]       *
C     using the ROMBERG-method.                                     *
C*********************************************************************
```

F 15.12 Adaptive Quadrature Methods

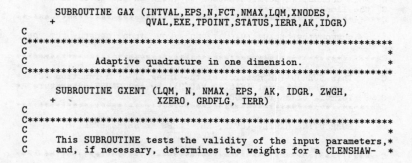

```
      SUBROUTINE GAX (INTVAL,EPS,N,FCT,NMAX,LQM,XNODES,
     +                QVAL,EXE,TPOINT,STATUS,IERR,AK,IDGR)
C
C*********************************************************************
C                                                                   *
C          Adaptive quadrature in one dimension.                    *
C*********************************************************************

      SUBROUTINE GXENT (LQM, N, NMAX, EPS, AK, IDGR, ZWGH,
     +                  XZERO, GRDFLG, IERR)
C
C*********************************************************************
C                                                                   *
C This SUBROUTINE tests the validity of the input parameters,*
C and, if necessary, determines the weights for a CLENSHAW-  *
```

```
C       CURTIS or GAUSS quadrature formula of degree IDGR.            *
C*****************************************************************

        SUBROUTINE GXQUAD (NODE, FCT, XZERO, ZWGH, LQM, GRDFLG,.
     +                     AK, IDGR, IERR2)
C
C*****************************************************************
C                                                              *
C       This SUBROUTINE determines the approximate integral of the  *
C       FUNCTION FCT(X) over the interval (A,B) by the quadrature   *
C       formula determined by LQM as follows:                  *
C           LQM = 1:   trapezoidal rule                        *
C           LQM = 2:   IDGR =2,...,20 :  I-point-formula of GAUSS  *
C           LQM = 3:   IDGR =2,4,6,...: CLENSHAW-CURTIS formula with *
C                                       IDGR + 1 weights        *
C           LQM = 4:   ROMBERG method                          *
C           LQM = 5:   IDGR =2,...,7 : NEWTON-COTES formulas with *
C                                       IDGR nodes             *
C*****************************************************************

        SUBROUTINE GXINS (NMAX,DATA,STATUS,N,T,N1,N2,ITNODE,IPOS)
C
C*****************************************************************
C                                                              *
C       This SUBROUTINE inserts two new sub-intervals into a linearly *
C       ordered list and and then re-sorts the list.          *
C       For this a binary search is used.                     *
C*****************************************************************

        SUBROUTINE GXDEL (NMAX, NCELLS, T, POS)
C
C*****************************************************************
C                                                              *
C       SUBROUTINE GXDEL erases the sub-interval indexed POS in the  *
C       linearly ordered list XNODES using the pointer vector T and  *
C       then re-sorts the vector T.                           *
C*****************************************************************

        SUBROUTINE GXDIV (NODE, R, L, FCT, XZERO, ZWGH, LQM, GRDFLG,
     +                    AK, IDGR, IERR2)
C
C*****************************************************************
C                                                              *
C       SUBROUTINE GXDIV subdivides the interval NODE into two equal  *
C       sized sub-intervals. For each of these two new sub-intervals *
C       the numerical value of the integral, the error and the new  *
C       interval endpoints are determined.                    *
C*****************************************************************

        SUBROUTINE GXACC (NMAX, DATA, STATUS, N, T ,NODE, ITNODE, K)
C
C*****************************************************************
C                                                              *
C       This SUBROUTINE permits access to the K-th sub-interval in  *
C       the linearly ordered list DATA.                        *
C*****************************************************************

        SUBROUTINE GXGAUS (VAL, XZERO, ZWGH, IDGR, GRDFLG, FCT, QV)
C
C*****************************************************************
C                                                              *
C       This SUBROUTINE determines the integral of a function FCT  *
C       over the interval [VAL(1),VAL(2)] by the GAUSS-quadrature  *
C       formula of degree IDGR                                *
C*****************************************************************

        SUBROUTINE GXRIEP (QVAL, EL, ELEPS, EPSEXE, EPSL, M,
     +                     EPS, EXE, QEXE)
C
C*****************************************************************
```

```
C                                                                    *
C       SUBROUTINE GXRIEP extrapolates the current quadrature        *
C       value using the new value of T(K,0) using RICHARDSON extra-*
C       polation and the EPSILON-algorithm. This results in a new   *
C       error and a new quadrature value for the sub-interval.      *
C********************************************************************

        SUBROUTINE GALEO (IDGR, FLAG, ALPHA, ZWGH)
C
C********************************************************************
C                                                                    *
C  The SUBROUTINE GALEO determines the nodes and weights of the     *
C  GAUSS-quadrature formula of degree IDGR.                         *
C********************************************************************

        SUBROUTINE GXPOLY (X, F, N, C)
C
C********************************************************************
C                                                                    *
C  This subroutine evaluates a polynomial of degree N at X          *
C  using the HORNER scheme. The value appears in F.                 *
C                                                                    *
C-------------------------------------------------------------------*

        SUBROUTINE GXPEGA (A, B, C, N, XSI)
C
C********************************************************************
C                                                                    *
C  This subroutine determines a zero of a polynomial of degree N,*
C  whose coefficients are given in the vector C(N) which lies       *
C  between A and B.                                                 *
C  The method used is a modified version of the PEGASUS method.     *
C                                                                    *
C-------------------------------------------------------------------*
```

F 16 Numerical Cubature

F 16.3 Newton-Cotes Cubature Formulas for Rectangles

```
        SUBROUTINE K4NECN(USERF,A,B,IP,C,D,IQ,METHOD,MOLD,CREC,
       +               ESTDIV,DIVIAT,WORK,IERR,IUFCLL)
C
C********************************************************************
C                                                                    *
C Cubature for rectangular regions using the NEWTON-COTES           *
C formulas:                                                          *
C********************************************************************

        DOUBLE PRECISION FUNCTION GRIDOT(I,J,WORK,METHOD,IPX,IQY)
C
C********************************************************************
C                                                                    *
C FUNCTION that determines the weights at the nodes.                *
C********************************************************************

        SUBROUTINE K4INIT(METHOD,WORK)
C
C********************************************************************
C                                                                    *
C Subroutine that initializes the constants for the various         *
```

```
C methods.                                                             *
C*********************************************************************
```

F 16.4 Newton-Cotes Cubature Formulas for Triangles

```
      SUBROUTINE K3NEC3(USERF,PX,PY,QX,QY,RX,RY,N,CTRI,IERR,
     +                  IUFCLL)
C
C*********************************************************************
C                                                                     *
C Cubature for triangular region using the three point NEWTON-        *
C COTES formulas:                                                     *
C*********************************************************************
```

F 16.5 Romberg Cubature for Rectangles and Triangles

```
      SUBROUTINE K4BURI(USERF,A,B,IP,C,D,IQ,N,CREC,DIVIAT,WORK,
     +                  IERR,IUFCLL)
C
C*********************************************************************
C                                                                     *
C Cubature over rectangular regions using the summed BULIRSCH-        *
C RICHARDSON method.                                                  *
C*********************************************************************

      SUBROUTINE K4BUST(USERF,A,B,IP,C,D,IQ,J,CBUST,IUFCLL)
C
C*********************************************************************
C                                                                     *
C Cubature over rectangular regions using the summed trapezoidal *
C rule and computation of the Jth term of the BULIRSCH sequence. *
C*********************************************************************

      SUBROUTINE BURIEX(BULI,WORK,N,M,VAL,ERREST)
C
C*********************************************************************
C                                                                     *
C RICHARDSON extrapolation for a given BULIRSCH sequence              *
C*********************************************************************

      DOUBLE PRECISION FUNCTION DENOM(J)
C
C*********************************************************************
C                                                                     *
C Determine the denominator for the step size in order to obtain *
C the Jth element of the BULIRSCH sequence                           *
C*********************************************************************

      SUBROUTINE K4RORI(USERF,A,B,IP,C,D,IQ,N,CREC,DIVIAT,WORK,
     +                  IERR,IUFCLL)
C
C*********************************************************************
C                                                                     *
C Cubature over rectangular regions using the summed ROMBERG-        *
C RICHARDSON method.                                                  *
C*********************************************************************

      SUBROUTINE RORIEX(ROMB,WORK,N,M,VAL,ERREST)
C
C*********************************************************************
C                                                                     *
```

```
      C RICHARDSON extrapolation for a given ROMBERG sequence.         *
      C*******************************************************************

            SUBROUTINE K4ROST(USERF,A,B,IP,C,D,IQ,J,CROST,IUFCLL)
      C
      C*******************************************************************
      C                                                                 *
      C Cubature over rectangular regions using the summed trapezoidal  *
      C rule in order to compute the Jth element of a ROMBERG sequence. *
      C*******************************************************************

            SUBROUTINE K3RORI(USERF,PX,PY,QX,QY,RX,RY,N,WORK,CTRI,
           +                  DIVIAT,IERR,IUFCLL)
      C
      C*******************************************************************
      C                                                                 *
      C Cubature over triangular regions using the summed 3-point       *
      C formula and ROMBERG-RICHARDSON extrapolation:                   *
      C*******************************************************************
```

F 16.6 Gauß Cubature Formulas for Rectangles

```
            SUBROUTINE K4GAUE(USERF,A,B,IP,C,D,IQ,METHOD,MOLD,CREC,
           +                  ESTDIV,DIVIAT,WORK,IERR,IUFCLL)
      C
      C*******************************************************************
      C                                                                 *
      C Cubature over rectangles via NEWTON-COTES formulas for GAUSSIAN *
      C nodes.                                                          *
      C*******************************************************************

            SUBROUTINE K4GINI(METHOD,WORK)
      C
      C*******************************************************************
      C                                                                 *
      C Subroutine that initializes the constants in K4GAUE depending   *
      C on method chosen.                                               *
      C*******************************************************************

            SUBROUTINE K4GAUV(USERF,X,NX,Y,NY,METHOD,MOLD,CREC,
           +                  ESTDIV,DIVIAT,WORK,IERR,IUFCLL)
      C
      C*******************************************************************
      C                                                                 *
      C Cubature over rectangles using NEWTON-COTES formulas for        *
      C GAUSSIAN nodes:                                                 *
      C*******************************************************************
```

F 16.7 Gauß Cubature Formulas for Triangles

```
            SUBROUTINE K3GAUN(USERF,PX,PY,QX,QY,RX,RY,N,METHOD,MOLD,
           +                  CTRI,WORK,IERR,IUFCLL)
      C
      C*******************************************************************
      C                                                                 *
      C Gaussian cubature over triangular regions:                      *
      C*******************************************************************

            SUBROUTINE K3GINI(METHOD,WORK)
      C
      C*******************************************************************
      C                                                                 *
      C SUBROUTINE that initializes the constants in accordance with    *
```

```
C the method.                                                          *
C**********************************************************************
```

F 16.8 Riemann Double Integrals using Bicubic Splines

```
      SUBROUTINE FIBICU (N,M,A,X,Y,VALUE)
C
C**********************************************************************
C                                                                    *
C This subroutine determines a double integral of a spline           *
C function over the rectangle [X(0), X(N)] x [Y(0), Y(M)],           *
C which is the complete domain of the spline.                        *
C**********************************************************************

      DOUBLE PRECISION FUNCTION FIBIC1 (N,M,A,I,J,XI,ETA)
C
C**********************************************************************
C                                                                    *
C The function FIBIC1 determines a double integral of a spline       *
C function over the rectangle [0, XI] x [0, ETA]. XI and ETA         *
C are relative coordinates in the rectangle I, J.                    *
C                                                                    *
C------------------------------------------------------------------*

      SUBROUTINE FIBIC2 (N,M,A,X,Y,XU,YU,XO,YO,VALUE,IERR)
C
C**********************************************************************
C                                                                    *
C The subroutine determines a double integral of a spline            *
C function over a rectangle [XU, XO] x [YU, YO].                     *
C                                                                    *
C------------------------------------------------------------------*
```

F 17 Initial Value Problems for Ordinary Differential Equations

F 17.3 One-Step Methods

```
      SUBROUTINE DEQSSP (X,Y,FCT,XEND,H,HMAX,ABSERR,RELERR,IND,
     1                   IERR,AUXF)
C
C**********************************************************************
C                                                                    *
C Numerical solution of an ordinary differential equation            *
C                                                                    *
C    Y' = F(X,Y) with given initial value  Y(X0) = Y0               *
C                                                                    *
C by a user specified one-step method.                               *
C Any one of the following methods may be chosen:                    *
C 1) EULER-CAUCHY polygonal method,                                  *
C 2) method of HEUN, and                                             *
C 3) classical RUNGE-KUTTA method.                                   *
C**********************************************************************
```

F 17.3.4.4 Embedding Formulas

```
      SUBROUTINE RKTRB (X,BETA,N,DES,Y,ABSERR,RELERR,IFLAG,WORK1,
     +                  WORK2,IWORK,IERR)
C
C*********************************************************************
C                                                                   *
C This program solves a system of at most 12 ordinary               *
C differential equations of first order by using a RUNGE-KUTTA      *
C embedding formula over the interval of integration                *
C I= [X0,BETA].                                                     *
C With the parameter IFLAG(1) one can chose the embedding formula*
C and IFLAG(2) assists in step size control.                        *
C*********************************************************************

      SUBROUTINE CIVPS (X,H,BETA,ABSERR,RELERR,N,FSAL,M,DES,Y,EPS,
     +                  XZI,QG,COEFF,MAXSTP,IERR)
C
C*********************************************************************
C                                                                   *
C This program solves a system of at most 12 ordinary               *
C differential equations of first order by using a RUNGE-KUTTA      *
C embedding formula over the interval of integration                *
C I= [X0,BETA]. The step size is automatically controlled.          *
C*********************************************************************

      SUBROUTINE HULL (X,H,BETA,ABSERR,RELERR,N,FSAL,M,DES,Y,EPS,
     +                 XZI,QG,COEFF,MAXSTP,IERR)
C
C*********************************************************************
C                                                                   *
C This program solves a system of at most 12 ordinary               *
C differential equations of first order by using a RUNGE-KUTTA      *
C embedding formula over the interval of integration                *
C I= [X0,BETA]. The step size is automatically controlled using     *
C the method of HULL.                                               *
C*********************************************************************

      SUBROUTINE RKSTEP (X,H,Y,N,M,K,DES,YHILO,COEFF,NOSTEP,FSAL,
     +                   XZI,IERR)
C
C*********************************************************************
C                                                                   *
C This program performs one integration step using the chosen       *
C RUNGE-KUTTA embedding formula.                                    *
C*********************************************************************

      SUBROUTINE STEP32 (X,H,Y,N,K,DES,YHILO,NOSTEP,XZI,IERR)
C
C*********************************************************************
C                                                                   *
C This subroutine performs one integration step using the           *
C RUNGE-KUTTA embedding formula composed of the RUNGE-KUTTA         *
C method of third order and the improved EULER-CAUCHY method of     *
C of second order.                                                  *
C*********************************************************************

      SUBROUTINE HSTART(DES,N,X,BETA,Y,RELERR,ABSERR,QG,DSMALL,
     +                  DLARGE,H)
C
C*********************************************************************
C                                                                   *
C HSTART computes the initial step size for solving an initial      *
C value problem numerically. The number of differential            *
C equations in the system is limited to 12.                         *
C*********************************************************************
```

```
      SUBROUTINE DMPREC (DSMALL,DLARGE)
C
C**********************************************************************
C                                                                    *
C This subroutine determines the machine constant in DOUBLE          *
C PRECISION and determines the largest representable DOUBLE          *
C PRECISION number for the machine used.                             *
C See the description of DLARGE !                                    *
C**********************************************************************

      SUBROUTINE COEFFI (M,IFLAG,COEFF,QG)
C
C**********************************************************************
C                                                                    *
C This subroutine determines the coefficients for the designated     *
C method.                                                            *
C**********************************************************************

      SUBROUTINE RKF43 (M,COEFF,QG)
C
C**********************************************************************
C                                                                    *
C This subroutine determines the coefficients for the embedding      *
C formula RKF4(3) (method of Fehlberg).                              *
C**********************************************************************

      SUBROUTINE RKF54 (M,COEFF,QG)
C
C**********************************************************************
C                                                                    *
C This subroutine determines the coefficients for the embedding      *
C formula RKF5(4) (method of Fehlberg).                              *
C**********************************************************************

      SUBROUTINE RK546M (M,COEFF,QG)
C
C**********************************************************************
C                                                                    *
C This subroutine determines the coefficients for the embedding      *
C formula RK5(4)6M (method of Dormand und Prince).                   *
C**********************************************************************

      SUBROUTINE RKE54 (M,COEFF,QG)
C
C**********************************************************************
C                                                                    *
C This subroutine determines the coefficients for the embedding      *
C formula RKE5(4) (method of England).                               *
C**********************************************************************

      SUBROUTINE HIHA5 (M,COEFF,QG)
C
C**********************************************************************
C                                                                    *
C This subroutine determines the coefficients for the embedding      *
C formula HIHA5 (method of Higham und Hall).                         *
C**********************************************************************

      SUBROUTINE RK547S (M,COEFF,QG)
C
C**********************************************************************
C                                                                    *
C This subroutine determines the coefficients for the embedding      *
C formula RK5(4)7S (method of Dormand und Prince).                   *
C**********************************************************************

      SUBROUTINE RK547M (M,COEFF,QG)
C
C**********************************************************************
C                                                                    *
```

```
C This subroutine determines the coefficients for the embedding  *
C formula RK5(4)7M (method of Dormand und Prince).               *
C****************************************************************

      SUBROUTINE RK547C (M,COEFF,QG)
C
C****************************************************************
C                                                                *
C This subroutine determines the coefficients for the embedding  *
C formula RK5(4)7C (method of Dormand und Prince).               *
C****************************************************************

      SUBROUTINE RK658M (M,COEFF,QG)
C
C****************************************************************
C                                                                *
C This subroutine determines the coefficients for the embedding  *
C formula RK6(5)8M (method of Dormand und Prince).               *
C****************************************************************

      SUBROUTINE RK658S (M,COEFF,QG)
C
C****************************************************************
C                                                                *
C This subroutine determines the coefficients for the embedding  *
C formula RK6(5)8S (method of Dormand und Prince).               *
C****************************************************************

      SUBROUTINE RK658C (M,COEFF,QG)
C
C****************************************************************
C                                                                *
C This subroutine determines the coefficients for the embedding  *
C formula RK6(5)8C (method of Dormand und Prince).               *
C****************************************************************

      SUBROUTINE RKV65 (M,COEFF,QG)
C
C****************************************************************
C                                                                *
C This subroutine determines the coefficients for the embedding  *
C formula RKV6(5) (method of Verner).                            *
C****************************************************************

      SUBROUTINE RKF65A (M,COEFF,QG)
C
C****************************************************************
C                                                                *
C This subroutine determines the coefficients for the embedding  *
C formula RKF6(5)A (method of Fehlberg).                         *
C****************************************************************

      SUBROUTINE RKF65B (M,COEFF,QG)
C
C****************************************************************
C                                                                *
C This subroutine determines the coefficients for the embedding  *
C formula RKF6(5)B (method of Fehlberg).                         *
C****************************************************************

      SUBROUTINE RKC65 (M,COEFF,QG)
C
C****************************************************************
C                                                                *
C This subroutine determines the coefficients for the embedding  *
C formula RKC6(5) (method of Calvo).                             *
C****************************************************************

      SUBROUTINE RKV65A (M,COEFF,QG)
C
```

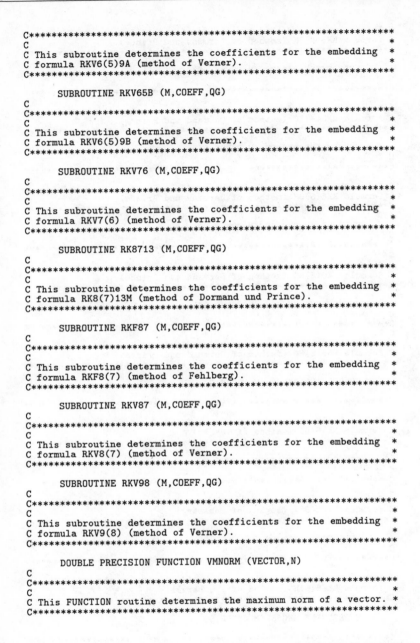

```
C*****************************************************************
C                                                               *
C This subroutine determines the coefficients for the embedding *
C formula RKV6(5)9A (method of Verner).                         *
C*****************************************************************

      SUBROUTINE RKV65B (M,COEFF,QG)
C
C*****************************************************************
C                                                               *
C This subroutine determines the coefficients for the embedding *
C formula RKV6(5)9B (method of Verner).                         *
C*****************************************************************

      SUBROUTINE RKV76 (M,COEFF,QG)
C
C*****************************************************************
C                                                               *
C This subroutine determines the coefficients for the embedding *
C formula RKV7(6) (method of Verner).                           *
C*****************************************************************

      SUBROUTINE RK8713 (M,COEFF,QG)
C
C*****************************************************************
C                                                               *
C This subroutine determines the coefficients for the embedding *
C formula RK8(7)13M (method of Dormand und Prince).             *
C*****************************************************************

      SUBROUTINE RKF87 (M,COEFF,QG)
C
C*****************************************************************
C                                                               *
C This subroutine determines the coefficients for the embedding *
C formula RKF8(7) (method of Fehlberg).                         *
C*****************************************************************

      SUBROUTINE RKV87 (M,COEFF,QG)
C
C*****************************************************************
C                                                               *
C This subroutine determines the coefficients for the embedding *
C formula RKV8(7) (method of Verner).                           *
C*****************************************************************

      SUBROUTINE RKV98 (M,COEFF,QG)
C
C*****************************************************************
C                                                               *
C This subroutine determines the coefficients for the embedding *
C formula RKV9(8) (method of Verner).                           *
C*****************************************************************

      DOUBLE PRECISION FUNCTION VMNORM (VECTOR,N)
C
C*****************************************************************
C                                                               *
C This FUNCTION routine determines the maximum norm of a vector.*
C*****************************************************************
```

F 17.3.5 Implicit Runge-Kutta Methods of Gaussian Type

```
      SUBROUTINE IRKDRV (DES,N,MMAX,IFLAG,LUNIN,LUNOUT,LUNPR,IERR,
     1                   EPSM,EPS,G,X0,XEND,Y0,YQ,
```

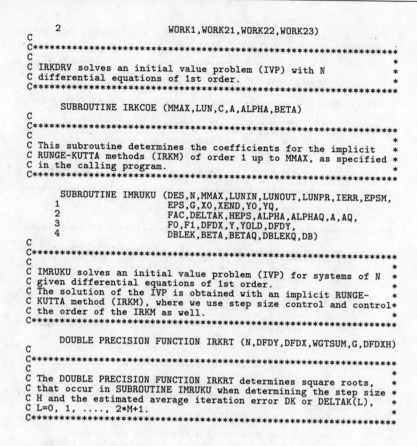

```
      2                        WORK1,WORK21,WORK22,WORK23)
C
C*********************************************************************
C                                                                   *
C IRKDRV solves an initial value problem (IVP) with N               *
C differential equations of 1st order.                              *
C*********************************************************************

      SUBROUTINE IRKCOE (MMAX,LUN,C,A,ALPHA,BETA)
C
C*********************************************************************
C                                                                   *
C This subroutine determines the coefficients for the implicit      *
C RUNGE-KUTTA methods (IRKM) of order 1 up to MMAX, as specified    *
C in the calling program.                                           *
C*********************************************************************

      SUBROUTINE IMRUKU (DES,N,MMAX,LUNIN,LUNOUT,LUNPR,IERR,EPSM,
     1                   EPS,G,XO,XEND,YO,YQ,
     2                   FAC,DELTAK,HEPS,ALPHA,ALPHAQ,A,AQ,
     3                   FO,F1,DFDX,Y,YOLD,DFDY,
     4                   DBLEK,BETA,BETAQ,DBLEKQ,DB)
C
C*********************************************************************
C                                                                   *
C IMRUKU solves an initial value problem (IVP) for systems of N     *
C given differential equations of 1st order.                        *
C The solution of the IVP is obtained with an implicit RUNGE-       *
C KUTTA method (IRKM), where we use step size control and control   *
C the order of the IRKM as well.                                    *
C*********************************************************************

      DOUBLE PRECISION FUNCTION IRKRT (N,DFDY,DFDX,WGTSUM,G,DFDXH)
C
C*********************************************************************
C                                                                   *
C The DOUBLE PRECISION FUNCTION IRKRT determines square roots,      *
C that occur in SUBROUTINE IMRUKU when determining the step size    *
C H and the estimated average iteration error DK or DELTAK(L),      *
C L=0, 1, ...., 2*M+1.                                              *
C*********************************************************************
```

F 17.3.7.2 Automatic Step Size Control, Adaptive Methods for Initial Value Problems

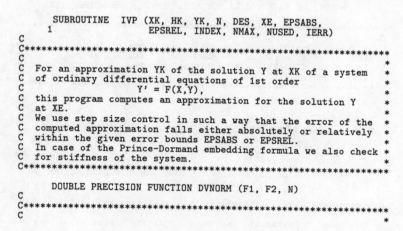

```
      SUBROUTINE  IVP (XK, HK, YK, N, DES, XE, EPSABS,
     1                 EPSREL, INDEX, NMAX, NUSED, IERR)
C
C*********************************************************************
C                                                                   *
C For an approximation YK of the solution Y at XK of a system       *
C of ordinary differential equations of 1st order                   *
C                  Y' = F(X,Y),                                     *
C this program computes an approximation for the solution Y         *
C at XE.                                                            *
C We use step size control in such a way that the error of the      *
C computed approximation falls either absolutely or relatively      *
C within the given error bounds EPSABS or EPSREL.                   *
C In case of the Prince-Dormand embedding formula we also check     *
C for stiffness of the system.                                      *
C*********************************************************************

      DOUBLE PRECISION FUNCTION DVNORM (F1, F2, N)
C
C*********************************************************************
C                                                                   *
```

```
C  This program finds the maximum norm of the difference F1 - F2 *
C  of two vectors F1 and F2 of length N.                          *
C                                                                 *
C-----------------------------------------------------------------*

      SUBROUTINE RUKU23 (X, H, Y, N, DES, Y2, Y3)
C
C*****************************************************************
C                                                                 *
C  Starting with an approximation Y at X, this program computes   *
C  approximations Y2 and Y3 at X + H using the RUNGE-KUTTA        *
C  embedding formula of 2nd and 3rd order for a system of         *
C  ordinary differential equations of 1st order                  *
C            Y' = F(X,Y).                                         *
C  The system of N ordinary differential equations of 1st order   *
C  must be provided by a SUBROUTINE DES.                          *
C*****************************************************************

      SUBROUTINE ENGL45 (X, H, Y, N, DES, Y4, Y5)
C
C*****************************************************************
C                                                                 *
C  Starting from an approximation Y at X, this program uses the   *
C  ENGLAND embedding formula of 4th and 5th order to find appro-  *
C  ximations Y4 and Y5 at X + H for the solution of the system of *
C  differential equations of 1st order                           *
C            Y' = F(X,Y)                                          *
C  The system contains N ordinary differential equations of 1st   *
C  order, with the right hand side F(X,Y) provided by the user    *
C  in a SUBROUTINE DES.                                           *
C*****************************************************************

      SUBROUTINE PRDO45 (X, H, Y, N, DES, Y4, Y5, ST1, ST2)
C
C*****************************************************************
C                                                                 *
C  Starting from an approximation Y at X, this program uses the   *
C  Prince-Dormand embedding formula of 4th and 5th order to find  *
C  apprximations Y4 and Y5 at X + H for the solution of the       *
C  system of differential equations of 1st order                 *
C            Y' = F(X,Y) .                                        *
C  The system contains N ordinary differential equations of 1st   *
C  order, with the right hand side F(X,Y) provided by the user    *
C  in a SUBROUTINE DES.                                           *
C  This program tests for stiffness in two ways.                  *
C*****************************************************************

      SUBROUTINE FRKFSY(A,DA,N,Y,DES,H,HMX,ABSERR,RELERR,IERR)
C
C*****************************************************************
C                                                                 *
C  A system of ordinary differential equations of 1st order is    *
C  integrated by applying the RUNGE-KUTTA-FEHLBERG method         *
C  [ of order O(H**5) ] with estimates for the local error and    *
C  step size control.                                             *
C*****************************************************************
```

F 17.4.3 The Predictor-Corrector Method of Adams-Moulton

```
      SUBROUTINE DESABM (X,Y,FCT,N,XEND,H,HMAX,ABSERR,RELERR,IND,
     1                   IERR,AUXF,LDF)
C
C*****************************************************************
C                                                                 *
C  Numerical solution of a system of ordinary differential        *
```

```
C   equations                                                              *
C                                                                          *
C       Y' = F(X,Y) with initial condition Y(X0) = Y0                      *
C                                                                          *
C   using the predictor-corrector method by ADAMS-BASHFORTH-              *
C   MOULTON. The starting values required are produced by the            *
C   RUNGE-KUTTA method with the same order as the A-B-M method.          *
C   An automatic step size control is used, that operates on the         *
C   principle of doubling or halving of the step size with sub-          *
C   sequent error estimation.                                             *
C**************************************************************************
```

F 17.5 Bulirsch-Stoer-Gragg Extrapolation

```
      SUBROUTINE DESEXT (X,Y,FCT,N,XEND,H,HMAX,ABSERR,RELERR,
     F                   IERR,AUXF,LDF)
C
C**************************************************************************
C                                                                          *
C   Numerical solution of a system of ordinary differential               *
C   equations                                                             *
C                                                                          *
C       Y' = F(X,Y)  with  given initial condition Y(X0) = Y0             *
C                                                                          *
C   using the extrapolations method of BULIRSCH-STOER.                    *
C   The maximal order of the extrapolation is determined depending*
C   the machine constant; the step size control lollows [HALL76]. *
C**************************************************************************
```

F 17.7.3 Gear's method for integrating stiff systems of DEs

```
      SUBROUTINE GEAR4 (XK, HK, YK, N, DES, XE, EPSABS,
     1                  EPSREL, NMAX, NUSED, IERR)
C
C**************************************************************************
C                                                                          *
C   Starting from an approximation YK for the solution Y of a             *
C   system of ordinary differential equations of first order             *
C                   Y' = F(X,Y)                                           *
C   at XK, this program computes an approximate solution YE at XE.*
C   Here we compute internally with a step size control, that            *
C   ensures the error of the computed solution to be less than the*
C   given absolute or relative error bounds EPSABS and EPSREL.           *
C   These bounds must be specified small enough for good results. *
C   The method used is the multistep method of Gear of fourth            *
C   order which is highly capable of solving stiff DEs. (Stiff DEs*
C   are those DE systems which have solution components of very          *
C   disparate growths.)                                                  *
C**************************************************************************
```

F 18 Boundary Value Problems for Ordinary Differential Equations

F 18.2 Reduction of Boundary Value Problems to Initial Value Problems

```
          SUBROUTINE  BVP  (A, B, H, YSTART, N, DEQ, BDCD, IPROC,
         1                  EPSIVP, EPSBC, IFMAX, ITMAX, ITER, IERR,
         2                  AUXF, LDF, WORK1, WORK21, WORK22, WORK23)
C
C*********************************************************************
C                                                                  *
C This program is designed to solve a general first order          *
C boundary value problem given in system form as                   *
C                                                                  *
C    Y' = F(X,Y)  for  A <= X <= B  with   R(Y(A), Y(B)) = 0.       *
C                                                                  *
C It uses the shooting method to determine an approximation        *
C YSTART for the correct initial value Y(A), which then can be     *
C used to find the solution Y using an initial value problem       *
C solver via the SUBROUTINE IVP. Here IPROC serves as a label      *
C to select the desired initial value problem solver of IVP.       *
C The non-linear system of equations that occurs in the           *
C shooting method is solved by  NEWTON's method.                   *
C*********************************************************************
```

B Bibliography

[ABRA72] ABRAMOWITZ, M.; STEGUN, I.A. (ed.): Handbook of Mathematical Functions, Dover Publications, New York, 10^{th} printing 1972.

[AHLB67] AHLBERG, J.H.; NILSON, E.N.; WALSH, J.L.: The theory of splines and their applications, Academic Press, New York, London 1967.

[AKIM70] AKIMA, H.: A New Method of Interpolation and Smooth Curve Fitting Based on Local Procedures, Journal of the Association for Computing Machinery, Vol. 17, (1970), p. 589 - 602.

[ANDE92] ANDERSON, E.; BAI, Z.; BISCHOF, C.; DEMMEL, J.; DONGARRA, J.; DUCROZ, J.; GREENBAUM, A.; HAMMARLING, S.; MCKENNEY, A.; OSTROUCHOV, S.; SORENSEN, D.: LAPACK User's Guide, SIAM, Philadelphia, 1992.

[ANDE73] ANDERSON, N.; BJÖRCK, A.: A new high order method of regula falsi type for computing a root of an equation, BIT 13 (1973), p.253-264.

[ATKI89] ATKINSON, K.E.: An introduction to numerical analysis, 2^{nd} ed., John Wiley, New York 1989.

[BALÁ91] BALÁZS, K.; KILGORE, T.: Interpolation and simultaneous mean convergence, Acta Math. Hung. 61 (1991), p. 65-74.

[BALÁ93] BALÁZS, K.; KILGORE, T.: Approximation of functions on $[-1, 1]$ and their derivatives by polynomial projection operators, Mathematische Nachrichten, 1993.

[BART87] BARTELS, R.H.; BEATTY, J.C.; BARSKY, B.A.: An introduction to the use of splines in computer graphics. Morgan Kaufmann Publ. Inc., Los Altos, California, 1987.

[BART94] BARTSCH, H.-J.: Taschenbuch mathematischer Formeln. 16^{th} edition, Fachbuchverlag, Leipzig 1994.

[BAUE65] BAUER, F. L.; HEINBOLD, J.; SAMELSON, K.; SAUER, R.:
Moderne Rechenanlagen, Teubner, Stuttgart 1965.

[BAUH70] BAUHUBER, F.: Diskrete Verfahren zur Berechnung der Null-
stellen von Polynomen, Computing $\underline{5}$ (1970), p.97-118.

[BERG86] BERG, L.: Lineare Gleichungssysteme mit Bandstruktur und ihr
asympotisches Verhalten, Hanser, München 1986.

[BEZI72] BÉZIER, P.: Numerical Control, Mathematics and Applications,
New York-London-Toronto 1972.

[BJÖR74] BJÖRCK, A.; DAHLQUIST, G.: Numerical Methods, Prentice
Hall, Englewood Cliffs, 1974.

[BÖHM84] BÖHM, W.; FARIN, G.; KOHMANN, J.: A survey of curve and
surface methods in CAGD. Computer Aided Geometric Design 1,
(1984) 1-60.

[BÖHM85] BÖHM, W.; GOSE, G.; KAHMANN, J.: Methoden der nu-
merischen Mathematik, Vieweg, Braunschweig-Wiesbaden 1985.

[BÖHM74] BÖHMER, K.: Spline-Funktionen, Teubner, Stuttgart 1974.

[BÖHM76] BÖHMER, K.; MEINARDUS, G.; SCHEMPP, W.: Spline func-
tions, Lecture Notes in Mathematics, Vol. 501, Springer, New
York, Berlin, Heidelberg 1976.

[BOOR62] BOOR, de C.: Bicubic Spline Interpolation, J. Math. Phys. $\underline{41}$
(1962), p. 215.

[BOOR78] BOOR, de C.; GOLUB, G.H.: Recent Advances in Numerical
Analysis, Academic Press, New York 1978.

[BOOR78/2] BOOR, de C.; PINKUS, A.: Proof of the conjecture of Bernstein
and Erdös concerning polynomial interpolation, J. Approx. Thy.
24 (1978), p. 289-303.

[BOOR79] BOOR, de C.: A Practical Guide to Splines, Springer, New York,
Berlin, Heidelberg 1979, 4^{th} printing 1987.

[BOTH83] BOTHA, J.F.; PINDER, G.F.: Fundamental Concepts in the Nu-
merical Solution of Differential Equations, John Wiley, New York
1983.

[BRAS77] BRASS, H.: Quadraturverfahren (Studia Mathematica, Skript 3)
Vandenhoeck & Ruprecht, Göttingen 1977

[BRAU83] BRAUN, M.: Differential Equations and their Applications, 3^{rd}
edition, Springer, New York, Berlin, Heidelberg 1983, $4.^{th}$ edition
1992.

[BRON85] BRONSTEIN, I.N.; SEMENDYAYEV, K.A.: Handbook of Mathematics, Verlag Harry Deutsch, Frankfurt a.M. 1985.

[BROS64] BROSOWSKI, B.; KRESS, R.: A Guide Book to Mathematics for Technologists and Engineers, Int. Series of Monographs on pure and applied Mathematics, Vol. 36, Pergamon Press, Oxford, London, New York, Paris 1964.

[BROW69] BROWN, K.M.: A quadratically convergent Newton–like method based upon Gaussian elimination, SIAM J. Num. Anal., 6 (1969), p. 560 - 569.

[BROW71] BROWN, K.M.; DENNIS, J.E., Jr., On the second order convergence of Brown's derivative-free method for solving simultaneous nonlinear equations, Yale University, Comp.Sci.Dept., Techn. Report, 1971, p. 71-77 .

[BULI66] BULIRSCH, R., STOER, J.: Numerical Treatment of Ordinary Differential Equations by Extrapolation Methods, Numerische Mathematik 8 (1966), p.1-13.

[BUNS85] BUNSE, W.; BUNSE-GERSTNER, A.: Numerische lineare Algebra, Teubner Studienbücher, Stuttgart 1985.

[BUTC87] BUTCHER, J.C.: The Numerical Analysis of Ordinary Differential Equations, John Wiley, New York 1987.

[CARN69] CARNAHAN, B.; LUTHER, H.A.; WILKES, J.O.: Applied Numerical Methods, John Wiley, New York 1969.

[CHAP88] CHAPRA, S.C.; CANALE, R.P.: Numerical Methods for Engineers, 2^{nd} Edition, McGraw-Hill, New York 1988.

[CHEN82] CHENEY, E. W.: Introduction to Approximation Theory, 2^{nd} edition, Chelsea, New York 1982.

[CLEN60] CLENSHAW, C.W.; CURTIS, A.R.: A method for numerical integration on a automatic computer, Numerische Mathematik 2 (1960), 197-205.

[CLIN79] CLINE, A.K.; MOLER, C.B.; STEWART, G.W.; WILKINSON, J.H.: An estimate for the condition number of a matrix. SIAM J. Numer. Anal. 16 (1979) p. 368-375.

[CLIN82] CLINE, A.K.; CONN, A.R.; VAN LOAN, C.: Generalizing the LINPACK condition estimator, in Hennart, J.P. ed. Numerical analysis. Lecture Notes in Mathematics, Vol. 909, Springer, New York, Berlin, Heidelberg, p.73-83.

[CLIN83] CLINE, A.K.; REW, P.K.: A set of counter-examples to three
 condition number estimators, SIAM J. Sci. Statist. Comput. 4
 (1983), p. 602-611.

[COLL63] COLLATZ, L.: Eigenwertaufgaben mit technischen Anwendungen
 2^{nd} ed. Akademische Verlagsgesellschaft, Leipzig 1963.

[COLL66] COLLATZ, L.: The Numerical Treatment of Differential Equa-
 tions, Springer, New York, Berlin, Heidelberg 1966.

[COLL68] COLLATZ, L.: Funktionalanalysis und numerische Mathematik,
 Springer, Berlin, Heidelberg, New York 1968.

[COLL73] COLLATZ, L.; ALBRECHT, J.: Aufgaben aus der Angewandten
 Mathematik I und II, Vieweg, Braunschweig 1972, 1973.

[COLL86] COLLATZ, L.: Differential Equations, John Wiley, New York
 1986.

[CONT80] CONTE, S.D.: Elementary Numerical Analysis. An algorithmic
 approach, McGraw-Hill, New York 1980.

[CRCS91] CRC Standard Mathematical Tables, CRC Press, Boca Raton
 1991.

[CUTH69] CUTHILL, E.; MCKEE, J.: Reducing the bandwith of sparse sym-
 metric matrices, ACM, New York 1969, S.157-172.

[CZIP58] CZIPSZER, J.; FREUD, G.: Sur ℓ' approximation d'une fonc-
 tion périodoque et de ses dérivées successives par un polynome
 trigonométrique et par ses dérivées successives, Acta Mathemat-
 ica (Stockholm), 99 (158), p. 33-51.

[DAHL74] DAHLQUIST, B.: Error Analysis for a Class of Methods for Stiff
 Nonlinear Initial Value Problems, Lecture Notes in Mathematics
 506, p.60-74, Springer, New York, Berlin, Heidelberg 1974.

[DAVI82] DAVIS, P.J.; RABINOWITZ, P.: Methods of Numerical Integra-
 tion, Academic Press, New York 1982.

[DEKK69] DEKKER, T.J.: Finding a zero by means of successive linear inter-
 polation in: Constructive aspects of the fundamental theorem of
 algebra, B. Dejon and P. Henrici, (eds.) Wiley-Interscience, New
 York 1969.

[DEKK84] DEKKER, K.; VERWER, J.G.: Stability of Runge-Kutta Meth-
 ods for Stiff Nonlinear Differential Equations, North-Holland, Am-
 sterdam 1984.

[DENN83] DENNIS, J.E.; SCHNABEL, R.B.: Numerical Methods for Un-
constrained Optimization and Nonlinear Equations, Prentice Hall,
Englewood Cliffs 1983.

[DÖRI69] DÖRING, B.: Über das Newtonsche Näherungsverfahren, Math.
Phys. Sem. ber. **16** (1969), p. 27-40.

[DONG79] DONGARRA, J.J.: LINPACK Users' Guide, SIAM Philadelphia
1979, 4^{th} printing 1984.

[DORM80] DORMAND, J.R.; PRINCE, P.J.: A family of embedded Runge-
Kutta formulae, J. of Computational and Applied Maths. Vol.6
(1980), p.19-26.

[DORM81] DORMAND, J.R.; PRINCE, P.J.: High order embedded Runge-
Kutta formulae, J. of Computational and Apllied Maths. Vol.7
(1981), p.67-75.

[DOWE71] DOWELL, M.; JARRATT, P.: A Modified Regula Falsi Method
for Computing the Root of an Equation, BIT 11 (1971), p.168-174.

[DOWE72] DOWELL, M.; JARRATT, P.: The "Pegasus" Method for Com-
puting the Root of an Equation, BIT 12 (1972), p.503-508.

[ENGE80] ENGELS, H.: Numerical Quadrature and Cubature, Academic
Press, London-New York-Toronto-Sydney-San Francisco 1980.

[ENGE87] ENGELN-MÜLLGES, G.; REUTTER, F.: Numerische Mathe-
matik für Ingenieure, BI-Wissenschaftsverlag, Mannheim-Wien-
Zürich, 5^{th} ed. 1987.

[ENGE96A] ENGELN-MÜLLGES, G.; UHLIG, F.: Numerical Algorithms
with C, Springer, New York, Berlin, Heidelberg 1996.

[ENGE96B] ENGELN-MÜLLGES, G.; NIEDERDRENK, K.: Fortran 90,
rororo Grundkurs Computerpraxis, Hrsg. Rudolf Hambusch,
Rowohlt-Verlag Hamburg 1996.

[ENRI76] ENRIGHT, W.H.; HULL, T.E.: Test Results for on Initial Value
Methods for Non-Stiff Ordinary Differential Equations, SIAM J.
Num. Anal. 13 (1976), p.944-950.

[ESSE75] ESSER, H.: Eine stets quadratisch konvergente Modifikation des
Steffensen-Verfahrens, Computing 14 (1975), p.367-369.

[ESSE77] ESSER, H.: Stabilitätsungleichungen für Diskretisierungen von
Randwertaufgaben gewöhnlicher Differentialgleichungen, Num.
Math. 28 (1977), p.69-100.

[ESSE80] ESSER, H.; NIEDERDRENK, K.: Nichtäquidistante Diskreti-
 sierungen von Randwertaufgaben, Num. Math. 35 (1980), p.465-
 478.

[FADD59] FADDEJEW, D.K.; FADDEJEWA, W.N.: Computational Meth-
 ods of Linear Algebra, Dover, New York 1959.

[FEHL60] FEHLBERG, E.: Neue genauere Runge-Kutta-Formeln für Differ-
 entialgleichungen zweiter Ordnung bzw. n-ter Ordnung, ZAMM
 40 (1960), p.252-259 and p.449-455.

[FEHL61] FEHLBERG, E.: Numerisch stabile Interpolationsformeln mit
 günstiger Fehlerfortpflanzung für Differentialgleichungen erster
 und zweiter Ordnung, ZAMM 41 (1961), p.101-110.

[FEHL66] FEHLBERG, E.: New high-order Runge-Kutta formulas with an
 arbitrarily small truncation error, ZAMM 46 (1966), p.1-16 (see
 also ZAMM 44 (1964), T 17 - T 29).

[FEHL69] FEHLBERG, E.: Klassische Runge-Kutta-Formeln fünfter und
 siebenter Ordnung mit Schrittweiten-Kontrolle, Computing 4
 (1969), p.93-106.

[FEHL70] FEHLBERG, E.: Klassische Runge-Kutta-Formeln vierter und
 niedriger Ordnung mit Schrittweiten-Kontrolle und ihre Anwen-
 dung auf Wärmeleitungsprobleme, Computing 6 (1970), p.61-71.

[FEHL75] FEHLBERG, E.: Klassische Runge-Kutta-Nyström-Formeln mit
 Schrittweiten-Kontrolle für Differentialgleichungen $\ddot{x} = f(t,x)$,
 Computing 10 (1972), p.305-315 and Klassische Runge-Kutta-
 Nyström-Formeln mit Schrittweiten-Kontrolle für Differentialgle-
 ichungen $\ddot{x} = f(t,x,\dot{x})$, Computing 14 (1975), p.371-387.

[FOLE84] FOLEY, T.A.: Three-stage Interpolation to scattered Data, Rocky
 Mountain J. of Math. Vol. 14, 1984.

[FORD77] FORD, J.A.: A Generalization of the Jenkins-Traub Method,
 Mathematics of Computation, Vol. 31 (1977), p.193-203.

[FORS67] FORSYTHE, G.E.; MOLER, C.B.: Computer Solution of Linear
 Algebraic Systems, Prentice Hall, Englewood Cliffs 1967.

[FORS77] FORSYTHE, G.E.; MALCOM, M.M.; MOLER, C.B.: Com-
 puter Methods for Mathematical Computations, Prentice-Hall,
 Englewood-Cliffs 1977.

[FRAN79] FRANKE, R.: A critical comparison of some methods for inter-
 polation of scattered data, Technical Report No. NPS 53-79-003,
 Naval Postgraduate School, Monterey, CA 1979.

[FRAN82] FRANKE, R.: Scatterend data Interpolation: Tests of some meth-
 ods. Mathematics of Computation 38 (1982), p.181-200.

[FRIE79] FRIED, I.: Numerical Solution of Differential Equations, Aca-
 demic Press, New York 1979,.

[FRIT80] FRITSCH, F.N.; CARLSON, R.E.: Monotonic piecewise cubic in-
 terpolation, SIAM J. Num. Anal. 17 (1980), p. 238 - 246.

[GEAR71/1] GEAR, C.W.: Numerical Initial Value Problems in Ordinary
 Differential Equations, Prentice-Hall, Englewood Cliffs 1971.

[GEAR71/2] GEAR, C.W.: The Automatic Integration of Stiff Ordinary Dif-
 ferential Equations, Comm. of the ACM, Vol. 14 (1971), p.176-179.

[GEAR71/3] GEAR, C.W.: DIFSUB for Solution of Ordinary Differential
 Equations (D 2), Comm. of the ACM, 14 (1971), p.185-190.

[GEAR74] GEAR, C. W.; TU, K. W.: The effects of variable mesh size on the
 stability of multistep methods, SIAM J. Num. Anal. 11 (1974), p.
 1025-1043.

[GEAR80] GEAR, C.W.: Runge-Kutta starters for multistep methods, ACM
 Transactions on Mathematical Software, Vol. 6 (1980), p. 263-279.

[GLAS66] GLASMACHER, W.; SOMMER, D.: Implizite Runge-Kutta-
 Formeln, Forschungsbericht des Landes NRW, Nr. 1763, West-
 deutscher Verlag, Opladen 1966.

[GOLU89] GOLUB, G.H.; VAN LOAN, C.F.: Matrix Computations, Balti-
 more, Maryland, The John Hopkins University Press, 2^{nd} edition
 1989, 4^{th} edition 1993.

[GOOS76] GOOS, G.; HARTMANIS, J. (edition): Lectures Notes in Com-
 puter Science, Vol. 6, EISPACK Guide, Springer, New York,
 Berlin, Heidelberg 1974, 2^{nd} ed. 1976.

[GOOS79] GOOS, G.; HARTMANIS, J. (edition): Lectures Notes in Com-
 puter Science, Vol. 76 Codes for Boundary-Value Problems in Or-
 dinary Differential Equations, Springer, New York, Berlin, Heidel-
 berg 1979.

[GORD69] GORDON, W.J.: Distributive Lattices and the Approximation of
 Multivariate Functions, in: J. Schoenberg (Ed.), Approximations
 with Special Emphasis on Spline Functions, Academic Press, 1969,
 p. 223-277.

[GRAG65] GRAGG, W.B.: On Extrapolation Algorithms for Ordinary Initial
 Value Problems, SIAM J. Num. Anal. Ser. B, 2 (1965), p. 384-403.

[GREV67] GREVILLE, T.N.E.: Theory and Applications of Spline Func-
 tions, Academic Press, New York-London 1967.

[GRIG77] GRIGORIEFF, R.D.: Numerik gewöhnlicher Differentialgleichun-
 gen, Band 1, Teubner, Stuttgart 1972, Band 2, Teubner, Stuttgart
 1977.

[HACK86] HACKBUSCH, W.: Theorie und Numerik elliptischer Differential-
 gleichungen, Teubner, Stuttgart 1986.

[HAGS82] HAGANDER, N.; SUNDBLAD, Y.: Aufgabensammlung Nu-
 merische Methoden, 2 vols., 2^{nd} edition, Oldenbourg, München
 1982.

[HAIR87] HAIRER, E.; NØRSETT, S.P.; WANNER, G.: Solving Ordinary
 Differential Equations I, Nonstiff Problems, 480 p., Springer, New
 York, Berlin, Heidelberg 1987.

[HAIR91] HAIRER, E.; WANNER, G.: Solving Differential Equations II,
 Stiff and Differential-Algebraic Problems, 601 p., Springer, New
 York, Berlin, Heidelberg 1991.

[HALL73] HALL, G.; ENRIGHT, W.H.; HULL, T.E.; SEDWICK, R.E.:
 Techn. Rep. #60, Dep. of Comp. Sc., U. of Toronto, 1973.

[HALL76] HALL, G.; WATT, J.M.: Modern Numerical Methods for Ordi-
 nary Differential Equations, Clarendon Press, Oxford 1976.

[HÄMM78] HÄMMERLIN, G.: Numerische Mathematik I, BI-Hskrpt.
 498/498a, BI-Wissenschaftsverlag, Mannheim-Wien-Zürich 1970,
 2^{nd} edition 1978.

[HÄMM91] HÄMMERLIN, G; HOFFMANN, K.H.: Numerical Mathematics,
 Springer, New York, Berlin, Heidelberg 1991.

[HART68] HART, J.F. et al: Computer Approximations, John Wiley, New
 York 1968.

[HENR68] HENRICI, P.: Discrete Variable Methods in Ordinary Differential
 Equations, John Wiley, New York, 1968.

[HENR64] HENRICI, P.: Elements of Numerical Analysis, John Wiley, New
 York 1964.

[HENR72] HENRICI, P.: Elements der Numerischen Analysis, Vol. 1, 2, BI
 HTB 551, 562, BI Verlag, Mannheim 1972.

[HILD78] HILDEBRAND, F.B.: Introduction to Numerical Analysis, 2^{nd}
 edition, McGraw-Hill, New York 1978.

[HIMM72] HIMMELBLAU, D.M.: Applied Nonlinear Programming, McGraw-Hill, New York 1972.

[HOSC89] HOSCHEK, J.; LASSER, D.: Grundlagen der geometrischen Datenverarbeitung, Teubner, Stuttgart 1989, 2^{nd} edition 1992.

[HOUS75] HOUSEHOLDER, A.S.: The Theory of Matrices in Numerical Analysis, Dover, New York 1975.

[HULL72] HULL, T. E.; ENRIGHT, W. H.; FELLEN, B. F.; SEDGWICK, R. E.: Comparing numerical methods for ordinary differential equations, SIAM J. Num. Anal. Vol. 9, (1972), p.603-637, Errata, ibidem, vol. 11 (1974), p. 681.

[HULL75] HULL, T.E.; ENRIGHT, W.H.; LINDBERG, B.: Comparing Numerical Methods for Stiff Systems of O.D.E.s, BIT 15 (1975), p.10-48.

[IGAR85] IGARASHI, M.: Practical stopping rule for finding roots of nonlinear equations, J. of Computational and Applied Mathematics 12, 13 (1985), p.371-380.

[IMSL87] IMSL International Mathematical and Statistical Library, Houston, Version 1.0, 1987.

[ISAA66] ISAACSON, E.; KELLER, H.B.: Analysis of Numerical Methods, Wiley, New York-London-Sydney 1966.

[JELT76] JELTSCH, R.: Stiff Stability and its relation to A_0- and A(0)-stability, SIAM J. Num. Anal. 13 (1976), p.8-17.

[JELT78] JELTSCH, R.: Stability on the imaginary axis and A-stability of linear multistep methods, BIT, 18 (1978), p.170-174.

[JENK70] JENKINS, M.A.; TRAUB, J.F.: A three-stage-algorithm for real polynomials using quadratic iteration, SIAM J. Num. Anal. Vol.7 (1970), p.545-566 (see also Numer. Math. 14 (1970) p.252-263).

[JOHN82] JOHNSON, L.W.; RIESS, R.D.: Numerical Analysis, 2^{nd} edition, Addison-Wesley, 1982.

[KAHA72] KAHANER, D.K.: Numerical quadrature by the ε-Algorithm, Math. Comp. 26 (1972), p. 689-693.

[KAHA83] KAHANER, D.K.; STOER, J.: Extrapolated adaptive quadrature, SIAM J. Sci. Stat. Comput., Vol. 4, (1983).

[KAPS81] KAPS, P.; WANNER, G.: A study of Rosenbrock-type methods of high order, Numer. Math. 38 (1981), p. 279-298.

[KELL68] KELLER, H.B.: Numerical Methods for Two Point Boundary
 Value Problems, Blaisdell, Waltham-Toronto-London 1968.

[KELL76] KELLER, H.B. (editor): Numerical Solution of Two Point Bound-
 ary Value Problems, SIAM Philadelphia, 1976.

[KIEL88] KIELBASINSKI, A.; SCHWETLICK, H.: Numerische lineare Al-
 gebra, VEB Deutscher Verlag der Wissenschaften, Berlin 1988.

[KILG77] KILGORE, T.: Optimization of the norm of the Lagrange inter-
 polation operator, Bull AMS 83 (1977), p. 1069-1071.

[KILG78] KILGORE, T.: A characterization of the Lagrange interpolating
 projection with minimal Tchebycheff norm, J. Approx. Thy. 24
 (1978), p. 273-288.

[KILG93] KILGORE, T.: An elementary simultaneous approximation theo-
 rem, Proc. AMS 118 (1993), to appear.

[KINC90] KINCAID, D.; CHENEY, W.: Numerical Analysis, Brooks/Cole,
 Pacific Grove, 1990.

[KING73] KING, R.F.: An Improved Pegasus-Method for Root Finding, BIT
 13 (1973), p.423-427.

[KIOU78] KIOUSTELIDIS, J.B.: Algorithmic Error Estimation for Approx-
 imate Solution of Nonlinear Systems of Equations, Computing 19
 (1978), p.313-320.

[KIOU79] KIOUSTELIDIS, J.B.: A Derivative-Free Transformation Preserv-
 ing the Order of Convergence of Iteration Methods in Case of
 Multiple Zeros, Num. Math. 33 (1979), p.385-389.

[KREI72] KREISS, H.O.: Difference Approximations for Boundary and
 Eigenvalue Problems for Ordinary Differential Equations, Math.
 of Comp. 26 (1972), p.605-624.

[KROG66] KROGH, F.T.: Predictor-Corrector-Methods of High Order with
 Improved Stability Characteristics, J. Ass. for Comp. Mach., Vol.
 13 (1966), p.374-385.

[KROG68] KROGH, F.T.: A Variable Step Variable Order Multistep Method
 for the Numerical Solution of Ordinary Differential Equations, Jet
 Propulsion Laboratory Pasadena/Cal. (Sect. Comp. and Anal.),
 May 1968, p. A91-A95.

[KRYL62] KRYLOV, V.I.: Approximate Calculation of Integrals, Macmillan,
 New York-London 1962.

[LANC85] LANCASTER , P.; TISMENETSKY, M.: The Theory of Matrices, 2^{nd} edition, Academic Press, New York 1985.

[LAPI71] LAPIDUS, L.; SEINFELD, J.H.: Numerical Solution of Ordinary Differential Equations, Academic Press, New York and London 1971.

[LIND77] LINDBERG, B.: Characterization of Optimal Stepsize Sequences for Methods for Stiff Differential Equations, SIAM J. Num. Anal. $\underline{14}$, (1977), p.859-887.

[LINI77] LINIGER, W.: Stability and Error Bounds for Multistep Solutions of Nonlinear Differential Equations, Proc. ISCAS-77 (IEEE International Symposium on Circuits and Systems), Phoenix, Ariz., April 25-27, 1977, p.277-280.

[LOAN92] VAN LOAN, C.: Matrix Framework for the Fast Fourier Transform, SIAM, Philadelphia, 1992.

[LORE86] LORENTZ, G. G.: Approximation of Functions, Chelsea, New York 1986.

[LUTH87] LUTHER, W.; NIEDERDRENK, K.; REUTTER, F.; YSERENTANT, H.: Gewöhnliche Differentialgleichungen, Reihe: Rechnerorientierte Ingenieurmathematik, Editor G. Engeln-Müllges, Vieweg, Braunschweig-Wiesbaden 1987.

[MAES84] MAESS, G.: Vorlesungen über Numerische Mathematik I, Akademie-Verlag, Berlin 1984, Birkhäuser, Basel, Stuttgart 1985.

[MAES88] MAESS, G.: Vorlesungen über Numerische Mathematik II, Akademie-Verlag, Berlin 1988.

[MART68] MARTIN, R.S.; J.H. WILKINSON: Similarity Reduction of a General Matrix to Hessenberg Form, Num. Math. $\underline{12}$ (1968), p.349-368.

[MART70] MARTIN, R.S.; PETERS, G. and WILKINSON, J. H.: The QR-Algorithm for Real Hessenberg Matrices, Num. Math. $\underline{14}$ (1970), p.219-231.

[McCA67] McCALLA, Th. R.: Introduction to Numerical Methods and Fortran Programming, John Wiley, New York 1967.

[McCR72] McCRACKEN, D.D.; DORN, W.S.: Numerical Methods with Fortran IV Case Studies, John Wiley, New York 1972.

[MEIN67] MEINARDUS, G.: Approximation of Functions: Theorie and Numerical Methods, Springer Tracts in Natural Philosophy, New York, Berlin, Heidelberg 1967.

[MEIN79] MEINARDUS, G.; MERZ, G.: Praktische Mathematik I, BI-Wissenschaftsverlag, Mannheim-Wien-Zürich 1979.

[MEING79] MEINGUET, B.J.: Multivariate interpolation at arbitrary points made simple, Z.A.M.P, Vol. $\underline{30}$ (1979), p. 292-304.

[MEIS81] MEIS, Th.; MARCOWITZ, U.: Numerical Solution of Partial Differential Equations, Applied Mathematical Sciences, Vol. 32, Springer, New York, Berlin, Heidelberg 1981.

[MENN77] MENNICKEN, R.; WAGENFÜHR, E.: Numerische Mathematik 1,2, Vieweg, Braunschweig-Wiesbaden 1977.

[MOOR80] MOORE, R.E.; KIOUSTELIDIS, J.B.: A simple test for accuracy of approximate solutions to nonlinear (or linear) systems, SIAM J. Num. Anal. $\underline{17}$ (1980), p. 521-529.

[MULL56] MULLER, D.E.: A Method for Solving Algebraic Equations using an Automatic Computer, Math. Tables Aids Comp. $\underline{10}$ (1956), p.208-215.

[NAGL89] NAGLE, R.K.; SAFF, E.B.: Fundamentals of Differential Equations, Cummings Publ. Comp., Calif. 1989.

[NIED84] NIEDERDRENK, K.: Die endliche Fourier- und Walsh-Transformation mit einer Einführung in die Bildverarbeitung, Hrsg. G. Engeln-Müllges, Vieweg-Verlag, 2^{nd} edition, Wiesbaden 1984, 3^{rd} edition 1993.

[NIED87] NIEDERDRENK, K.; YSERENTANT, H.: Funktionen einer Veränderlichen, Reihe: Rechnerorientierte Ingenieurmathematik, Editor G. Engeln-Müllges, Vieweg, Braunschweig-Wiesbaden 1987.

[NIEM87] NIEMEYER, H.: Lineare Algebra, Reihe: Rechnerorientierte Ingenieurmathematik, Editor G. Engeln-Müllges, Vieweg-Verlag, Braunschweig-Wiesbaden 1987.

[NIEM91] NIEMEYER, H.: Funktionen von mehreren Veränderlichen, analytische und numerische Behandlung, Reihe: Rechnerorientierte Ingenieurmathematik, Editor G. Engeln-Müllges, Vieweg-Verlag, Braunschweig-Wiesbaden 1991.

[NIET70] NIETHAMMER, W.: Über- und Unterrelaxation bei linearen Gleichungssystemen, Computing $\underline{5}$ (1970), p.303-311.

[NOBL65] NOBLE, B.: Numerical Methods, Interscience, New York 1965.

[NORT85] NORTON, V.: Finding a bracketed zero by larkin's method of rational interpolation, ACM Transactions on Mathematical Software, Vol. 11, (1985), p.120-134.

[ORTE70] ORTEGA, J.M.; RHEINBOLDT, W.C.: Iterative Solution of Nonlinear Equations in Several Variables, Academic Press, New York 1970.

[ORTE90] ORTEGA, J.M.: Numerical Analysis, a Second Course, SIAM, Philadelphia 1990.

[PARL69] PARLETT, B.N.; REINSCH, C.: Balancing a matrix for calculation of eigenvalues and eigenvectors, Num. Math. 13 (1969), p.293-304.

[PARL80] PARLETT, B.N.: The Symmetric Eigenvalue Problem, Prentice Hall, Englewood Cliffs, New York 1980.

[PATR75] PATRICK, M.L.; SAARI, D.G.: A globally convergent algorithm for determining approximate real zeros of a class of functions, BIT 15 (1975), p.296-303.

[PETE70] PETERS, G.; WILKINSON, J.H.: Eigenvectors of real and complex matrices by LR and QR triangularizations , Num. Math. 16 (1970), p.181-204.

[PIEGL95] PIEGL, L.; TILLER, W.: The Nurbs Book, Springer, New York, Berlin, Heidelberg 1995.

[PIEPE95] PIEPER, W.M.: Recursive Multidimensional Integration, Curtin University Perth, WA, Australia on leave from the Fachhochschule Giessen-Friedberg, Germany, email: wilhelm.pieper@mni.fh-giessen.de, 1995.

[PRES86] PRESS, W.H.; FLANNERY, B.P.; TENKOLSKY, S.A.; VETTERLING, W.T.: Numerical Recipes, Cambridge Univ. Press, 1986, 1988.

[QUAD83] QUADPACK, A Subrative Package for Automatic Integration, Springer, New York, Berlin, Heidelberg 1983.

[RALS60] RALSTON, A., WILF, H.S.: Mathematical Methods for Digital Computers, 3 vol., Wiley, New York 1960-1977.

[RALS67] RALSTON, A.; WILF, H.S.: Mathematical Methods for Digital Computers, John Wiley, New York 1967.

[RALS78] RALSTON, A.; RABINOWITZ, P.: A First Course in Numerical Analysis, International Student Edition, McGraw-Hill, Kogokusha, 2^{nd} edition 1978.

[RAVI83] RAVIART, P.-A.; THOMAS, J.M.: Introduction á l'Analyse Numérique des Équations aux Dérivées Partielles, Masson, Paris 1983.

[REIN71] REINSCH, C.: Smoothing by spline functions I, Num. Math. 10 (1967), p.177-183; II.: Num. Math. 16 (1971), p.451-454.

[RENN81] RENNER, G.; POCHOP, V.: A new method for local smooth interpolation, Eurographics 81, J.L. Encarnacao (ed.), p. 137-147.

[RENN82] RENNER, G.: A method of shape description for mechanical engineering practice, Computers in Industry 3 (1982), p.137-142.

[RICE77] RICE, J.E. (ed.): Math. Software I, Academic Press, New York, London 1970/71, III 1977.

[RICE83] RICE, John R.: Numerical Methods, Software and Analysis, McGraw-Hill, New York 1983.

[RITT69] RITTER, K.: Two Dimensional Splines and their Extremal Properties, ZAMM 49 (1969), p.597-608.

[RUTI52] RUTISHAUSER, H.: Über die Instabilität von Methoden zur Integration gewöhnlicher Differentialgleichungen, Z.A.M.P 3 (1952), p.65-74.

[RUTI60] RUTISHAUSER, H.: Bemerkungen zur numerischen Integration gewöhnlicher Differentialgleichungen n-ter Ordnung, Num. Math. 2 (1960), p.263-279 (see also Z.A.M.P 6 (1955), p.497-498).

[SCHE89] SCHEDEL, U.: Sparse Matrices, Numerical Aspects with Applications for Scientists and Engineers, Holstead Press, New York 1989.

[SCHM76] SCHMEISSER, G.; SCHIRMEIER, H.: Praktische Mathematik, de Gruyter, Berlin-New York 1976.

[SCHU76/1] SCHUMAKER, L.L.: Two-stage spline methods for fitting surfaces, Approximation Theory Bonn, 1976, Editors: A. Dold u. B. Eckmann, Springer, New York, Berlin, Heidelberg 1976.

[SCHUM76/2] SCHUMAKER, L.L.: Fitting surfaces to scattered data, Approximation Theory II, Editors: G.G. Lorenz, C.K. Chui u. L.L. Schumaker, Academic Press, New York 1976.

[SCHW72] SCHWARZ, H.R.; STIEFEL, E.; RUTISHAUSER, H.: Numerik symmetrischer Matrizen, Teubner, Stuttgart 1968, 2^{nd} edition 1972.

[SCHW84] SCHWARZ, H.R.: Methoden der finiten Elemente unter Berücksichtigung der technischen Praxis, Teubner, Stuttgart, 2^{nd} edition 1984, 3^{rd} edition 1991.

[SCHW88] SCHWARZ, H.R.: FORTRAN-Programme zur Methode der finiten Elemente, Teubner, Stuttgart 1988, 3^{rd} edition 1991..

[SCHW89] SCHWARZ, H.R.: Numerical Analysis: A Comprehensive Introduction, John Wiley, New York 1989.

[SCHWE79] SCHWETLICK, H.: Numerische Lösung nichtlinearer Gleichungssysteme, VEB Verlag der Wissenschaften, Berlin 1979.

[SHAH70] SHAH, J.M.: Two-dimensional-polynomial splines, Num. Math. 15 (1970), p.1-14.

[SHAM73] SHAMPINE, L.F.; ALLEN, R.C. Jr.: Numerical Computing: An Introductuion, Saunders, Philadelphia, London, Toronto 1973.

[SHAM75] SHAMPINE, L.F.; GORDON, M.K.: Computer Solution of Ordinary Differential Equations, W. H. Freeman, San Francisco 1975.

[SHEP68] SHEPARD, D.: A two dimensional interpolation function for irregularly-spaced data, ACM National Conference (1968), p.517-524.

[SHRA85] SHRAGER, R.I.: A rapid robust rootfinder. Mathematics of Computation Vol. 44, (1985), p.151-165.

[SINW82] SINWEL, H. F., RUNCK, P. O.: Konstanten im Approximationssatz von Jackson-Timan-Teljakowskii, in Numerical Methos in Approximation Theory, vol. 6, ISNM 59, Birkhäuser, Basel, Boston: p. 199-208, 1982.

[SOMM67] SOMMER, D.: Neue implizite Runge-Kutta-Formeln und deren Anwendungsmöglichkeiten, Ph.D. thesis, RWTH Aachen 1967.

[SPÄT71] SPÄTH, H.: Zweidimensionale glatte Interpolation, Computing 4 (1969), p.178-182 (see also Computing 7 (1971), p.364-369).

[SPÄT73] SPÄTH, H.: Algorithmen für elementare Ausgleichsmodelle, Oldenbourg, München-Wien 1973.

[SPÄT74/1] SPÄTH, H.: Algorithmen für multivariable Ausgleichsmodelle, Oldenbourg, München-Wien 1974.

[SPÄT74/2] SPÄTH, H.: Splines for Curves and Surfaces, Utilitas Mathematica, Winnipeg, 1974.

[SPEL85] SPELLUCCI, P.; TÖRNIG, W.: Eigenwertberechnung in den In-
 genieurwissenschaften, Teubner, Stuttgart 1985.

[STEP79] STEPLEMAN, R.S.; WINARSKY, N.D.: Adaptive numerical dif-
 ferentiation, Mathematics of Computations, $\underline{33}$, (1979), p. 1257-
 1264.

[STET73] STETTER, J.: Analysis of Discretization Methods for Ordinary
 Differential Equations, Springer, New York, Berlin, Heidelberg
 1973.

[STET76] STETTER, H.J.: Numerik für Informatiker, Oldenbourg,
 München-Wien 1976, 2^{nd} edition 1986.

[STEU79] STEUTEN, G.: Realisierung von Splinemethoden zur Konstruk-
 tion glatter Kurven und Flächen auf dem Bildschirm, Diplom the-
 sis, RWTH Aachen 1979.

[STEW73] STEWART, G. W.: Introduction to Matrix Computations, Aca-
 demic Press, New York 1973.

[STIE63] STIEFEL, E. L.: An Introduction to Numerical Mathematics,
 Academic Press, New York 1963.

[STOE91] STOER, J.; BULIRSCH, R.: Introduction to Numerical Analysis,
 2^{rd} edition, Springer, New York, Berlin, Heidelberg 1991/92.

[STRO66] STROUD, A.H.; SECREST, D.: Gaussian Quadrature Formulas,
 Prentice-Hall, Englewood Cliffs, New Jersey, 1966.

[STRO71] STROUD, A.H.: Approximate Calculation of Multiple Integrals,
 Prentice-Hall, Englewood Cliffs, New Jersey, 1971.

[STRO74] STROUD, A.H.: Numerical Quadrature and Solution of Ordinary
 Differential Equations, Springer, New York, Berlin, Heidelberg
 1974.

[STUM82] STUMMEL, F.; HAINER, K.: Praktische Mathematik, Teubner,
 Stuttgart 1970, 2^{nd} edition 1982.

[TERW95] TERWEGE, B.: Analyse verschiedener Methoden zur Erzeugung
 von Spline-Flächen, Research work FH Aachen (Editor G. Engeln-
 Müllges 1995)

[TIKH77] TIKHONOV, A.N.; ARSENIN, V.Y.: Solutions of Ill-Posed Prob-
 lems, Winston & Sons, New York-Toronto-London-Sydney 1977.

[TÖRN79] TÖRNIG, W.: Numerische Mathematik für Ingenieure und
 Physiker, Springer, Berlin, Heidelberg, New York 1979, Vol. 1: Nu-
 merische Methoden der Algebra; Vol. 2: Eigenwertprobleme und
 numerische Methoden der Analysis.

[TRAU66] TRAUB, J.F.: A class of globally convergent iteration functions
 for the solution of polynomal equations, Math. of Comp. 20 (1966),
 p.113-138.

[TRAU81] TRAUB, J.F.: Iterative Methods for the Solution of Equations,
 Prentice-Hall, Englewood Cliffs, New Jersey 1964, 2^{nd} edition
 1981.

[VERN78] VERNER, J.H.: Explizit Runge-Kutta methods with estimates of
 the local truncation error, SIAM J. Num. Anal. 15 (1978), p. 772-
 790.

[WATK91] WATKINS, D. S.: Fundamentals of Matrix Computations, John
 Wiley, New York 1991.

[WEIS90] WEISSINGER, J.: Spärlich besetzte Gleichungssys-
 teme. Eine Einführung mit BASIC- und PASCAL-Programmen,
 BI-Wissenschaftsverlag, Mannheim 1990.

[WERN79] WERNER, H.; SCHABACK, R.: Praktische Mathematik II,
 Springer, Berlin, Heidelberg, New York 1972, 2^{nd} edition 1979.

[WERN82] WERNER, H.: Praktische Mathematik I, Springer, Berlin, Hei-
 delberg, New York 1970, 3^{rd} edition 1982.

[WERN86] WERNER, H.; ARNDT, H.: Gewöhnliche Differentialgleichun-
 gen, Eine Einführung in Theorie und Praxis, Springer, Berlin,
 Heidelberg, New York 1986.

[WILK61] WILKINSON, J.H.: Error analysis of direct methods of matrix
 inversion, J. Assoc. Comput. March 8 (1961) p. 281-330.

[WILK64] WILKINSON, J.H.: Rounding Errors in Algebraic Processes,
 Prentice Hall, Englewood Cliffs, 1964.

[WILK65] WILKINSON, J.H.: The Algebraic Eigenvalue Problem, Claren-
 don Press, Oxford 1965.

[WILK71] WILKINSON, J.H.: REINSCH, C.: Handbook for Automatic
 Computation, Springer, New York, Berlin, Heidelberg 1971.

[WILK85] WILKINSON, J. H.: The perfidious polynomial, MAA Studies in
 Mathematics, Vol. 24, Studies in Numerical Analysis, G. H. Golub
 ed., 1984, p. 1-28.

[WODI77] WODICKA, R.: Kombinierte interpolatorische Polynomsplines,
 Report of the Instituts für Geometrie und Praktische Mathematik,
 RWTH Aachen, 1977.

[WODI91] WODICKA, R.: Ergänzungen zu Akima's Steigungsformel, Mitteilungen aus dem Mathem. Seminar Giessen, Heft 203, Selbstverlag des Mathematischen Instituts, Giessen 1991.

[YOUN71] YOUNG, D.M.: Iterative Solution of Large Linear Systems, Academic Press, New York and London 1971.

[ZIEL86] ZIELKE, G.: Report on test matrices for generalized inverses, Computing $\underline{36}$ (1986), p.105-162.

[ZURM65] ZURMÜHL, R.: Praktische Mathematik für Ingenieure und Physiker, Springer, Berlin, Heidelberg, New York, 5^{th} edition 1965.

[ZURM84] ZURMÜHL, R.; FALK, S.: Matrizen und ihre Anwendungen für Angewandte Mathematiker, Physiker und Ingenieure, Springer, Berlin, Heidelberg, New York 1984.

LITERATURE FOR OTHER TOPICS

Numerical Treatment of Partial Differential Equations

AMES, W.F. (Ed.): Nonlinear partial differential equations in engineering, Academic Press, New York, London, Vol.1 1965, Vol.2 1972.

AMES, W.F.: Numerical Methods for partial differential equations, Academic Press, New York, London, 1977.

ANSORGE, R.: Differenzenapproximationen partieller Anfangswertaufgaben, Teubner, Stuttgart 1978.

COLLATZ, L.: Differentialgleichungen, Teubner, Stuttgart, 7^{th} edition 1990.

FARLOW, S.J.: Partial Differential Equations for Scientists and Engineers, John Wiley, New York, 1982.

FORSYTHE, G.E.; WASOV, W.: Finite-difference methods for partial differential equations, New York 1960, 4^{th} edition 1967.

GREENSPAN, D.: Introductory numerical analysis of elliptic boundary value problems, Harper & Row, New York 1965.

HABERMAN, R.: Elementary Applied Partial Differential Equations, Prentice-Hall, Englewood Cliffs, New Jersey 1983.

HACKBUSCH, W.: Theorie und Numerik partieller Differentialgleichungen, Teubner, Stuttgart 1984.

HACKBUSCH, W.: Multi-Grid Methods and Applications, Springer, New York, Berlin, Heidelberg 1980.

JANENKO, N.N.: Die Zwischenschrittmethode zur Lösung mehrdimensionaler Probleme der mathematischen Physik, Springer, Berlin, Heidelberg, New York 1969.

LAPIDUS, L.; PINDER, G.F.: Numerical Solution of Partial Differential Equations in Science and Engineering, Academic Press, New York, 1982.

MITCHELL, A.R.: Computational Methods in Partial Differential Equations, John Wiley, New York 1976.

RICHTER, W.: Partielle Differentialgleichungen: Einführung in Theorie und Praxis der partiellen Differentialgleichungen mit FORTRAN-Programmen, Spektrum Akademischer Verlag, Heidelberg, Berlin, Oxford 1995.

SMITH, G.D.: Numerical Solution of Partial Differential Equations: Finite Difference Methods, 3^{rd} edition, Clarendon Press, Oxford 1985.

TÖRNIG, W.; GIPSER, M.; KASPAR, B.: Numerische Lösung von partiellen Differentialgleichungen der Technik, Teubner, Stuttgart 1985.

TÖRNIG, W.: Numerische Mathematik für Ingenieure und Physiker, Vol. 2, Springer, Berlin, Heidelberg, New York 1979.

WENDLAND, W.L.: Elliptic Systems in the Plane, London, San Francisco, Melbourne, 1979.

Finite Element Method

ACIZ, A.K. (ed.): The Mathematical Foundations of the Finite Elemente Method with Applications to Partial Differential Equations, Academic Press, New York 1972.

BATHE, K.J.; WILSON, E.: Finite Element Procedures in Engineering Analysis, Prentice Hall, Englewood Cliffs (New Jersey) 1982.

BATHE, K.J.: WILSON, E. L. Numerical Methods in Finite Element Analysis, Prentice Hall, Englewood Cliffs, 1976.

BLANK, R.E.; DUPONT, T.F.: An Optimal Order Process for Solving Finite Element Equations, Center for Numerical Analysis University of Texas at Austin, CNA-158, May 1980.

BOOR, de, C. (Ed.): Mathematical Aspects of Finite Element in Partial Differential Equations, Academic Press, New York, London 1974.

BREBBIA, C.A.; CONNOR, J.J.: Fundamentals of Finite Element Techniques for Structural Engineers, Butterworths, London 1973.

CIARLET, Ph.G.: The Finite Element Method for Elliptic Problems, North-Holland Public, Amsterdam 1978.

CUVELIER, C.; SEGAL, A.; v. STEENHOVEN, A.: Finite Element Methods and Navier-Stokes Equations, Kluwer Academic, Boston 1986.

FENNER, R.T.: Finite Element Methods for Engineers, Macmillan, London 1975.

FIX, G.; STRANG, G.: An analysis of the finite element method, Prentice Hall, Englewood Cliffs 1973.

GALLAGHER, R.H.: Finite-Element-Analysis, Springer, Berlin, Heidelberg, New York 1976.

GAWEHN, W.: Finite-Elemente-Methode, Vieweg, Braunschweig, Wiesbaden, 3^{rd} edition 1988.

GIRAULT, V.; RAVIARD, P.-A.: Finite Elements Approximation of the Navier-Stokes Equations, Springer, New York, Berlin, Heidelberg 1986.

GOERING, H.M.; ROOS, H. J.; TOBISKA, L.: Finite-Elemente-Methode, Springer, Berlin, Heidelberg, New York 1985.

HAHN, H.G.: Methode der Finiten Elemente in der Festigkeitslehre, Akademische Verlagsgesellschaft, Wiesbaden, 2^{nd} edition 1982.

HINTON, E.; OWEN, D.R.J.: Finite Element Programming, Academic Press, London 1980.

HUGHES, T.J.R.: Finite Element Method: Linear Static and Dynamic Finite Element Analysis, Prentice-Hall, Englewood Cliffs 1987.

LINK, M.: Finite Elemente in der Statik und Dynamik, Teubner, Stuttgart, 2^{nd} edition 1989.

LIVESLEY, R.K.: Finite Elements: An Introduction for Engineers, Cambridge University Press, London, New York 1983.

ODEN, J.T.; REDDY, J.N.: An Introduction to the Mathematical Theory of Finite Elements, John Wiley, New York 1976.

RICHTER, W.: Numerical Solution of Partial Differential Equations with the Finite Element Method, Beatrix Perera-Verlag, Stuttgart/Colombo 1990.

SCHWARZ, H.R.: Methoden der finiten Elemente unter Berücksichtigung der technischen Praxis, Teubner, Stuttgart, 2^{nd} edition 1984, 3^{rd} edition 1991.

SCHWARZ, H.R.: FORTRAN-Programme zur Methode der finiten Elemente, Teubner, Stuttgart 1988, 3^{rd} edition 1991.

SEWELL, G.: Analysis of Finite Element Method: PDE/PROTRAN, Springer, New York, Berlin, Heidelberg, Tokyo 1985.

WENDLAND, W.L.; STEPHAN, E.: Mathematische Grundlagen der finiten Elemente Methoden, Lang, Frankfurt 1982.

WHITE, R.E.: An Introduction to the Finite Element Method with Applications to Nonlinear Problems, New York, 1985.

ZIENKIEWICZ, O.C.: The Finite Element Method in Engineering Science, McGraw Hill, London 1971.

ZIENKIEWICZ, O.C.; MORGAN, K.: Finite Elements and Approximation, John Wiley, New York 1983.

C Index

Printing: Saladruck, Berlin
Binding: Buchbinderei Lüderitz & Bauer, Berlin